Synthesis, Characterization and Applications of Sustainable Advanced Nanomaterials

Synthesis, Characterization and Applications of Sustainable Advanced Nanomaterials

Editor

Thomas Dippong

MDPI • Basel • Beijing • Wuhan • Barcelona • Belgrade • Manchester • Tokyo • Cluj • Tianjin

Editor
Thomas Dippong
Chemistry and Biology
Technical University of
Cluj Napoca
Baia Mare
Romania

Editorial Office
MDPI
St. Alban-Anlage 66
4052 Basel, Switzerland

This is a reprint of articles from the Special Issue published online in the open access journal *Materials* (ISSN 1996-1944) (available at: www.mdpi.com/journal/materials/special_issues/sustainable_advanced_nanomaterials).

For citation purposes, cite each article independently as indicated on the article page online and as indicated below:

LastName, A.A.; LastName, B.B.; LastName, C.C. Article Title. *Journal Name* **Year**, *Volume Number*, Page Range.

ISBN 978-3-0365-6913-0 (Hbk)
ISBN 978-3-0365-6912-3 (PDF)

© 2023 by the authors. Articles in this book are Open Access and distributed under the Creative Commons Attribution (CC BY) license, which allows users to download, copy and build upon published articles, as long as the author and publisher are properly credited, which ensures maximum dissemination and a wider impact of our publications.

The book as a whole is distributed by MDPI under the terms and conditions of the Creative Commons license CC BY-NC-ND.

Contents

About the Editor vii

Thomas Dippong
Synthesis, Physicochemical Characterization and Applications of Advanced Nanomaterials
Reprinted from: *Materials* **2023**, *16*, 1674, doi:10.3390/ma16041674 1

Thomas Dippong, Oana Cadar and Erika Andrea Levei
Effect of Transition Metal Doping on the Structural, Morphological, and Magnetic Properties of $NiFe_2O_4$
Reprinted from: *Materials* **2022**, *15*, 2996, doi:10.3390/ma15092996 11

Firuta Goga, Rares Adrian Bortnic, Alexandra Avram, Mioara Zagrai, Lucian Barbu Tudoran and Raluca Anca Mereu
The Effect of Ni^{2+} Ions Substitution on Structural, Morphological, and Optical Properties in $CoCr_2O_4$ Matrix as Pigments in Ceramic Glazes
Reprinted from: *Materials* **2022**, *15*, 8713, doi:10.3390/ma15248713 25

Agnieszka Feliczak-Guzik
Nanomaterials as Photocatalysts—Synthesis and Their Potential Applications
Reprinted from: *Materials* **2022**, *16*, 193, doi:10.3390/ma16010193 41

Marwène Oumezzine, Cristina Florentina Chirila, Iuliana Pasuk, Aurelian Catalin Galca, Aurel Leca and Bogdana Borca et al.
Magnetocaloric and Giant Magnetoresistance Effects in La-Ba-Mn-Ti-O Epitaxial Thin Films: Influence of Phase Transition and Magnetic Anisotropy
Reprinted from: *Materials* **2022**, *15*, 8003, doi:10.3390/ma15228003 63

Nikola Lenar, Robert Piech and Beata Paczosa-Bator
The New Reliable pH Sensor Based on Hydrous Iridium Dioxide and Its Composites
Reprinted from: *Materials* **2022**, *16*, 192, doi:10.3390/ma16010192 73

Kai-Chun Hsu, Chung-Lun Yu, Heng-Jyun Lei, Subramanian Sakthinathan, Po-Chou Chen and Chia-Cheng Lin et al.
Modification of Electrospun CeO_2 Nanofibers with $CuCrO_2$ Particles Applied to Hydrogen Harvest from Steam Reforming of Methanol
Reprinted from: *Materials* **2022**, *15*, 8770, doi:10.3390/ma15248770 87

Roman Atanasov, Rares Bortnic, Razvan Hirian, Eniko Covaci, Tiberiu Frentiu and Florin Popa et al.
Magnetic and Magnetocaloric Properties of Nano- and Polycrystalline Manganites $La_{(0.7-x)}Eu_xBa_{0.3}MnO_3$
Reprinted from: *Materials* **2022**, *15*, 7645, doi:10.3390/ma15217645 101

Liliana Cepoi, Inga Zinicovscaia, Tatiana Chiriac, Ludmila Rudi, Nikita Yushin and Dmitrii Grozdov et al.
Modification of Some Structural and Functional Parameters of Living Culture of *Arthrospira platensis* as the Result of Selenium Nanoparticle Biosynthesis
Reprinted from: *Materials* **2023**, *16*, 852, doi:10.3390/ma16020852 117

Tatjana Glaskova-Kuzmina, Leons Stankevics, Sergejs Tarasovs, Jevgenijs Sevcenko, Vladimir Špaček and Anatolijs Sarakovskis et al.
Effect of Core–Shell Rubber Nanoparticles on the Mechanical Properties of Epoxy and Epoxy-Based CFRP
Reprinted from: *Materials* 2022, 15, 7502, doi:10.3390/ma15217502 133

Jie Che, Xin Jiang, Yangchun Fan, Mingfeng Li, Xuejuan Zhang and Daojiang Gao et al.
A Novel Dual-Emission Fluorescence Probe Based on CDs and Eu^{3+} Functionalized UiO-66-$(COOH)_2$ Hybrid for Visual Monitoring of Cu^{2+}
Reprinted from: *Materials* 2022, 15, 7933, doi:10.3390/ma15227933 149

Jan Krajczewski, Robert Ambroziak, Sylwia Turczyniak-Surdacka and Małgorzata Dziubałtowska
WO_3 Nanopores Array Modified by Au Trisoctahedral NPs: Formation, Characterization and SERS Application
Reprinted from: *Materials* 2022, 15, 8706, doi:10.3390/ma15238706 161

Dipendu Saha, Gerassimos Orkoulas and Dean Bates
One-Step Synthesis of Sulfur-Doped Nanoporous Carbons from Lignin with Ultra-High Surface Area, Sulfur Content and CO_2 Adsorption Capacity
Reprinted from: *Materials* 2023, 16, 455, doi:10.3390/ma16010455 175

Ioan Petean, Gertrud Alexandra Paltinean, Emanoil Pripon, Gheorghe Borodi and Lucian Barbu Tudoran
Silver Depreciation in 3-Polker Coins Issued during 1619–1627 by Sigismund III Vasa King of Poland
Reprinted from: *Materials* 2022, 15, 7514, doi:10.3390/ma15217514 187

Cristina Iosif, Stanca Cuc, Doina Prodan, Marioara Moldovan, Ioan Petean and Anca Labunet et al.
Mechanical Properties of Orthodontic Cements and Their Behavior in Acidic Environments
Reprinted from: *Materials* 2022, 15, 7904, doi:10.3390/ma15227904 203

Alexandru Turza, Violeta Popescu, Liviu Mare and Gheorghe Borodi
Structural Aspects and Intermolecular Energy for Some Short Testosterone Esters
Reprinted from: *Materials* 2022, 15, 7245, doi:10.3390/ma15207245 219

Agnieszka Stróż, Joanna Maszybrocka, Tomasz Goryczka, Karolina Dudek, Patrycja Osak and Bożena Łosiewicz
Influence of Anodizing Conditions on Biotribological and Micromechanical Properties of Ti–13Zr–13Nb Alloy
Reprinted from: *Materials* 2023, 16, 1237, doi:10.3390/ma16031237 235

Agnieszka Stróż, Thomas Luxbacher, Karolina Dudek, Bartosz Chmiela, Patrycja Osak and Bożena Łosiewicz
In Vitro Bioelectrochemical Properties of Second-Generation Oxide Nanotubes on Ti–13Zr–13Nb Biomedical Alloy
Reprinted from: *Materials* 2023, 16, 1408, doi:10.3390/ma16041408 253

Julija Volmajer Valh, Tanja Pušić, Mirjana Čurlin and Ana Knežević
Extending the Protection Ability and Life Cycle of Medical Masks through the Washing Process
Reprinted from: *Materials* 2023, 16, 1247, doi:10.3390/ma16031247 269

About the Editor

Thomas Dippong

Thomas Dippong (Associate Professor, Doctor at the Technical University of Cluj Napoca) is a chemical engineer with a Ph.D. and hability in chemistry. His current research activities are related to obtained and characterization nanoparticles for various applications, as part of ongoing research in partnership with the Technical University of Cluj-Napoca within the field of ferrite embedded in silica matrix. He is an expert in analytical chemistry, organic/inorganic chemistry, thermal treatment, instrumental analysis, and synthesis of nanomaterials. Dr Dippong has published 133 peer-reviewed publications (77 papers in high ranked scientific ISI-Thomson journals (42 Q1, 12 Q2, and 23 Q3), 56 in other national and international journals), 1309 citations, h-index: 31 (WoS) and has given 34 lectures at international conferences (ICTAC 14 Brazil, ESTAC Brasov, JTACC Budapest, CEEC-TAC5 Roma, etc). He has also published two books at international publishing houses and 15 books at national publishing houses. He has been the contract manager for projects and is currently an active member of four other projects. Dr. Dippong has reviewed 400 scientific articles for 78 ISI-Thomson journals. He is the guest editor of seven Special Issues of five prestigious Q1 ISI-Thomson journals.

Editorial

Synthesis, Physicochemical Characterization and Applications of Advanced Nanomaterials

Thomas Dippong

Faculty of Science, Technical University of Cluj-Napoca, 76 Victoriei Street, 430122 Baia Mare, Romania; dippong.thomas@yahoo.ro

Abstract: This Special Issue highlights the last decade's progress regarding new nanostructured materials. In this regard, the development of nanoscale syntheses and innovative characterization tools that resulted in the tailored design of nanostructured materials with versatile abilities in many applications were investigated. Various types of engineered nanostructures, usually metal nanoparticles or nanoporous metal oxides, have been synthesized for various applications. This Special Issue covers the state-of-the-art of advanced nanoparticles in many disciplines (chemistry, pharmacy, nanomedicine, agriculture, catalysis, and environmental science). The crystallite sizes depended on the annealing temperature and type of doping ion. A combination of rigid and soft particles could simultaneously enhance both the tensile properties and the fracture toughness, which could not be achieved by the single-phase particles independently. The surface charge and in vitro corrosion resistance are key parameters characterizing biomaterials in the interaction of the implant with the biological environment. Solar energy in the presence of a photocatalyst can be effectively converted into electricity/fuel, break down chemical and microbial pollutants, and help water purification. The saturation magnetization, remanent magnetizations, coercivity, and anisotropy were found to depend on the doping ion, annealing temperature, and particle size. The efficiency of the photocatalysis reaction depends on several factors, including light absorption capacity/light intensity, the type of photocatalyst used, the concentration of a photocatalyst and contaminant particles, the pH of the reaction medium, etc. The variety of color pigments and coloring properties of the targeted application in the ceramic industry was also of interest.

Keywords: advanced metal oxide nanoparticles; synthesis; photocatalysts; sustainable processes; energy conversion; nanosensors; smart nanostructures and nanodevices for virus detection

Citation: Dippong, T. Synthesis, Physicochemical Characterization and Applications of Advanced Nanomaterials. *Materials* **2023**, *16*, 1674. https://doi.org/10.3390/ma16041674

Received: 8 February 2023
Revised: 9 February 2023
Accepted: 13 February 2023
Published: 17 February 2023

Copyright: © 2023 by the author. Licensee MDPI, Basel, Switzerland. This article is an open access article distributed under the terms and conditions of the Creative Commons Attribution (CC BY) license (https:// creativecommons.org/licenses/by/ 4.0/).

With the rapid development of nanotechnology, nanomaterials have recently attracted the attention of the scientific community due to their unique structural, morphological, optical, electrical, thermal, and magnetic characteristics [1–3]. These enhanced properties are caused by their high surface-to-volume ratio due to their size falling in the 1–100 nm range [1–3]. Nanomaterials can be metallic-based nanoparticles (ferrites, chromates, aluminates, bismutates, etc.) or carbon oxides (carbon nanotubes, graphenes, graphene oxides, etc.). The tailoring of the shape, size, and size distribution of nanoparticles, and the properties of hybrid nanoparticles, is achieved through different synthesis routes by modifying parameters such as the pH, concentration of reactants, dopants, or stirring speed. Some of these methods are complex, involving the use of reduction agents with little to no impact on the environment and needing a longer reaction time or a high processing temperature to complete crystallization [1–3]. These tailored properties of nanoparticles make them suitable candidates for technological applications in photocatalysis, photoluminescence, biosensors, catalysis, humidity sensors, permanent magnets, magnetic drug delivery, magnetic liquids, magnetic refrigeration, ceramic pigments, microwave absorbents, corrosion protection, water decontamination, photocatalysis, antimicrobial agents or biomedicine (hyperthermia) [1–3]. Multifunctional magnetic nanocomposites are among those heterogeneous nanosized systems where at least one phase component is magnetic and can act as an intermediate of either the actuation or the response of the overall system. The main advantage of heterogeneous nanosystems is the possibility of combining and inter-influencing the electronic properties of the constituent interfaced nanophases.

The Special Issue, entitled *Synthesis, Characterization and Application of Sustainable Advanced Nanomaterials*, includes 18 original research works and focuses on highlighting the progress, challenges, and future directions in the area of the synthesis and characterization of nanomaterials and nanostructures with multiple applications in chemistry, physics, biology, and medicine [1–3].

A sol–gel route followed by thermal treatment used to produce $NiFe_2O_4$ doped with transition metal ions (Zn^{2+}, Mn^{2+}, Co^{2+}) was reported by Dippong et al. [4]. The TG/DTA curves of samples dried at 40 °C indicated the formation and decomposition of metallic precursors to ferrites in single or two stages, with comparable mass losses [4]. The functional groups identified by Fourier-transform infrared spectroscopy confirmed the decomposition of metal nitrates, the formation and decomposition of precursors, and the formation of the SiO_2 matrix [4]. The X-ray diffraction indicated that the sol–gel synthesis produced single-phase crystalline ferrites in the case of the Zn^{2+}- and Co^{2+}-doped Ni-ferrites. By doping with Mn^{2+}, several secondary phases derived from the SiO_2 matrix (cristobalite, quartz, and Fe_2SiO_4) accompanied the crystalline spinel ferrite. The XRD parameters were influenced by the crystallite size, lattice strain, defects, annealing temperature, and doping ions. The gradual increase of the lattice parameters suggested the uniform distribution of doping metal ions in the $NiFe_2O_4$ lattice. The unit cell volume increases by doping with Mn^{2+} ion and decreases by doping with Zn^{2+} and Co^{2+} ions. By contrast, the X-ray and bulk densities and porosity decrease by doping with Mn^{2+} and increase with doping Zn^{2+} and Co^{2+} ions [4]. The $NiFe_2O_4$ particle size increases by doping with Mn^{2+} and decreases by doping with Zn^{2+} and Co^{2+} ions, respectively. The doping of $NiFe_2O_4$ with Zn^{2+}, Mn^{2+}, and Co^{2+} leads to a decrease of the saturation magnetization and remanent magnetization, whereas the coercivity decreases at 700 °C and increases at 1000 °C. The obtained magnetic transition metal dopped-Ni ferrite nanoparticles are possible candidates for various medical applications such as controlled drug delivery, cancer therapy, biosensing, and magnetic resonance imaging [4].

In this Special Issue, Goga et al. [5], focused on the Ni substitution of cobalt ions in a $Co_xCr_2O_4$ matrix (x = 0, 0.25, 0.5, 0.75, and 1.00) by using a sol–gel synthesis route, was performed. The X-ray diffraction (XRD) studies reveal a spinel-type Face–Centered Cubic structure and a secondary Cr_2O_3 phase when $x \leq 0.75$ and a Body–Centered Tetragonal structure when x = 1. The structural characterization is consistent with the Ni^{2+} substitution of Co^{2+} ions, thereby a decrease of unit cell parameter and the unit cell volume was observed with an increase of x. The increase of Ni^{2+} substitution in the matrix has a pronounced increase in the size of the crystallites from 39.9 nm (for x = 0) to 99.42 nm (for x = 0.75) [5]. FT-IR indicated two strong absorption bands corresponding to the metal–oxygen stretching from tetrahedral and octahedral sites, characteristic of a spinel structure. The UV-VIS absorption bands assigned to the $Co_{(1-x)}Ni_xCr_2O_4$ spinel confirm the Ni^{2+} ions at the A site and the Cr^{3+} ions at the B sites. Adjusting the nickel content in the $CoCr_2O_4$ matrix, the color of the pigment can be easily controlled. Lighter shades of ceramic glazes can be obtained when embedding in glossy glaze and darker shades can be obtained when embedding in a matte glaze, confirming the spinels' applicability as pigments in the ceramic tile industry [5]. SEM and TEM microscopy evidenced the powder morphology and the tendency of nanoparticles to agglomerate. The elemental EDX distribution of Ni Kα1 in the glossy ceramic tile confirms the homogeneous and uniform pigment distribution in the glaze after firing. All the samples present distinct stable colors and good structural and morphological properties that recommend them to be used as ceramic pigments [5].

In this Special Issue, Feliczak-Guzik [6] reviewed nanomaterials as photocatalysts—synthesis and their potential applications. The increasing demand for energy and environmental degradation are current serious problems. Therefore, searching for new, efficient, and stable photocatalysts with high application potential is a point of great interest. Over the years, research on the synthesis of photocatalysts has evolved considerably from transition metal oxides (e.g., TiO_2/ZnO) to much more advanced materials [6]. The photocatalysts should be characterized by the ability to absorb radiation from a wide spectral range of light, the appropriate position of the semiconductor energy bands to the redox reaction potentials, and the long diffusion path of charge carriers, besides the thermodynamic, electrochemical, and photoelectrochemical stabilities. The light absorption capacity/light intensity, the type of photocatalyst used, the concentration of a photocatalyst and contaminant particles, and reaction medium's pH are key factors to determining the optimal amount of these factors for a given photocatalyst and type of pollutant [6]. Therefore, efforts are being performed to increase the efficiency of photo processes by changing the electron structure, surface morphology, and crystal structure of semiconductors. The photocatalysis process has been studied for a long time on a laboratory scale; however, its large-scale application is greatly hampered (by blocking light penetration in thick coatings, leaching effects, and difficulty in recovering the photocatalyst). In natural systems, the decomposition rate of pollutants is not limited by a time regime, unlike in industrial installations, where the technical challenge of photocatalytic processes, especially in heterogeneous systems, is unsatisfactory reaction kinetics [6].

Oumezzine et al. [7] reported the magnetocaloric and giant magnetoresistance effects in La-Ba-Mn-Ti-O epitaxial thin films: the influence of phase transition and magnetic anisotropy. Magnetic perovskite films have promising properties for energy-efficient spintronic devices and magnetic refrigeration. Here, an epitaxial ferromagnetic $La_{0.67}Ba_{0.33}Mn_{0.95}Ti_{0.05}O_3$ thin film was grown on a $SrTiO_3$ single crystal substrate by pulsed laser deposition. High-resolution X-ray diffraction proved the high crystallinity of the film with tetragonal symmetry. The magnetic, magnetocaloric effect and magnetoresistance of the $La_{0.67}Ba_{0.33}Mn_{0.95}Ti_{0.05}O_3$ film with a thickness of 97 nm have been studied at different directions of the applied magnetic field to the sample plane. The $La_{0.67}Ba_{0.33}Mn_{0.95}Ti_{0.05}O_3$ epilayer exhibits a second-order ferromagnetic phase transition around 234 K together with a metal–semiconductor transition close to this Curie temperature (T_C) and an in-plane magnetic uniaxial easy axis. Further, a gradual metal-to-paramagnetic semiconductor transition at higher temperature finishes at 245 K. The magnetic entropy variation under 5 T induction of a magnetic field applied parallel to the film surface reaches a maximum of 17.27 mJ/cm^3 K, and the relative cooling power is 1400 mJ/cm^3 K for the same applied magnetic field [7]. Another important finding is that the $La_{0.67}Ba_{0.33}Mn_{0.95}Ti_{0.05}O_3$ epitaxial thin film has a giant magnetoresistance as high as 82% at a temperature close to the T_C, which may be interesting for electromagnetic applications.

In this Special Issue, Lenar et al. [8] present a new reliable pH sensor based on hydrous iridium dioxide and its composites. The addition of a conducting polymer to the composite material changed the wetting properties of the material, making it highly hydrophobic, which consequently contributed to the stability of the potentiometric response during the water-layer test. Three $hIrO_2$-based materials were prepared and applied as solid-contact layers in pH-selective electrodes with polymeric membranes [8]. The material included a standalone hydrous iridium oxide; a composite material of hydrous iridium oxide, carbon nanotubes, and triple composite material composed of hydrous iridium oxide; carbon nanotubes; and poly(3-octylthiophene-2,5-diyl) [8]. Each component contributed differently to the sensors' performance—the addition of carbon nanotubes increased the electrical capacitance of the sensor. Oppositely, the addition of the conducting polymer allowed it to increase the contact angle of the material, changing its wetting properties and enhancing the stability of the potentiometric response. The hydrous iridium oxide-contacted electrodes exhibit linear responses in a wide linear range of pH (2–11) and stable potentiometric responses (the lowest potential drift of 0.036 mV/h is attributed to the electrode with triple composite material). The response towards hydrogen ions turned out to be repeatable and reversible within this range of pH values, and neither redox nor light sensitivity were detected. No presence of a water layer was detected in the solid-contact electrodes with IrO_2^--based materials [8].

The study of Hsu et al. [9], focusing on the modification of electrospun CeO_2 nanofibers with $CuCrO_2$ particles applied to hydrogen harvest from steam reforming of methanol, was also included in this Special Issue. Hydrogen is an alternative renewable energy source for addressing the energy crisis and climate change. $CuCrO_2$ particles were attached to the surfaces of electrospun CeO_2 nanofibers to form CeO_2-$CuCrO_2$ nanofibers; the catalyst was produced and used for steam reforming of methanol. The $CuCrO_2$ particles did not readily adhere to the surfaces of the CeO_2 nanofibers, so a trace amount of SiO_2 was added to the surfaces to make them hydrophilic. After the SiO_2 modification, the CeO_2 nanofibers were immersed in a Cu-Cr-O precursor and annealed in a vacuum to form CeO_2-$CuCrO_2$ nanofibers. The specific surface area of the CeO_2-$CuCrO_2$ nanofibers is 15.06 m^2/g [9]. According to the findings, the increased hydrogen production rate can be ascribed to the stronger catalytic activity, larger surface area, lower reactor temperature, and higher methanol flow rate of the CeO_2-$CuCrO_2$ nanofiber catalyst. According to the H_2 production performance, the CeO_2-$CuCrO_2$ nanofibers can be a better catalyst for commercial H_2 production and are suitable for fuel cell vehicles without high-temperature activation [9].

The study of Atanasov et al. [10] compares the structural and magnetic properties of the nano- and polycrystalline manganites $La_{(0.7-x)}Eu_xBa_{0.3}MnO_3$, which are potential magnetocaloric materials to be used in domestic magnetic refrigeration close to room temperature. The sol–gel method produced nano-scale particles, showing an average size of 30–70 nm. Both systems are single-phase, with rhombohedral lattice symmetry. Iodometry was used to estimate the oxygen content in samples, showing a lower concentration of Mn^{4+} ions, leading to the lowest oxygen content value of $O_{2.97\pm0.02}$ for the x = 0.4 sample. To reduce this temperature below 300 K, the La^{3+} ions were partially replaced by Eu ions. In nano-sized manganites, the reduction of T_C is accompanied by a broad magnetic transition, extending the magnetic cooling effect to a larger temperature range [10]. The magnetic measurements revealed single magnetic phases, low magnetic anisotropy, and very small coercivity for both systems. The bulk samples with x < 0.4 show a metallic–insulator transition at a temperature T_p lower than magnetic transition temperature T_c. All samples show a negative magnetoresistance. A modified Arrott plot analysis revealed that bulk samples' critical exponents were in the tricritical mean field model range and in the 3D Heisenberg model range for nanocrystalline samples. The maximum magnetic entropy change of 4.2 J/kgK was observed for the x = 0.05 bulk sample for $\mu_0 \Delta H$ = 4 T [10]. Since the temperature range (δT_{FWHM}) for nano-sized samples $La_{0.7}Ba_{0.3}MnO_3$ and $La_{0.65}Eu_{0.05}Ba_{0.3}MnO_3$ covers a wide range including room temperature, they may be used in multistep refrigeration processes. The magnetocaloric effect was found to be larger and close to room temperature for the bulk samples, while for the nano-samples, it was lower, but extended on a large temperature range. This wide range of effective nanoparticle cooling and high entropy change in bulk material can be combined for suitable commercial cooling [10].

Cepoi et al. [11] presented that *Arthrospira platensis* easily tolerates the presence of high concentrations of selenium (up to 125 mg/L) in the medium, growth, and biomass accumulation, being within the limits of the values characteristic for the control biomass. The biosynthesis of selenium nanoparticles has become particularly important due to the environmentally friendly character of the process and the special properties of the obtained particles. For selenium concentrations up to 50 mg/L, the amount of biomass accumulated during the cultivation cycle increased by up to 18% compared to the control [11]. The content of lipids and carbohydrates in biomass increased with the increasing sodium selenite concentration added to the nutrient medium. The content of protein and phycobilin also increased, and the dose-dependent character of this relationship was maintained up to a concentration of sodium selenite of 175 mg/L [11]. With an increase in the content of lipids, the level of malonic dialdehyde in the cells also increased. Most of the bioaccumulated selenium was determined in the protein (47.5% of the accumulated selenium) and the lipid (24.1%) fractions, the ultrastructural changes in the cells during biosynthesis and the change in the expression of some genes involved in stress response reactions [11]. In the protein fraction, selenium nanoparticles with a size of 2–8 nm were formed. Thus, the expression level of iron-superoxide dismutase and heat-shock protein increased, which may be associated with the need to manage the increased flow of reactive oxygen species and to stabilize the proteins subjected to the action of the xenobiotics. Selenium also caused ultrastructural changes in *Arthrospira platensis* expressed in the damage and disorder of thylakoids, the detachment of the cytoplasmic membrane from the cell wall, the change in the density of the cell wall, and the formation of carbon reserves in the cells, indicating the negative effects of selenium ions [11]. Thus, *Arthrospira platensis* tolerates high concentrations of selenium, accumulates significant amounts of this element, and carries out the biosynthesis of selenium nanoparticles, which are mainly located in the protein and lipid fractions. The process is accompanied by biochemical, ultrastructural, and gene expression changes associated with the response of spirulina to stress conditions [11].

Glaskova-Kuzmina et al. [12] reported on the effect of core–shell rubber nanoparticles on the tensile properties, fracture toughness, and glass transition temperature of the epoxy and epoxy-based carbon fiber reinforced polymer. The Hansen model was applied to describe the elastic modulus of the epoxy possessing a certain fraction of the core–shell rubber nanoparticles and pores. Three additives containing core–shell rubber nanoparticles were used for the research, resulting in a filler fraction of 2–6 wt.% in the epoxy resin. The effect of the core–shell rubber nanoparticles on the tensile properties of the epoxy resin was notable, leading to a reduction of 10–20% in the tensile strength and elastic modulus and an increase of 60–108% in the fracture toughness for the highest filler fraction. No considerable distinction in the fracture toughness among the additives was detected, thereby proving that the small (100 nm) and large (300 nm) core–shell rubber nanoparticles were equally efficient [12]. The glass transition temperature of the epoxy was gradually improved by 10–20 °C with the increase of core–shell rubber nanoparticles for all of the additives, which could be attributed to the high crosslink density and toughening effect of rubber modifiers, thereby testifying to their dissolution in the continuous epoxy phase. The possible combination of rigid and soft particles could be a compromise to simultaneously improve both the tensile properties and the fracture toughness, which cannot be achieved by the single-phase particles independently [12].

Che et al. [13] presented a novel dual-emission fluorescence probe based on carbon dots and an Eu^{3+} functionalized UiO-66-$(COOH)_2$ hybrid for visual monitoring of Cu^{2+}. The carbon dots-UiO-66-$(COOH)_2$ exhibits outstanding selectivity, excellent sensitivity, and good anti-interference for ratiometric sensing Cu^{2+} in water. The linear range is 0–200 µM, and the detection limit is 0.409 µM [13]. The carbon dots-UiO-66-$(COOH)_2$ silicon plate achieves rapid and selective detection of Cu^{2+} and the change in fluorescence color can be observed by the naked eye [13]. These results reveal that the carbon dots-UiO-66-$(COOH)_2$ hybrid can be employed as a simple, rapid, and sensitive fluorescent probe to detect Cu^{2+} [13]. The possible sensing mechanism of this dual-emission fluorescent probe is discussed in detail [13]. The result reveals that adding Cu^{2+} would affect the energy transfer between the ligand and Eu^{3+}, which would quench the luminescence of Eu^{3+}. This finding indicates that carbon dots-UiO-66-$(COOH)_2$ material can be employed as a fluorescent probe to rapidly and efficiently detect Cu^{2+} in aqueous solutions [13].

Krajczewski et al. [14] reported on the WO_3 nanopores array modified by Au trisoctahedral NPs: formation, characterization and SERS application. The WO_3 nanopores array was obtained by an anodization method in an aqueous solution with the addition of F^- ions. Several factors affecting the final morphology of the samples were tested, such as potential, time, and F^- concentrations. Using smaller trisoctahedron Au NPs as seeds for the growth of larger nanoparticles permits easy tuning of the size of particles while maintaining the well-defined trisoctahedron shape. The nanopore's size increased with the increasing potential [14]. The XPS measurements do not show any contamination by F^- on the surface, typical for WO_x samples formed by an anodization method in the range of 0.5–1 h. Such a layer was successfully modified by anisotropic gold trisoctahedral NPs of various sizes. The UV-Vis spectroscopy showed shifting of SPR into longer wavelengths with the successive growth of nanoparticles. The WO_3—Au trisoctahedron-modified nanoarray was successfully used as a SERS platform. The highest enhancement was observed for the Au NPs with a 94 nm diameter [14].

Saha et al. [15] effectively synthesized sulfur-doped nanoporous carbon with an ultrahigh surface area from lignin by one-step carbonization with the help of sodium thiosulfate as a sulfurizing agent and potassium hydroxide as an activating agent to create porosity. Lignin is the second-most available biopolymer in nature. Lignin was employed as the carbon precursor for the one-step synthesis of sulfur-doped nanoporous carbons and has several applications in scientific and technological sectors. The peak deconvolution results of XPS confirmed that the nanoporous carbons possess sulfur contents of 1 to 12.6 at.%, and the key functionalities include S=C, S-C=O, and SO_x. The nanoporous carbons' porosity analysis revealed that the BET-specific surface areas of the carbons are in 741–3626 m^2/g and a total pore volume of 0.5–1.74 cm^3/g [15]. The surface area of 3626 m^2/g is one of the highest for carbon-based materials reported in the literature [15]. Pure-component adsorption isotherms of CO_2, CH_4, and N_2 were measured on all the porous carbons at 298 K, with a pressure up to 760 torrs [15]. The carbon with the highest BET surface area demonstrated the highest CO_2 uptake of more than 10.89 mmol/g, at 298 K and 760 torr, which is one of the highest for porous carbon-based materials, compared to previous studies [15]. Ideally, the adsorbed solution theory was employed to calculate the selectivity for CO_2/N_2, CO_2/CH_4, and CH_4/N_2, from the pure-component isotherm data, and some of the carbons reported a very high selectivity value.

Petean et al. [16] presented the silver depreciation in 3-Polker coins issued from 1619–1627 by Sigismund III Vasa, King of Poland, in the context of the "Kipper- und Wipperzeit" financial crisis generated by the 30-years war, using non-destructive investigation methods such as X-ray diffraction and Scanning Electron Microscopy coupled with Energy Dispersion Spectroscopy (EDS) elemental analysis. It was characterized by a strong debasing of the silver title of the coins issued by the countries involved in the war. Silver coins issued by Poland were generally considered safer. Some historical references mention forgeries of this monetary type issued in copper plated with a thin silver foil. Using modern material investigation techniques, the authors aimed to find the precise situation of the officially issued 3-Polker by the Poland mints. A significant achievement of this research is the SEM–EDS elemental maps recorded for each coin that reveal the silver alpha phase grains and Ag-Cu eutectic grains without metallographic analysis [16]. These methods allow proper investigation of the coins and preserve their integrity, a necessary factor for valuable museum artifacts. The findings reveal important facts for historians: the 3-Polker coins issued by Sigismund III Vasa, King of Poland, from 1619–1627 evidenced a certain depreciation of the silver title from about 84.3% to a range of 63.2–74.6% for the coins issued between 1621–1625 [16]. It is a mild decrease in the silver title compared to the historical data regarding the currency affected by the Kipper- und Wipperzeit crisis. The findings reveal that the silver title in 3-Polker coins was restored to the normal value between 1626 and 1627. The author concludes that the 3-Polker issued in the official Poland mints, even those affected by silver depreciation, was considered good money (being hoarded) and definitely could not be the rich copper debased coins mentioned in some of the medieval sources [16].

Iosif et al. [17] reported on the mechanical properties of orthodontic cements and their behavior in acidic environments and investigated the mechanical properties and morphology of three categories of orthodontic cements: resin composites (BracePaste); resin-modified glass ionomer (Fuji Ortho) and resin cement (Transbond) exposed to acidic environments such as Coca Cola™ and Red Bull™. Their mechanical properties, such as compressive strength, diametral tensile strength, and flexural strength, were correlated with the samples' microstructures, liquid absorption, and solubility in liquid [17]. The findings suggest that Transbond resin cement presents the best compression strength and BracePaste features the best flexural strength. The elastic modulus is very important considering the solicitations induced by chewing forces. The BracePaste has the best value of the elastic modulus, followed by Fuji Ortho. Therefore, each material has strong points that are useful for personalized orthodontic treatment according to the patient's requirements. Acid soft drinks and energy drinks are very popular among young patients with orthodontic brackets. The acidic components within these soft drinks (phosphoric acid in Coca-Cola™ and citric acid in Red Bull™) can erode the bonding layer and affect the bracket's stability. Atomic force microscopy reveals the nanostructural alteration of the investigated orthodontic materials, such as roughness increasing and nano-filler particles acid erosion [17]. It was found that these parameters strongly influence the orthodontic material behavior (e.g., BracePaste roughness decreasing under acid exposure proves an excellent resistance to in-depth erosive penetration), a fact that must be considered when the orthodontic treatment is prescribed to the patient. BracePaste is recommended for long-term orthodontic treatment for patients who regularly consume acidic beverages, Fuji Ortho is recommended for short-term orthodontic treatment for patients who regularly consume acidic beverages, and Transbond is recommended for orthodontic treatment over an average time period for patients who do not regularly consume acidic beverages.

The study by Turza et al. [18] showed the structural aspects and intermolecular energy of some short testosterone esters. Testosterone (17β-Hydroxyandrost-4-en-3-one) is the primary male anabolic–androgenic steroid. A single crystal X-ray diffraction technique was employed to elucidate the crystal structures of three short testosterone esters: propionate, phenylpropionate, and isocaproate. They were shown to belong to the non-centrosymmetric orthorhombic $P2_12_12_1$, and monoclinic $P2_1$ space groups. Structural features were described and evaluated in terms of Hirshfeld surfaces, and crystal energies were further compared with the base native form (without ester) and with the acetate ester [18]. The investigation of crystals in the solid state via computational methods yielded that, in all crystals, the crystal stability and formation of supramolecular self-assemblies are governed by dominant dispersion effects. Although the C-H-O hydrogen bonds are present in all compounds, they play a less noticeable role [18]. Total crystal lattice energies are greater in absolute terms with the increase in ester chain length. The core steroidal rings depict similar conformations in all prodrugs, with the six-membered A rings in intermediate sofa-half-chair geometries, B and C rings showing chair-like conformations, and five-membered D rings showing intermediate envelope-half-chair conformations. The molecular overlap indicates a good match of backbone skeleton rings representing the native part of the ester's structures and the differences occurring in the carbon tails orientation. From a pharmaceutical point of view, their solubility is correlated with ester length, which implies the added ester functionalities. The shortest acetate ester possesses the lowest solubility, while the longest isocaproate ester is approximately four-fold greater. Phenylpropionate and propionate forms show similar values and are between the other two [18].

The influence of anodizing conditions on the biotribological and micromechanical properties of Ti–13Zr–13Nb alloy was reported by Stróz et al. [19]. The porous oxide nanotubes' layers of various geometries and lengths on the Ti–13Zr–13Nb alloy surface can be produced by anodizing to improve osseointegration, which shows that Vickers microhardness determined under variable loads changed depending on the type of electrolyte and applied voltage–time parameters of electrochemical oxidation. By anodizing, first-generation, second-generation, and third-generation oxide nanotubes layers were produced on the Ti–13Zr–13Nb alloy surface. Vickers microhardness decreased from 181(5) to 252(6) and from 254(3) to 221(3) with the increasing load for second-generation and third-generation oxide nanotube layers, respectively, compared to the alloy substrate [19]. The kinetic coefficient of friction determined based on the friction coefficient took the smallest value of 0.86(8) for the second-generation oxide nanotubes' layer. The highest coefficient of kinetic friction of 0.94(1) was characterized by the surface of the first-generation oxide nanotubes' layer [19]. Based on the results obtained, a three-body abrasion wear mechanism was proposed for biotribological wear of the Ti–13Zr–13Nb alloy before and after anodizing in Ringer's solution. Based on the biotribological tests carried out in Ringer's solution in a reciprocating motion in the ball-on-flat system for the Ti–13Nb–13Zr alloy before and after anodizing, it was found that the non-anodized alloy was characterized by the highest wear resistance for which the average material volume consumption. Wear scars' analysis of the ZrO_2 ball was performed using optical microscopy. It was found that the composition of the electrolyte with the presence of fluoride ions was an essential factor influencing the micromechanical and biotribological properties of the obtained oxide nanotubes' layers [19].

Stróż et al. [20] reported in vitro bioelectrochemical properties of second-generation oxide nanotubes on a Ti–13Zr–13Nb biomedical alloy. In a neutral aqueous KCl solution, the second-generation oxide nanotubes layer moves the isoelectric point from 4.2 for the non-anodized Ti–13Zr–13Nb alloy, which is typical for the surface without a functional group to pH of 5.4, which is characteristic for the amorphous oxide phase. Comparison of the influence of different electrolytes such as KCl, PBS, and artificial blood on the zeta potential at pH of 7.4 for the Ti–13Zr–13Nb alloy before and after anodizing revealed a strong reaction of calcium anions with amorphous surfaces [20]. The complex ions in artificial blood have demonstrated a stronger affinity to the hydrophobic surface before anodizing than hydrophilic ones after electrochemical oxidation. The increase in the corrosion resistance of the anodized Ti–13Zr–13Nb electrode in PBS compared with the non-anodized Ti–13Zr–13Nb electrode was due to the presence of a stable second-generation oxide nanotubes layer. The zeta potential method used in these in vitro studies could not be used in vivo due to technical limitations, determination of the breakdown potential of the second-generation oxide nanotubes layer on the Ti–13Zr–13Nb alloy in PBS was not possible due to the technical limitations of the potentiostat to the tested potential range of 10 V. Knowledge about the kinetics of drug release from the obtained oxide nanotubes will facilitate the development of personalized implants that are carriers of tissue-forming and therapeutic substances, supporting the process of osseointegration of the implant in the human body [20].

In the last Special Issue article, Valh [21] investigated extending the protection ability and life cycle of medical masks through the washing process. Numerous challenges and the pandemic period of SARS-CoV-2 affecting people's respiratory systems have raised specific questions and doubts about the extent to which consumer laundry detergents can reasonably ensure the level of disinfection during washing. Reusing decontaminated disposable medical face masks could contribute to reducing the environmental burden of discarded masks. The hydrophobicity of medical masks determined through the static contact angle depends on the number of cycles carried out. The static contact angle of the samples after the first cycle is lower than after the fifth cycle in all procedures. The barrier properties of the medical mask were analyzed before and after the first and fifth washing cycles indirectly by measuring the contact angle of the liquid droplets with the front and back surface of the mask and further by measuring the air permeability and determining the antimicrobial resistance. Images of ultrapure water drops on the surface confirm the hydrophobicity of the front/back of the medical mask before and after washing. The additional analysis included FT-IR, pH of the material surface and aqueous extract, and the determination of residual substances—surfactants—in the aqueous extract of washed versus unwashed medical masks, while their aesthetic aspect was examined by measuring their spectral characteristics [21,22]. The results showed that household washing had a more substantial impact on the change of some functional properties, primarily air permeability, than laboratory washing [22]. The disinfectant agent, didecyldimethylammonium chloride, contributes to the protective ability and supports the idea that washing medical masks under controlled conditions can preserve barrier properties and enable reusability [21].

I am aware that the diversity and innovation of new compounds and tools rapidly developing in multidisciplinary research related to nanomaterials based on metals cannot all be collected in a single volume. However, this collection will contribute to the interest of research in this area, providing our readers with a broad and updated scenario. All these published studies will offer a new approach for future studies to create important advances in materials science and engineering.

In conclusion, as the Editors of this Special Issue, we would like to thank all the authors and reviewers who contributed to this Special Issue with innovative ideas and constructive reviewers' comments. We are grateful for the consistent support from the *Materials* Editorial Office. We are sure that this Special Issue will provide our readers with a platform to understand the novel real-world synthesis and characterization of innovative nanomaterials and nanostructures with their pivotal roles in diverse applications.

Funding: This research received no external funding.

Conflicts of Interest: The authors declare no conflict of interest.

References

1. Dippong, T.; Levei, E.A.; Cadar, O. Formation, structure and magnetic properties of $MFe_2O_4@SiO_2$ (M = Co, Mn, Zn, Ni, Cu) nanocomposites. *Materials* **2021**, *14*, 1139. [CrossRef] [PubMed]
2. Dippong, T.; Cadar, O.; Levei, E.A.; Deac, I.G.; Borodi, G.; Barbu-Tudoran, L. Influence of polyol structure and molecular weight on the shape and properties of $Ni_{0.5}Co_{0.5}Fe_2O_4$ nanoparticles obtained by sol-gel synthesis. *Ceram. Int.* **2019**, *45*, 7458–7467. [CrossRef]
3. Dippong, T.; Levei, E.A.; Cadar, O.; Deac, I.G.; Lazar, M.; Borodi, G.; Petean, I. Effect of amorphous SiO_2 matrix on structural and magnetic properties of $Cu_{0.6}Co_{0.4}Fe_2O_4/SiO_2$ nanocomposites. *J. Alloy. Compd.* **2020**, *849*, 156695. [CrossRef]
4. Dippong, T.; Cadar, O.; Levei, E.A. Effect of Transition Metal Doping on the Structural, Morphological, and Magnetic Properties of $NiFe_2O_4$. *Materials* **2022**, *15*, 2996. [CrossRef] [PubMed]
5. Goga, F.; Bortnic, R.A.; Avram, A.; Zagrai, M.; Barbu Tudoran, L.; Mereu, R.A. The Effect of Ni^{2+} Ions Substitution on Structural, Morphological, and Optical Properties in $CoCr_2O_4$ Matrix as Pigments in Ceramic Glazes. *Materials* **2022**, *15*, 8713. [CrossRef]
6. Feliczak-Guzik, A. Nanomaterials as Photocatalysts—Synthesis and Their Potential Applications. *Materials* **2022**, *16*, 193. [CrossRef]
7. Oumezzine, M.; Chirila, C.F.; Pasuk, I.; Galca, A.C.; Leca, A.; Borca, B.; Kuncser, V. Magnetocaloric and Giant Magnetoresistance Effects in La-Ba-Mn-Ti-O Epitaxial Thin Films: Influence of Phase Transition and Magnetic Anisotropy. *Materials* **2022**, *15*, 8003. [CrossRef] [PubMed]
8. Lenar, N.; Piech, R.; Paczosa-Bator, B. The New Reliable pH Sensor Based on Hydrous Iridium Dioxide and Its Composites. *Materials* **2023**, *16*, 192. [CrossRef]
9. Hsu, K.-C.; Yu, C.-L.; Lei, H.J.; Sakthinathan, S.; Chen, P.-C.; Lin, C.C.; Chiu, T.W.; Nagaraj, K.; Fan, L.; Lee, Y.-H. Modification of Electrospun CeO_2 Nanofibers with $CuCrO_2$ Particles Applied to Hydrogen Harvest from Steam Reforming of Methanol. *Materials* **2022**, *15*, 8770. [CrossRef]
10. Atanasov, R.; Bortnic, R.; Hirian, R.; Covaci, E.; Frentiu, T.; Popa, F.; Deac, I.G. Magnetic and Magnetocaloric Properties of Nano- and Polycrystalline Manganites $La_{(0.7-x)}Eu_xBa_{0.3}MnO_3$. *Materials* **2022**, *15*, 7645. [CrossRef]
11. Cepoi, L.; Zinicovscaia, I.; Chiriac, T.; Rudi, L.; Yushin, N.; Grozdov, D.; Tasca, I.; Kravchenko, E.; Tarasov, K. Modification of Some Structural and Functional Parameters of Living Culture of Arthrospira platensis as the Result of Selenium Nanoparticle Biosynthesis. *Materials* **2023**, *16*, 852. [CrossRef] [PubMed]
12. Glaskova-Kuzmina, T.; Stankevics, L.; Tarasovs, S.; Sevcenko, J.; Špacek, V.; Sarakovskis, A.; Zolotarjovs, A.; Shmits, K.; Aniskevich, A. Effect of Core–Shell Rubber Nanoparticles on the Mechanical Properties of Epoxy and Epoxy-Based CFRP. *Materials* **2022**, *15*, 7502. [CrossRef] [PubMed]
13. Che, J.; Jiang, X.; Fan, Y.; Li, M.; Zhang, X.; Gao, D.; Ning, Z.; Li, H. A Novel Dual-Emission Fluorescence Probe Based on CDs and Eu^{3+} Functionalized $UiO-66-(COOH)_2$ Hybrid for Visual Monitoring of Cu^{2+}. *Materials* **2022**, *15*, 7933. [CrossRef] [PubMed]
14. Krajczewski, J.; Ambroziak, R.; Turczyniak-Surdacka, S.; Dziubałtowska, M. WO3 Nanopores Array Modified by Au Trisoctahedral NPs: Formation, Characterization and SERS Application. *Materials* **2022**, *15*, 8706. [CrossRef] [PubMed]
15. Saha, D.; Orkoulas, G.; Bates, D. One-Step Synthesis of Sulfur-Doped Nanoporous Carbons from Lignin with Ultra-High Surface Area, Sulfur Content and CO_2 Adsorption Capacity. *Materials* **2023**, *16*, 455. [CrossRef] [PubMed]
16. Petean, I.; Paltinean, G.A.; Pripon, E.; Borodi, G.; Barbu Tudoran, L. Silver Depreciation in 3-Polker Coins Issued during 1619–1627 by Sigismund III Vasa King of Poland. *Materials* **2022**, *15*, 7514. [CrossRef]
17. Iosif, C.; Cuc, S.; Prodan, D.; Moldovan, M.; Petean, I.; Labunet, A.; Barbu Tudoran, L.; Badea, I.C.; Man, S.C.; Badea, M.E.; et al. Mechanical Properties of Orthodontic Cements and Their Behavior in Acidic Environments. *Materials* **2022**, *15*, 7904. [CrossRef]
18. Turza, A.; Popescu, V.; Mare, L.; Borodi, G. Structural Aspects and Intermolecular Energy for Some Short Testosterone Esters. *Materials* **2022**, *15*, 7245. [CrossRef]
19. Stróż, A.; Maszybrocka, J.; Goryczka, T.; Dudek, K.; Osak, P.; Łosiewicz, B. Influence of Anodizing Conditions on Biotribological and Micromechanical Properties of Ti–13Zr–13Nb Alloy. *Materials* **2023**, *16*, 1237. [CrossRef]
20. Stróż, A.; Luxbacher, T.; Dudek, K.; Chmiela, B.; Osak, P.; Łosiewicz, B. In Vitro Bioelectrochemical Properties of Second-Generation Oxide Nanotubes on Ti–13Zr–13Nb Biomedical Alloy. *Materials* **2023**, *16*, 1408. [CrossRef]
21. Valh, J.V.; Pušić, T.; Curlin, M.; Knežević, A. Extending the Protection Ability and Life Cycle of Medical Masks through the Washing Process. *Materials* **2023**, *16*, 1247. [CrossRef] [PubMed]
22. Dippong, T.; Mihali, C.; Năsui, D.; Berinde, Z.; Butean, C. Assessment of water physicochemical parameters in the Strimtori-Firiza reservoir in N-W Romania. *Water Environ. Res.* **2018**, *90*, 220–233. [CrossRef] [PubMed]

Disclaimer/Publisher's Note: The statements, opinions and data contained in all publications are solely those of the individual author(s) and contributor(s) and not of MDPI and/or the editor(s). MDPI and/or the editor(s) disclaim responsibility for any injury to people or property resulting from any ideas, methods, instructions or products referred to in the content.

Article

Effect of Transition Metal Doping on the Structural, Morphological, and Magnetic Properties of NiFe$_2$O$_4$

Thomas Dippong [1], Oana Cadar [2] and Erika Andrea Levei [2,*]

[1] Faculty of Science, Technical University of Cluj-Napoca, 76 Victoriei Street, 430122 Baia Mare, Romania; dippong.thomas@yahoo.ro

[2] INCDO-INOE 2000, Research Institute for Analytical Instrumentation, 67 Donath Street, 400293 Cluj-Napoca, Romania; oana.cadar@icia.ro

* Correspondence: erika.levei@icia.ro

Abstract: Sol-gel route followed by thermal treatment was used to produce NiFe$_2$O$_4$ doped with transition metal ions (Zn^{2+}, Mn^{2+}, Co^{2+}). The structural, morphological, and magnetic properties of the doped NiFe$_2$O$_4$ were compared with those of virgin NiFe$_2$O$_4$. The metal-glyoxylates' formation and decomposition as well as the thermal stability of the doped and virgin ferrites were assessed by thermal analysis. The functional groups identified by Fourier-transform infrared spectroscopy confirmed the decomposition of metal nitrates, the formation and decomposition of precursors, and the formation of the SiO$_2$ matrix. The X-ray diffraction indicated that the sol-gel synthesis produced single-phase crystalline ferrites in case of virgin, Zn^{2+} and Co^{2+}-doped Ni-ferrites. By doping with Mn^{2+}, several secondary phases derived from the SiO$_2$ matrix accompanied the crystalline spinel ferrite. The crystallite sizes depended on the annealing temperature and type of doping ion. The gradual increase of lattice parameters suggested the uniform distribution of doping metal ions in the NiFe$_2$O$_4$ lattice. The saturation magnetization, remanent magnetizations, coercivity, and anisotropy were found to depend on the doping ion, annealing temperature, and particle size. The high saturation magnetization values of the obtained nanocomposites make them suitable for a wide range of applications in the field of sensors development and construction.

Keywords: nickel ferrite; nanoparticle; divalent metal doping; magnetic properties

1. Introduction

Spinel ferrites are the topic of numerous studies due to their magnetic nature and crystalline structure. Small changes of the particle size, composition or presence of surface effects give them unique magnetic features [1]. Nanosized spinel ferrites received a huge amount of interest due to their low cost, excellent chemical stability, moderate saturation magnetization, high surface area, high wear resistance, low density, low thermal expansion coefficient, and low toxicity to both human health and environment [2–4]. These ferrites are promising candidates for a broad range of applications in the industry (magnetic recording media, photoelectric devices, sensors, magnetic pigments, photocatalysts in dye degradation, controlled signal transformation, storage devices, batteries, solar cells) and biomedicine (controlled drug delivery, tumor treatment, magnetic resonance imaging, biomagnetic separation, cellular therapy, tissue repair, cell separation, and biosensing) [1–8].

The NiFe$_2$O$_4$ is a soft magnetic, semiconducting material with ferromagnetic properties, prominent electrical resistivity, low conductivity, low eddy current loss, high chemical stability, catalytic behavior, etc. [1,5,9,10]. NiFe$_2$O$_4$ possesses an inverse spinel structure, with the Fe^{3+} ions distributed equally between the tetrahedral (A) and octahedral (B) sites while, the Ni^{2+} ions occupy the octahedral (B) sites [1,5,11].

The ferrite structure and properties are sensitive to the synthesis method, additive substitution, and calcination process [12–14]. Doping with transition metal ions, such as Co^{2+},

Ni^{2+}, or Zn^{2+}, is an effective way to improve and control the structure and consequently the optical, electrical, dielectric, and magnetic properties of nanosized $NiFe_2O_4$ [1,7,8,15–17]. The doping with transition metal ions into spinel ferrite structure changes the cations' distribution between the tetrahedral (A) and octahedral (B) sites, leading to different magnetic properties. The dopant ion may also change the energy of the grain boundaries, acting as a driving force of the grain growth [1]. The electrical resistivity can also be improved by doping the host matrices with smaller divalent cations or by controlling their microstructures [8]. Zn^{2+} doping disturbs the cation distribution, enhances the dielectric and magnetic properties [18]. The substitution of $NiFe_2O_4$ with magnetic divalent transition metal ions like Mn^{2+} led to appealing magnetic and electrical features [2,12,14,19,20]. By adjusting the Mn-to-Ni ratio in the ferrite, the magnetic properties of the ferrite can be controlled [2]. The Ni^{2+} ions' addition overcomes the grain formation, leading to low surface roughness [13,21].

The physico-chemical properties of nanosized ferrites are highly influenced by the synthesis route, dopant ion nature and amount, as well as the presence of structural order–disorder effects [1,5]. The annealing temperature influences the grain boundary migration and grain boundary diffusion, which further determines the grain shape, grain size, core density, and microstructure [4]. The synthesis route is a key factor to obtain high-purity nanoferrites [1]. Several methods for producing nanoferrites, such as sol-gel, co-precipitation, refluxing, hydrothermal, mechano-chemical, solid-state, precursor, auto combustion, microwave plasma, microemulsion, mechanical alloying, etc. are described in the literature [1,5,7]. Among these, to produce ferrite nanocomposites, the sol-gel method is one of the most-used approaches due to its simplicity, low cost, low processing temperature, and good control over the structure, physico-chemical properties, surface properties, and magnetic behavior [22]. To obtain spinel ferrites by the sol-gel method, nitrate salts are frequently used, as they act as water-soluble, low-temperature oxidizing agents [23]. Solvothermal synthesis allows the large-scale production of ferrites with controlled size and shape by choosing the appropriate aqueous or non-aqueous solvent mixture, by varying the synthesis temperature, pressure, and reaction time [3]. The microwave-assisted synthesis of ferrites has a lower yield than hydrothermal or thermal-decomposition methods [3]. The co-precipitation method is another frequently used method to produce nanoparticles with a specific shape and size [22]. The major disadvantage associated with the ferrite production by co-precipitation is the poor crystallinity of the resulting NPs, that may be enhanced by subsequent heat treatment [3]. Auto-combustion is a simple and low-cost process that requires a short reaction time and low energy consumption [4,24]. The ferrites prepared by this method have homogeneous chemical composition, high-purity, and good sinterability [24]. In the modified sol-gel method, the reactants are mixed with tetraethyl orthosilicate (TEOS), the sol is exposed to air until the gelation of the silica (SiO_2) network, the gels are thermally treated to obtain carboxylate precursors that are further thermally decomposed into the oxidic systems. This method is versatile, simple, and effective in producing pure nanoparticles, but has the drawback of having the presence of amorphous phases at low annealing temperatures and of secondary crystalline phases at high annealing temperatures [25]. Among different coating materials, mesoporous SiO_2 is non-toxic and biocompatible, allows the control of the particle growth, minimizes the nanoparticles agglomeration, improves their stability, enhances the magnetic guidability and bio-compatibility, and favors the conjugation with functional groups, [26–29].

The paper aims to investigate the structural, morphological and magnetic properties of virgin $NiFe_2O_4$ and $NiFe_2O_4$ doped with transition divalent metal ions Zn^{2+} ($Zn_{0.15}Ni_{0.85}Fe_2O_4$), Mn^{2+} ($Mn_{0.15}Ni_{0.85}Fe_2O_4$), and Co^{2+} ($Co_{0.15}Ni_{0.85}Fe_2O_4$) embedded in a SiO_2 matrix produced by sol-gel route, followed by thermal treatment at various temperatures. This study is of particular interest due to the lack of information on the effect of dopant nature (Zn^{2+}, Mn^{2+} and Co^{2+}) on the size and magnetic properties of mixed $M_{0.15}Ni_{0.85}Fe_2O_4$ (M=Co, Mn and Zn) type ferrites embedded in SiO_2 matrix. Because the oxidic phases at low temperatures are poorly crystalline or even amorphous, the desired surface properties and crystallinity can be achieved by using specific annealing conditions. Besides, the reactivity of the amorphous phases allows their participation in a variety of chemical transformations. In this regard, the X-ray diffraction (XRD) parameters were compared for different annealing temperatures to get important structural information. The thermal (TG-DTA) analysis and Fourier transform infrared (FT-IR) spectroscopy depicted the formation and decomposition of metallic glyoxylate precursors, the stability of the produced ferrites and formation SiO_2 matrix. A special emphasis was given to the evolution of magnetic properties (saturation magnetization (M_S), remanent magnetization (M_R), coercivity (H_C), and anisotropy (K)) with the increase of annealing temperature and the type of doping ion.

2. Materials and Methods

2.1. Reagents

Iron (III) nitrate nonahydrate ($Fe(NO_3)_3 \cdot 9H_2O$, 98%), nickel (II) nitrate hexahydrate ($Ni(NO_3)_2 \cdot 6H_2O$, 99%), zinc (II) nitrate hexahydrate ($Zn(NO_3)_2 \cdot 6H_2O$, 98%), manganese (II) nitrate tetrahydrate ($Mn(NO_3)_2 \cdot 4H_2O$, 98%), cobalt (II) nitrate hexahydrate ($Co(NO_3)_2 \cdot 6H_2O$, 98%), 1,2 ethanediol (1,2-ED, 99%), tetraethyl orthosilicate (TEOS, 99%) and ethanol 96% (Merck, Darmstadt, Germany) were used in the synthesis.

2.2. Synthesis

$NiFe_2O_4$ and $M-NiFe_2O_4$ embedded in SiO_2 ($M_{0.15}Ni_{0.85}Fe_2O_4$, M=Co, Mn and Zn) nanocomposites, containing 70 wt.% ferrite and 30 wt.% SiO_2, were prepared by modified sol-gel method by using a M/Ni/Fe molar ratio of 0.15/0.85/2. A schematic diagram of the synthesis method is given in Figure 1. To prepare the sols, the metal nitrates were mixed with 1,2-ED, TEOS and ethanol by using a NO_3^-/ED/TEOS molar ratio of 1/1/0.50. The resulting sols were stirred continuously for 30 min and maintained in open air, at room temperature until gelation occurs. The formed gel embedded a homogenous mixture of metal nitrates and 1,2-ED. As the production of high-purity gels with high crystallites size is favored by a thermal pretreatment before annealing [5], the obtained gels were grinded, dried at 40 and 200 °C, and annealed at 400 °C (5 h), 700 °C (5 h) and 1000 °C (5 h), respectively, by using a LT9 muffle furnace (Nabertherm, Lilienthal, Germany).

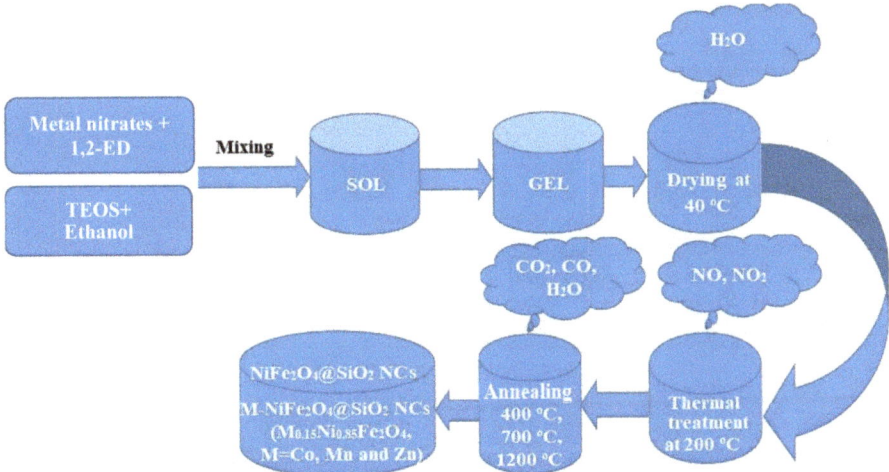

Figure 1. Schematic diagram of virgin and doped $NiFe_2O_4$ ($Zn-NiFe_2O_4$, $Mn-NiFe_2O_4$, $Co-NiFe_2O_4$) synthesis.

By heating the gels at 200 °C the redox reactions between the nitrates and 1,2-ED take place in the pores of the SiO_2 matrix resulting a mixture of Fe(III), Ni(II), and M(II) glyoxylates. The mixtures of glyoxylates around 300 °C decompose into metal oxides that reacts at temperatures above 300 °C and forms the ferrites. SiO_2 has de role of a spacer between the nanoparticles, reducing the particle agglomeration [14,29].

2.3. Characterization

The thermal behavior was investigated by thermogravimetric (TG) and differential thermal analysis (DTA) by using a Q600 SDT (TA Instruments, Newcastle, DE, USA) thermal analyzer, in air up to 1000 °C, at 5 °C/min. The FT-IR spectra were recorded by using a Spectrum BX II (Perkin Elmer, Waltham, MA, USA) Fourier-transform infrared spectrometer on pellets containing 1% (w/w) sample in KBr. The X-ray diffraction patterns were recorded by using a D8 Advance (Bruker, Karlsruhe, Germany) diffractometer, operating at room temperature, 40 kV, and 40 mA with CuKα radiation (λ = 1.54060 Å). The Co/Ni/Fe ($Co_{0.15}Ni_{0.85}Fe_2O_4@SiO_2$), Mn/Ni/Fe ($Mn_{0.15}Ni_{0.85}Fe_2O_4@SiO_2$), and Zn/Ni/Fe ($Zn_{0.15}Ni_{0.85}Fe_2O_4@SiO_2$) molar ratios were verified by inductively coupled plasma optical emission spectrometry (ICP-OES) by using a Perkin Elmer Optima 5300 DV (Norwalk, CT, USA) spectrometer, after microwave digestion with aqua regia. The nanoparticles morphology was studied by transmission electron microscopy (TEM) and scanning electron microscopy (SEM) on samples deposited from suspension onto carbon-coated copper grids by using an HD-2700 (Hitachi, Tokyo, Japan) transmission electron microscope and a SU8230 (Hitachi, Tokyo, Japan) scanning electron microscope. A cryogen-free vibrating-sample magnetometer (Cryogenic Limited, London, UK) was used for the magnetic measurements.

3. Results and Discussion

3.1. Thermal Analysis

The TG/DTA curves of virgin and doped $NiFe_2O_4$ samples dried at 40 °C are presented in Figure 2. The DTA curve shows three processes: (I) loss of moisture and physically adsorbed water suggested by the endothermic effects at 64–95 °C, (II) formation of metal-glyoxylate precursors shown by the exothermic effects at 116–182 °C and (III) decomposition of glyoxylate precursors into ferrites as indicated by the exothermic effect at 260–315 °C.

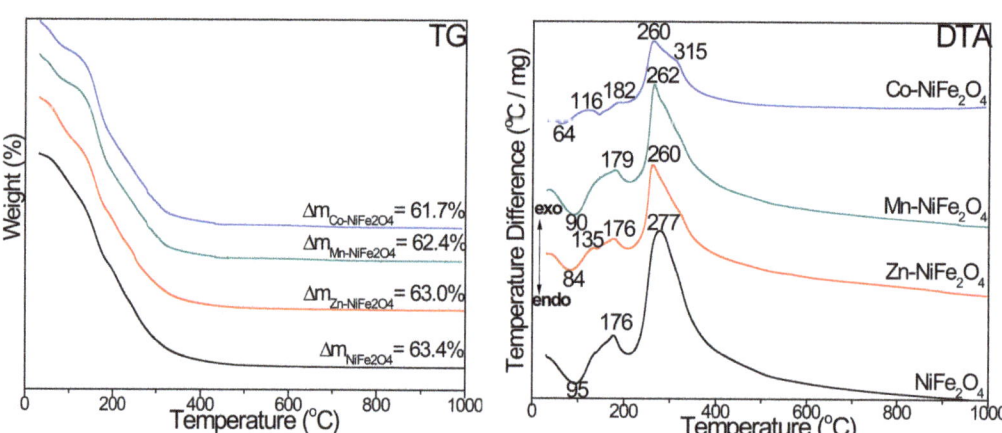

Figure 2. TG and DTA curves of virgin and doped $NiFe_2O_4$ ($Zn-NiFe_2O_4$, $Mn-NiFe_2O_4$, $Co-NiFe_2O_4$) dried at 40 °C.

In case of virgin NiFe$_2$O$_4$ and Mn-doped NiFe$_2$O$_4$, both the glyoxylate precursors formation and decomposition into NiFe$_2$O$_4$ and Mn$_{0.15}$Ni$_{0.85}$Fe$_2$O$_4$ take place in single stages. The total mass loss is 63.4% for NiFe$_2$O$_4$ and 62.4% for Mn$_{0.15}$Ni$_{0.85}$Fe$_2$O$_4$, respectively. Zn-doped NiFe$_2$O$_4$ shows the same three processes, but the formation of metal-glyoxylate precursors takes place in two stages: the Ni- and Zn-glyoxylates are formed at 135 °C, whereas the Fe- glyoxylate at 176 °C. The total mass loss shown on the TG curve is 63%. In case of Co-doped NiFe$_2$O$_4$, both the formation and the decomposition of the glyoxylate precursors occur in two stages: Co- and Ni- glyoxylates are formed at 166 °C and decomposed at 260 °C, whereas Fe- glyoxylate is formed at 182 °C and decomposed at 315 °C. During the metal glyoxylates decomposition, the resulted Fe$_2$O$_3$ reacts with Co$_3$O$_4$ and NiO to form Co$_{0.15}$Ni$_{0.85}$Fe$_2$O$_4$ [7,8]. The TG curve indicate a total mass loss of 61.7%. Thus, between 260 and 277 °C, the virgin Ni-ferrite, as well as the Zn- and Mn-doped ferrites are formed, whereas the Co-doped ferrite is formed at 315 °C. The mass losses are comparable, the highest mass loss being recorded for the virgin Ni-ferrite and the lowest mass loss for Co doped Ni-ferrite.

3.2. Fourier-Transform Infrared Spectroscopy

The FT-IR spectra offers data on the presence of different functional groups, molecular geometry and inter-molecular interactions [1]. In samples heated at 40 °C, the FT-IR spectra (Figure 3) display an intense band at 1384 cm^{-1} specific to nitrate groups [25,30], which disappears for samples heated at 200 °C, indicating the nitrates decomposition.

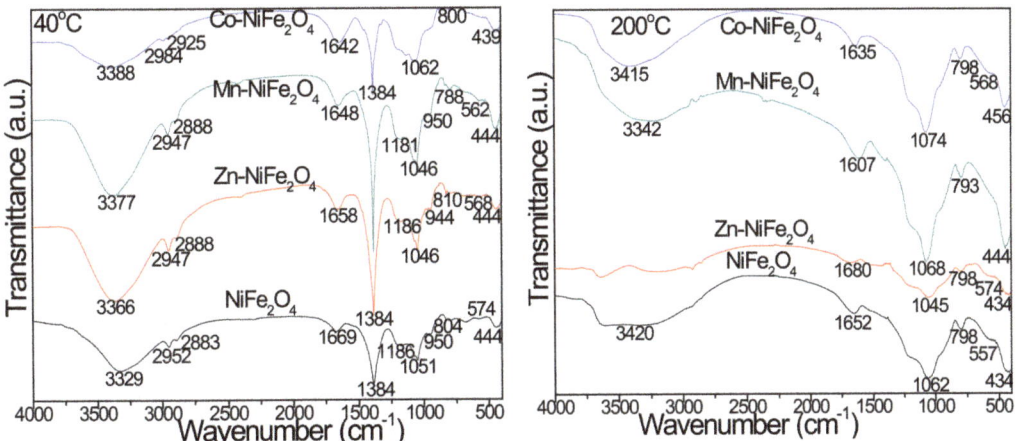

Figure 3. FT-IR spectra of virgin and doped NiFe$_2$O$_4$ (Zn-NiFe$_2$O$_4$, Mn-NiFe$_2$O$_4$, Co-NiFe$_2$O$_4$) heated at 40 and 200 °C.

The bands at 2984–2952 and 2888–2925 cm^{-1} are attributed to C-H bond-asymmetric and symmetric stretching in 1,2ED. The band at 1669–1642 cm^{-1} is assigned to the O-H stretching and bending in both 1,2ED and adsorbed water. The band at 3388–3329 cm^{-1} is assigned to the O-H stretching and intermolecular hydrogen bonds in 1,2ED at 40 °C, and the band at 950–944 cm^{-1} is assigned to -OH stretching and Si-OH deformation vibration following the hydrolysis of -Si (OCH$_2$CH$_3$)$_4$ groups in TEOS [25,30]. In samples heated to 200 °C, the vibration of C=O in COO$^-$ groups indicated by the bands at 1680–1607 cm^{-1}, confirms the coordination of carboxylate groups by metal ions and the formation of a chelated complex [25,30]. The band at 574–557 cm^{-1} is assigned to tetrahedral M-O bonds and cyclic Si-O-Si structures vibrations, whereas the band at 456–434 cm^{-1} is assigned to the octahedral M-O and Si-O bonds vibration [25,30]. The formation of the SiO$_2$ matrix is confirmed by the Si-O bond vibration at 439–456 cm^{-1}, Si-O-Si cyclic structures vibration at

574–557 cm^{-1}, Si-O-Si chains symmetric stretching and bending at 788–810 cm^{-1}, and Si-O-Si bonds stretching vibration at 1046–1074 cm^{-1} with shoulder at 1181–1186 cm^{-1} [25,30].

The FT-IR spectra of virgin and doped NiFe$_2$O$_4$ (Zn-NiFe$_2$O$_4$, Mn-NiFe$_2$O$_4$, Co-NiFe$_2$O$_4$) annealed at high temperatures (Figure 4) show the presence of the characteristic bands for SiO$_2$ matrix: O-H bonds vibration in Si-OH group (3465–3346 cm^{-1}), H-O-H bending (1697–1615 cm^{-1}), Si-O-Si stretching (1094–1070 cm^{-1}), Si-O chains symmetric stretching and bending in the SiO$_4$ tetrahedron (803–794 cm^{-1}), Si-O bonds vibration (47–450 cm^{-1}) and Si-O-Si cyclic structures vibration (613–568 cm^{-1}) [5,6,10,11,15]. The absorption band at 470–450 cm^{-1} can be attributed also to the M-O stretching vibration at the octahedral (B) site, whereas that at 613–568 cm^{-1} to the M-O stretching vibration at the tetrahedral site [1,6,25,30], indicating the formation of ferrites with cubic structure [1]. The doping of NiFe$_2$O$_4$ with larger size and higher atomic weight divalent ions forces the migration of Fe^{3+} ions to the octahedral (B) sites leading to a decrease of the tetrahedral vibration frequency and an increase of the octahedral vibration frequency [9].

3.3. X-ray Diffraction

The XRD patterns of virgin and doped NiFe$_2$O$_4$ annealed at 400, 700, and 1000 °C are presented in Figure 4. The samples annealed at 400 °C display the diffraction peaks corresponding to the reflection planes of (220), (311), (222), (400), (422), (511), and (440), confirming the presence of low-crystallized single phase NiFe$_2$O$_4$ (JCPDS card no 89-4927) [31]), with no detectable impurity phases [1,25]. By increasing the annealing temperature (700 and 1000 °C), in case of NiFe$_2$O$_4$, Zn-(Zn-NiFe$_2$O$_4$) and Co-doped (Co-NiFe$_2$O$_4$) NiFe$_2$O$_4$ single phase ferrites are obtained. The increase of the diffraction lines' intensity indicates the increase of crystallinity and particle size [5]. In case of Mn-doped NiFe$_2$O$_4$ (Mn-NiFe$_2$O$_4$), both at 700 and 1000 °C, cristobalite (JCPDS card no. 89-3434 [31], quartz (JCPDS card 85-0457 [31]) and Fe$_2$SiO$_4$ (JCPDS card no.87-0315 [31]) are also identified as secondary phases. The presence of secondary phases could be explained by the higher mobility of cations and strain variation induced by the annealing process, that also slightly shifts the 2θ positions and broadens the peaks, concomitantly with the increase of crystallite sizes [31]. The formation of Fe$_2$SiO$_4$ could be attributed to the reducing conditions generated during the carboxylate precursors decomposition that partially reduce the Fe^{3+} ions to Fe^{2+} ions within the SiO$_2$ matrix pores, which further reacts with SiO$_2$ to form Fe$_2$SiO$_4$ [12,13,32].

The variation of the oxygen atoms' positions results in structural distortion of the FeO$_6$, FeO$_4$, and NiO$_6$ complexes that highly disturb the NiFe$_2$O$_4$ lattice, leading to structural changes with high impact on the physico-chemical properties [5]. In case of doping with Mn^{2+} ions, the diffraction peak situated near 2θ = 35° are slightly shifted. Some possible explanations could be the Mn^{2+} ions that enter in the octahedral (B) sites as well as the larger radius of Mn^{2+} (0.80 Å) than of Ni^{2+} (0.72 Å) [15]. The crystallite size (D) calculated from the most intense diffraction peaks (311), lattice constant (a), unit cell volume (V), bulk density (d$_p$), X-ray density (d$_{XRD}$), porosity (P), and hopping length in tetrahedral (L$_A$) and octahedral (L$_B$) sites [6,8,33–35] are shown in Table 1. XRD parameters are influenced not only by the crystallite size, lattice strain and defects, but also by the annealing temperature and doping ions [6]. The sharpening and narrowing of the diffraction peaks suggest the crystallite size become more obvious with the annealing temperature [16]. At high annealing temperatures (1000 °C), a significant agglomeration takes place without subsequent recrystallization, supporting the formation of a single crystal instead of a polycrystal structure [5,36].

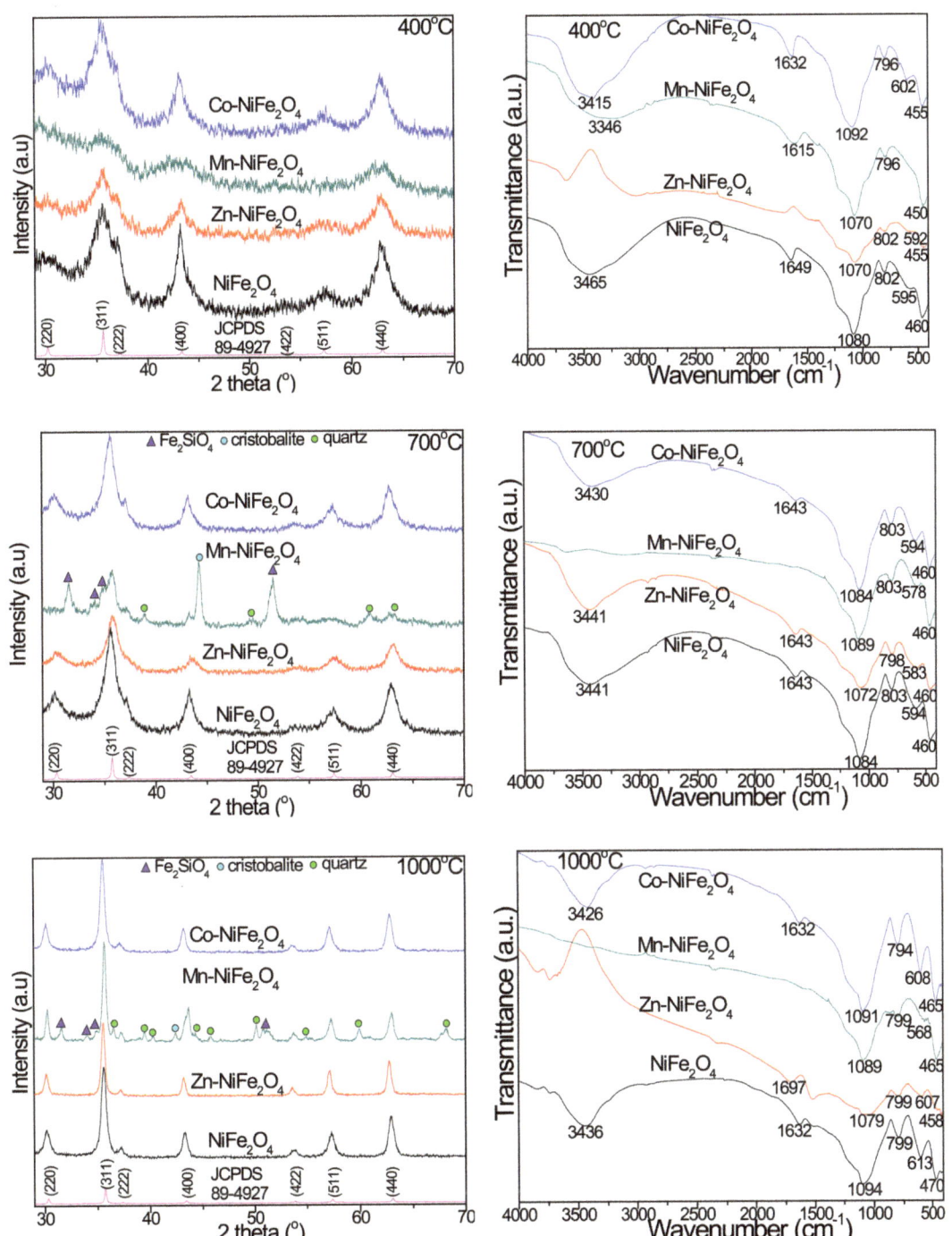

Figure 4. XRD patterns and FT-IR spectra of virgin and doped $NiFe_2O_4$ ($Zn-NiFe_2O_4$, $Mn-NiFe_2O_4$, $Co-NiFe_2O_4$) samples annealed at 400, 700, and 1000 °C.

Table 1. Crystallite size (D), lattice constant (a), unit cell volume (V), bulk density (d_p), X-ray density (d_{XRD}), porosity (P), hopping length in tetrahedral sites (L_A) and in octahedral sites (L_B) and M/Ni/Fe molar ratio of virgin and doped $NiFe_2O_4$ ($Zn-NiFe_2O_4$, $Mn-NiFe_2O_4$, $Co-NiFe_2O_4$) NCs.

Temp (°C)	Sample	D (nm)	A (Å)	V (Å³)	d_p (g/cm³)	d_{XRD} (g/cm³)	P (%)	L_A (Å)	L_B (Å)	M/Ni/Fe Molar Ratio
400	$NiFe_2O_4$	11.5	8.4008	592.9	4.528	5.251	13.76	3.638	2.970	0/0.96/2.04
	$Zn-NiFe_2O_4$	4.4	8.3584	583.9	4.602	5.355	14.06	3.619	2.955	0.97/0.97/2.03
	$Mn-NiFe_2O_4$	16.5	8.4135	595.6	4.508	5.215	13.56	3.643	2.975	0.96/0.95/2.02
	$Co-NiFe_2O_4$	5.8	8.3462	581.4	4.582	5.356	14.45	3.614	2.951	0.97/0.96/2.03
700	$NiFe_2O_4$	18.2	8.4058	593.9	4.496	5.243	14.25	3.640	2.972	0/0.96/2.04
	$Zn-NiFe_2O_4$	6.7	8.3676	585.9	4.552	5.337	14.71	3.623	2.958	0.98/0.98/2.01
	$Mn-NiFe_2O_4$	24.6	8.4231	597.6	4.452	5.198	14.35	3.647	2.978	-
	$Co-NiFe_2O_4$	9.5	8.3923	591.1	4.488	5.267	14.79	3.634	2.967	0.99/0.98/2.01
1000	$NiFe_2O_4$	27.6	8.4182	596.6	4.411	5.219	15.48	3.645	2.976	0/0.99/2.00
	$Zn-NiFe_2O_4$	8.7	8.3824	589.9	4.478	5.301	15.52	3.630	2.964	0.99/1.00/2.00
	$Mn-NiFe_2O_4$	38.4	8.4295	599.0	4.401	5.186	15.13	3.650	2.980	-
	$Co-NiFe_2O_4$	20.2	8.4095	594.7	4.368	5.237	16.59	3.641	2.973	0.00/0.99/2.01

The lattice constant (a) increases, whereas the X-ray density (d_{XRD}) decreases with increasing crystallite size. Some possible explanations could be the surface tension decrease caused by the size effect and the expansion of unit cell by replacing Ni^{2+} with Zn^{2+}, Co^{2+}, and Mn^{2+} ions [6,17,33]. Considering the small difference between the atomic weight of Ni^{2+} and Mn^{2+} ions, the d_{XRD} variation may be attributed to the changes of the lattice constant (a) [37]. The lattice constant (a) shows a linear behavior and it follows Vegard's law. The differences between the lattice parameter of investigated samples were attributed to the different ionic radii of Fe^{3+} (tetra: 0.49; octa: 0.64 Å), Zn^{2+} (tetra: 0.60; octa: 0.74 Å), Ni^{2+} (tetra: 0.54; octa: 0.78 Å), Mn^{2+} (tetra: 0.58; octa: 0.69 Å), and Co^{2+} (tetra: 0.58; octa: 0.74 Å) [17,33,35]. The decrease of porosity (P) with the increase of annealing temperature may be a consequence of the rapid densification during the annealing process [6,17,33].

3.4. Chemical Analysis

The M/Ni/Fe molar ratio calculated based on Co, Mn, Zn, Ni and Fe concentrations measured by ICP-OES confirmed the theoretical elemental composition of the obtained NCs (Table 1). In all cases, the best fit of experimental and theoretical data was remarked for samples annealed at 1000 °C. In case of Mn-dopped $NiFe_2O_4$ annealed at 700 and 1000 °C, the Mn/Ni/Fe molar ratio could not be calculated based on the metal concentrations, due to the presence of Fe_2SiO_4 as secondary phase.

3.5. Transmission and Scanning Electron Microscopy

The TEM images (Figure 5) reveal irregularly shaped particles that form agglomerates. As a result of the doping with Zn^{2+} and Co^{2+} ions, a decrease of the particle size from 29 nm ($NiFe_2O_4$) to 10 nm ($Zn-NiFe_2O_4$) and 21 nm ($Co-NiFe_2O_4$) was observed, whereas by doping with Mn^{2+} ion, the particle size increases to 43 nm ($Mn-NiFe_2O_4$).

The variation of particle size by doping may be determined by the different kinetics of metal oxides' formation reaction, the different particle growth rate or the presence of structural disorder and strain in the lattice due to different ionic radii [14,37]. The different particle arrangement could be attributed to the formation of well-delimited grains that form solid boundaries.

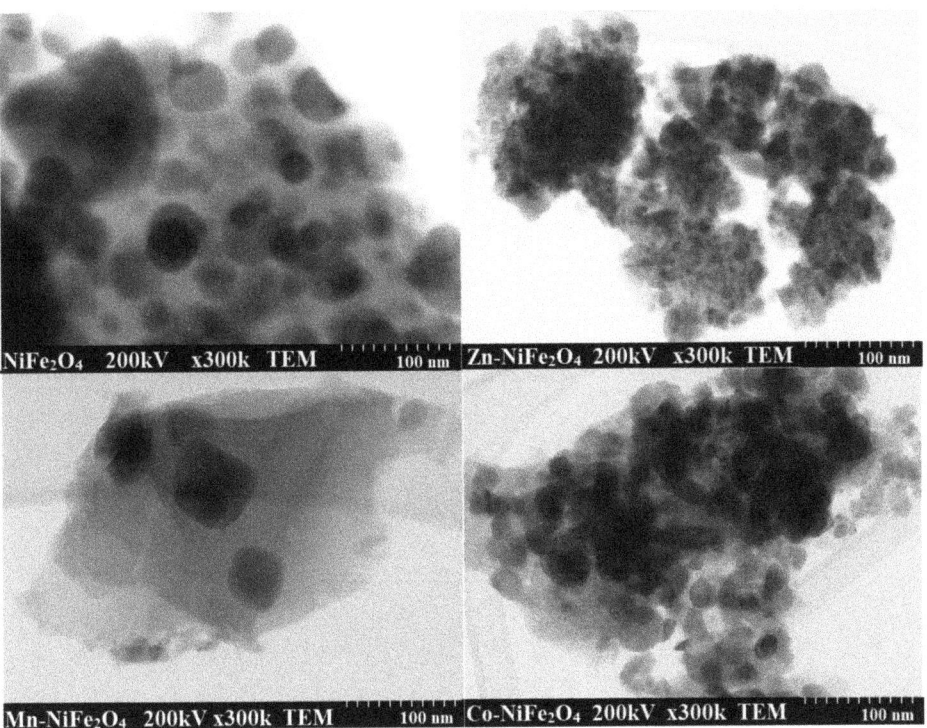

Figure 5. TEM images of virgin and doped $NiFe_2O_4$ (Zn-$NiFe_2O_4$, Mn-$NiFe_2O_4$, Co-$NiFe_2O_4$) NCs annealed at 1000 °C.

The particle agglomeration is frequently observed in case of NCs synthesized by chemical routes and is caused most probably by the assembling tendency of small particles, magnetic nature, and weak surface interaction due to Van der Waals forces [8,9,25,33]. The internal heat energy produced during the annealing may also lead to the agglomeration of particles due to interfacial surface tensions [8,25].

The differences obtained between particle and crystallite size result most probably due to the interference of amorphous matrix and of large-size nanoparticles that highly influence the diffraction patterns, by the large fraction of the total number of atoms contained [8]. The crystal-growth rate increase could be attributed to volume expansion and supersaturation reduction of the system at high annealing temperatures, which further leads to increase of the amorphous Fe oxides solubility and crystallization of $M_{0.15}Ni_{0.85}Fe_2O_4$ when Mn, Zn, and Co diffuse into the crystal structure of $NiFe_2O_4$. When the nucleation rate exceeds the growth rate, small and homogenously distributed particles are obtained. At high annealing temperatures, these particles may join together due to coalescence, formation of crystalline clusters, and joint cementation [8,14,25,37].

The SEM images (Figure 6) indicates agglomerations of homogenous, clearly delimited particles typical of ferrite materials containing magnetic elements [25]. The particles in Zn and Co doped $NiFe_2O_4$ have a homogenous microstructure with closely packed, irregularly shaped small particles, whereas those in Mn doped $NiFe_2O_4$ are bigger and more loosely packed.

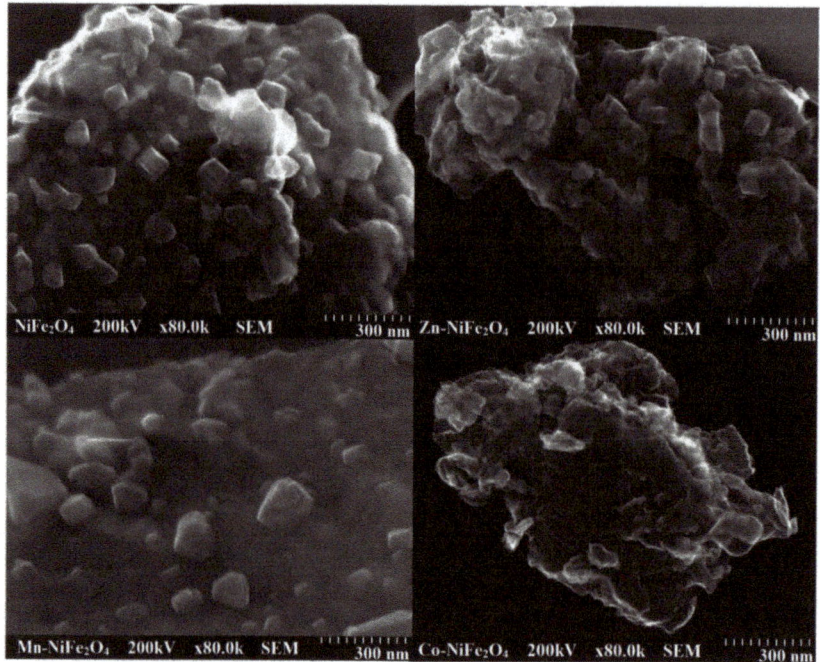

Figure 6. SEM images of virgin and doped $NiFe_2O_4$ (Zn-$NiFe_2O_4$, Mn-$NiFe_2O_4$, Co-$NiFe_2O_4$) NCs annealed at 1000 °C.

3.6. Magnetic Properties

All samples display superparamagnetic behavior with well-defined hysteresis loops (Figure 7), but important differences in the magnetic parameters are induced by the doping ions. Small particles contain fewer domain walls and require higher demagnetization force, whereas large particles have a higher probability of domain formation [9].

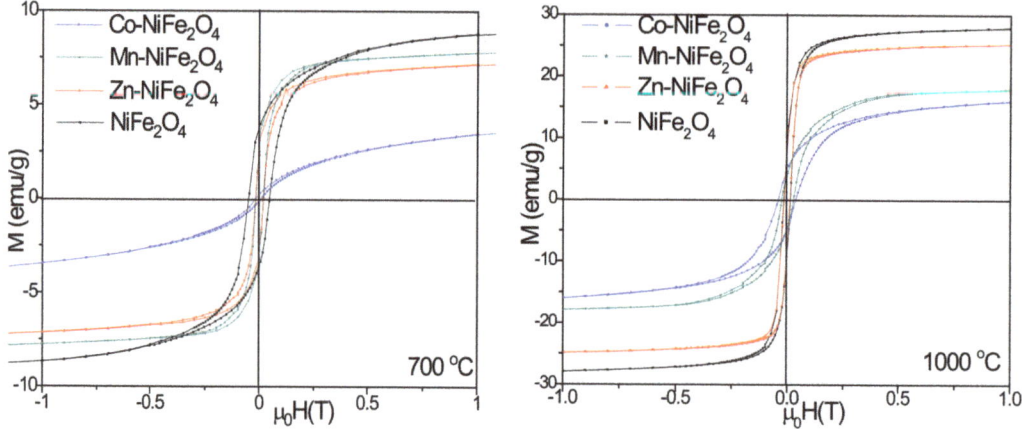

Figure 7. Magnetic hysteresis loops of virgin and doped $NiFe_2O_4$ (Zn-$NiFe_2O_4$, Mn-$NiFe_2O_4$, Co-$NiFe_2O_4$) NCs annealed at 700 and 1000 °C.

The saturation magnetization (M_S), remanent magnetization (M_R), coercivity (H_C), squareness (Sq), and anisotropy (K) of NCs annealed at 700 and 1000 °C are shown in Table 2. The doping of $NiFe_2O_4$ with Zn^{2+}, Mn^{2+}, or Co^{2+} ions lead to a decrease of the M_S and M_R after annealing at 700 and 1000 °C. Above the single-domain critical size, the competition between the increasing magnetostatic energy and the domain-wall energy favors the domain-wall formation and the single-domain particle splits into multi-domain [9].

Table 2. Saturation magnetization (M_S), remanent magnetization (M_R), coercivity (H_C), squareness (S), and anisotropy (K) of virgin and doped $NiFe_2O_4$ (Zn-$NiFe_2O_4$, Mn-$NiFe_2O_4$, Co-$NiFe_2O_4$) NCs.

Sample	M_S (emu/g)		M_R (emu/g)		H_C (Oe)		Sq		$K \cdot 10^3$ (erg/cm^3)	
	700	1000	700	1000	700	1000	700	1000	700	1000
$NiFe_2O_4$	9.4	31.2	3.91	7.48	48	18	0.416	0.240	0.283	0.354
Zn-$NiFe_2O_4$	7.4	25.7	3.24	6.65	20	17	0.438	0.259	0.274	0.306
Mn-$NiFe_2O_4$	8.1	17.8	3.12	2.49	18	24	0.390	0.140	0.268	0.318
Co-$NiFe_2O_4$	3.8	16.2	0.26	3.84	11	32	0.068	0.237	0.026	0.326

The low magnetization value of Co-$NiFe_2O_4$ is due to the incomplete crystallization and small-sized crystallites, which generate structural disorder on the nanoparticles surface. As the particles surface behaves as an inactive layer, its magnetization become negligible [5]. Some possible explanations for the variation of Ms in case of doped $NiFe_2O_4$ could be: (*i*) occupation of the octahedral sites by Zn^{2+} ions, (*ii*) random incomplete A–O–B linkages resulting in the replacement of non-magnetic ions by magnetic ions in the spinel, and (*iii*) the presence of non-collinear magnetic structures [16]. The magnetization caused by domain wall movement needs less energy than the domain rotation. The number of domain walls increases with increasing particle size. In case of Zn^{2+} doping, the wall movement contribution to magnetization is higher than that of the domain rotation [33]. Moreover, the presence of impurity phases with antiparallel magnetic ordering to the ferrite ordering reduces the M_S. The doping with Co^{2+} ions having higher magnetic moment than Ni^{2+} ions result in a decrease of M_S, as Ni^{2+} ion may occupy both the tetrahedral (A) and octahedral (B) sites [36].

By doping $NiFe_2O_4$, the H_C decreases at 700 °C and increases at 1000 °C as a consequence of increased spin disorder in the surface layer and smaller particle size [1,32]. The H_C value of 48 Oe of virgin $NiFe_2O_4$ decreases to 11 Oe in case of Co doping and annealing at 700 °C, most probably due to agglomerates' formation which leads to the increase of average particles size above the critical single domain of $NiFe_2O_4$ particles and further leads to a multidomain magnetic structure [5]. Moreover, by annealing at low temperatures, the grain growth occurs, weakening the domain wall pinning effects at the grain boundary [5].

To calculate the magnetic anisotropy constant (K) of the samples, we assumed that the spinel ferrite particles have spherical shape. The K value of virgin $NiFe_2O_4$ is larger than that of doped $NiFe_2O_4$. The magnetic anisotropy of particles behaves as energy barrier and stops the switching of the magnetization's direction to the easy axis [38,39]. At a certain temperature, the thermal activation overcomes the magnetic anisotropy energy barrier and the magnetization direction of the particles change, indicating a super-paramagnetic behavior [38,39]. A conceivable explanation could be the presence of a magnetically disordered surface layer, where a competition of exchange interactions between surface spins exists. Moreover, the magnetic disorder may originate in uneven magnetic interactions of the surface spins, arbitrarily oriented grains of different sizes and disordered vacancies [38,39].

4. Conclusions

The structural, morphological and magnetic properties of virgin and Zn-, Mn-, and Co-doped $NiFe_2O_4$ embedded in SiO_2 matrix obtained through a modified sol-gel route and thermal treatment were investigated. The FT-IR spectra evidenced the formation of metallic precursors and of SiO$_2$ matrix. The TG/DTA curves of samples dried at 40 °C indicated the formation and decomposition of metallic precursors to ferrites in single or two stages, with comparable mass losses. The XRD analysis revealed single-phase ferrites for virgin, Zn- and Co-doped NiFe$_2$O$_4$, and the presence of secondary crystalline phases derived from the SiO$_2$ matrix (cristobalite, quartz, and Fe$_2$SiO$_4$) in case of Mn-doped NiFe$_2$O$_4$. XRD parameters were influenced not only by the crystallite size, lattice strain, and defects, but also by the annealing temperature and doping ions. The lattice constant and unit cell volume increase by doping with Mn^{2+} ion and decrease by doping with Zn^{2+} and Co^{2+} ions. By contrast, X-ray and bulk densities, and porosity decrease by doping with Mn^{2+} and increase with doping Zn^{2+} and Co^{2+} ions. The NiFe$_2$O$_4$ particle size increases by doping with Mn^{2+} and decrease by doping with Zn^{2+} and Co^{2+} ions, respectively. The doping of NiFe$_2$O$_4$ with Zn^{2+}, Mn^{2+} and Co^{2+} leads to a decrease of the saturation magnetization and remanent magnetization, whereas the coercivity decreases at 700 °C and increases at 1000 °C. The obtained magnetic transition metal dopped-Ni ferrite nanoparticles are possible candidates for various medical applications like controlled drug delivery, cancer therapy, biosensing, and magnetic resonance imaging.

Author Contributions: T.D., conceptualization, methodology, investigation, validation, writing-original draft, visualization, supervision; O.C. and E.A.L., methodology, investigation, writing-review and editing. All authors have read and agreed to the published version of the manuscript.

Funding: T.D. was financed from the own funds of the Technical University of Cluj-Napoca. O.C. and E.A.L. were financed by the Ministry of Research, Innovation and Digitalization through Program 1—Development of the National Research & Development System, Subprogram 1.2—Institutional performance—Projects that finance the RDI excellence, contract no. 18PFE/30.12.2021.

Institutional Review Board Statement: Not applicable.

Informed Consent Statement: Not applicable.

Data Availability Statement: Not applicable.

Acknowledgments: The authors would like to express their gratitude to Lucian Barbu Tudoran for the TEM measurements.

Conflicts of Interest: The authors declare no conflict of interest.

References

1. Abushad, A.; Arshad, M.; Naseem, S.; Ahmed, H.; Ansari, A.; Chakradhary, V.K.; Husain, S.; Khan, W. Synthesis and role of structural disorder on the optical, magnetic and dielectric properties of Zn doped NiFe$_2$O$_4$ nanoferrites. *J. Mol. Struct.* **2022**, *1253*, 132205.
2. Suresh, J.; Trinadh, B.; Babu, B.V.; Reddy, P.V.S.S.S.N.; Mohan, B.S.; Krishna, A.R.; Samatha, K. Evaluation of micro-structural and magnetic properties of nickel nano-ferrite and Mn^{2+} substituted nickel nano-ferrite. *Phys. B Condens. Matter.* **2021**, *620*, 413264. [CrossRef]
3. Rama, G.; Dhiman, P.; Kumar, A.; Vo, D.V.N.; Sharma, G.; Sharma, S.; Naushad, M. Recent advances on nickel nano-ferrite: A review on processing techniques, properties and diverse applications. *Chem. Eng. Res. Des.* **2021**, *175*, 182–208.
4. Samieemehr, M.; Arab, A.; Kiani, E. Influence of two-step sintering on power loss and permeability dispersion of MnZnNi ferrite. *J. Magn. Magn. Mater.* **2022**, *553*, 169269. [CrossRef]
5. Pottker, W.E.; Ono, R.; Cobos, M.A.; Hernando, A.; Araujo, J.F.D.F.; Bruno, A.C.O.; Lourenço, S.A.; Longo, E.; La Porta, F.A. Influence of order-disorder effects on the magnetic and optical properties of NiFe$_2$O$_4$ nanoparticles. *Ceram. Int.* **2018**, *44*, 17290–17297. [CrossRef]
6. Samson, V.A.; Bernadsha, S.B.; Xavier, R.; Rueshwin, C.S.T.; Prathap, S.; Madhavan, J.; Raj, M.V.A. One pot hydrothermal synthesis and characterization of NiFe$_2$O$_4$ nanoparticles. *Mater. Today Proc.* **2022**, *50*, 2665–2667. [CrossRef]
7. Barvinschi, P.; Stefanescu, O.; Dippong, T.; Sorescu, S.; Stefanescu, M. CoFe$_2$O$_4$/SiO$_2$ nanocomposites by thermal decomposition of some complex combinations embedded in hybrid silica gels. *J. Therm. Anal. Calorim.* **2013**, *112*, 447–453. [CrossRef]
8. Dippong, T.; Levei, E.A.; Goga, F.; Petean, I.; Avram, A.; Cadar, O. The impact of polyol structure on the formation of Zn$_{0.6}$Co$_{0.4}$Fe$_2$O$_4$ spinel-based pigments. *J. Sol-Gel Sci. Technol.* **2019**, *92*, 736–744. [CrossRef]
9. Vinosha, P.A.; Xavier, B.; Krishnan, S.; Das, S.J. Investigation on zinc substituted highly porous improved catalytic activity of NiFe$_2$O$_4$ nanocrystal by co-precipitation method. *Mater. Res. Bull.* **2018**, *101*, 190–198. [CrossRef]

10. Patil, K.; Phadke, S.; Mishra, A. A study of structural and dielectric properties of Zn^{2+} doped $MnFe_2O_4$ and $NiFe_2O_4$ spinel ferrites. *Mater. Today Proc.* **2021**, *46*, 2226–2228. [CrossRef]
11. Ugendar, K.; Babu, V.H.; Reddy, V.R.; Markaneyulu, G. Cationic ordering and magnetic properties of rare-earth doped $NiFe_2O_4$ probed by Mössbauer and X-ray spectroscopies. *J. Magn. Magn. Mater.* **2019**, *484*, 291–297. [CrossRef]
12. Maaz, K.; Duan, J.L.; Karim, S.; Chen, Y.H.; Zhai, P.F.; Xu, L.J.; Yao, H.J.; Liu, J. Fabrication and size dependent magnetic studies of $Ni_xMn_{1-x}Fe_2O_4$ (x = 0.2) cubic nanoplates. *J. Alloys Compd.* **2016**, *684*, 656–662. [CrossRef]
13. Abdallah, H.M.I.; Moyo, T. Superparamagnetic behavior of $Mn_xNi_{1-x}Fe_2O_4$ spinel nanoferrites. *J. Magn. Magn. Mater.* **2014**, *361*, 170–174. [CrossRef]
14. Dippong, T.; Levei, E.A.; Deac, I.G.; Petean, I.; Borodi, G.; Cadar, O. Sol-gel synthesis, structure, morphology and magnetic properties of $Ni_{0.6}Mn_{0.4}Fe_2O_4$ nanoparticles embedded in SiO_2 matrix. *Nanomaterials* **2021**, *11*, 3455. [CrossRef]
15. Luo, T.; Hou, X.; Liang, Q.; Zhang, G.; Chen, F.; Xia, Y.; Ru, Q.; Yao, L.; Wu, Y. The influence of manganese ions doping on nanosheet assembly $NiFe_2O_4$ for the removal of Congo red. *J. Alloys Compd.* **2018**, *763*, 771–780. [CrossRef]
16. Chakradhary, V.K.; Ansari, A.; Jaleel Akhtar, M. Design, synthesis and testing of high coercivity cobalt doped nickel ferrite nanoparticles for magnetic applications. *J. Magn. Magn. Mater.* **2019**, *469*, 674–680. [CrossRef]
17. Chand, P.; Vaish, S.; Kumar, P. Structural, optical and dielectric properties of transition metal (MFe_2O_4; M= Co, Ni and Zn) nanoferrites. *Phys. B* **2017**, *524*, 53–63. [CrossRef]
18. Bhame, S.D.; Joy, P.A. Enhanced strain sensitivity in magnetostrictive spinel ferrite $Co_{1-x}Zn_xFe_2O_4$. *J. Magn. Magn. Mater.* **2018**, *447*, 150–154. [CrossRef]
19. Marinca, T.F.; Chicinaș, I.; Isnard, O.; Neamțu, B.V. Nanocrystalline/nanosized manganese substituted nickel ferrites—$Ni_{1-x}Mn_xFe_2O_4$ obtained by ceramic-mechanical milling route. *Ceram. Int.* **2016**, *42*, 4754–4763. [CrossRef]
20. Shobana, M.K.; Sankar, S. Structural, thermal and magnetic properties of $Ni_{1-x}Mn_xFe_2O_4$ nanoferrites. *J. Magn. Magn. Mater.* **2009**, *321*, 2125–2128. [CrossRef]
21. Hassadee, A.; Jutarosaga, T.; Onreabroy, W. Effect of zinc substitution on structural and magnetic properties of cobalt ferrite. *Procedia Eng.* **2012**, *32*, 597–602. [CrossRef]
22. Reddy, M.P.; Zhou, X.; Yann, A.; Du, S.; Huang, Q.; Mohamed, A. Low temperature hydrothermal synthesis, structural investigation and functional properties of $Co_xMn_{1-x}Fe_2O_4$ ($0 \leq x \leq 1$) nanoferrites. *Superlattices Microst.* **2015**, *81*, 233–242. [CrossRef]
23. Desai, H.B.; Hathiya, L.J.; Joshi, H.H.; Tanna, A.R. Synthesis and characterization of photocatalytic $MnFe_2O_4$ nanoparticles. *Mater. Today Proc.* **2020**, *21*, 1905–1910. [CrossRef]
24. Salunkhe, A.B.; Khot, V.M.; Phadatare, M.R.; Thorat, N.D.; Joshi, R.S.; Yadav, H.M.; Pawar, S.H. Low temperature combustion synthesis and magnetostructural properties of Co-Mn nanoferrites. *J. Magn. Magn. Mater.* **2014**, *352*, 91–98. [CrossRef]
25. Dippong, T.; Levei, E.A.; Lengauer, C.L.; Daniel, A.; Toloman, D.; Cadar, O. Investigation of thermal, structural, morpho-logical and photocatalytic properties of $Cu_xCo_{1-x}Fe_2O_4$ nanoparticles embedded in SiO_2 matrix. *Mater. Charact.* **2020**, *163*, 110268. [CrossRef]
26. Philipse, A.P.; van Bruggen, M.P.B.; Pathmamanoharan, C. Magnetic silica dispersions: Preparation and stability of sur-face-modified silica particles with a magnetic core. *Langmuir* **1994**, *10*, 92–99. [CrossRef]
27. Asghar, K.; Qasim, M.; Das, D. Preparation and characterization of mesoporous magnetic $MnFe_2O_4@mSiO_2$ nanocomposite for drug delivery application. *Mater. Today Proc.* **2020**, *26*, 87–93. [CrossRef]
28. Vestal, C.R.; Zhang, Z.J. Synthesis and magnetic characterization of Mn and Co spinel ferrite-silica nanoparticles with tunable magnetic core. *Nano Lett.* **2003**, *3*, 1739–1743. [CrossRef]
29. Dippong, T.; Levei, E.A.; Deac, I.G.; Petean, I.; Cadar, O. Dependence of structural, morphological and magnetic proper-ties of manganese ferrite on Ni-Mn substitution. *Int. J. Mol. Sci.* **2022**, *23*, 3097. [CrossRef]
30. Torkian, S.; Ghasemi, A.; Razavi, R.S. Cation distribution and magnetic analysis of wideband microwave absorptive $Co_xNi_{1-x}Fe_2O_4$ ferrites. *Ceram. Int.* **2017**, *43*, 6987–6995. [CrossRef]
31. Joint Committee on Powder Diffraction Standards. *Powder Diffraction File*; International Center for Diffraction Data: Swarthmore, PA, USA, 1999.
32. Mathubala, G.; Manikandan, A.; Arul Antony, S.; Ramar, P. Photocatalytic degradation of methylene blue dye and magnetooptical studies of magnetically recyclable spinel $Ni_xMn_{1-x}Fe_2O_4$ (x = 0.0–1.0) nanoparticles. *J. Mol. Struct.* **2016**, *113*, 79–87. [CrossRef]
33. Džunuzović, A.S.; Ilić, N.I.; Vijatović Petrović, M.M.; Bobić, J.D.; Stojadinović, B.; Dohčević-Mitrović, Z.; Stojanović, B.D. Structure and properties of Ni-Zn ferrite obtained by auto-combustion method. *J. Magn. Magn. Mater.* **2015**, *374*, 245–251. [CrossRef]
34. Ati, M.A.; Othaman, Z.; Samavati, A. Influence of cobalt on structural and magnetic properties of nickel ferrite nanoparticles. *J. Mol. Struct.* **2013**, *1052*, 177–182. [CrossRef]
35. Gaffour, A.; Ravinder, D. Characterization of nano-structured nickel-cobalt ferrites synthesized by citrate-gel auto combustion method. *Int. J. Sci. Eng. Res.* **2014**, *4*, 73–79.
36. Sontu, U.B.; Rao, N.; Reddy, V.R.M. Temperature dependent and applied field strength dependent magnetic study of cobalt nickel ferrite nano particles: Synthesized by an environmentally benign method. *J. Magn. Magn. Mater.* **2018**, *352*, 398–406. [CrossRef]
37. Jadhav, J.; Biswas, S.; Yadav, A.K.; Jha, S.N.; Bhattacharyya, D. Structural and magnetic properties of nanocrystalline Ni-Zn ferrites: In the context of cationic distribution. *J. Alloys Compd.* **2017**, *696*, 28–41. [CrossRef]

38. Ozçelik, B.; Ozçelik, S.; Amaveda, H.; Santos, H.; Borrell, C.J.; Saez-Puche, R.; de la Fuente, G.F.; Angurel, L.A. High speed processing of NiFe$_2$O$_4$ spinel using a laser furnace. *J. Mater.* **2020**, *6*, 661–670. [CrossRef]
39. Shen, W.; Zhang, l.; Zhao, B.; Du, Y.; Zhou, X. Growth mechanism of octahedral like nickel ferrite crystals prepared by modified hydrothermal method and morphology dependent magnetic performance. *Ceram. Int.* **2018**, *44*, 9809–9815. [CrossRef]

Article

The Effect of Ni²⁺ Ions Substitution on Structural, Morphological, and Optical Properties in CoCr₂O₄ Matrix as Pigments in Ceramic Glazes

Firuta Goga [1], Rares Adrian Bortnic [2], Alexandra Avram [1], Mioara Zagrai [3], Lucian Barbu Tudoran [4] and Raluca Anca Mereu [1,*]

1. Faculty of Chemistry and Chemical Engineering, Babeş-Bolyai University, 11 Arany Janos Street, 400028 Cluj-Napoca, Romania
2. Faculty of Physics, Babes-Bolyai University, 1 Kogalniceanu Street, 400084 Cluj-Napoca, Romania
3. National Institute for Research and Development of Isotopic and Molecular Technologies, 67-103 Donath Street, 400293 Cluj-Napoca, Romania
4. Electron Microscopy Center, Faculty of Biology and Geology, Babes-Bolyai University, 400006 Cluj-Napoca, Romania
* Correspondence: raluca.mereu@ubbcluj.ro

Abstract: The structural, morphological, and optical properties of Ni²⁺ ions substitution in CoCr₂O₄ matrix as ceramic pigments were investigated. The thermal decomposition of the dried gel was performed aiming to understand the mass changes during annealing. The X-ray diffraction (XRD) studies reveal a spinel-type Face–Centered Cubic structure and a secondary Cr₂O₃ phase when $x \leq 0.75$ and a Body–Centered Tetragonal structure when $x = 1$. Fourier Transform Infrared Spectroscopy (FT–IR) indicated two strong absorption bands corresponding to the metal–oxygen stretching from tetrahedral and octahedral sites, characteristic of spinel structure. Ultraviolet–Visible (UV–Vis) spectra exhibited the electronic transitions of the Cr²⁺ Cr³⁺ and Ni²⁺ ions. From the UV–Vis data, the CIE color coordinates, (x, y) of the pigments were evaluated. The morphology was examined by Scanning Electron Microscopy (SEM) and Transmission Electron Microscopy (TEM) showing the agglomeration behavior of the particles. The stability, coloring properties and potential ceramic applications of studied pigments were tested by their incorporation in matte and glossy tile glazes followed by the application of obtained glazes on ceramic tiles. This study highlights the change in pigment color (from turquoise to a yellowish green) with Ni²⁺ ions substitution in the CoCr₂O₄ spinel matrix.

Keywords: ceramic pigments; spinel; chromite; optical properties; coloring properties

1. Introduction

The color of ceramic tiles, followed by the design and the technical properties, is one of the most appreciated characteristics by the customer. Thereby, there is a real need in diversifying the ceramic color palette. In addition, the use of the newer technologies for obtaining easier, faster, and qualitative colored models on tiles keeps the research in the field of ceramic pigments open. Nowadays, the demand is still on in finding new materials suitable for the challenge created by the digital decoration ink-jet printing for ceramic tiles. This application allows the use of colloidal suspensions of pigments to improve decorative features of ceramic parts and other ceramic tiles [1].

The ceramic pigments are inorganic materials which have high thermal and chemical stability. These properties allow them to be subjected to the processing conditions without losing their coloring characteristics.

Spinel compounds have a huge contribution in the field of ceramic pigments to obtain different colors like red, pink, brown, gray, or blue.

Spinels with the general formula AB₂O₄ represent a large family of inorganic compounds which are now used in different industrial branches as ceramic pigments [2–5], magnetic materials [6,7], catalytic materials [8], supercapacitors [9,10], due to their excellent chemical, thermal, mechanical, magnetic properties, and high melting point. Although spinels have a complex composition, they are intrinsically colored and present several advantages, when compared to the metal oxides as colorants in ceramics and enamels.

The properties of the oxide materials and consequently of spinels, are strongly correlated with the synthesis method and thermal behavior, which affects the interspersion of the components and thus defines the morphology [11]. This also influences the coloring properties of pigments.

In general, the synthesis methods employed depend especially on the final use of the material and are based on the relation between structure—morphology and properties. Consequently, there are different methods utilized in the synthesis of spinel structures, the conventional ones (co-precipitation and hydro/solvo-thermal) and non-conventional ones (sol-gel and derivatives or auto-combustion) [12–14]. Among them, the sol-gel synthesis offers great versatility to prepare low-cost materials with a rigorous control from the compositional and microstructural point of view. The sol-gel method is based on inorganic polymerization reactions and so, starting from molecular precursors an oxide can be easily obtained via hydroxylation-condensation reactions [1,12].

Spinel pigments can be obtained by the addition of chromophores, transition metal ions into inert matrices such as oxides or by calcination of metal-organic complexes [15]. They have the general formula AB_2O_4 where the A-sites are tetrahedral coordinated occupied by divalent ions as A^{2+} (like Co, Ni, Mn, Zn), and the B-sites are octahedral coordinated and occupied by trivalent ions B^{3+} (like Cr or Al) [16].

In the literature, there are studies regarding spinel oxides such as $CoCr_2O_4$, $MgCr_2O_4$, $Co_{(1-x)}Mg_xCr_2O_4$ [17,18] and $NiCr_2O_4$ [19] which have been synthesized by different methods and are investigated for different applications due to their thermal stability, as previously noted.

Nowadays, for the blue pigment, cobalt-based materials are used [1]. There are few drawbacks for the use of cobalt, i.e., abundance, extraction, geopolitical instabilities, and the need to be used in the electronic or other industries.

In the present context, the aim of this work is to synthesize and to investigate new competitive pigments for novel technologies with as little cobalt as possible. Considering that, we studied the nickel substitution in the cobalt chromite matrix. For this investigation, an easy to use and environmentally friendly modified sol-gel method was employed for the synthesis of $Co_{(1-x)}Ni_xCr_2O_4$ (x = 0, 0.25, 0.5, 0.75 and 1).

Structural, morphological, and optical properties were studied as a function of the nickel substitution in the spinel matrix focusing on the application as ceramic pigments.

The proposed modified sol-gel method involves metallic salts combined with the sucrose and pectin as organic precursors. The method was successfully utilized in the synthesis of other spinel oxide compounds with nanometric dimensions [20].

More than obtaining these chromite spinel pigments through an easy, cost-effective method for the first time, the aim of this work was to test their potential applicability in the ceramic tile industry. For this purpose, all synthetized pigments were embedded in two different types of glazes to also test their efficacy as ceramic pigments. The novelty of this work is to investigate the viability of all synthesized pigments to be successfully used in matte and glossy glazes applied on tiles. The coloring properties were successfully demonstrated for all samples embedded in both types of glazes.

2. Materials and Methods

2.1. Materials

The following precursors were used for the synthesis of pigments: $Co(NO_3)_2 \cdot 6H_2O$ (99.5% purity, Merck, Darmstadt, Germany Merk), $Ni(CH_3COO)_2 \cdot 4H_2O$ (99.5% purity, Merck, Darmstadt, Germany) and $(NH_4)_2Cr_2O_7$ (99.5% purity, Merck, Darmstadt, Germany). All the reagents used in this experiment are of high analytical grade and they were used as purchased, without further purification. HNO_3 (65%, Merck, Darmstadt, Germany) was used as a pH regulator. The sucrose and pectin used to support the condensation reaction were commercial food grade.

2.2. Synthesis of Pigments

Concentrated precursor solutions were prepared using the metal salts ($Co(NO_3)_2 \cdot 6H_2O$, $Ni(CH_3COO)_2 \cdot 4H_2O$ and $(NH_4)_2Cr_2O_7$) dissolved in distilled water. Separately, another solution containing the sucrose with the ratio sucrose: grams of final oxide (wt.:wt.) of 2:1 was prepared. The sol-gel process starts when mixing the precursor solutions with sucrose solution and pectin to form a gel. The addition of sucrose and pectin to the precursor solution containing the metal cations forms a polymer matrix in which the metal cations are distributed through the polymeric network structure [20]. This mechanism, and the role played by sucrose and pectin in the formation of the oxide structures, is discussed in more detail in reference [20].

The hydrolysis process occurs during the vigorous magnetic stirring (1000 RPM) of the metal solutions under strict temperature control (at 40–45 °C) and with a pH correction to around 1–1.5. After stirring, the obtained sol is left to age for 24 h at 80 °C to ensure the formation of the gel lattice, with the elimination of the water present in the pores and the final formation of a porous structure. The thermal treatment of the dried gels was performed in an electric furnace, in porcelain crucibles. The furnace temperature had an increase rate of 300 °C/h, with an isotherm plateau of 30 min, at 1000 °C.

2.3. Analysis Methods

The mass loss and the phase transformations occurring during the heating of the dried gel were investigated by thermal analyses (TG-DTA) employing a SDTQ600 TA Instruments (TA Instruments New Castle, DE, USA) thermal analyzer.

The thermal analysis was performed on a sample of about 10 mg placed in an alumina crucible and non-isothermally heated from 30 °C to 1000 °C at a heating rate of 10 °C/min in dynamic flow-air.

The structural characterization was carried out at room temperature by powder X-ray diffraction using a Bruker D8 Advance AXS diffractometer (Karlsruhe, Germany) with Cu Kα radiation in the 20–80° 2θ region. Crystallite size, cell arrangements and phase fractions were calculated by Rietveld refinement analysis using FullProf Suit Software (FullProff suite July–2017).

The FT-IR absorption spectra were recorded with a JASCO FTIR 6200 spectrometer in the 400–1500 cm^{-1} spectral range, with a standard resolution of ± 2 cm^{-1}.

Scanning Electron Microscopy (SEM, Chiyoda, Tokyo, Japan) and Transmission Electron Microscopy (TEM, Chiyoda, Tokyo, Japan) were performed employing a combined electron scanning (SE, Chiyoda, Tokyo, Japan) and transmission (TE) Hitachi HD-2700 electron microscope (Chiyoda, Tokyo, Japan) operated at a maximum acceleration voltage of 200 kV. The energy–dispersive X-ray spectrometry (EDX) was used to obtain the images with the nickel distribution in the glossy ceramic glaze.

The absorption spectra of the powder samples were obtained using a Perkin–Elmer Lambda 45 UV/Vis spectrometer (Waltham, MA, USA) with integrating sphere using the pellet technique. The powder samples mixed with $BaSO_4$ were uniaxial press in a pellet matrix using a load force of 10 tons/cm^2 to form transparent disks with diameter of 13 mm, and 2 mm thick. The spectra were recorded in the 200–850 nm wavelength range, with the wavelength accuracy ± 2 nm.

The stability of pigments was checked by incorporating them in both matte and glossy tile glaze, using a 2% pigment addition. Thermal treatment for glaze melting was performed at 1200 °C for 6 h at maximum temperature. The glazes were characterized using a color spectrophotometer spectro–guide series (BYK Gardner, Los Angeles, CA, USA).

3. Results

3.1. Thermal Analysis

The thermal decomposition of dried precursor gel was investigated to understand the details of the decomposition process, but most importantly, to determine the temperature at which nucleation and crystallization takes place. This is imperative from a technological point of view and the potential application of the studied pigments in the industry. The thermogravimetric (TG) and differential thermal analysis (DTA) thermograms (20–1000 °C) of the $CoCr_2O_4$ dried gel are presented in Figure 1.

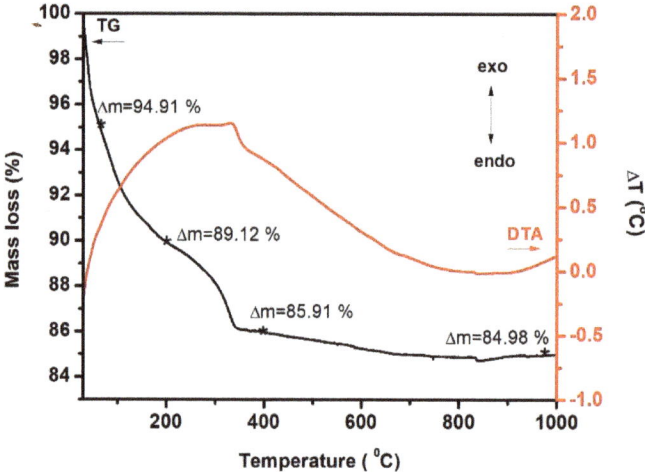

Figure 1. The TG and DTA curves for $CoCr_2O_4$ dried gel.

The decomposition of the dried gel follows two steps:
- the removal of the residual water (the absorbed and the coordinated water) with a mass loss of about 5.09% (30–61 °C);
- the weight loss of the organic fragment with formation of volatile compounds, with a total mass loss of approximately 9.13% (61–1000 °C).

A significant mass loss takes place up to 400 °C, after which no significant losses are recorded. Overall, the total mass loss at 1000 °C was 15.02% and is assigned to the drying process of the gel, in which much of the organic precursor has been removed. Prior to 900 °C, a small mass loss of 1.03% is attributed to the volatile compounds formed and remained inside the material pores. The DTA curve presents a big exothermic peak which unfolds in the 20–800 °C range confirm the two listed decomposition steps. The peak corresponding to the elimination of the organic fragments can be described by the following temperatures: T_{onset} = 318.35 °C, T_{peak} = 331.02 °C and T_{end} = 352.76 °C.

The dried gels were subjected to thermal treatment at 1000 °C for 30 min to ensure the compound crystallization.

3.2. Structural Characterization of Spinels

Figure 2 presents the XRD patterns of the Ni^{2+} ions substitution in $CoCr_2O_4$ spinel matrix after the thermal treatment in air atmosphere at 1000 °C for 30 min.

Both $NiCr_2O_4$ and $CoCr_2O_4$ oxides crystallize into normal spinel structures in the AB_2O_4 form with Co^{2+} and Ni^{2+} ions occupying the tetrahedral A sites and Cr^{3+} ions occupying octahedral B sites.

The XRD patterns present only intense and sharp peaks characteristic to the crystalline phases.

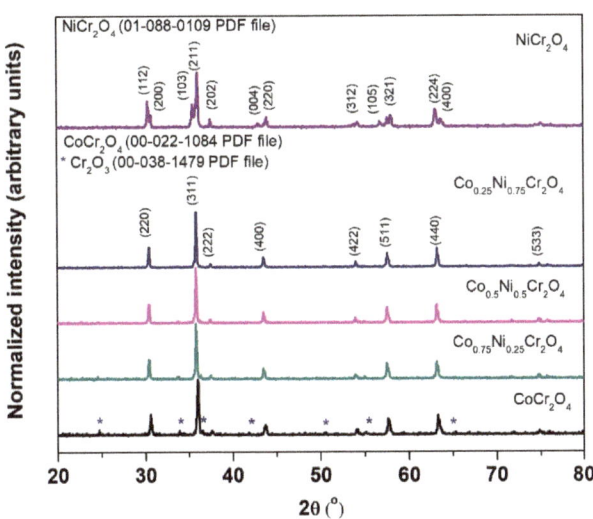

Figure 2. XRD patterns for $Co_{(1-x)}Ni_xCr_2O_4$ pigments annealed at 1000 °C in air.

The $Co_{(1-x)}Ni_xCr_2O_4$ spinel (when $x \neq 1$) crystallizes in Face-Centered Cubic structure and is indexable to the space group Fd-3m (no. 227). This is consistent with the standard values of Face-Centered Cubic phase (00-022-1084 PDF files). On the other hand, $NiCr_2O_4$ crystalizes in the Body-Centered Tetragonal structure, the space group I41/amd (no. 141). So, a transition from the cubic spinel (when $x < 1$) to a Body Centered Tetragonal structure (when $x = 1$) is observed.

For samples with $x < 1$ the XRD data analysis indicated the formation of two crystalline phases: a well crystallized phase $Co_{(1-x)}Ni_xCr_2O_4$ spinel with similar structure to $CoCr_2O_4$ (00-022-1084 PDF files) as the major phase, and Cr_2O_3 (00-038-1479 PDF files) as a secondary phase. Crystallinity dependent properties of crystals are attributed to their crystallite size. Larger crystallites develop sharper peaks on the XRD pattern for each crystal plane. The width of a peak is related to its crystallite size [21]. The formation of Cr_2O_3 impurity phase is common during the synthesis process of $NiCr_2O_4$ and is often reported in the literature [22,23].

The XRD peaks at 2θ values observed in the $NiCr_2O_4$, pattern, match the (112), (200), (103), (211), (202), (004), (220), (312), (106), (321), (224) and (400) crystalline planes of the Body-Centered Tetragonal spinel structure (01-088-0109 PDF files). No other secondary peaks were detected, suggesting that the Ni^{2+} substitution occurred.

The Rietveld refinement fittings were carried out to investigate the changes in the crystal structure by using FullProf software. The experimental recorded data points are represented with black circles, the structural model fit is represented with a red solid line, the structural model fit is represented by the blue solid line and the vertical green bars represent the Bragg positions. Up until the best possible convergence, the refinement was continued. The good fit is evidenced in Figure 3 where the small variance in the difference curve can be observed.

The average crystallite sizes, the lattice parameters and the average strain and standard deviation calculated with the FullProf software are listed in Tables 1 and 2. Thereby, the average crystallite sizes increase with the increase in concentration of Ni^{2+} ions in $Co_{(1-x)}Ni_xCr_2O_4$ spinel starting from 39.9 nm when $x = 0$ and reaching 99.42 nm when $x = 0.75$. The average crystallite size of $NiCr_2O_4$ (when $x = 1$) was 58.98 nm.

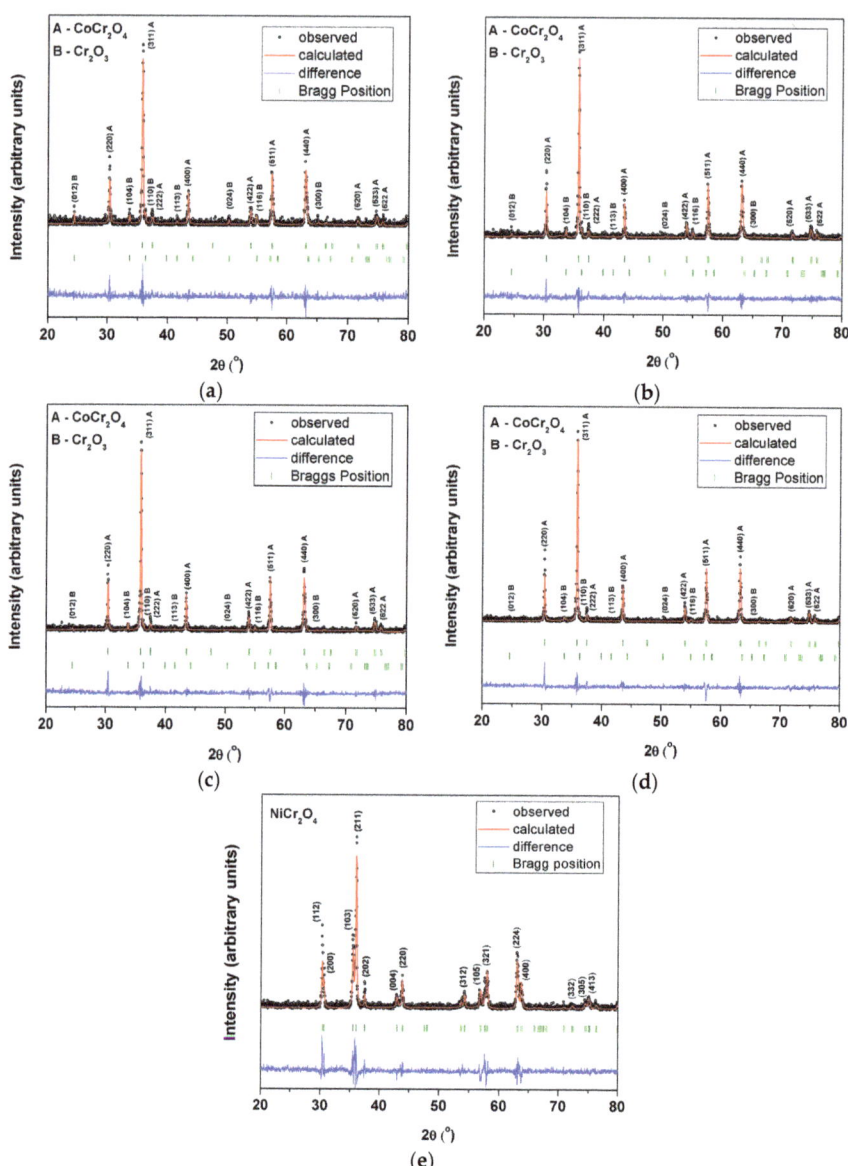

Figure 3. Rietveld analysis of powder-diffraction pattern of (**a**) CoCr$_2$O$_4$ (**b**) Co$_{0.75}$Ni$_{0.25}$Cr$_2$O$_4$ (**c**) Co$_{0.5}$Ni$_{0.5}$Cr$_2$O$_4$ (**d**) Co$_{0.25}$Ni$_{0.75}$Cr$_2$O$_4$ (**e**) NiCr$_2$O$_4$.

The evolution of the average crystallite size, cell parameter, density, and unit cell volume as function of nickel concentration are represented in Figure 4. Thereby, the lattice constant value of Co$_{(1-x)}$Ni$_x$Cr$_2$O$_4$ spinel when x ≤ 0.75 structure was measured to be 8.33 Å, and it is in concordance with the literature data [24]. This large value of the lattice constant was assigned to the disordering of cations in the spinel structure of CoCr$_2$O$_4$ and due to the exchange of tetrahedral A-site Co^{2+} ions with octahedral B-site Cr^{3+} ions [2].

Table 1. Calculated crystallographic parameters.

Ni Concentration (%)	Average Apparent Size and Standard Deviation (nm)	Cell Parameter a (Å)	Cell Parameter c (Å)	Average Maximum Strain and Standard Deviation
0	39.902 (9.847)	8.33325		1.9764 (0.0007)
0.25	65.478 (26.594)	8.33072		1.9764 (0.0007)
0.5	92.199 (52.736)	8.32525		1.9520 (0.0005)
0.75	99.428 (61.333)	8.31979		1.9764 (0.0007)
1	58.983 (13.616)	5.84012	8.42349	5.3902 (0.0046)

Table 2. Calculated phase distribution.

Ni Concentration (%)	Mass (%) Phase 1	Mass (%) Phase 2 (Cr_2O_3)	Density Phase 1 (g/cm^3)	Density Phase 2 (g/cm^3)
0	90.91 (2.25)	9.09 (0.85)	5.209	5.229
0.25	93.22 (2.23)	6.78 (0.74)	5.209	5.210
0.5	97.12 (2.32)	2.88 (0.50)	5.224	5.248
0.75	95.36 (2.00)	4.65 (0.71)	5.234	5.250
1	-	-	5.241	-

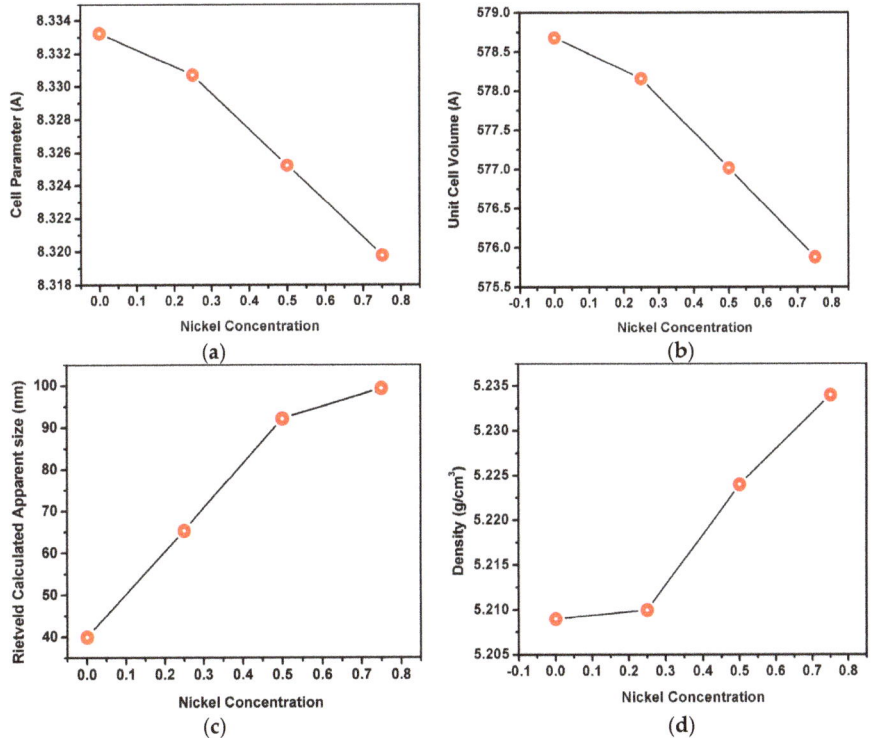

Figure 4. Variation of (a) cell parameter, (b) unit cell volume with the nickel concentration, (c) Rietveld Calculated Apparent size and density (d).

When the Ni^{2+} content is increased, the lattice parameter and volume unit decrease (Figure 4a,b). This decrease is assigned to the difference of ionic radius of Ni^{2+} and Co^{2+}, i.e., with nickel having a smaller radius. Similarly, the density increases with the increase of nickel concentration (Figure 4d).

Considering the targeted application as pigments, the crystallite size plays a determining role in the coloration capacity as being conditioned/size dependent on the specific surface.

3.3. FT–IR Spectroscopy

The FT–IR spectra of $Co_{(1-x)}Ni_xCr_2O_4$ (x = 0, 0.25, 0.5, 0.75 and 1) samples are presented in Figure 5. Two strong absorption bands centered at 516, 620 cm^{-1} with a shoulder at 666 cm^{-1} are observed. These absorption bands are the spinel characteristic peaks and depend on the vibration of the cations at the B site [25–27].

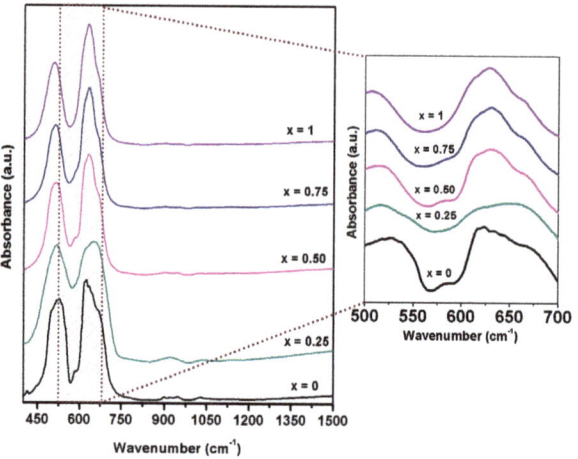

Figure 5. FT-IR spectra of $Co_{(1-x)}Ni_xCr_2O_4$ (x = 0, 0.25, 0.5, 0.75 and 1).

As can be observed in FTIR spectra, the peak centered at 525 cm^{-1} from the sample with x = 0 shifts to a lower frequency, at 504 cm^{-1}, with the increasing nickel content x = 1 (in $NiCr_2O_4$). The crystal theory of the field stabilization energy advises that the Cr^{3+} and Co^{2+} are incorporated into the octahedral and tetrahedral interstices site and compared to the Cr^{3+} and Ni^{2+}, which by occupying the octahedral interstices site own a smaller ligand gravity and splitting energy of d orbital due to the smaller amount of electric charge will cause the peak to shift to lower frequencies [25,26,28]. The second peak at 620 cm^{-1} also shifts towards lower frequencies with the increase of nickel content and has the same origin as the peak from 516 cm^{-1} [25]. Minor changes in the observed band shape can be assigned to the particle size, a similar behavior being observed in $CoCr_2O_4$ samples presented in the literature [29].

3.4. Morphological Characteristics

SEM and TEM micrographs of $Co_{(1-x)}Ni_xCr_2O_4$ samples, with x = 0, 0.25, 0.5, 0.75 and 1, are presented in Figure 6. All samples have the tendency to agglomerate their small particles generating irregular and larger aggregates with various rhombohedral-like shapes. Additionally, no significant differences in the spinel microstructure with the increase of Ni^{2+} ion content was observed, suggesting a similar morphology for all compositions. The estimated crystallite sizes, according to TEM analysis, confirm the calculated values from XRD.

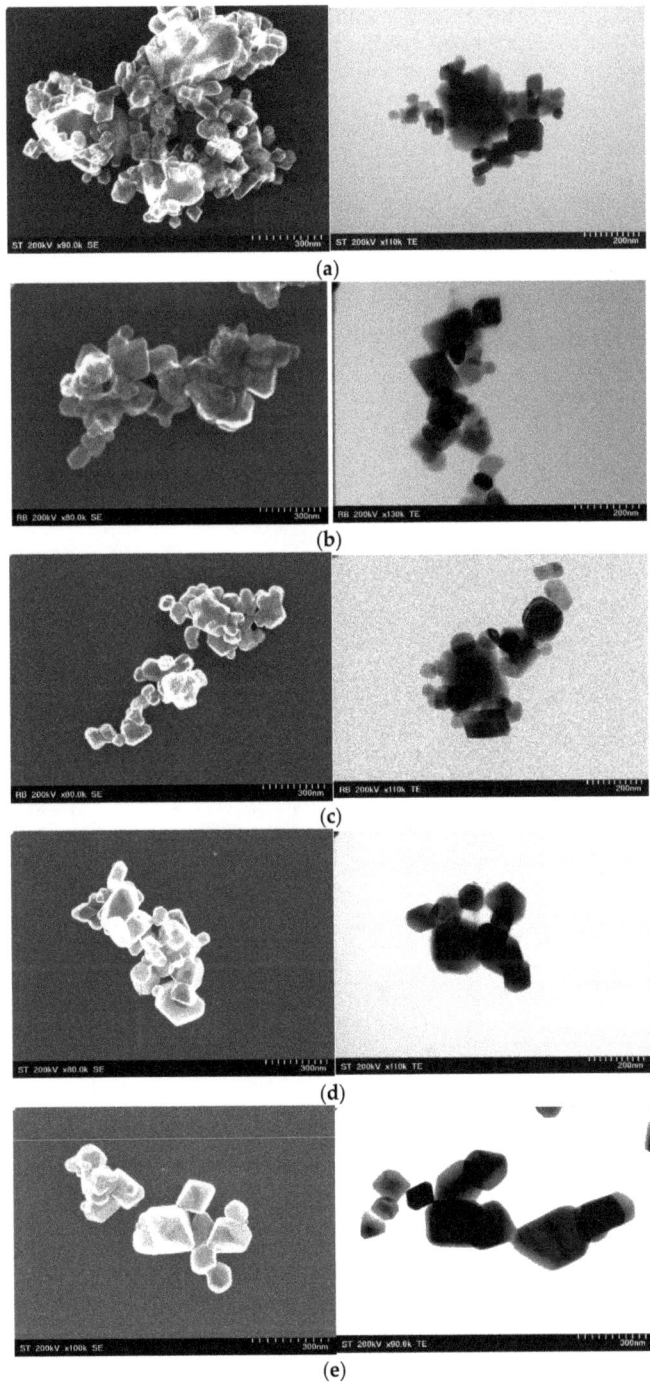

Figure 6. SEM and TEM images of (**a**) $CoCr_2O_4$ (**b**) $Co_{0.75}Ni_{0.25}Cr_2O_4$ (**c**) $Co_{0.5}Ni_{0.5}Cr_2O_4$ (**d**) $Co_{0.25}Ni_{0.75}Cr_2O_4$ (**e**) $NiCr_2O_4$.

3.5. UV-Vis Spectroscopy

The Ni^{2+} substitution of Co^{2+} in $CoCr_2O_4$ matrix was investigated by UV-VIS spectroscopy. The optical absorption spectra of the $Co_{(1-x)}Ni_xCr_2O_4$ (x = 0, 0.25, 0.50, 0.75 and 1) recorded in the wavelength range from 250–850 nm are presented in Figure 7. The recorded spectra of the synthesized nanoparticles are very similar to the spectra of $CoCr_2O_4$ [29] and, also of $NiCr_2O_4$ [29] as reported in the literature. The band around 259 nm can be attributed to the charge-transfer transitions between O^{2-} and Cr^{3+} [17,29].

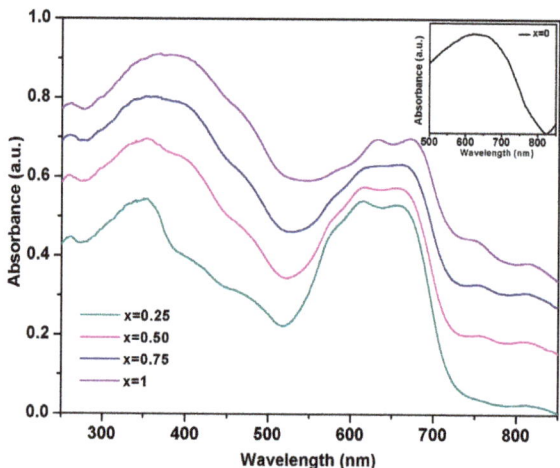

Figure 7. UV–Vis spectra of $Co_{(1-x)}Ni_xCr_2O_4$ pigment (x = 0, 0.25, 0.5, 0.75 and 1).

The incorporation of nickel ions in the $CoCr_2O_4$ matrix leads to the broadening of the band centered around 350 nm, band assigned to a charge transfer of Ni^{2+} cation [25,30] and to a decrease in intensity of the bands in the 574–700 nm wavelength range. Additionally, the absorption bands of $Co_{(1-x)}Ni_xCr_2O_4$ samples centered at 574 nm, 612 nm, and 660 nm shifts with 10 nm to higher wavelengths and the band centered at 760 nm shifts to lower wavelengths with the increase of Ni^{2+} content.

The shoulder present in all the spectra at around 460 nm is attributed to the intrinsic d-d transition of Cr^{3+} ions [31–33]. A decrease in the intensity of the absorption bands at wavelengths higher than 500 nm was observed with the increase of Ni^{2+} ions in the matrix. The absorption bands at about 574 nm, 612 nm are frequently observed in cobalt chromite oxides and can be attributed to the spin-allowed electronic transition $^4A_2(F) \rightarrow {}^4T_1(P)$ of Co^{+2} ions in A site. The absorption band from 612 nm it is also assigned to the transition $^4A_{2g} \rightarrow {}^4T_{2g}$ of Cr^{3+} ion at the B site [17,25]. The band from 660 nm and 750 nm corresponds to $^4A_2 \rightarrow {}^4T_2$ and $^3A_2 \rightarrow {}^3T_1$ transitions and are due to the d-d transition of Ni^{2+} ions in the tetrahedral coordinated O^{2-} environment [20,32]. The bands from about 750 nm and 820 nm become more visible with the increase of Ni^{2+} ions in the matrix and confirm the incorporation of Ni^{2+} in tetrahedral coordination [34].

3.6. CIE Diagram of the Obtained Pigments

The UV-Vis data were used to determinate the CIR color coordinates, (x, y) of the pigment samples. The obtained color parameters are listed in Table 3 and represented in the CIE diagram illustrated in Figure 8. As can observed, the color coordinates slightly vary from (0.3865, 0.3295) when x = 0.25 to (0.3118, 0.2762) when x = 1 and ranging in between for x = 0.5 and 0.75. Increasing the nickel concentration in the matrix leads to a shift from a pink-turquoise to a greenish-blue color. This observation is correlated with the literature data [10] and the UV-Vis absorption spectra described previously.

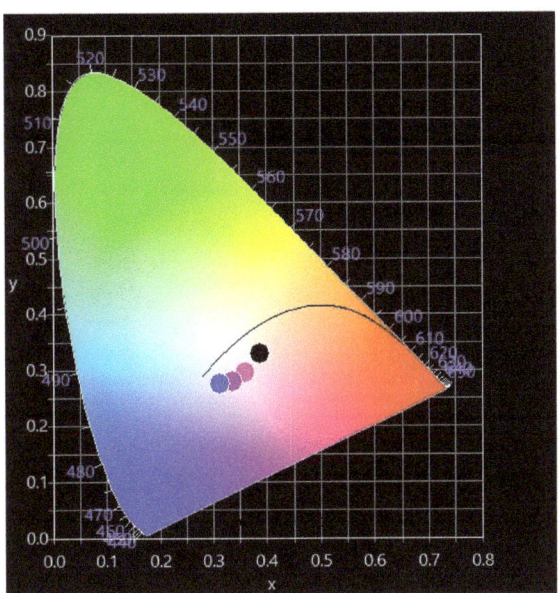

Figure 8. Chromaticity diagram. The place of the pigment nanopowders.

Table 3. Calculated values of the pigment nanopowder color.

Sample $Co_{(1-x)}Ni_xCr_2O_4$ (% Ni)	Color Coordinates (x, y)		
	x	y	Representation in the CIE Diagram
x = 0.25	0.3865	0.3295	dark circle
x = 0.50	0.3585	0.2957	pink
x = 0.75	0.3361	0.2801	purple
x = 1	0.3118	0.2762	blue

The pigments were embedded in both matte and glossy tile glazes to test their applications in the ceramic tile industry. Thereby, the homogeneous glasses were prepared by incorporating spinel nanopowders into ceramic glazes which were further applied on ceramic substrates and were finally subjected to thermal treatments.

Figure 9 presents a more comprehensive view of the pigments before and after their embedding. As can be observed, the pigment powders exhibit bluish tones that tend to become greener with the progression of Ni^{2+} substitution up to a yellowish green with the full Co^{2+} replacement.

The color of pigment powders and those embedded in glazes, in terms of color parameters (L^* = lightness, a^* = red-green axe values, b^* = yellow-blue axe values, G^* = gloss) are presented in Table 4. A major difference that can be observed is that of the 'L' parameter, that ranges between 31.09 and 33.49 for the matte glaze and from 40.69 to 45.72 for the glossy one. The gloss parameter ('G') is lower for matte glazes (50.5–58.1) and higher for glossy ones (83.8–86.8), due to the nature of the glaze itself. This can be explained by the fact that the latter reflects light due to its smoother surface, whereas the former tends to scatter it. Therefore, pigments embedded in glossy glazes will naturally become 'brighter' in nature.

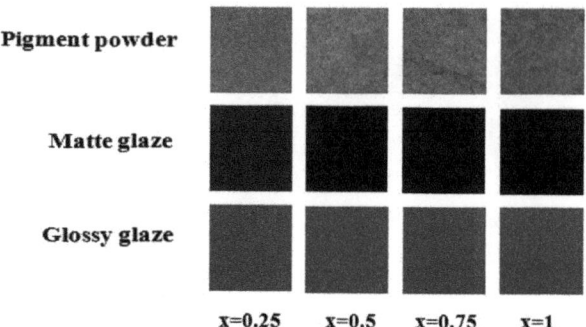

Figure 9. The $Co_{(1-x)}Ni_xCr_2O_4$ pigment Nano powders before after their embedding in matte and glossy glaze.

Table 4. The CIELab coordinates of pigments embedded in opaque tile glazes.

Pigment		Matte Glaze				Glossy Glaze			
CIELab Coordonates	L*	a*	b*	G	L*	a*	b*	G
$Co_{0.75}Ni_{0.25}Cr_2O_4$	32.32	−9.78	−3.2	50.5	40.69	−10.63	5.65	86.1
$Co_{0.5}Ni_{0.5}Cr_2O_4$	31.31	−6.67	1.2	57.6	42.03	−9.45	0.15	83.8
$Co_{0.25}Ni_{0.75}Cr_2O_4$	31.09	−4.07	3.82	58.1	43.34	−6.63	4.98	86.7
$NiCr_2O_4$	31.71	−0.62	6.74	52.7	45.72	−1.63	11.10	86.8

The a* and b* coordinates of the glazes increase with increasing nickel substitution in pigment which impacts the color by his chromophore capacity and also by the impact on particle size. Generally, this ranging can also be correlated with different synthesis methods, with different factors such as synthesis conditions (temperature, stoichiometry, and pH) or particle size. For example, the growth in b* value coordinate was correlated with the increase in particle size in $CoAl_2O_4$ for blue pigment [1]. In glazes, b* values are highly reduced to a range of −6.7 and −0.4, also with greener hues [1,35].

An increase in a* (red-green) and b* (yellow-blue) values can also be observed with an increase in Ni substitution. Higher a* values tend to lead to yellowish-brown hues and higher b* values to reddish-brown ones. This can be better observed by the color progression in Figure 9. Once embedded in the glazes, the color of the pigment becomes darker and more intense due to thermal treatment.

The experimental results indicate that the synthesized $Co_{(1-x)}Ni_xCr_2O_4$ spinel ($0.25 \leq x \leq 1$) oxide nanopowders can be successfully used as ceramic pigments due to their coloration capacity and thermal resistance. The thermal and chemical stability of the $Co_{(1-x)}Ni_xCr_2O_4$ spinel pigment in the glazes is high enough to obtain a uniform color distribution of the product and is not affected by melting in the firing process.

3.7. EDX Mapping

The nickel substitution in the spinel was evidenced after embedding in glossy ceramic tile by using the Energy-Dispersive X-ray Spectrometry (EDX) from SEM. Figure 10 presents the EDS element layered image including Ni Kα1 of the ceramic glossy tile cross-section as a function of nickel substitution in the spinel. A uniform distribution of the nickel in the cross-section of the glaze is observed for all samples and the partial and total substitution of nickel is evidenced in Figure 10.

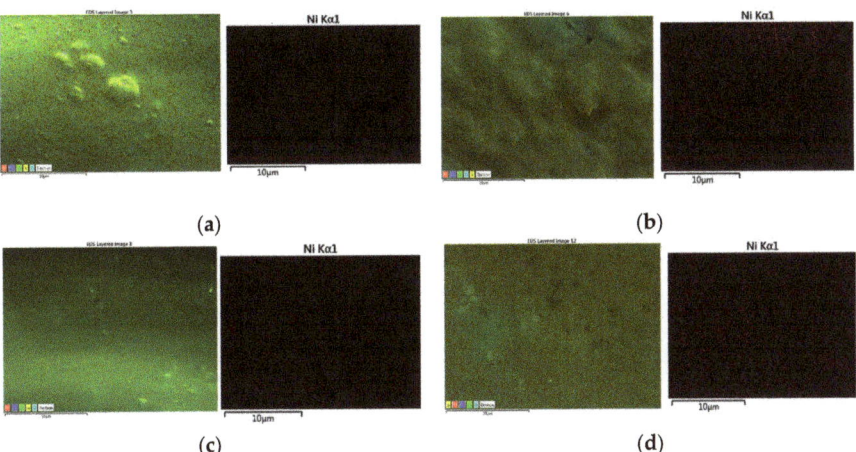

Figure 10. EDX element layered image including Ni Kα1 of the ceramic glossy tile cross-section as a function of nickel substitution in the chromite spinel $(Co_{(1-x)}Ni_xCr_2O_4)$ where (**a**) x = 0.25, (**b**) x = 0.5, (**c**) x = 0.75 and (**d**) x = 1.

4. Conclusions

The paper reports structural, morphological, and optical properties of new ceramic pigments-based spinel structure, obtained by sol-gel route. The structural characterization is consistent with the substitution of chromophores Co^{2+} ions with the Ni^{2+} ions in the $CoCr_2O_4$ matrix. The XRD analysis revealed the formation of $Co_{(1-x)}Ni_xCr_2O_4$ spinel-type Face-Centered Cubic structure as a principal phase and Cr_2O_3 as a secondary one when x ≤ 0.75. A decrease in unit cell parameter and the unit cell volume was achieved with an increase of x. When x = 1, $NiCr_2O_4$ crystallizes in the Body-Centered Tetragonal spinel as a single phase when compared to lower 'x' values when a secondary phase appears. The increase of nickel substitution in the matrix has a pronounced increase in the size of the crystallites from 39.9 nm (for x = 0) to 99.42 nm (for x = 0.75). FT-IR spectra confirmed the spinel structure formation and the elimination of all the organic fragments. The UV-Vis absorption spectra presented the bands corresponding nickel ions located at the A site and the chromium ions located at the B sites. Adjusting the nickel content in the $CoCr_2O_4$ matrix the color of the pigment can be easily controlled, as it can also be seen in the CIE diagram chromaticity. SEM and TEM microscopy confirmed evidence of the powder morphology and the tendency of nanoparticles to agglomerate. The CIELab coordinates of the pigments embedded in glossy and matte tile glazes reveal the color ranging of the two glazes. Additionally, the elemental EDX distribution of Ni Kα1 confirms the homogeneous and uniform distribution of the pigment in the glossy glaze after firing. The obtained pigments can be successfully applied in glaze tiles and ceramics.

Author Contributions: Conceptualization and methodology, validation, investigation, data curation, supervision, F.G.; writing—original draft preparation, R.A.M.; validation, investigation, data curation, supervision, writing—review and editing, R.A.M., A.A., R.A.B. and M.Z., investigation, data curation, L.B.T. All authors have read and agreed to the published version of the manuscript.

Funding: This research received no external funding.

Institutional Review Board Statement: Not applicable.

Informed Consent Statement: Not applicable.

Data Availability Statement: Not applicable.

Acknowledgments: This work was partially supported by a grant of the Ministry of Research, Innovation and Digitization: Nucleu-Program, project number PN 19 35 01 01.

Conflicts of Interest: The authors declare no conflict of interest.

References

1. Enríquez, E.; Reinosa, J.; Fuertes, V.; Fernández, J. Advances and challenges of ceramic pigments for inkjet printing. *Ceram. Int.* **2022**, *48*, 31080–31101. [CrossRef]
2. Mindru, I.; Gingasu, D.; Marinescu, G.; Patron, L. *Design de Nanomateriale Oxidice cu Structura Spinelica. De la Sinteza la Aplicatii*, 1st ed.; Matrixrom: Bucharest, Romania, 2018; pp. 1–148. (In Romanian)
3. Livage, J.; Sanchez, C.; Henry, M.; Doeuff, S. The chemistry of the sol-gel process. *Solid State Ionics* **1989**, *32–33*, 633–638. [CrossRef]
4. Ilosvai, A.M.; Dojcsak, D.; Váradi, C.; Nagy, M.; Kristály, F.; Fisercz, B.; Vanyorek, L. Sonochemical Combined Synthesis of Nickel Ferrite and Cobalt Ferrite Magnetic Nanoparticles and Their Application in Glycan Analysis. *Int. J. Mol. Sci.* **2022**, *23*, 5081. [CrossRef] [PubMed]
5. Chen, T.-W.; Tamilalagan, E.; Chen, S.M.; Akilarasan, M.; Maheshwaran, S.; Liu, X. An Ultra-Sensitive Electrochemical Sensor for the Detection of Carcinogen Oxidative Stress 4-Nitroquinoline N-Oxide in Biologic Matrices Based on Hierarchical Spinel Structured $NiCo_2O_4$ and $NiCo_2S_4$; A Comparative Study. *Int. J. Mol. Sci.* **2020**, *21*, 2373. [CrossRef] [PubMed]
6. Younis, M.; Saleem, M.; Atiq, S.; Naseem, S. Magnetic phase transition and magneto-dielectric analysis of spinel chromites: MCr_2O_4 (M = Fe, Co and Ni). *Ceram. Int.* **2018**, *44*, 10229–10235. [CrossRef]
7. Dippong, T.; Levei, E.A.; Deac, I.G.; Petean, I.; Cadar, O. Dependence of Structural, Morphological and Magnetic Properties of Manganese Ferrite on Ni-Mn Substitution. *Int. J. Mol. Sci.* **2022**, *23*, 3097. [CrossRef]
8. Zhou, X.; Zhang, X.; Zou, C.; Chen, R.; Cheng, L.; Han, B.; Liu, H. Insight into the Effect of Counterions on the Chromatic Properties of Cr-Doped Rutile TiO_2-Based Pigments. *Mater* **2022**, *15*, 2049. [CrossRef]
9. Tena, M.Á.; Mendoza, R.; Trobajo, C.; García-Granda, S. Cobalt Minimisation in Violet $Co_3P_2O_8$ Pigment. *Materials* **2022**, *15*, 1111. [CrossRef] [PubMed]
10. Mohanty, P.; Prinsloo, A.R.E.; Doyle, B.P.; Carleschi, E.; Sheppard, C.J. Structural and magnetic properties of $(Co_{1-x}Ni_x)Cr_2O_4$ (x = 0.5, 0.25) nanoparticles. *AIP Adv.* **2018**, *8*, 056424. [CrossRef]
11. Anju, Y.R.S.; Pötschke, P.; Pionteck, J.; Krause, B.; Kuritka, I.; Vilcáková, J.; Škoda, D.; Urbánek, P.; Machovský, M.; Masar, M.; et al. $Cu_xCo_{1-x}Fe_2O_4$ (x = 0.33, 0.67, 1) Spinel Ferrite Nanoparticles Based Thermoplastic Polyurethane Nanocomposites with Reduced Graphene Oxide for Highly Efficient Electromagnetic Interference Shielding. *Int. J. Mol. Sci.* **2022**, *23*, 2610. [CrossRef]
12. Novikov, V.A.; Xanthopoulou, G.G.; Amosov, A.P. Solution Combustion Synthesis of Nanostructured $NiCr_2O_4$ Spinel and Its Catalytic Activity in CO Oxidation. *Int. J. Self-Propagating High-Temp. Synth.* **2021**, *30*, 246–250. [CrossRef]
13. Wang, C.; Zhou, E.; He, W.; Deng, X.; Huang, J.; Ding, M.; Wei, X.; Liu, X.; Xu, X. $NiCo_2O_4$-Based Supercapacitor Nanomaterials. *J. Nanomater.* **2017**, *7*, 41. [CrossRef] [PubMed]
14. Xu, X.; Gao, J.; Hong, W. Ni-based chromite spinel for high-performance supercapacitors. *RSC Adv.* **2016**, *6*, 29646–29653. [CrossRef]
15. Dippong, T.; Levei, E.A.; Goga, F.; Petean, I.; Avram, A.; Cadar, O. The impact of polyol structure on the formation of $Zn_{0.6}Co_{0.4}Fe_2O_4$ spinel-based pigments. *J. Sol-Gel Sci. Technol.* **2019**, *92*, 736–744. [CrossRef]
16. Lazau, I.; Pacurariu, C.; Ecsedi, Z.; Ianos, R. *Metode Neconvenționale Utilizate în Sinteza Compușilor Oxidici*, 1st ed.; Politehnica: Timisoara, Romania, 2006; pp. 1–500. (In Romanian)
17. Chavarriaga, E.; Lopera, A.; Bergmann, C.; Alarcónd, J. Effect of the substitution of Co^{2+} by Mg^{2+} on the color of the $CoCr_2O_4$ ceramic pigment synthesized by solution combustion. *Bol. Soc. Esp. Ceram. Vidr.* **2020**, *59*, 176–184. [CrossRef]
18. Grazenaite, E.; Pinkas, J.; Beganskiene, A.; Kareiva, A. Sol–gel and sonochemically derived transition metal (Co, Ni, Cu, and Zn) chromites as pigments: A comparative study. *Ceram. Int.* **2016**, *42*, 9402–9412. [CrossRef]
19. Enhessari, M.; Salehabadi, A.; Khanahmadzadeh, S.; Arkat, K.; Nouri, J. Modified Sol-Gel Processing of $NiCr_2O_4$ Nanoparticles; Structural Analysis and Optical Band Gap. *High Temp. Mater. Process.* **2017**, *36*, 121–125. [CrossRef]
20. Suciu, C.; Hoffmann, A.C.; Vik, A.; Goga, F. Effect of calcination conditions and precursor proportions on the properties of YSZ nanoparticles obtained by modified sol–gel route. *J. Chem. Eng.* **2008**, *138*, 608–615. [CrossRef]
21. Abdullah, M.M.; Rajab, F.M.; Al-Abbas, S.M. Structural and optical characterization of Cr_2O_3 nanostructures: Evaluation of its dielectric properties. *AIP Adv.* **2014**, *4*, 027121. [CrossRef]
22. Mohanty, P.; Sheppard, C.J.; Prinsloo, A.R.E.; Roos, W.D.; Olivi, L.; Aquilanti, G. Effect of cobalt substitution on the magnetic properties of nickel chromite. *J. Magn. Magn. Mater.* **2018**, *451*, 20–28. [CrossRef]
23. Ptak, M.; Maczka, M.; Gagor, A.; Pikul, A.; Macalik, L.; Hanuza, J. Temperature-dependent XRD, IR, magnetic, SEM and TEM studies of Jahn–Teller distorted $NiCr_2O_4$ powders. *J. Solid State Chem.* **2013**, *201*, 270–279. [CrossRef]
24. Castiglioni, G.L.; Minelli, G.; Portab, P.; Vaccari, A. Synthesis and Properties of Spinel-Type Co–Cu–Mg–Zn–Cr Mixed Oxides. *J. Solid State Chem.* **2000**, *152*, 526–532. [CrossRef]
25. Gao, Y.; Chang, H.; Wu, Q.; Wang, H.-Y.; Pang, Y.-B.; Liu, F.; Zhu, H.-J.; Yun, Y.-h. Optical properties and magnetic properties of antisite-disordered $Ni_{1-x}Co_xCr_2O_4$ spinels. *Trans. Nonferrous Met. Soc. China* **2017**, *27*, 863–867. [CrossRef]

26. Wang, Z.; Saxena, S.; Lazor, P.; O'Neill, H. An in-situ Raman spectroscopic study of pressure induced dissociation of spinel $NiCr_2O_4$. *J. Phys. Chem. Solids* **2003**, *64*, 425–431. [CrossRef]
27. Ahmadyari-Sharamin, M.; Hassanzadeh-Tabrizi, S. Polyacrylamide gel synthesis, characterization, and optical properties of $Co_{1-x}Ni_xCr_2O_4$ spinel nanopigment. *J. Sol-Gel Sci. Technol.* **2021**, *99*, 534–545. [CrossRef]
28. Han, A.; Ye, M.; Zhang, Z.; Liao, J.; Li, N. Crystal structure and optical properties of $CoCr_2O_4$–$NiCr_2O_4$ solid solutions prepared by low-temperature combustion synthesis method. *Adv. Mater. Res.* **2013**, *616–618*, 1877–1881. [CrossRef]
29. Mączka, M.; Ptak, M.; Kurnatowska, M.; Hanuza, J. Synthesis, phonon and optical properties of na-nosized $CoCr_2O_4$. *Mater. Chem. Phys.* **2013**, *138*, 682–688. [CrossRef]
30. El-Kemary, M.; Nagy, N.; El-Mehasseb, I. Nickel oxide nanoparticles: Synthesis and spectral studies of interactions with glucose. *Mater. Sci. Semicond. Process.* **2013**, *16*, 1747–1752. [CrossRef]
31. Liang, S.-t.; Zhang, H.-l.; Luo, M.-t.; Luo, K.-j.; Li, P.; Xu, H.-b.; Zhang, Y. Colour performance investigation of a Cr_2O_3 green pigment prepared via the thermal decomposition of CrOOH. *Ceram. Int.* **2014**, *40*, 4367–4373. [CrossRef]
32. Singh, J.; Kumar, R.; Verma, V.; Kumar, R.K. Role of Ni^{2+} substituent on the structural, optical and magnetic properties of chromium oxide ($Cr_{2-x}Ni_xO_3$) nanoparticles. *Ceram. Int.* **2020**, *46*, 24071–24082. [CrossRef]
33. Li, L.; Yan, Z.F.; Lu, G.Q.; Zhu, Z.H. Synthesis and structure characterization of chromium oxide prepared by solid thermal decomposition reaction. *J. Phys. Chem. B* **2006**, *110*, 178–183. [CrossRef] [PubMed]
34. Tripathi, V.K.; Nagarajan, R. Influencing Optical and Magnetic Properties of $NiCr_2O_4$ by the Incorporation of Fe(III) for Cr(III) Following Epoxide Gel Synthesis. *J. Electron. Mater.* **2019**, *48*, 1139–1146. [CrossRef]
35. Costa, G.; Ribeiro, M.; Hajjaji, W.; Seabra, M.; Labrincha, J.; Dondi, M.; Cruciani, G. Ni-doped hibonite ($CaAl_{12}O_{19}$): A new turquoise blue ceramic pigment. *J. Eur. Ceram. Soc.* **2009**, *29*, 2671–2678. [CrossRef]

Review

Nanomaterials as Photocatalysts—Synthesis and Their Potential Applications

Agnieszka Feliczak-Guzik

Faculty of Chemistry, Adam Mickiewicz University in Poznań, Uniwersytetu Poznańskiego 8, 61-614 Poznań, Poland; agaguzik@amu.edu.pl

Abstract: Increasing demand for energy and environmental degradation are the most serious problems facing the man. An interesting issue that can contribute to solving these problems is the use of photocatalysis. According to literature, solar energy in the presence of a photocatalyst can effectively (i) be converted into electricity/fuel, (ii) break down chemical and microbial pollutants, and (iii) help water purification. Therefore, the search for new, efficient, and stable photocatalysts with high application potential is a point of great interest. The photocatalysts must be characterized by the ability to absorb radiation from a wide spectral range of light, the appropriate position of the semiconductor energy bands in relation to the redox reaction potentials, and the long diffusion path of charge carriers, besides the thermodynamic, electrochemical, and photoelectrochemical stabilities. Meeting these requirements by semiconductors is very difficult. Therefore, efforts are being made to increase the efficiency of photo processes by changing the electron structure, surface morphology, and crystal structure of semiconductors. This paper reviews the recent literature covering the synthesis and application of nanomaterials in photocatalysis.

Keywords: nanomaterials; semiconductors; photocatalyst; photocatalysis; plasmonic properties of metals

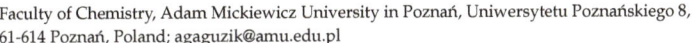

1. Heterogeneous Photocatalysis

According to the International Union of Pure and Applied Chemistry, heterogeneous photocatalysis is a reaction in which a photocatalyst initiates the process after absorption of the exciting radiation and the photocatalyst occurs in a different thermodynamic phase than the reactants [1,2]. Among the materials most commonly used as photocatalysts are solid semiconductors, primarily transition metal oxides. The specific resistance of semiconductors at room temperature ranges from 10^{-2} Ωcm to 10^9 Ωcm and is strongly temperature dependent. Due to their conductive properties, semiconductors are intermediate between dielectrics (insulators) and conductors (metals) [2]. In semiconductor materials, the valence band (VB) is fully occupied, while the conduction band (CB) is completely empty at absolute zero temperature. The energy of the excited band (Eg) in such materials is in the range of 0–4 eV, and the Fermi level lies between the conduction band and the valence band. For semiconductors the forbidden band in a macroscopic scale assumes constant values.

Heterogeneous photocatalysis includes a wide range of chemical reactions, for example: partial or complete oxidation, dehydrogenation, hydrogen transfer, oxygen and deuterium isotopic exchange, metal deposition, water detoxification, and removal of gaseous pollutants [2].

This type of photocatalysis involves the following steps: (i) adsorption of the substrates involved in the reaction on the surface of the photocatalyst, (ii) absorption of radiation quanta of appropriate energy by the applied photocatalyst, (iii) generation of reactive electron-hole pairs, and (iv) electron and hole reactions with the adsorbed compounds or their recombination. To begin with, excitation of the photocatalyst, which is a semiconductor, involves absorption of radiation with an energy equal to, or greater than, the energy gap, followed by excitation of an electron from the valence band to the conduction band. Then, the resulting individuals react with the surrounding components: the electron causes reduction (photoreduction) and the hole causes oxidation (photooxidation) of the compounds adsorbed on the applied photocatalyst, as shown in Figure 1 [3].

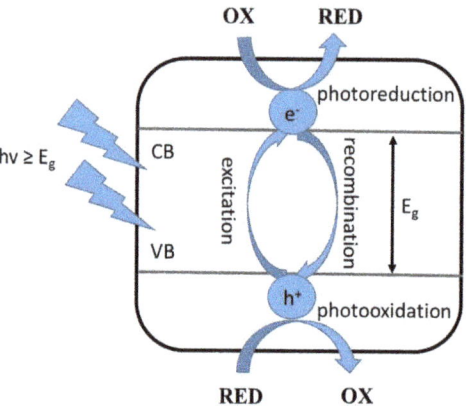

Figure 1. Excitation of a semiconductor photocatalyst, where: OX—oxidized compound, RED—reduced compound, VB—valence band, CB—conduction band, Eg—the energy of the excited band, hv—the energy of the radiation quantum [3].

Absorption of light by a semiconductor generates the formation of an electron-hole pair, increasing the concentrations of electrons and holes above their equilibrium concentration by Δn^* and Δp^*, respectively. This occurs until the recombination reaction slows down the process. The free energy of these individual charge carriers is expressed by the so-called quasi-Fermi level [4]. For the minority charges, the position of the quasi-Fermi level differs significantly from that of the Fermi level of an unlit semiconductor, while the difference is small for the majority charges.

It should be added here that the absorption of light depends primarily on the distance from the surface of the semiconductor. This means that the distribution of excess charges is not uniform. As the distance from the semiconductor surface increases, the probability of a minority charge appearance decreases. However, the biggest changes will occur near the surface itself, where the minority charges accumulate. The change in charge density has a major impact on all processes occurring in this area [4].

On the basis of the compositions of the initial materials, Yang et al. grouped photocatalysts into six categories: (i) traditional semiconductor, (ii) molecular, (iii) plasmonic, (iv) 2D, (v) quantum dots, and (vi) traditional semiconductor-based photovoltaic assisted [5]. Of the aforementioned, plasmonic nanostructures, which can be used to enhance light absorption by semiconductors or to drive direct photocatalysis with visible light on their surface, are currently attracting considerable attention. Interest in this type of materials, may also be due to the fact that photocatalytic processes carried out using semiconductors show some limitations, such as: low absorption coefficient, limited wavelength range for light, and low selectivity towards a specific chemical reaction pathway.

Plasmonic nanostructures can confine electromagnetic energy in free space to nanometer-sized regions and convert it into various forms, including confined and scattering fields, high-energy "hot" electrons, and holes, or heat and thermal radiation. These nanostructures are designed in principle to mainly express one of such energy transformations; their properties depend on where the nanostructures are to be used [6].

The Localized Surface Plasmon Resonance (LSPR) generated in plasmonic structures can lead to enhancement or formation of linear and nonlinear optics phenomena (e.g., spontaneous emission, nonlinear absorption or Raman scattering) [7]. LSPR occurs when an electromagnetic wave with a frequency identical to the vibrational frequency of localized surface plasmons falls on a plasmonic nanoparticle [8]. The main feature of the localized resonance of surface plasmons is that its frequency can be varied by selecting both the size, shape, and position of the nanostructures, as well as the type of matrix and material of which they are made. This makes it possible to control the resonance and adjust it to those wavelengths to be used in the planned applications.

2. The Properties of Photocatalysts

Photocatalysts, compared to traditional catalysts, operate on a different principle. Table 1 compares the properties of traditional catalysts with those of photocatalysts [9].

Table 1. Comparison of the properties of traditional catalysts with photocatalysts [9].

Conventional Catalysts	Photocatalysts
- have a certain number of active centers with which the chemical reaction takes place and which, with the time of using the material, can be poisoned	- usually are semiconducting compounds (they can exist in a self-supporting form or can be deposited on a carrier), which can be excited by radiation from the UV-Vis range; for a semiconductor to photocatalyze a particular chemical reaction, the potential of such a process should be within the limits of the photocatalyst's band gap, between the potential of the valence band and conduction
- formation of transition products as a result of the reaction of substrates with a catalyst	- generate reactive electron-hole pairs across the excited surface, which then interact with substrates
- modification of the course of the reaction, affecting the reduction of activation energy and increasing the rate of product formation	- do not react directly with reagents
- selective activity	- non-selective activity

The course of the catalytic, and photocatalytic processes, according to Ohtanie using a catalyst, and a photocatalyst is shown in Figure 2 [10]. Most researchers use the term 'photocatalytic activity', but in almost all cases the meaning is the same as absolute or relative reaction rate. The reason for this may be to get others to think of 'photocatalytic reaction rate' as one of the properties of a photocatalyst (i.e., photocatalysts have an individual activity, whereas "reaction rate" is controlled by activity under given reaction conditions). In catalysis, "catalytic activity" (Figure 2a) was used to show the properties (or performance of the catalyst) because the 'active site' on the catalyst is responsible for the catalytic reaction. The reaction rate per active site can be estimated and should be the "catalytic activity". On the other hand, there are no active sites on the photocatalyst, in the same sense as in thermal catalysis, i.e., the rate of the catalytic reaction is predominantly governed by the number of active sites, and the reaction rate strongly depends on various factors, e.g.: the intensity of the irradiated light that initiates the photocatalytic reaction (Figure 2b). Given that the dark side of a photocatalyst or suspension has no effect on the photocatalytic reaction, the use of the term 'active site' is inappropriate, and therefore a relationship between photocatalytic activity and active sites cannot be expected. In the kinetic analysis of general chemical reactions, the rate constant is estimated and compared. Given that photoexcited electrons (e^-) and positive holes (h^+) induce a redox reaction, the rate constant of these active species can be estimated. Since e- and h+ recombine with each other, the overall rate of the photocatalytic reaction also depends on this recombination rate. Assuming that k(redox) and k(recombination) are the rate constants of the reaction rates occurring by e^- and h^+ and their recombination, respectively, i.e., in the simplest kinetic model, the ratio k(redox)/k(recombination) should be a measure of the intrinsic photocatalytic activity [10].

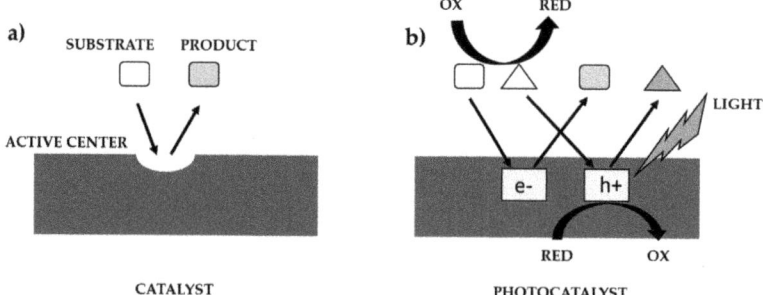

Figure 2. Comparison of catalysis process (**a**) with photocatalysis (**b**) [10].

Great influence on the properties of a given semiconducting material applicable in photocatalysis has its size. As the size of the crystal is minimized, the ratio of the number of atoms on its surface to those inside the crystal increases, which causes a change not only in the surface properties of the semiconductor but also in the entire material. It affects, among other things: the melting point of the material and the electron properties - if the size of the semiconductor decreases to a few nanometers, separate energy levels are created instead of continuous energy bands (this is called the quantum size effect) [11].

In nanoparticles, the electron and hole are closer to each other than in bulk semiconductors, which has the effect of increasing the Coulombic interactions between them. This also affects the size of the energy gap. Increasing the energy gap allows the absorption threshold to shift toward shorter wavelengths as the particle size decreases. Consequently, this leads to an increase in the molar absorption coefficient, which is related to the overlap of the wave functions of the charge carriers. This has become the basis for the use of nanomaterials (nano semiconductors) in catalysis, luminescence, optoelectronic devices, and solar cells [12].

3. Synthesis of Photocatalysts

Over the past few years, the synthesis of nanomaterials has been a dominant trend in many fields of science and technology [13]. Nanomaterials of metallic or semiconducting nature that can be excited by radiation in the UV-Vis range have become very popular [3]. These materials differ in characteristics from micrometric-sized materials because nanometer-sized particles exhibit new and unique magnetic, electrical, optical, and catalytic properties [3]. A good photocatalyst should be characterized by: (i) the ability to absorb radiation from a wide spectral range of light, (ii) the appropriate position of the energy bands of the semiconductor about the redox reaction potentials, (iii) high mobility and long diffusion path of charge carriers, (iv) thermodynamic, electrochemical, and photoelectrochemical stability [14]. Moreover, for the reactions involving the resulting photocarriers to occur efficiently, it is necessary to effectively prevent recombination by separating electron-hole pairs, and then their transport to the semiconductor surface. Meeting these requirements by semiconductors is very difficult. Therefore, efforts are being made to increase the efficiency of photo processes by changing the electron structure, surface morphology, and crystal structure of semiconductors [10]. Broadening the range of radiation absorption can be achieved by, among other things, doping the cationic and/or anionic subgrid, or introducing plasmonic metal nanoparticles. An increase in the degree of crystallinity of the resulting materials by reducing the concentration of defects leads to a reduction in the number of recombination centers. Reducing the size of particles, while increasing the specific surface area, has a beneficial effect on the efficient diffusion of charge carriers to the surface. The selection of the shape of nanocrystals by adjusting the proportion of selected crystallographic planes differing in surface energy allows the adsorption of only selected particles, thus providing selectivity of photocatalytic processes [10].

3.1. Synthesis of Semiconductors

The optical and electrical properties of semiconducting materials are strongly related to the distribution of energy bands, or more precisely, to the energy of the excited band. It determines the threshold energy that electrons must have at the moment of transition from the valence band to the conduction band. Semiconducting nanocrystals can be considered as a multi-atomic molecule in which the electron orbitals are formed: the highest occupied molecular orbital (HOMO) and the lowest unoccupied molecular orbital (LUMO). In the ground (non-excited) state, the HOMO orbital is filled, while the LUMO is unfilled. Upon excitation, electrons in the semiconducting materials can switch from HOMO to LUMO. This transition is analogous to the transitions of electrons from the highest occupied molecular orbital to the lowest unoccupied molecular orbital of organic compounds. When electrons are excited from the valence band to the conduction band, a gap of positive charge, called an "electron hole," remains in the valence band. At a later stage, as a result of recombination of electrons and holes, the energy released in this process is emitted in the form of a photon (radiative recombination) or is transferred to another charge carrier during non-radiative recombination [15].

According to the number of charge carriers formed as a result of excitation, we can distinguish between intrinsic semiconductors (the concentration of free electrons of a semiconductor is equal to the concentration of holes) and doped semiconductors, in which the introduction of a doping material generates carriers of one type. Hence, we distinguish between n-type (over-doped) semiconductors, in which there is electron overshoot, and p-type (undoped) semiconductors, which are characterized by hole overshoot (the number of holes is greater than the number of electrons in the conduction band). Intrinsic semiconductors include materials made of a single type of atoms, such as B, Ge, Si, Se, S, Sb, Te, or I. Germanium, silicon, and selenium are known as intrinsic semiconductors, while the other elements are most often used as dopants or as components of the so-called complex (non-self-contained semiconductors) semiconductor mate-materials. The group of compound semiconductor materials includes chemical compounds of two, three, or more chemical elements, of which the most common are semiconductors of two-element compounds of the type A(III)B(V), A(II)B(V), A(II)B(VI), or A(IV)B(VI). Depending on the elements that make up the semiconductors in question, they can exhibit both p-type (ZnTe) and n-type (ZnSe) conductivity. A(II)B(VI) type semiconductors, which include ZnS, ZnSe, ZnTe, CdS, CdTe, HgSe, HgTe, and HgS, are used as luminophores in the visible light range for the production of optical fibers and photovoltaic components [16]. Depending on the type of absorption occurring in semiconductor materials, a distinction is made between the semiconductors with a straight or oblique interband gap. For the semiconductors with a straight gap, the bottom of the conduction band and the top of the valence band occur for the same value of k (lattice vector), which determines the position of the unfilled state in the band. For the semiconductors with a straight transition for effective absorption of light there must be a probability of meeting of two particles—an electron and a photon. In oblique transitions, three particles—an electron, a photon, and a phonon—should meet. This means that the absorption coefficient for straight transitions takes a higher value than for oblique transitions [17]. Therefore, semiconductors with straight transitions are used as luminophores; they have high luminescence efficiency [18].

3.1.1. Nanoscale Semiconductors

At the nanoscale, the physicochemical properties of materials change in a fundamental way compared to those of their bulk counterparts [15], which is related to the so-called quantum entrapment effect [19]. It occurs as the particle size decreases below the Bohr radius of the exciton, causing the electron in the nanocrystal to behave as if in a three-dimensional box of potential. As the Bohr radius of the exciton increases, the energy of the excited band decreases [20]. The quantum confinement effect plays a key role in the properties of nanocrystals, which is related to the change in the energy of the excited band. In semiconductor nanocrystals, analogous to organic molecules, a photon can be released or absorbed at the transition of charge carriers between the quantum levels of the valence band and the conduction band. The wavelength of absorption or luminescence can be controlled by changing the size of the quantum dots, which are often called "artificial atoms" [21].

Doping semiconductor nanocrystals affects their electrical, optical or magnetic properties. It may cause an increase in conductivity (an increase in the number of electrons or holes), the formation of a new energy level, which in turn contributes to the appearance of completely new optoelectronic properties of nanocrystals [22]. In A(II)B(VI) type semiconductors, doping with atoms of other elements has the greatest impact on their luminescent properties. For example, CdS quantum dots (4.2 nm) doped with Mn^{2+} ions show a blue shift in photoluminescence spectra and exhibit orange photoluminescence, compared to the undoped CdS nanocrystals. This phenomenon is caused by the additional 6A1-4T1 transition of Mn^{2+} ions, resulting in an increase in the quantum yield of luminescence reaching up to 41%. The observed properties of doped nanocrystals compared to undoped ones occur due to a shift in the energy levels in the nanocrystals caused by the introduction of the dopant [23,24]. Depending on the desired characteristic of the nanomaterials, selected dopants are introduced. Thus, doping zinc selenide nanocrystals with silver atoms causes a shift of the emission maximum from the blue light region to the green light region [25], and doping ZnSe NCs with Cr^{2+} atoms into the green light region, and contribute to changing the electrical properties of the nanocrystals themselves, which has been successfully applied in the production of lasers [26]. The combination of the relative ease of fabrication of semiconductor nanomaterials with the ability to adjust the position and magnitude of their bandgap energy has produced promising materials with a wide range of applications, including optoelectronics, photonics, catalysis, photovoltaics, various sensors, and biomedicine [27–29].

3.1.2. Titanium(IV) oxide (TiO_2)

Titanium(IV) oxide is a semiconductor material with high efficiency in various photocatalytic reactions. It exhibits high chemical and photochemical stability.

However, besides the above-mentioned advantages, has some limitations, viz:

- the possibility of its wide application is limited by, e.g., the fact that it absorbs only UV radiation (387.5 nm), which needs the use of the light sources of high cost of exploitation [30];
- a decrease in the efficiency of the photocatalytic reaction, which is related to the phenomenon of recombination of photo-excited charge carriers (electrons (e^-) and holes (h^+)) [31];
- low selectivity especially in photo-oxidation reactions of organic compounds [32].

In view of above-mentioned limitations, work has begun on the synthesis of new photocatalysts on titanium(IV) oxide matrix, whose photocatalytic activity would be in the visible radiation range (>400 nm). This would significantly expand the applicability of heterogeneous photocatalysis in environmental protection, either by using the main part of the sunlight spectrum or by using a light source with lower irradiance. Currently, the goal of most of the work carried out in the world is to obtain a visible-light-activated photocatalyst, which would be obtained by modification of TiO_2. Reactions carried out under hydrothermal conditions made it possible to obtain semiconductor materials with different morphological structures, for example, during the synthesis of anatase (TiO_2), it was important to use appropriate substances to control the formation of crystal morphology during hydrothermal synthesis, including fluorine compounds [33]. However, these compounds at higher temperatures can undergo transformations to highly toxic compounds with corrosive properties. This poses quite a limitation to the applications of this method [34]. An alternative to hydrothermal methods, for the synthesis of anatase, can be the process of crystallization in the gas phase. In this process, it is possible to obtain anatase crystals with a decahedral structure (decahedral anatase particles) [34]. Amano and colleagues showed that rapid heating, up to 1200 °C, and cooling of a mixture of titanium(IV) chloride and oxygen promoted the formation of anatase in the form of decahedral-shaped crystals [35]. However, this method did not permit for controlled and continuous dosing of titanium(IV) chloride ($TiCl_4$) into the reactor system, which was connected to a coaxial flow of reaction gases. This was a key element in the controlled preparation of anatase crystals with different morphological parameters. Hence, Janczarek and co-workers [36] have developed a method for precise dosing of titanium(IV) chloride vapor into the tubular reactor space, combined with a constant flow of oxygen. This solution made it possible to obtain a product characterized by well-defined properties with high efficiency and reproducibility.

The efficiency of heterogeneous photocatalysis using TiO_2 as a photocatalyst depends primarily on the polymorphic variety of the material. Polymorphic varieties of titanium oxide include rutile, anatase, and brucite [37]. The most desirable form of TiO_2 is anatase, which is characterized by a large specific surface area, a high degree of surface hydroxylation, and a bandgap energy of Eg = 3.23 eV (384 nm). Rutile (Eg = 3.02 eV (411 nm)) is less effective in photocatalytic processes, which is due to the presence of differences in the recombination rates of electron-hole pairs; the recombination time between the electron (e^-) and the hole (h^+) for rutile is shorter than their migration time to the surface. Besides, rutile has fewer active sites and hydroxyl groups on the surface compared to anatase [38].

TiO_2-based materials with enhanced UV activity or activity under visible light can also be obtained by: (i) the addition of transition metal ions, e.g., Mn, Nb, V, Fe, Au, Ag [39,40]; (ii) preparation of a reduced form of TiO_2 [41]; (iii) sensitization of TiO_2 with dyes [42] and with semiconductors with a smaller Eg bandwidth [43], (iv) doping with non-metals, e.g.: nitrogen [44], carbon [45], or phosphorus [46].

The mechanism of excitation of the photocatalyst depends on how the material is modified. The main types of excitation of TiO_2 under the influence of radiation from the visible range include [41,47,48]: (a) the appearance of a new energy state associated with the presence of an oxygen vacancy, (b) dye sensitization, where the dye is a sensitizer, (c) dye sensitization, where the dye is both a sensitizer and a degradant, and (d) the formation of a new energy level below the conduction band associated with the presence of metal cations.

Another method of titanium(IV) oxide modification is its doping with metals or non-metals, such as boron, tungsten, or precious metals. The doping with boron can enhance the photocatalytic activity of titanium(IV) oxide under visible light.

The introduction of boron into the TiO_2 structure inhibits the growth of crystal size, can affect the phase transformation of anatase to rutile, and can increase the specific surface area of photocatalysts [49–53].

Tungsten oxide, on the other hand, due to the width of its excited band (2.8 eV), can be used as an admixture of titanium dioxide, thus causing an increase in its photocatalytic activity in the visible light range [54–58].

Modification with noble metals (primarily gold, silver, or platinum) can enhance the activity of titanium dioxide in the visible light range since nanoparticles of noble metals such as silver and gold exhibit the ability to absorb visible radiation, which is a result of the existence of a surface plasmon. This enables them to absorb light in the visible and near-infrared range, which favors their potential use for activating titanium dioxide with solar radiation. In addition, they can capture charge carriers (e^-/h^+), and thus cause a reduction in the rate of the recombination process of electron-hole pairs, which is associated with an increase in the quantum yield of the reaction [59].

It has been observed that the photocatalytic activity of TiO_2 modified with noble metals depends, among other things, on the size of the metal particles. The size of the obtained noble metal nanoparticles is affected by the reaction temperature, the reducing reagents used, the type of stabilizer used, and other factors [60–63].

3.1.3. Zinc Oxide (ZnO)

One of the commonly used photocatalysts, along with TiO_2, is zinc oxide ZnO. It is a material of increasing interest due to its ability to form various nanostructures such as nanowires, nanobelts, nanoscratches, nanospheres, nanofibers, and nanotetrapods. Currently, however, nanowires of zinc oxide are of the greatest interest, especially when arranged in layers oriented in perpendicular to the conducting substrate. Nanowires deposited in this way are characterized by a high diameter-to-height ratio, which means that the total surface area of the deposited ZnO can be up to 100 times greater than the geometric surface area on which this deposition occurs. Consequently, a large amount of photosensitive material can be deposited on the ZnO surface, resulting in a high light absorption efficiency value. The ordered nanowire layers are used, for example, in lasers, electroluminescent devices, sensors, photocatalytic systems, and third-generation solar cells [64,65].

In the synthesis of ZnO, a key process is the preparation of zinc hydroxide. There are several natural forms of $Zn(OH)_2$, denoted as: α-, β-, γ-, δ-, ε- $Zn(OH)_2$. The latter is the most stable. Usually, during deposition, the α- form is deposited first, which under aging changes to the ε- form [66].

The main crystalline form of zinc oxide is wurtzite, the form that is thermodynamically stable under normal atmospheric conditions. It is a system consisting of $4O^{2-}$ and Zn^{2+} ions arranged in a characteristic manner. A characteristic feature of ZnO is the presence of polar and non-polar crystal planes [67].

3.1.4. Comparison of the Properties of TiO_2 and ZnO

Despite the promising properties of zinc oxide, titanium(IV) oxide is still the most commonly used photocatalyst. This is largely related to the higher chemical stability of TiO_2. Titanium(IV) oxide has a similar energy gap to that of ZnO (3.2 eV) and a similar energy band pattern. In addition, TiO_2 also has the advantage of higher electrical permeability than ZnO, which allows it to better retain electrons and inhibit the recombination process [68]. An advantageous feature of ZnO over TiO_2 is also that it is a straight energy gap semiconductor (unlike TiO_2, whose energy gap type depends on the crystallographic form). In addition, ZnO exhibits a higher electron mobility than TiO_2 (200 cm^2 /Vs for ZnO and 10 cm^2 /Vs for TiO_2) [69]. This results in a lower resistance of ZnO. The ease of fabrication and the low cost of the process may also be in favor of ZnO over TiO_2.

3.2. Plasmonic Materials

Plasmonics is concerned with the studies of plasmons that are the quasiparticles made of quanta of plasma oscillations at the characteristic plasma frequency ω_p, as a result of the action of an electromagnetic wave on quasi-free carriers originating from the conduction band of a metal or semiconductor [70]. As a result of the electromagnetic field, the quasi-submissive carriers move away from the positively charged atomic nucleus and then return to their previous state when the field no longer acts, due to the attractive Coulombic forces [71].

The two main groups of methods for obtaining plasmonic materials include the so-called top-down methods (building from the top down) and bottom-up methods (building the material from scratch, atom by atom, or particle by particle) [72,73].

Top-down methods include lithographic techniques, which include lithographic nanoprinting, soft lithographic methods, or methods based on the use of a scanning tunneling microscope (STM) and 3D Direct Laser Writing [72,73].

3.2.1. Lithographic Techniques

Electron Beam Lithography (EBL) and Focus Ion Beam (FIB) lithography favor obtaining the desired nanostructure in two steps, i.e.: hardening the resist with an electron or ion beam, and etching the nanostructure by deep plasma etching, e.g., using the Reactive Ion Etching (RIE) technique. The resulting nanostructures have very high resolution (on the order of a few nanometers). The photolithography technique, on the other hand, is based on the use of a light beam, in which a specially prepared mask—a metallic plate with appropriately selected holes through which the light beam is passed—is additionally used to obtain the desired nanostructure [74].

In addition to the above-mentioned lithographic techniques, soft lithography techniques are also used, including Nanoimprint Lithography (NIL) and Room Temperature Nanoimprint Lithography (RTNIL). Using room-temperature lithographic nanoimprinting, among other things, the optically active, planar, chiral photonic metamaterials are obtained [75]. The technique is based on duplicating a nanostructure on a polymer stamp, which is formed by pouring a polymeric material onto a suitably prepared template. The template (usually a quartz substrate) is obtained by micromachining or modern lithography techniques.

3.2.2. Techniques Based on Scanning Tunneling Microscopes

Plasmonic nanostructures can also be obtained using scanning tunneling microscopes. They control the conditions for layer growth. Obtaining a given structure is made possible by a needle that mimics the given structure while scanning the electrically conductive material [76].

Bottom-up methods include those using self-assembly, direct nanoparticle doping methods, and gas-phase deposition techniques [77].

3.2.3. Nanoparticle Direct Doping (NPDD)

The use of nanoparticle direct doping permits obtaining desired materials through a chemical process. The advantage of this method is the deagglomerated state of nanoparticles [77], which is vital because agglomerates of plasmonic nanoparticles are useless in plasmonics, as they exhibit no or weak resonance phenomena. The main advantages of this method, in addition to those mentioned above, is the speed of obtaining composites, the preservation of the original form of the dopant in the composite by controlling the size and shape of the introduced nanoparticles, and the possibility of obtaining composites doped with particles, both metallic and non-metallic [77].

An example of the materials obtained by this method is precast glass doped with silver nanoparticles (0.15 wt%) [77].

3.2.4. Techniques Based on Self-Assembly

Self-organization techniques employ the mechanisms of natural self-assembly, e.g., the self-assembly of block copolymers. These techniques permit obtaining periodic domain nanostructures by microscopic phase separation [75]. The advantage of self-assembly is that it provides ordered structures with different morphologies [78].

On the other hand, the method of self-organization to single layers of nanoparticles involves dynamic evaporation of nanoparticles on the surface of liquid-air separation. This leads to the nucleation of islands of nanoparticles, followed by the growth of a monolayer [79]. Volume plasmonic materials can be obtained by the self-organization of liquid crystals, which yields a three-dimensional synthetic material that exhibits strong resonances in the visible range [76].

3.2.5. Gas-Phase Deposition Techniques

There are two main varieties of gas-phase deposition: Physical Vapor Deposition (PVD) and Chemical Vapor Deposition (CVD). In the PVD methods (the so-called "clean technology" as no harmful chemicals are required), the applied coating exhibits an adhesive nature, and its properties depend on the purity of the substrate. The process involves obtaining vapors of the material, which are then transported to the surface where they condense, and the coating grows. With this method, only flat or simple shapes can be coated as the process requires rotation of the coated parts. The CVD methods, on the other hand, involve the introduction of gas substrates into a chamber, where the appropriate chemical reactions take place on the substrate, leading to the formation of a coating. With this method, it is also possible to coat three-dimensional parts, as the process does not require rotation of the workpieces. However, when using this method, there may be a problem with the difficulty of balancing the compound decomposition reactions throughout the volume of the working chamber, which is associated with the formation of less pure layers [74].

Among the plasmonic materials currently used, silver and gold predominate. Silver has the lowest losses for the visible and near-infrared light range [80], which permitted the use of this plasmonic material to obtain super-lenses and hyper senses [81], or to increase the efficiency of solar panels made of layers of amorphous silicon [82]. Gold, compared to silver, exhibits higher chemical stability under natural conditions, which permits the use of gold layers in plasmonic biosensors [83].

The materials alternative to expensive noble metals may be plasmonic materials belonging to the nitride group, e.g., zirconium nitride ZrN, titanium nitride TiN, hafnium nitride HfN and tantalum nitride TaN [84]; doped transparent conducting oxides such as aluminum-doped zinc oxide Al: ZnO, zinc oxide doped with gallium Ga:ZnO, tin oxide doped with antimony oxide Sb_2O_3:SnO_2 and indium oxide doped with tin Sn:In_2O_3 [85], graphene [86], and the metals copper, aluminum, chromium, or iridium [87].

3.2.6. Plasmonic Properties of Metals

Surface plasmons in metals have many fascinating properties, which enables their applications in optics, sensorics, photonics, and nonlinear fields. Recently, the plasmonic properties of some metals (e.g., Au, Ag, Cu, Al, Mg, Pt and Rh) have been widely studied both experimentally and theoretically, which is related to the fact that for the development of efficient synthesis of nanoporous metals, the elucidation of their basic plasmonic properties is crucial. The plasmonic properties of nanoporous metals can be tuned by using different strategies for their preparation (these compounds are obtained by synthetic routes) [88], including (i) templating, which permits a precise control of the size and structure of porous metallic structures, (ii) dealloying, which permits production of structures characterized by open nanopores, tunable pore sizes, structural properties and multifunctionality, and (iii) colloidal chemistry [6]. Very often nanoporous metallic nanoparticles are produced by using a combination of lithographic techniques (Section 3.2.1) and dealloying methods. An example is the synthesis of nanoporous gold nanoparticles, by depositing gold and silver on a substrate made of a silicon wafer or a glass slide [89,90].

On the other hand, a nanoporous silver structure was obtained by Yang and coworkers using a silver halide electroreduction process, which permitted getting a material with tunable pore size [91]. Nanoporous silver films of variable composition have been produced by Shen and O'Carroll using non-lithographic and heat-assisted methods [92].

Besides Au and Ag nanoparticles, aluminum has also been widely studied as a UV plasmonic material. In most cases, the preparation of suitable Al structures required several processing steps, ranging from chemical synthesis of Al nanoparticles to nano-lithography for nanostructured films [93,94]. Garoli and co-workers have described the fabrication of nanoporous aluminum structures from Al_2Mg_3 alloy by means of a galvanic exchange reaction [95].

Jiang et al. have proposed the preparation of nanoporous Mg in a two-step process [96]. In the first step, Ti(Nb,Ta,V,Fe)$_{50}$Cu$_{50}$ alloys were melted in liquid Mg to synthesize inter-penetrating phase composites. The Ti-rich phase was then etched by selective dissolution in 15 M aqueous HF solution for several minutes in an ultrasonic bath, followed by cleaning in deionized water and alcohol. On the other hand, Liu et al. synthesized nanoporous magnesium for hydrogen generation using physical vapor deposition, starting with Mg powders of large granulation [97].

3.3. Other Materials

Recently, several new porous materials have been obtained for use in photocatalysis, including metal-organic frameworks (MOFs, Figure 3a), covalent organic frameworks (COFs, Figure 3b), hydrogen-bonded organic frameworks (HOFs, Figure 3c), and porous organic polymers (POPs, Figure 3d) [98–101]. MOFs composed of metal ions/clusters and organic linkers through coordination bonds, exhibit some unique features, which include periodic and well-defined structure, high specific surface area, structural diversity, and customizability. However, in addition to these advantages, they typically exhibit relatively low chemical stability and poor conductivity, which hinders their practical application [102,103]. COFs are a group of fully engineered crystalline materials obtained by polymerizing organic building blocks through strong covalent bonds [104–107]. HOFs consist of organic molecules linked through hydrogen bonding, exhibiting a specific structure and low density. They exhibit poor chemical stability, which also limits their application [108,109]. POPs are highly stable porous materials linked through strong covalent bonds based on organic molecules. Due to their undefined structure and irregular pores, it is difficult to gain adequate knowledge of the structure-activity relationship [110,111].

Of the aforementioned materials, COFs not only combine their advantages, but also offset their disadvantages, and as a result, they are attracting increasing scientific interest and are used in, e.g.,: gas adsorption and separation, detection, and catalysis [112–116].

The particularly desirable solid-state behavior of COFs makes them promising materials for photocatalytic applications [117–119]. In 2008. Wan et al. [120] described a boronic ester-based COF that exhibited solid-state behavior confirmed by a linear I-V profile. Stegbauer and co-workers published a paper in which hydrazone-bound COF was used for the first time as a photocatalyst for hydrogen evolution under visible light radiation [121]. The publication of these results has stimulated a rapid increase in the application of these materials in photocatalytic research, for example: in photocatalytic CO_2 reduction, organic transformation, and in pollutant degradation [122–124].

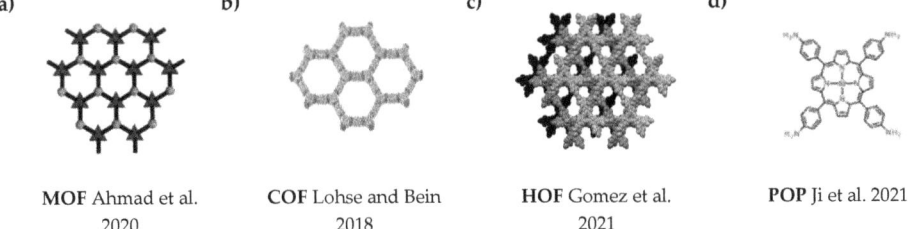

Figure 3. The structures of porous materials: (**a**) MOF [125], (**b**) COF [126], (**c**) HOF [127], (**d**) POP [110].

4. Application of Nanomaterials as Photocatalysts

Due to the use of solar radiation and other low-cost light sources, photocatalytic processes using photocatalysts show great application potential in four main areas: environmental, energy, biomedical, and chemical synthesis [128].

Heterogeneous photocatalysis is a relatively low-cost and sustainable technology for the treatment of many pollutants found in air and water, including organic compounds and heavy metals. Japan, the US, India, and China are major users of this technology, as evidenced in part by the number of research publications in the field. Like any method, it has both advantages and disadvantages. In general, the main advantages of conventional photocatalysts include chemical and physical stability, low cost, and environmental friendliness. The main limitations are the lack of solar sensitivity and lower efficiency, which encourages researchers to design strategies based on multiple photocatalysts to overcome the problems, such as recombination and low solar sensitivity, and further expand the processing capacity.

4.1. Photodegradation of Pollutants

Recently, increasing attention is paid to the search for safe and environmentally friendly methods of removing contaminants of organic compounds and inorganic compounds (these impurities are usually oxidized to simpler, less harmful compounds). Among others, photocatalytic degradation in the presence of suitable photocatalysts, often titanium(IV) oxide nanoparticles, is used to remove contaminants from the aqueous and gas phases. The first degradation of biphenyl and chlorobiphenyls in the presence of titanium(IV) oxide was performed by Carey and colleagues [129]. Since then, heterogeneous photocatalysis has gained importance as a potential method for the removal of environmental pollutants, such as the degradation of organic compounds [130], inorganic compounds [131], removal of odors from confined spaces [132], and destruction of bacteria in the presence of weak UV radiation [104].

Currently, the method of photodegradation in the presence of TiO_2 and solar radiation is used to purify water by removing benzene, toluene, ethylbenzene, and xylene, among others [133,134].

Figure 4 shows the mechanism of semiconductor excitation and photodegradation of impurities [3]. As a result of the excitation of the semiconductor with a radiation quantum of the same or greater energy than the energy gap of the photocatalyst, an electron is excited from the VB band to the CB band. This leads to the formation of a reactive electron-hole pair, which reacts with environmental pollutants. The electron-hole exhibits oxidizing properties and can react directly with the contaminants present in the environment, while the electron reacts with molecular oxygen, generating a superoxide anion radical, which in aqueous environments is immediately converted to $HO_2 \bullet$ and finally to do OH•. The resulting radicals rapidly oxidize impurities. If the process takes place in the aqueous phase, hydroxyl radicals can additionally be generated, which are also involved in the photodegradation of compounds. The final product of photodegradation is carbon dioxide and water, although other carbon-containing compounds at lower oxidation levels may be formed [135].

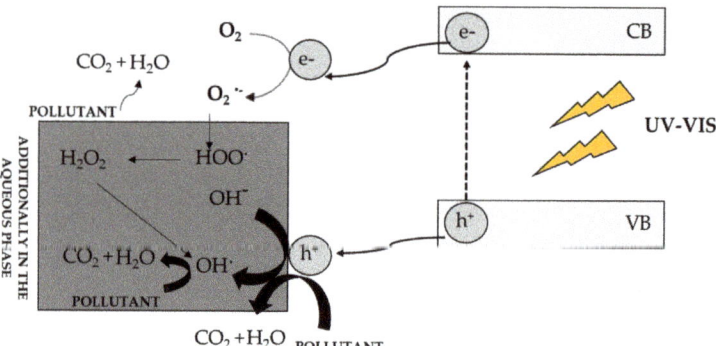

Figure 4. Mechanism of photocatalyst excitation and photodegradation of contaminants in the aqueous and gas phases [3].

Examples of photocatalysts used to remove contaminants are:
- Nanoparticles: titanium(IV) oxide (removal of acetone [136], methyl blue, methyl orange [137]), SnO_2 (removal of benzene [138]), ZnO (removal of methyl blue [139]; removal of methyl orange [140–142], ZnS (removal of rhodamine [143]), Ag^0 (removal of methyl blue [144]);
- Nanotubes: titanium(IV) oxide (removal of formaldehyde [145]), Ta_2O_5 (removal of toluene [146]);
- Nanocomposites: $TiO_2/CoxOy$ (phenol removal [147]);
- Nanowires: CdS (methyl orange removal [148]).

From among gas pollutants, nitrogen oxide and nitrogen dioxide are particularly toxic, the latter poses a particular threat to human health. Short-term exposure to a gas with a high concentration of NO_2 leads to irritation of the upper respiratory tract, and long-term exposure can provoke chronic respiratory diseases, including cancer. Nitric oxide is several times less harmful to human health, but it oxidizes to nitrogen dioxide on contact with the air. One solution that can contribute to reducing the concentration of nitrogen oxides in the air is photocatalytic concrete. This material is the most widely used human construction material, so its use in the context of reducing the concentration of nitrogen oxides seems a promising idea. Embedded in the structure of photocatalytic concrete are molecules of a photocatalyst, which, as a result of absorption of solar radiation, initiate a chemical reaction or affect the rate of an already occurring reaction. The most commonly used photocatalyst in concrete technology is titanium (IV) oxide in the form of anatase [149].

4.2. Heterogeneous Photocatalysis in the Removal of Heavy Metals

As a result of heterogeneous photocatalysis, metals and metalloid elements are converted to less toxic (lower valence) forms and/or are permanently deposited on the semiconductor surface [150,151]. The reaction of photocatalytic removal of heavy metals from the aqueous phase begins with excitation of the semiconductor by absorption of energy equal to or exceeding the energy of its excited band. Doping of zinc oxide with selenium [152] and silver [153] enhances the photocatalytic performance of the nanocomposite and results in a significant reduction or complete removal of heavy metals, such as: Cd, Ni, Pb, Zn, Cu, Cr from the aqueous environment. Modification of titanium(IV) oxide surface with silica [154], graphene [155], organic acids [156], and sulfur compounds [157] also allows the elimination of heavy metals from the reaction mixture.

The metals present in trace amounts in the natural environment such as mercury (Hg), chromium (Cr), lead (Pb) and others are considered highly hazardous to human health. Environmental applications of heterogeneous photocatalysis include the removal of heavy metals such as (Hg), chromium (Cr), lead (Pb), cadmium (Cd), lead (Pb), arsenic (As), nickel (Ni), and copper (Cu). The photoreduction capability has also been used to recover expensive metals such as gold (Au), platinum (Pt), and silver (Au) from industrial wastewater [158].

4.3. CO_2 Photoconversion and Water Photodecomposition

In addition to the above-mentioned applications, photocatalysts are now increasingly used in the transformation of carbon(IV) oxide to fuels or other compounds. All over the globe, there is an increasing demand for energy at an alarming rate, which is related to both continuous population growth and economic and technological development [159]. Figure 5 shows that more than 80% of the generated energy comes from burning fossil fuels, which include oil, natural gas, and solid fuels (mainly coal and lignite) [160].

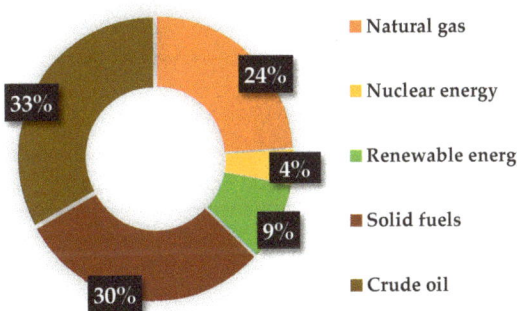

Figure 5. Main energy resources and their percentage contribution to energy production [161].

Fossil fuels show great importance in the global economy, for this reason, they are referred to as strategic raw materials, however, their massive use around the world is associated with concerns about the sufficiency of energy for future generations [162]. At present, fossil fuel resources are heavily depleted [159]. Moreover, in addition to concerns about the sufficiency of fossil resources, attention has always been paid to their adverse environmental impact. The use of energy derived from the combustion of fossil fuels is associated with the emission of significant amounts of carbon(IV) oxide into the atmosphere, which in turn exacerbates the global warming problem [163]. According to forecasts, the energy demand will continue to increase, so it is necessary to find a solution that reconciles the ever-increasing demand for energy with the need to protect the environment [164]. The solution to this problem, in the main, is the development of energy from renewable resources, especially in the direction of obtaining alternative substitutes for transportation fuels. After all, one of the main directions of consumption of energy produced from fossil raw materials is the transportation sector, consuming as much as 60% of its total amount [164]. When obtaining an alternative energy source, the reduction of carbon(IV) oxide in the atmosphere should also be considered. Such an opportunity is provided by the photoconversion of carbon(IV) oxide to light hydrocarbons. Most photocatalysts in use reduce CO_2 to carbon(II) oxide, methane, methanol, formic acid, or higher hydrocarbons. The energy required for this process is electromagnetic radiation of the desired wavelength. To date, this reaction has primarily used oxides and sulfides of transition metals, such as CdS [165], $ZnIn_2S_4$ [166], TiO_2 [167], Bi_2WO_6 [168]. Yet, due to these materials' low stability and lifetime, the search is constantly underway for a semiconductor material that both absorbs visible light and has an optimally located conduction band potential. The mechanism of photoconversion of carbon dioxide is based on the use of semiconductor materials to excite reactions involving solar radiation. The initiation of the redox reaction is a result of the photoexcitation of an electron (it is important that the refractory gap of the semiconductor material is smaller than the photon energy). The excited electrons are transferred from the valence band (VB) to the conduction band (CB). The generated charge carriers can move to the surface of the semiconductor material and react with substances adsorbed on the surface, such as CO_2. Holes and electrons can recombine or combine in trap states. Table 2 shows examples of photocatalysts used in the photoconversion of carbon(IV) oxide reaction to hydrocarbons under UV radiation in the ultraviolet range and VIS radiation in the visible range.

Table 2. Examples of photocatalysts used in the photoconversion of carbon(IV) oxide reaction to hydrocarbons under UV radiation in the ultraviolet range and VIS radiation in the visible range.

Material	Type of Radiation	Ref.
TiO_2 nanoparticles	UV-VIS	[169]
Ag/TiO_2 nanoparticles	UV-VIS	[170]
Au/TiO_2 nanoparticles	UV-VIS	[171]
$I-TiO_2$ nanoparticles	UV	[172]
Co/Co_3O_4 nanoparticles	UV-VIS	[173]
$AgBr/TiO_2$ nanocomposite	UV	[174]
$ZnO/ZnTe$ nanocomposite	VIS	[175]
$C-Na_2TiO_3$ nanowires	UV-VIS	[176]
$N-Na_2TiO_3$ nanowires	UV-VIS	[177]
$N-Na_2TiO_6$ nanowires	UV-VIS	[177]

One of the main alternative fuels that can replace existing fossil fuels is hydrogen. It is of great interest because of its characteristics; it is abundant, has a high heat of combustion, its combustion produces only water, and, most importantly, is environmentally friendly [178–180]. The element, in the presence of oxygen, burns with a nearly colorless, light blue flame with a relatively high propagation speed (2.7 m/s). The possibility of spontaneous ignition of a hydrogen-air mixture depends mainly on its concentration, for example, at 293 K the mixture can spontaneously ignite if the volume concentration of hydrogen is between 4 and 75%. Mixtures with particularly explosive properties are obtained in the hydrogen concentration range of 18–65%. Handling gaseous hydrogen is dangerous due to its flammability and explosiveness and requires extreme caution. At present, about 48% of hydrogen is generated by steam reforming of methane at elevated temperatures 30% is generated from oil, 18% from coal, and 4% from water electrolysis [181]. Among the best known methods for obtaining hydrogen is natural gas reforming [182]; coal or coke gasification [183], plasma technology [181], water electrolysis [183], photoelectrolysis [184], and biological methods [181]. This indicates that it can be extracted from a variety of feedstocks in many ways, although minimizing the cost of its production is currently a challenge for scientists and engineers. Currently, hydrogen is used as a fuel to power spacecraft, in fuel cells to provide heat and electricity, and to power some automobiles [185].

Photocatalytic reactions are the basis of technologies for obtaining a clean, environmentally safe energy carrier, which is hydrogen, not only from water but also from organic substances using the renewable energy source, which is the Sun. Photocatalytic hydrogen production is carried out in two ways: (i) photocatalytic splitting of water and (ii) photocatalytic reforming of organic compounds [186]. The first approach is based on the ability of water to reduce and oxidize by reacting with photogenerated electrons and positively charged "holes", during the irradiation of semiconductors, in the presence of selected cocatalysts. The second approach is based on the ability of some organic species to donate electrons to the positive holes of the illuminated photocatalyst and be oxidized, generating proton ions, while photogenerated electrons reduce the latter to produce hydrogen in the presence of proper co-catalysts [187].

Table 3 shows examples of photocatalysts that can be used in the decomposition of water under the influence of UV radiation in the ultraviolet range and VIS radiation in the visible range.

Table 3. Examples of photocatalysts that can be used in the decomposition of water under the influence of UV radiation in the ultraviolet range and VIS radiation in the visible range.

Material	Type of Radiation	Ref.
TiO_2 nanoparticles	UV	[188]
Au/TiO_2 nanoparticles	UV	[188]
Pt-TiO_2 nanoparticles	UV	[189]
TiO_2/F/Pt nanocomposite	UV	[190]
La:$NaTaO_3$ nanocomposite	UV-VIS	[190]
RuO_2/La-$NaTaO_3$ nanocomposite	UV-VIS	[190]
SnO_2 nanowires	VIS	[191]
SnO_2/SnS_2 nanowires	VIS	[191]

5. Conclusions

This paper reviews the literature covering the synthesis and exemplary applications of nanomaterials as photocatalysts. These materials, metallic and semiconducting in nature, which can be excited by radiation in the UV-Vis range can potentially be used, among other applications, for the removal of pollutants from the environment (mainly on a laboratory scale). Over the years, research on the synthesis of photocatalysts has evolved considerably; from the use of transition metal oxides (e.g., TiO_2/ZnO) to much more advanced materials. The research advances favor the development of heterogeneous photocatalysts, in which the absorption of light is shifted from the UV to the visible regime (sunlight) to utilize sunlight/white light for photocatalysis. The efficiency of the photocatalysis reaction depends on a number of factors, including: light absorption capacity/light intensity, the type of photocatalyst used, the concentration of a photocatalyst and contaminant particles, the pH of the reaction medium and others. Therefore, it is very important to determine the optimal amount of these factors for a given photocatalyst and type of pollutant. The process of photocatalysis has been studied for a long time on a laboratory scale, but unfortunately, its large-scale application is greatly hampered, for example, by blocking light penetration in thick coatings, leaching effect, difficulty in recovering the photocatalyst, etc. In addition, in natural systems, the rate of decomposition of pollutants is not limited by a time regime, unlike in industrial installations, where the technical challenge of photocatalytic processes, especially in heterogeneous systems, is unsatisfactory reaction kinetics.

Researchers' interest in this topic is growing year by year, as evidenced by the steadily increasing number of papers in this thematic area. However, future work should focus on finding solutions for large-scale production, commercialization, and applications. Obtaining efficient, low-cost, stable, visible-light-activated photocatalysts continues to be a challenge in the field of photocatalysis. However, in my opinion, it is worthwhile to continue this work and to develop a highly efficient photocatalyst, which would be used, among other things, to obtain a pure energy carrier, such as hydrogen. Although photocatalytic reactions for obtaining hydrogen have been extensively studied, their overall efficiency towards hydrogen evolution is low. In many cases, satisfactory photocatalytic activity is only achieved when irradiated with UV light. For this reason, the search for new materials that are active and stable in the presence of sunlight is of great interest.

Funding: This research received no external funding.

Institutional Review Board Statement: Not applicable.

Informed Consent Statement: Not applicable.

Data Availability Statement: Not applicable.

Conflicts of Interest: The author declares no conflict of interest.

References

1. Braslavsky, S.E.; Braun, A.M.; Cassano, A.E.; Emeline, A.V.; Litter, M.I.; Palmisano, L.; Parmon, V.N.; Serpone, N. Glossary of terms used in photocatalysis and radiation catalysis (IUPAC Recommendations 2011). *Pure Appl. Chem.* **2011**, *83*, 931–1014. [CrossRef]
2. Herrmann, J.M. Heterogeneous photocatalysis: Fundamentals and applications to the removal of various types of aqueous pollutants. *Catal. Today* **1999**, *53*, 115–129. [CrossRef]
3. Baluk, M.A.; Mykowska, E.; Gołaszewska, D. Photoactive nanomaterials in environmental application. *Kwadrans Dla Chem.* **2020**, 33–45.
4. Gerischer, H. The Principles of Photoelectrochemical Energy Conversion. In *Photovoltaic and Photoelectrochemical Solar Energy Conversion*; Plenum Press: New York, NY, USA, 1981.
5. Yang, X.; Wang, D. Photocatalysis: From fundamental principles to materials and applications. *ACS Appl. Energy Mater.* **2018**, *1*, 6657–6693. [CrossRef]
6. Koya, A.N.; Zhu, X.; Ohannesian, N.; Yanik, A.A.; Alabastri, A.; Proietti Zaccaria, R.; Krahne, R.; Shih, W.C.; Garoli, D. Nanoporous metals: From plasmonic properties to applications in enhanced spectroscopy and photocatalysis. *ACS Nano* **2021**, *15*, 6038–6060. [CrossRef] [PubMed]
7. Sui, M.; Kunwar, S.; Pandey, P.; Lee, J. Strongly confined localized surface plasmon resonance (LSPR) bands of Pt, AgPt, AgAuPt nanoparticles. *Sci. Rep.* **2019**, *9*, 16582–16596. [CrossRef]
8. Scroccarello, A.; della Pelle, F.; del Carlo, M.; Compagnone, D. Optical plasmonic sensing based on nanomaterials integrated in solid supports. A critical review. *Anal. Chim. Acta* **2022**, *1237*, 340594–340610. [CrossRef] [PubMed]
9. Lerner, L. 7 Things You May Not Know about Catalysis, Argonne National Laboratory. Available online: https://www.anl.gov/article/7-things-you-may-not-know-about-catalysis (accessed on 16 December 2022).
10. Ohtani, B. Photocatalysis A to Z-what we know and what we do not know in a scientific sense. *J. Photochem. Photobiol.* **2010**, *11*, 157–178. [CrossRef]
11. Kelsall, R.W.; Hamley, I.W.; Geoghegan, M. *Nanoscale Science and Technology*; Wiley & Sons: Chichester, UK, 2005.
12. Trindade, T.; O'Brien, P.; Pickett, N.L. Nanocrystalline semiconductors: Synthesis, properties, and perspectives. *Chem. Mater.* **2001**, *13*, 3843–3858. [CrossRef]
13. Shankar, K.; Basham, J.I.; Allam, N.K.; Varghese, O.K.; Mor, G.K.; Feng, X.; Paulose, M.; Seabold, J.A.; Choi, K.S.; Grimes, C.A. Recent advances in the use of TiO_2 nanotube and nanowire arrays for oxidative photoelectrochemistry. *J. Phys. Colloid Chem.* **2009**, *113*, 6327–6359. [CrossRef]

14. Li, J.; Wu, N. Semiconductor-based photocatalysts and photoelectrochemical cells for solar fuel generation: A review. *Catal. Sci. Technol.* **2015**, *5*, 1360–1384. [CrossRef]
15. Smith, A.M.; Nie, S. Semiconductor nanocrystals: Structure, properties, and band gap engineering. *Acc. Chem. Res.* **2010**, *43*, 190–200. [CrossRef] [PubMed]
16. Rao, C.N.R.; Muller, A.; Cheetham, A.K. *The Chemistry of Nanomaterials: Synthesis, Properties and Application*; Wiley-VCH: Weinheim, Germany, 2004.
17. Prasad, P.N. *Nanophotonics*; Wiley & Sons: Chichester, UK, 2004.
18. Gaponenko, S.V. *Introduction to Nanophotonics*; Cambridge University Press: Cambridge, UK, 2010.
19. Xia, Y.; Yang, P. Chemistry and physics of nanowires. *Adv. Mater.* **2003**, *15*, 351–352. [CrossRef]
20. Ubaid, K.A.; Zhang, X.; Sharma, V.K.; Li, L. Fate and risk of metal sulfide nanoparticles in the environment. *Environ. Chem. Lett.* **2020**, *18*, 97–111. [CrossRef]
21. Kastner, M.A. Artificial atoms. *Phys. Today* **1993**, *46*, 24–31. [CrossRef]
22. Norris, D.J.; Efros, A.L.; Erwin, S.C. Doped Nanocrystals. *Science* **2008**, *319*, 1776–1779. [CrossRef] [PubMed]
23. Tynkevych, O.; Karavan, V.; Vorona, I.; Filonenko, S.; Khalavka, Y. Synthesis and properties of water-soluble blue-emitting Mn-alloyed CdTe quantum dots. *Nanoscale Res. Lett.* **2018**, *13*, 132–138. [CrossRef]
24. Khan, I.; Saeed, K.; Khan, I. Nanoparticles: Properties, applications and toxicities. *Arab. J. Chem.* **2019**, *12*, 908–931. [CrossRef]
25. Nguyen, V.K.; Pham, D.K.; Tran, N.Q.; Dang, L.H.; Nguyen, N.H.; Nguyen, T.V.; Nguyen, T.H.; Luong, T.B. comparative studies of blue-emitting zinc selenide nanocrystals doped with Ag, Cu, and Mg towards medical applications. *Crystals* **2022**, *12*, 625. [CrossRef]
26. Feng, G.; Yang, C.; Zhou, S. Nanocrystalline Cr^{2+}-doped ZnSe nanowires laser. *Nano Lett.* **2013**, *13*, 272–275. [CrossRef]
27. Malode, S.J.; Shanbhag, M.M.; Kumari, R.; Dkhar, D.S.; Chandra, P.; Shetti, N.P. Biomass-derived carbon nanomaterials for sensor applications. *J. Pharm. Biomed. Anal.* **2023**, *222*, 115102–115126. [CrossRef]
28. Wang, Z.; Wang, M.; Wang, X.; Hao, Z.; Han, S.; Wang, T.; Zhang, H. Photothermal-based nanomaterials and photothermal-sensing: An overview. *Biosens. Bioelectron.* **2023**, *220*, 114883–114895. [CrossRef]
29. Singh, N.; Kim, J.; Kim, J.; Lee, K.; Zunbul, Z.; Lee, I.; Kim, E.; Chi, S.G.; Kim, J.S. Covalent organic framework nanomedicines: Biocompatibility for advanced nanocarriers and cancer theranostics applications. *Bioact. Mater.* **2023**, *21*, 358–380. [CrossRef]
30. Pelaez, M.; Nolan, N.T.; Pillai, S.C.; Seery, M.K.; Falaras, P.; Kontos, A.G.; Dunlop, P.S.M.; Hamilton, J.W.J.; Byrne, J.A.; O'Shea, K. A review on the visible light active titanium dioxide photocatalysts for environmental applications. *Appl. Catal. B* **2012**, *125*, 331–349. [CrossRef]
31. Schneider, J.; Matsuoka, M.; Takeuchi, M.; Zhang, J.; Horiuchi, Y.; Anpo, M.; Bahnemann, D.W. Understanding TiO_2 photocatalysis: Mechanisms and Materials. *Chem. Rev.* **2014**, *114*, 9919–9986. [CrossRef]
32. Ghosh-Mukerji, S.; Haick, H.; Schvartzman, M.; Paz, Y. Selective photocatalysis by means of molecular recognition. *J. Am. Chem. Soc.* **2001**, *123*, 10776–10777. [CrossRef]
33. Yang, H.G.; Sun, C.H.; Qiao, S.Z.; Zou, J.; Liu, G.; Smith, S.C.; Cheng, H.M.; Lu, G.Q. Anatase TiO_2 single crystals with a large percentage of reactive facets. *Nature* **2008**, *453*, 638–641. [CrossRef]
34. Ahonen, P.P.; Moisala, A.; Tapper, U.; Brown, D.P.; Jokiniemi, J.K.; Kauppinen, E.I. Gas-phase crystallization of titanium dioxide nanoparticles. *J. Nanoparticle Res.* **2002**, *4*, 43–52. [CrossRef]
35. Amano, F.; Prieto-Mahaney, O.O.; Terada, Y.; Yasumoto, T.; Shibayama, T.; Ohtani, B. Decahedral single-crystalline particles of anatase titanium(IV) oxide with high photocatalytic activity. *Chem. Mater.* **2009**, *21*, 2601–2603. [CrossRef]
36. Janczarek, M.; Kowalska, E.; Ohtani, B. Decahedral-shaped anatase titania photocatalyst particles: Synthesis in a newly developed coaxial-flow gas-phase reactor. *Chem. Eng. J.* **2016**, *289*, 502–512. [CrossRef]
37. Carp, O.; Huisman, C.L.; Reller, A. Photoinduced reactivity of titanium dioxide. *Prog. Solid State Chem.* **2004**, *32*, 33–177. [CrossRef]
38. Banerjee, S.; Gopal, J.; Raj, B. Physics and chemistry of photocatalytic titanium dioxide: Visualization of bactericidal activity using atomic force microscopy. *Res. Commun.* **2006**, *10*, 1378–1385.
39. Anpo, M. Use of Visible Light. Second-generation titanium oxide photocatalysts prepared by the application of an advanced metal ion-implantation method. *Pure Appl. Chem.* **2000**, *72*, 1787–1792. [CrossRef]
40. Zaleska-Medynska, A.; Lezner, M.; Grabowska, E.; Zaleska, A. Preparation and photocatalytic activity of iron-modified titanium dioxide photocatalyst. *Physicochem. Probl. Miner. Process.* **2012**, *48*, 193–200.
41. Nakamura, I.; Negishi, N.; Kutsuna, S.; Ihara, T.; Sugihara, S.; Takeuchi, K. Role of oxygen vacancy in the plasma-treated TiO_2 photocatalyst with visible light activity for NO removal. *J. Mol. Catal. A Chem.* **2000**, *161*, 205–212. [CrossRef]
42. Chatterjee, D.; Mahata, A. Demineralization of organic pollutants on the dye modified TiO_2 semiconductor particulate system using visible light. *Appl. Catal. B* **2001**, *33*, 119–125. [CrossRef]
43. Hirai, T.; Suzuki, K.; Komasawa, I. Preparation and photocatalytic properties of composite CdS nanoparticles-titanium dioxide particles. *J. Colloid Interface Sci.* **2001**, *244*, 262–265. [CrossRef]
44. Irie, H.; Watanabe, Y.; Hashimoto, K. Nitrogen-concentration dependence on photocatalytic activity of TiO_2-XN_x powders. *J. Phys. Chem. B* **2003**, *107*, 5483–5486. [CrossRef]
45. Li, Y.; Hwang, D.S.; Lee, N.H.; Kim, S.J. Synthesis and characterization of carbon-doped titania as an artificial solar light sensitive photocatalyst. *Chem. Phys. Lett.* **2005**, *404*, 25–29. [CrossRef]

46. Korösi, L.; Dékány, I. Preparation and investigation of structural and photocatalytic properties of phosphate modified titanium dioxide. *Colloids Surf. A Physicochem. Eng. Asp.* **2006**, *280*, 146–154. [CrossRef]
47. Lobedank, J.; Bellmann, E.; Bendig, J. Sensitized Photocatalytic Oxidation of Herbicides Using Natural Sunlight. *J. Photochem. Photobiol.* **1997**, *108*, 89–93. [CrossRef]
48. Li, F.B.; Li, X.Z. The enhancement of photodegradation efficiency using Pt-TiO$_2$ catalyst. *Chemosphere* **2002**, *10*, 1103–1111. [CrossRef]
49. Geng, H.; Yin, S.; Yang, X.; Shuai, Z.; Liu, B. Geometric and electronic structures of the boron-doped photocatalyst TiO$_2$. *J. Condens. Matter Phys.* **2006**, *18*, 87–96. [CrossRef]
50. Wu, Y.; Xing, M.; Zhang, J.; Chen, F. Effective visible light-active boron and carbon modified TiO$_2$ photocatalyst for degradation of organic pollutant. *Appl. Catal. B* **2010**, *97*, 182–189. [CrossRef]
51. Ohtani, B. Preparing Articles on Photocatalysis—beyond the illusions, misconceptions, and speculation. *Chem. Lett.* **2008**, *37*, 217–229. [CrossRef]
52. Xu, T.; Song, C.; Liu, Y.; Han, G. Band structures of TiO$_2$ doped with N, C and B. *J. Zhejiang Univ. Sci. B* **2006**, *7*, 299–303. [CrossRef]
53. Chen, D.; Yang, D.; Wang, Q.; Jiang, Z. Effects of Boron Doping on Photocatalytic Activity and Microstructure of Titanium Dioxide Nanoparticles. *Ind. Eng. Chem. Res.* **2006**, *45*, 4110–4116. [CrossRef]
54. Chai, S.Y.; Kim, Y.J.; Lee, W.I. Photocatalytic WO$_3$/TiO$_2$ nanoparticles working under visible light. *J. Electroceram.* **2006**, *17*, 909–912. [CrossRef]
55. Akurati, K.K.; Vital, A.; Dellemann, J.P.; Michalow, K.; Graule, T.; Ferri, D.; Baiker, A. Flame-Made WO$_3$/TiO$_2$ nanoparticles: Relation between surface acidity, structure and photocatalytic activity. *Appl. Catal. B* **2008**, *79*, 53–62. [CrossRef]
56. Michalow, K.A.; Vital, A.; Heel, A.; Graule, T.; Reifler, F.A.; Ritter, A.; Zakrzewska, K.; Rekas, M. Photocatalytic Activity of W-Doped TiO$_2$ Nanopowders. *J. Adv. Oxid. Technol.* **2008**, *11*, 56–64. [CrossRef]
57. Iliev, V.; Tomova, D.; Rakovsky, S.; Eliyas, A.; Puma, G.L. Enhancement of photocatalytic oxidation of oxalic acid by gold modified WO$_3$/TiO$_2$ Photocatalysts under UV and Visible Light Irradiation. *J. Mol. Catal. A Chem.* **2010**, *327*, 51–57. [CrossRef]
58. Kubacka, A.; Colón, G.; Fernández-García, M. N- and/or W-(Co)doped TiO$_2$-Anatase catalysts: Effect of the calcination treatment on photoactivity. *Appl. Catal. B* **2010**, *95*, 238–244. [CrossRef]
59. Sclafani, A.; Herrmann, B.J.M. Influence of metallic silver and of platinum-silver bimetallic deposits on the photocatalytic activity of titania (anatase and rutile) in organic and aqueous media. *J. Photochem. Photobiol. A Chem.* **1998**, *113*, 181–188. [CrossRef]
60. Zhang, F.; Jin, R.; Chen, J.; Shao, C.; Gao, W.; Li, L.; Guan, N. High photocatalytic activity and selectivity for nitrogen in nitrate reduction on Ag/TiO$_2$ catalyst with fine silver clusters. *J. Catal.* **2005**, *232*, 424–431. [CrossRef]
61. Hou, X.G.; Ma, J.; Liu, A.D.; Li, D.J.; Huang, M.D.; Deng, X.Y. Visible light active TiO$_2$ films prepared by electron beam deposition of noble metals. *Nucl. Instrum. Methods Phys. Res. B* **2010**, *268*, 550–554. [CrossRef]
62. Kowalska, E.; Remita, H.; Colbeau-Justin, C.; Hupka, J.; Belloni, J. Modification of titanium dioxide with platinum ions and clusters: Application in photocatalysis. *J. Phys. Chem. C* **2008**, *112*, 1124–1131. [CrossRef]
63. Hwang, S.; Lee, M.C.; Choi, W. Highly Enhanced photocatalytic oxidation of CO on titania deposited with Pt nanoparticles: Kinetics and mechanism. *Appl. Catal. B* **2003**, *46*, 49–63. [CrossRef]
64. Zarębska, K.; Kwiatkowski, M.; Gniadek, M.; Skompska, M. Electrodeposition of Zn(OH)$_2$, ZnO thin films and nanosheet-like Zn seed layers and influence of their morphology on the growth of ZnO nanorods. *Electrochim. Acta* **2013**, *98*, 255–262. [CrossRef]
65. Skompska, M.; Zarębska, K. Electrodeposition of ZnO nanorod arrays on transparent conducting substrates-a review. *Electrochim. Acta* **2014**, *127*, 467–488. [CrossRef]
66. Peulon, S.; Lincot, D. Mechanistic study of cathodic electrodeposition of zinc oxide and zinc hydroxychloride films from oxygenated aqueous zinc chloride solutions. *J. Electrochem. Soc.* **1998**, *145*, 864–875. [CrossRef]
67. Baruah, S.; Dutta, J. Hydrothermal growth of ZnO nanostructures. *Sci. Technol. Adv. Mater.* **2009**, *10*, 13001–13019. [CrossRef]
68. Hodes, G. Comparison of dye- and semiconductor-sensitized porous nanocrystalline liquid junction solar cells. *J Phys. Chem. C* **2008**, *112*, 17778–17787. [CrossRef]
69. Briscoe, J.; Dunn, S. Extremely thin absorber solar cells based on nanostructured semiconductors. *Mater. Sci. Technol.* **2011**, *27*, 1741–1756. [CrossRef]
70. Barnes, W.L.; Dereux, A.; Ebbesen, T.W. Surface plasmon subwavelength optics. *Nature* **2003**, *424*, 824–830. [CrossRef]
71. Maier, S.A. *Plasmonics: Fundamentals and Applications*; Springer: New York, NY, USA, 2007.
72. Liu, N.; Guo, H.; Fu, L.; Kaiser, S.; Schweizer, H.; Giessen, H. Three-dimensional photonic metamaterials at optical frequencies. *Nat. Mater.* **2008**, *7*, 31–37. [CrossRef]
73. Nagpal, P.; Lindquist, N.C.; Oh, S.-H.; Norris, D.J. Ultrasmooth Patterned Metals for Plasmonics and Metamaterials. *Science* **2009**, *325*, 594–597. [CrossRef]
74. Lindquist, N.C.; Nagpal, P.; McPeak, K.M.; Norris, D.J.; Oh, S.H. Engineering metallic nanostructures for plasmonics and nanophotonics. *Rep. Prog. Phys.* **2012**, *75*, 36501–36562. [CrossRef]
75. Chen, Y.; Tao, J.; Zhao, X.; Cui, Z.; Schwanecke, A.S.; Zheludev, N.I. Nanoimprint lithography for planar chiral photonic meta-materials. *Microelectron. Eng.* **2005**, *78–79*, 612–617. [CrossRef]
76. Korzeb, K.; Gajc, M.; Pawlak, D. Przegląd Metod Otrzymywania Materiałów Plazmonicznych Oraz Wybranych Alternatywnych Materiałów. *Electron. Mater.* **2014**, *42*, 18–30.

77. Gajc, M.; Surma, H.B.; Klos, A.; Sadecka, K.; Orlinski, K.; Nikolaenko, A.E.; Zdunek, K.; Pawlak, D.A. Nanoparticle direct doping: Novel method for manufacturing three-dimensional bulk plasmonic nanocomposites. *Adv. Funct. Mater.* **2013**, *23*, 3443–3451. [CrossRef]
78. Mai, Y.; Eisenberg, A. Self-assembly of block copolymers. *Chem. Soc. Rev.* **2012**, *41*, 5969–5985. [CrossRef]
79. Bigioni, T.P.; Lin, X.M.; Nguyen, T.T.; Corwin, E.I.; Witten, T.A.; Jaeger, H.M. Kinetically driven self assembly of highly ordered nanoparticle monolayers. *Nat. Mater.* **2006**, *5*, 265–270. [CrossRef]
80. West, P.R.; Ishii, S.; Naik, G.V.; Emani, N.K.; Shalaev, V.M.; Boltasseva, A. Searching for better plasmonic materials. *Laser Photonics Rev.* **2010**, *4*, 795–808. [CrossRef]
81. Liu, Z.; Lee, H.; Xiong, Y.; Sun, C.; Zhang, X. Far-field optical hyperlens magnifying sub-diffraction-limited objects. *Science* **2007**, *315*, 1686. [CrossRef]
82. Ferry, V.E.; Verschuuren, M.A.; Li, H.B.T.; Verhagen, E.; Walters, R.J.; Schropp, R.E.I.; Atwater, H.A.; Polman, A.; Li, H.; van der Werf, C.H.M.; et al. Light trapping in ultrathin plasmonic solar cells. *Opt. Express* **2010**, *18*, 238–247. [CrossRef]
83. Homola, J. Present and future of surface plasmon resonance biosensors. *Anal. Bioanal. Chem.* **2003**, *377*, 528–539. [CrossRef]
84. Feigenbaum, E.; Diest, K.; Atwater, H.A. Unity-order index change in transparent conducting oxides at visible frequencies. *Nano Lett.* **2010**, *10*, 2111–2116. [CrossRef]
85. Minami, T. New N-type transparent conducting oxides. *MRS Bull.* **2000**, *25*, 38–44. [CrossRef]
86. Katsnelson, M.I.; Novoselov, K.S.; Geim, A.K. Chiral tunnelling and the klein paradox in graphene. *Nat. Phys.* **2006**, *2*, 620–625. [CrossRef]
87. Tassin, P.; Koschny, T.; Kafesaki, M.; Soukoulis, C.M. A Comparison of graphene, superconductors and metals as conductors for metamaterials and plasmonics. *Nat. Photonics* **2012**, *6*, 259–264. [CrossRef]
88. Rebbecchi, T.A.; Chen, Y. Template-based fabrication of nanoporous metals. *J. Mater. Res.* **2018**, *33*, 2–15. [CrossRef]
89. Raj, D.; Palumbo, M.; Fiore, G.; Celegato, F.; Scaglione, F.; Rizzi, P. Sustainable nanoporous gold with excellent SERS performances. *Mater. Chem. Phys.* **2023**, *293*, 126883–126895. [CrossRef]
90. Arnob, M.M.P.; Artur, C.; Misbah, I.; Mubeen, S.; Shih, W.C. 10×-Enhanced heterogeneous nanocatalysis on a nanoporous gold disk array with high-density hot spots. *ACS Appl. Mater. Interfaces* **2019**, *11*, 13499–13506. [CrossRef]
91. Seok, J.Y.; Lee, J.; Yang, M. Self-generated nanoporous silver framework for high-performance iron oxide pseudocapacitor anodes. *ACS Appl. Mater. Interfaces* **2018**, *10*, 17223–17231. [CrossRef]
92. Shen, Z.; O'Carroll, D.M. Nanoporous silver thin films: Multifunctional platforms for influencing chain morphology and optical properties of conjugated polymers. *Adv. Funct. Mater.* **2015**, *25*, 3302–3313. [CrossRef]
93. Yang, W.; Zheng, X.-G.; Wang, S.-G.; Jin, H.-J. Nanoporous aluminum by galvanic replacement: Dealloying and inward-growth plating. *J. Electrochem. Soc.* **2018**, *165*, C492–C496. [CrossRef]
94. Vargas-Martínez, J.; Estela-García, J.E.; Suárez, O.M.; Vega, C.A. Fabrication of a porous metal via selective phase dissolution in Al-Cu alloys. *Metals* **2018**, *8*, 378. [CrossRef]
95. Garoli, D.; Schirato, A.; Giovannini, G.; Cattarin, S.; Ponzellini, P.; Calandrini, E.; Zaccaria, R.P.; D'amico, F.; Pachetti, M.; Yang, W. Galvanic replacement reaction as a route to prepare nanoporous aluminum for UV plasmonics. *Nanomaterials* **2020**, *10*, 102. [CrossRef]
96. Jiang, B.; Li, C.; Dag, Ö.; Abe, H.; Takei, T.; Imai, T.; Hossain, M.S.A.; Islam, M.T.; Wood, K.; Henzie, J. Mesoporous metallic rhodium nanoparticles. *Nat. Commun.* **2017**, *8*, 1–8. [CrossRef]
97. Liu, J.; Wang, H.; Yuan, Q.; Song, X. A Novel material of nanoporous magnesium for hydrogen generation with salt water. *J. Power Sources* **2018**, *395*, 8–15. [CrossRef]
98. Cao, J.; Yang, Z.; Xiong, W.; Zhou, Y.; Wu, Y.; Jia, M.; Zhou, C.; Xu, Z. Ultrafine metal species confined in metal–organic frameworks: Fabrication, characterization and photocatalytic applications. *Coord. Chem. Rev.* **2021**, *439*, 213924–213944. [CrossRef]
99. Wang, H.; Wang, H.; Wang, Z.; Tang, L.; Zeng, G.; Xu, P.; Chen, M.; Xiong, T.; Zhou, C.; Li, X. Covalent organic framework photocatalysts: Structures and applications. *Chem. Soc. Rev.* **2020**, *49*, 4135–4165. [CrossRef]
100. Xia, C.; Kirlikovali, K.O.; Nguyen, T.H.C.; Nguyen, X.C.; Tran, Q.B.; Duong, M.K.; Nguyen Dinh, M.T.; Nguyen, D.L.T.; Singh, P.; Raizada, P. The emerging covalent organic frameworks (COFs) for solar-driven fuels production. *Coord. Chem. Rev.* **2021**, *446*, 214117–214142. [CrossRef]
101. Zhang, Z.; Jia, J.; Zhi, Y.; Ma, S.; Liu, X. Porous Organic Polymers for Light-Driven Organic Transformations. *Chem. Soc. Rev.* **2022**, *51*, 2444–2490. [CrossRef]
102. Zhou, H.C.J.; Kitagawa, S. Metal-Organic Frameworks (MOFs). *Chem. Soc. Rev.* **2014**, *43*, 5415–5418. [CrossRef]
103. Wang, L.; Zhang, Y.; Chen, L.; Xu, H.; Xiong, Y. Solar energy conversion: 2D polymers as emerging materials for photocatalytic overall water splitting. *Adv. Mater.* **2018**, *30*, 1870369–1870380. [CrossRef]
104. Evans, A.M.; Parent, L.R.; Flanders, N.C.; Bisbey, R.P.; Vitaku, E.; Kirschner, M.S.; Schaller, R.D.; Chen, L.X.; Gianneschi, N.C.; Dichtel, W.R. Seeded growth of single-crystal two-dimensional covalent organic frameworks. *Science* **2018**, *361*, 52–57. [CrossRef]
105. Feng, X.; Ding, X.; Jiang, D. Covalent organic frameworks. *Chem. Soc. Rev.* **2012**, *41*, 6010–6022. [CrossRef]
106. Zhao, X.; Liang, R.R.; Jiang, S.Y.; Ru-Han, A. Two-dimensional covalent organic frameworks with hierarchical porosity. *Chem. Soc. Rev.* **2020**, *49*, 3920–3951.
107. Wang, Z.; Zhang, S.; Chen, Y.; Zhang, Z.; Ma, S. Covalent organic frameworks for separation applications. *Chem. Soc. Rev.* **2020**, *49*, 708–735. [CrossRef]

108. Luo, J.; Wang, J.W.; Zhang, J.H.; Lai, S.; Zhong, D.C. Hydrogen-bonded organic frameworks: Design, structures and potential applications. *CrystEngComm* **2018**, *20*, 5884–5898. [CrossRef]
109. Lin, R.B.; He, Y.; Li, P.; Wang, H.; Zhou, W.; Chen, B. Multifunctional porous hydrogen-bonded organic framework materials. *Chem. Soc. Rev.* **2019**, *48*, 1362–1389. [CrossRef] [PubMed]
110. Ji, W.; Wang, T.X.; Ding, X.; Lei, S.; Han, B.H. Porphyrin- and phthalocyanine-based porous organic polymers: From synthesis to application. *Coord. Chem. Rev.* **2021**, *439*, 213875–213904. [CrossRef]
111. Hetemi, D.; Pinson, J. Surface functionalisation of polymers. *Chem. Soc. Rev.* **2017**, *46*, 5701–5713. [CrossRef]
112. Yadav, P.; Yadav, M.; Gaur, R.; Gupta, R.; Arora, G.; Srivastava, A.; Goswami, A.; Gawande, M.B.; Sharma, R.K. Chemistry of magnetic covalent organic frameworks (MagCOFs): From synthesis to separation applications. *Mater. Adv.* **2022**, *3*, 1432–1458. [CrossRef]
113. Ren, X.; Li, C.; Liu, J.; Li, H.; Bing, L.; Bai, S.; Xue, G.; Shen, Y.; Yang, Q. The fabrication of Pd single atoms/clusters on COF layers as Co-catalysts for photocatalytic H_2 evolution. *ACS Appl. Mater. Interfaces* **2022**, *14*, 6885–6893. [CrossRef]
114. Zhang, W.; Chen, L.; Dai, S.; Zhao, C.; Ma, C.; Wei, L.; Zhu, M.; Chong, S.Y.; Yang, H.; Liu, L. Reconstructed covalent organic frameworks. *Nature* **2022**, *604*, 72–79. [CrossRef]
115. Wang, X.; Sun, L.; Zhou, W.; Yang, L.; Ren, G.; Wu, H.; Deng, W.Q. Iron single-atom catalysts confined in covalent organic frameworks for efficient oxygen evolution reaction. *Cell Rep. Phys. Sci.* **2022**, *3*, 100804–100817. [CrossRef]
116. Ahmed, I.; Jhung, S.H. Covalent organic framework-based materials: Synthesis, modification, and application in environmental remediation. *Coord. Chem. Rev.* **2021**, *441*, 213898–214015. [CrossRef]
117. Wang, G.B.; Li, S.; Yan, C.X.; Zhu, F.C.; Lin, Q.Q.; Xie, K.H.; Geng, Y.; Dong, Y. Covalent organic frameworks: Emerging high-performance platforms for efficient photocatalytic applications. *J. Mater. Chem. A Mater.* **2020**, *8*, 6957–6983. [CrossRef]
118. Huang, W.; Luo, W.; Li, Y. Two-dimensional semiconducting covalent organic frameworks for photocatalytic solar fuel production. *Mater. Today* **2020**, *40*, 160–172. [CrossRef]
119. Xie, J.; Shevlin, S.A.; Ruan, Q.; Moniz, S.J.A.; Liu, Y.; Liu, X.; Li, Y.; Lau, C.C.; Guo, Z.X.; Tang, J. Efficient visible light-driven water oxidation and proton reduction by an ordered covalent triazine-based framework. *Energy Environ. Sci.* **2018**, *11*, 1617–1624. [CrossRef]
120. Wan, S.; Guo, J.; Kim, J.; Ihee, H.; Jiang, D. A belt-shaped, blue luminescent, and semiconducting covalent organic framework. *Angew. Chem. Int. Ed. Engl.* **2008**, *47*, 8826–8830. [CrossRef] [PubMed]
121. Stegbauer, L.; Schwinghammer, K.; Lotsch, B.V. A hydrazone-based covalent organic framework for photocatalytic hydrogen production. *Chem. Sci.* **2014**, *5*, 2789–2793. [CrossRef]
122. Nguyen, H.L.; Alzamly, A. Covalent organic frameworks as emerging platforms for CO_2 photoreduction. *ACS Catal.* **2021**, *11*, 9809–9824. [CrossRef]
123. Tian, M.; Wang, Y.; Bu, X.; Wang, Y.; Yang, X. An ultrastable olefin-linked covalent organic framework for photocatalytic decarboxylative alkylations under highly acidic conditions. *Catal. Sci. Technol.* **2021**, *11*, 4272–4279. [CrossRef]
124. Kan, X.; Wang, J.C.; Chen, Z.; Du, J.Q.; Kan, J.L.; Li, W.Y.; Dong, Y. Synthesis of metal-free chiral covalent organic framework for visible-light-mediated enantioselective photooxidation in water. *J. Am. Chem. Soc.* **2022**, *144*, 6681–6686. [CrossRef]
125. Ahmad, M.; Luo, Y.; Wöll, C.; Tsotsalas, M.; Schug, A. Design of metal-organic framework templated materials using high-throughput computational screening. *Molecules* **2020**, *25*, 4875. [CrossRef]
126. Lohse, M.S.; Bein, T. Covalent organic frameworks: Structures, synthesis, and applications. *Adv. Funct. Mater.* **2018**, *28*, 1705553–1705624. [CrossRef]
127. Gomez, E.; Hisaki, I.; Douhal, A. Synthesis and photobehavior of a new dehydrobenzoannulene-based hof with fluorine atoms: From solution to single crystals observation. *Int. J. Mol. Sci.* **2021**, *22*, 4803. [CrossRef]
128. Kisch, H. *Semiconductor Photocatalysis: Principles and Applications*; Wiley & Sons: Chichester, UK, 2015.
129. Carey, J.H.; Lawrence, J.; Tosine, H.M. Photodechlorination of PCB's in the presence of titanium dioxide in aqueous suspensions. *Bull. Environ. Contam. Toxicol.* **1976**, *16*, 697–701. [CrossRef] [PubMed]
130. Liu, H.; Gao, Y. Photocatalytic decomposition of phenol over a novel kind of loaded photocatalyst of TiO_2/activated carbon/silicon rubber film. *React. Kinet. Catal. Lett.* **2004**, *83*, 213–219. [CrossRef]
131. Barakat, M.A.; Chen, Y.T.; Huang, C.P. Removal of toxic cyanide and Cu(II) ions from water by illuminated TiO_2 catalyst. *Appl. Catal. B* **2004**, *53*, 13–20. [CrossRef]
132. Bum Kim, S.; Tae Hwang, H.; Chang Hong, S. Photocatalytic degradation of volatile organic compounds at the gas-solid interface of a TiO_2 photocatalyst. *Chemosphere* **2008**, *48*, 437–444. [CrossRef] [PubMed]
133. Maness, P.C.; Smolinski, S.; Blake, D.M.; Huang, Z.; Wolfrum, E.J.; Jacoby, W.A. Bactericidal activity of photocatalytic TiO_2 reaction: Toward an understanding of its killing mechanism. *Appl. Environ. Microbiol.* **1999**, *65*, 4094–4098. [CrossRef]
134. Hoffmann, M.R.; Martin, S.T.; Choi, W.; Bahnemann, D.W.; Keck, W.M. Environmental Applications of Semiconductor Photocatalysis. *Chem. Rev.* **1995**, *95*, 69–96. [CrossRef]
135. Chatterjee, D.; Dasgupta, S. Visible Light Induced Photocatalytic Degradation of Organic Pollutants. *J. Photochem. Photobiol.* **2005**, *6*, 186–205. [CrossRef]
136. Bettoni, M.; Candori, P.; Falcinelli, S.; Marmottini, F.; Meniconia, S.; Rol, C.; Sebastiani, G.V. Gas phase photocatalytic efficiency of TiO_2 powders evaluated byacetone photodegradation. *J. Photochem. Photobiol. A Chem.* **2013**, *268*, 1–6. [CrossRef]
137. Haque, F.Z.; Nandanwar, R.; Singh, P. Evaluating photodegradation properties of anatase and rutile TiO_2 nanoparticles for organic compounds. *Optik* **2017**, *128*, 191–200. [CrossRef]

138. Chen, S.; Sun, Z.; Zhang, L.; Xie, H. Photodegradation of gas phase benzene by SnO_2 nanoparticles by direct hole oxidation mechanism. *Catalysts* **2020**, *10*, 117. [CrossRef]
139. Chakrabarti, S.; Dutta, B.K. Photocatalytic degradation of model textile dyes in wastewater using ZnO as semiconductor catalyst. *J. Hazard. Mater.* **2004**, *112*, 269–278. [CrossRef] [PubMed]
140. Lee, H.J.; Kim, J.H.; Park, S.S.; Hong, S.S.; Lee, G.D. Degradation kinetics for photocatalytic reaction of methyl orange over Al-doped ZnO nanoparticles. *J. Ind. Eng. Chem.* **2015**, *25*, 199–206. [CrossRef]
141. del Gobbo, S.; Poolwong, J.; D'Elia, V.; Ogawa, M. Simultaneous controlled seeded-growth and doping of ZnO nanorods with aluminum and cerium: Feasibility assessment and effect on photocatalytic activity. *Cryst. Growth Des.* **2020**, *20*, 5508–5525. [CrossRef]
142. Gerawork, M. Photodegradation of methyl orange dye by using zinc oxide—copper oxide nanocomposite. *Optik* **2020**, *216*, 164864–164870. [CrossRef]
143. Osuntokun, J.; Ajibade, P.A.; Onwudiwe, D.C. Synthesis and photocatalytic studies of ZnS nanoparticles from heteroleptic complex of Zn(II) 1-cyano-1-carboethoxy-2,-2-ethylenedithiolato diisopropylthiourea and its adducts with N-donor ligands. *Superlattices Microstruct.* **2016**, *100*, 605–618. [CrossRef]
144. Kamarudin, N.S.; Jusoh, R.; Setiabudi, H.D.; Sukor, F. Photodegradation of Methylene Blue Using Phyto-Mediated Synthesis of Silver Nanoparticles: Effect of Calcination Treatment. *Mater. Today Proc.* **2018**, *5*, 21981–21989. [CrossRef]
145. Sahrin, N.T.; Nawaz, R.; Kait, C.F.; Lee, S.L.; Wirzal, M.D.H. Visible light photodegradation of formaldehyde over TiO_2 nanotubes synthesized via electrochemical anodization of titanium foil. *Nanomaterials* **2020**, *10*, 128. [CrossRef]
146. Baluk, M.A.; Kobylanski, M.P.; Lisowski, W.; Trykowski, G.; Klimczuk, T.; Mazierski, P.; Zaleska-Medynska, A. Fabrication of durable ordered Ta_2O_5 nanotube arrays decorated with Bi_2S_3 quantum dots. *Nanomaterials* **2019**, *9*, 1347. [CrossRef]
147. Kobylański, M.P.; Juchno, Z.; Baluk, M.A.; Zaleska-Medynska, A. Deposition of gold nanoparticles on titanium nanopits. *Appl. Biosci.* **2018**, *1*, 21–22.
148. Hanifehpour, Y.; Soltani, B.; Amani-Ghadim, A.R.; Hodayi, H.; Min, B.K.; Joo, S.W. Novel visible light photocatalyst based on holmium-doped cadmium sulfide: Synthesis, characterization and kinetics study. *J. Inorg. Organomet. Polym. Mater.* **2017**, *27*, 1–12. [CrossRef]
149. Skalska, K.; Malankowska, A.; Balcerzak, J.; Gazda, M.; Nowaczyk, G.; Jurga, S.; Zaleska-Medynska, A. NOx photooxidation over different noble metals modified TiO_2. *Catalysts* **2022**, *12*, 857. [CrossRef]
150. Ethaib, S.; Al-Qutaifia, S.; Al-Ansari, N.; Zubaidi, S.L. Function of nanomaterials in removing heavy metals for water and wastewater remediation: A review. *Environments* **2022**, *9*, 123. [CrossRef]
151. Biswas, A.; Chandra, B.P.; Prathibha, C. Highly efficient and simultaneous remediation of heavy metal ions (Pb(II), Hg(II), As(V), As(III) and Cr(VI)) from water using Ce intercalated and ceria decorated titanate nanotubes. *Appl. Surf. Sci.* **2023**, *612*, 155841–155856. [CrossRef]
152. Krishnegowda, J.; Shivanna, S.; Shyni, L.S.; Jagadish, K.; Srikantaswamy, S.; Abhilash, M.R. Photocatalytic Degradation and Removal of Heavy Metals in Pharmaceutical Waste by Selenium Doped ZnO Nano Composite Semiconductor. *J. Res.* **2016**, *5*, 47–54.
153. Nagaraju, G.; Prashanth, S.A.; Shastri, M.; Yathish, K.V.; Anupama, C.; Rangappa, D. Electrochemical heavy metal detection, photocatalytic, photoluminescence, biodiesel production and antibacterial activities of Ag–ZnO nanomaterial. *Mater. Res. Bull.* **2017**, *94*, 54–63. [CrossRef]
154. Machida, S.; Kato, R.; Hasegawa, K.; Gotoh, T.; Katsumata, K.I.; Yasumori, A. photoreduction of copper ions using silica–surfactant hybrid and titanium(IV) oxide under sulfuric acid conditions. *Materials* **2022**, *15*, 5132. [CrossRef]
155. Li, Y.; Cui, W.; Liu, L.; Zong, R.; Yao, W.; Liang, Y.; Zhu, Y. Removal of Cr(VI) by 3D TiO_2-graphene hydrogel via adsorption enriched with photocatalytic reduction. *Appl. Catal. B* **2016**, *199*, 412–423. [CrossRef]
156. Wang, N.; Zhu, L.; Deng, K.; She, Y.; Yu, Y.; Tang, H. Visible light photocatalytic reduction of Cr(VI) on TiO_2 in situ modified with small molecular weight organic acids. *Appl. Catal. B* **2010**, *95*, 400–407. [CrossRef]
157. Hafeez, M.; Afyaz, S.; Khalid, A.; Ahmad, P.; Khandaker, M.U.; Sahibzada, M.U.K.; Ahmad, I.; Khan, J.; Alhumaydhi, F.A.; Emran, T. Synthesis of cobalt and sulphur doped titanium dioxide photocatalysts for environmental applications. *J. King Saud Univ. Sci.* **2022**, *34*, 102028–102034. [CrossRef]
158. Gao, X.; Meng, X. Photocatalysis for heavy metal treatment: A review. *Processes* **2021**, *9*, 1729. [CrossRef]
159. Wang, Y.; He, T.; Liu, K.; Wu, J.; Fang, Y. From biomass to advanced bio-fuel by catalytic pyrolysis/hydro-processing: Hydrodeoxygenation of bio-oil derived from biomass catalytic pyrolysis. *Bioresour. Technol.* **2012**, *108*, 280–284. [CrossRef] [PubMed]
160. Arun, N.; Sharma, R.V.; Dalai, A.K. green diesel synthesis by hydrodeoxygenation of bio-based feedstocks: Strategies for catalyst design and development. *Renew. Sustain. Energy Rev.* **2015**, *48*, 240–255. [CrossRef]
161. Global Energy Consumption 2013. Available online: http://www.euanmearns.com/wp-content/uploads/2014/06/global_energy_2013.png (accessed on 16 December 2022).
162. Viju, C.; Kerr, W.A. Taking an option on the future: Subsidizing biofuels for energy security or reducing global warming. *Energy Policy* **2013**, *56*, 543–548. [CrossRef]
163. Joselin Herbert, G.M.; Unni Krishnan, A. Quantifying environmental performance of biomass energy. *Renew. Sustain. Energy Rev.* **2016**, *59*, 292–308. [CrossRef]
164. Su, Y.; Zhang, P.; Su, Y. An overview of biofuels policies and industrialization in the major biofuel producing countries. *Renew. Sustain. Energy Rev.* **2015**, *50*, 991–1003. [CrossRef]

165. Kuehnel, M.F.; Orchard, K.L.; Dalle, K.E.; Reisner, E. Selective photocatalytic CO_2 reduction in water through anchoring of a molecular Ni catalyst on CdS nanocrystals. *J. Am. Chem. Soc.* **2017**, *139*, 7217–7223. [CrossRef]
166. Yu, B.; Zhou, Y.; Li, P.; Tu, W.; Li, P.; Tang, L.; Ye, J.; Zou, Z. Photocatalytic reduction of CO_2 over Ag/TiO_2 nanocomposites prepared with a simple and rapid silver mirror method. *Nanoscale* **2016**, *8*, 11870–11874. [CrossRef]
167. Kim, J.; Kwon, E.E. Photoconversion of carbon dioxide into fuels using semiconductors. *J. CO2 Util.* **2019**, *33*, 72–82. [CrossRef]
168. Kong, X.Y.; Tan, W.L.; Ng, B.J.; Chai, S.P.; Mohamed, A.R. Harnessing Vis–NIR broad spectrum for photocatalytic CO_2 reduction over carbon quantum dots-decorated ultrathin Bi_2WO_6 nanosheets. *Nano Res.* **2017**, *10*, 1720–1731. [CrossRef]
169. Nasir, A.; Khalid, S.; Yasin, T.; Mazare, A. A Review on the progress and future of TiO_2/graphene photocatalysts. *Energies* **2022**, *15*, 6248. [CrossRef]
170. Hong, D.; Lyu, L.M.; Koga, K.; Shimoyama, Y.; Kon, Y. Plasmonic Ag@TiO_2 core-shell nanoparticles for enhanced CO_2 photoconversion to CH_4. *ACS Sustain. Chem. Eng.* **2019**, *7*, 18955–18964. [CrossRef]
171. Karthick Raj, A.G.; Murugan, C.; Pandikumar, A. Efficient photoelectrochemical reduction of carbon dioxide into alcohols assisted by photoanode driven water oxidation with gold nanoparticles decorated titania nanotubes. *J. CO2 Util.* **2021**, *52*, 101684–101694. [CrossRef]
172. Klein, M.; Zielinska-Jurek, A.; Janczarek, M.; Cybula, A.; Zielińska-Jurek, A.; Zaleska, A. Carbon dioxide photoconversion. The effect of titanium dioxide immobilization conditions and photocatalyst type. *Physicochem. Probl. Miner. Process.* **2012**, *48*, 159–167.
173. Choi, J.Y.; Lim, C.K.; Park, B.; Kim, M.; Jamal, A.; Song, H. Surface activation of cobalt oxide nanoparticles for photocatalytic carbon dioxide reduction to methane. *J. Mater. Chem. A* **2019**, *7*, 15068–15072. [CrossRef]
174. Abou Asi, M.; He, C.; Su, M.; Xia, D.; Lin, L.; Deng, H.; Xiong, Y.; Qio, R.; Li, X.Z. Photocatalytic reduction of CO_2 to hydrocarbons using AgBr/TiO_2 nanocomposites under visible light. *Catal. Today* **2011**, *175*, 256–263. [CrossRef]
175. Ehsan, M.F.; He, T. In situ synthesis of ZnO/ZnTe common cation heterostructure and its visible-light photocatalytic reduction of CO_2 into CH_4. *Appl. Catal. B Environ.* **2015**, *166–167*, 345–352. [CrossRef]
176. Parayil, S.K.; Razzaq, A.; Park, S.M.; Kim, H.R.; Grimes, C.A.; In, S. Photocatalytic conversion of CO_2 to hydrocarbon fuel using carbon and nitrogen co-doped sodium titanate nanotubes. *Appl. Catal. A Gen.* **2015**, *498*, 205–213. [CrossRef]
177. Yong, L.; Tian, Z.; Zhao, Z.; Liu, Q.; Kou, J.; Chen, X.; Gao, J.; Yan, S.; Zou, Z. High-yield synthesis of ultrathin and uniform Bi_2WO_6 square nanoplates benefitting from photocatalytic reduction of CO_2 into renewable hydrocarbon fuel under visible. light. *ACS Appl. Mater. Interfaces* **2011**, *3*, 3594–3601.
178. Andrews, J.; Shabani, B. Re-envisioning the role of hydrogen in a sustainable energy economy. *Int. J. Hydrog. Energy* **2012**, *37*, 1184–1203. [CrossRef]
179. Dincer, I.; Acar, C. Review and evaluation of hydrogen production methods for better sustainability. *Int. J. Hydrog. Energy* **2015**, *40*, 11094–11111. [CrossRef]
180. Dodds, P.E.; Staffell, I.; Hawkes, A.D.; Li, F.; Grünewald, P.; McDowall, W.; Ekins, P. Hydrogen and fuel cell technologies for heating: A review. *Int. J. Hydrog. Energy* **2015**, *40*, 2065–2083. [CrossRef]
181. Zuttel, A.; Borgschulte, A.; Schlapbach, L. *Hydrogen as a Future Energy Carrier*; Wiley-VCH: Weinheim, Germany, 2008.
182. Vrieling, E.G.; Beelen, T.P.M.; van Santen, R.A.; Gieskes, W.W.C. Nanoscale uniformity of pore architecture in diatomaceous silica: A combined small and wide angle x-ray scattering study. *J. Phycol.* **2000**, *36*, 146–159. [CrossRef]
183. Vrieling, E.G.; Beelen, T.P.M.; Sun, Q.Y.; Hazelaar, S.; van Santen, R.A.; Gieskes, W.W.C. Ultrasmall, small, and wide angle X-ray scattering analysis of diatom biosilica: Interspecific differences in fractal properties. *J. Mater. Chem.* **2004**, *14*, 1970–1975. [CrossRef]
184. Shan, A.Y.; Ghazi, T.I.M.; Rashid, S.A. Immobilisation of titanium dioxide onto supporting materials in heterogeneous photocatalysis: A review. *Appl. Catal. A* **2010**, *389*, 1–8. [CrossRef]
185. Sahu, D.R.; Hong, L.Y.; Wang, S.C.; Huang, J.L. Synthesis, analysis and characterization of ordered mesoporous TiO_2/SBA-15 matrix: Effect of calcination temperature. *Microporous Mesoporous Mat.* **2009**, *117*, 640–649. [CrossRef]
186. Momirlan, M.; Veziroglu, T.N. The properties of hydrogen as fuel tomorrow in sustainable energy system for a cleaner planet. *Int. J. Hydrog. Energy* **2005**, *30*, 795–802. [CrossRef]
187. Christoforidis, K.C.; Fornasiero, P. Photocatalytic Hydrogen Production: A Rift into the Future Energy Supply. *ChemCatChem* **2017**, *9*, 1523–1544. [CrossRef]
188. Bamwenda, G.; Tsubota, S.; Nakamura, T.; Haruta, M. Photoassisted hydrogen production from a water-ethanol solution: A comparison of activities of Au·TiO_2 and Pt·TiO_2. *J. Photochem. Photobiol. A* **1995**, *89*, 177–189. [CrossRef]
189. Jiaguo, Y.; Qi, L.; Jaroniec, M. Hydrogen Production by Photocatalytic Water Splitting over Pt/TiO_2 Nanosheets with Exposed (001) Facets. *J. Phys. Chem. C* **2010**, *114*, 13118–13125.
190. Torres-Martínez, L.; Gomez, R.; Vázquez-Cuchillo, O.; Juárez-Ramírez, I.; Cruz-Lopez, A.; Alejandre-Sandovala, J. Enhanced photocatalytic water splitting hydrogen production on RuO_2/La:$NaTaO_3$ prepared by sol–gel method. *Catal. Commun.* **2010**, *12*, 268–272. [CrossRef]
191. Yue-Ying, L.; Wang, J.G.; Sun, H.H.; Hua, W.; Liu, X.R. Heterostructured SnS_2/SnO_2 nanotubes with enhanced charge separation and excellent photocatalytic hydrogen production. *Int. J. Hydrog. Energy* **2018**, *43*, 14121–14129.

Disclaimer/Publisher's Note: The statements, opinions and data contained in all publications are solely those of the individual author(s) and contributor(s) and not of MDPI and/or the editor(s). MDPI and/or the editor(s) disclaim responsibility for any injury to people or property resulting from any ideas, methods, instructions or products referred to in the content.

Article

Magnetocaloric and Giant Magnetoresistance Effects in La-Ba-Mn-Ti-O Epitaxial Thin Films: Influence of Phase Transition and Magnetic Anisotropy

Marwène Oumezzine [1,*], Cristina Florentina Chirila [2], Iuliana Pasuk [2], Aurelian Catalin Galca [2], Aurel Leca [2], Bogdana Borca [2,*] and Victor Kuncser [2,*]

1. Laboratoire de Physico-Chimie des Matériaux, Université de Monastir, Monastir 5019, Tunisia
2. National Institute of Materials Physics, 077125 Magurele, Romania
* Correspondence: oumezzine@hotmail.co.uk (M.O.); bogdana.borca@infim.ro (B.B.); kuncser@infim.ro (V.K.)

Abstract: Magnetic perovskite films have promising properties for use in energy-efficient spintronic devices and magnetic refrigeration. Here, an epitaxial ferromagnetic $La_{0.67}Ba_{0.33}Mn_{0.95}Ti_{0.05}O_3$ (LBMTO-5) thin film was grown on $SrTiO_3(001)$ single crystal substrate by pulsed laser deposition. High-resolution X-ray diffraction proved the high crystallinity of the film with tetragonal symmetry. The magnetic, magnetocaloric and magnetoresistance properties at different directions of the applied magnetic field with respect to the *ab* plane of the film were investigated. An in-plane uni-axial magnetic anisotropy was evidenced. The LBMTO-5 epilayer exhibits a second-order ferromagnetic-paramagnetic phase transition around 234 K together with a metal–semiconductor transition close to this Curie temperature (T_C). The magnetic entropy variation under 5 T induction of a magnetic field applied parallel to the film surface reaches a maximum of 17.27 mJ/cm^3 K. The relative cooling power is 1400 mJ/cm^3 K (53% of the reference value reported for bulk Gd) for the same applied magnetic field. Giant magnetoresistance of about 82% under 5 T is obtained at a temperature close to T_C. Defined as the difference between specific resistivity obtained under 5 T with the current flowing along the magnetic easy axis and the magnetic field oriented transversally to the current, parallel and perpendicular to the sample plane, respectively, the in-plane magneto-resistance anisotropy in 5 T is about 9% near the T_C.

Keywords: perovskite manganite; epitaxial thin films; magnetoresistance; magnetocaloric effect; anisotropy

1. Introduction

Significant attention has been given in recent years to the search for new-generation device materials for magnetic storage technology and spintronics. Strongly correlated materials consisting of thin manganite films with a perovskite structure have been in the research spotlight because of their combination of spin, charge, orbit and lattice degrees of freedom [1,2]. Nevertheless, epitaxial thin films of this type demonstrate great potential for multifunctional device applications, such as magnetic refrigeration [3–7], spintronics and faster reading devices [8,9]. Such applications can be improved by the astonishing electronic specificities concomitant with giant magnetoresistance (GMR) and a large magnetocaloric effect (MCE). Moreover, the magnetic anisotropy and anisotropic magnetoresistance (AMR) properties of these materials have also attracted attention [10,11]. MCE research today is limited to bulk and single crystal materials, but studies on epitaxial rare-earth oxide thin films for MCE are very challenging. Thus, new fundamental studies will contribute to a deep understanding of the MCE of thin films. The AMR effect on perovskite manganites has been studied by a few research groups, indicating a peak in temperature dependence of the AMR near the Curie temperature T_C [12]. From the scientific point of view, fabrication of thin films of high structural quality and convenient and reliable control of their magneto-transport properties is crucial for making performant magneto-resistive devices. Note

that many studies on mixed valence perovskite manganites have been focused on the partial substitution of manganese (electron-doped) with various metallic elements owing to significant changes in their magnetic and magneto-transport properties. Among the most cited microscopic mechanisms underlying the fascinating observations of macroscopic physical properties is the double-exchange (DE) interaction between Mn^{3+}/Mn^{4+} and the Jahn–Teller-distorted ions Mn^{3+} [13,14]. Hence, the ferromagnetic ground state with metallic conduction arises from the hopping of the itinerant e_g electron between neighboring Mn^{3+} and Mn^{4+} ions. Among the perovskite manganites, $La_{0.67}Ba_{0.33}MnO_3$ (LBMO) has attracted particular attention due to its high Curie temperature of 345 K. Unlike bulk materials, few reports on the substitution of manganese with various metallic elements in manganite epilayers are available [15–17]. The substitution with the non-magnetic Ti^{4+} (d^0) of the magnetic ion Mn^{4+} will causes a sudden break of the ferromagnetic Mn^{3+}-O-Mn^{4+} interactions without any ferromagnetic compensation, allowing the fine tuning of T_C toward lower temperatures [18]. Using pulsed laser deposition (PLD), high-quality epitaxial thin films of La-Ba-Mn-Ti-O can be obtained, with pronounced properties such as MCE [19] and GMR [15]. Due to a strong shape anisotropy of thin films, it can be anticipated that the easy axis of magnetization will lie in the plane of the films, if the strain and surface/interface anisotropies are negligible. In this manuscript, the $La_{0.67}Ba_{0.33}MnO_3$ system with substituent Ti cations at the Mn-site was chosen (5% of Mn is replaced by Ti). The film was successfully epitaxially grown on a $SrTiO_3$(001) single crystalline substrate by PLD.

Here, we report a study of the temperature dependence of magnetization (M), magnetic entropy change (ΔS_M), resistivity (ρ) and magnetoresistance (MR) under different directions of the applied magnetic field versus the *ab* plane of the epitaxial film. An increased response of such magneto-functionalities was evidenced for the in-plane orientations of the applied magnetic field.

2. Materials and Methods

The epitaxial $La_{0.67}Ba_{0.33}Mn_{0.95}Ti_{0.05}O_3$ (referred to as LBMTO-5) thin film is grown by pulsed-laser deposition (PLD) on $SrTiO_3$ (STO) with a (001) orientation. The detailed LBMTO-5 target material and LBMTO-5 film fabrication method is described in detail in a previous work [19]. HRXRD (High Resolution X-Ray Diffraction) patterns are measured with Cu $K_{\alpha 1}$ radiation (1.5406 Å wavelength) using a D8 Discover diffractometer from Bruker AXS (Billerica, Massachusetts (MA), United States) in the 2θ-ω, φ-scan, and reciprocal-space mapping modes (RSMs). These structural characterizations permit us to determine the pseudocubic out-of-plane and in-plane lattice parameters of the films and to confirm the quality of the epitaxy. The magnetic properties were investigated by using a SQUID-Superconducting Quantum Interference Device magnetometer (MPMS 7T, Quantum Design, San Diego, California (CA), USA). According to our previous experience on the optimization of the magnetocaloric effects, the isothermal magnetizations were measured with the applied magnetic field along the in-plane directions. Temperature- and angle-dependent resistivity were investigated under various magnetic fields by using a Physical Property Measurements System (PPMS 14 T, Quantum Design, San Diego, California (CA), USA). In these measurements, the current was always flowing along the easy axis of magnetization specific to the rectangular-shaped sample, whereas the magnetic field was applied perpendicular to the sample plane or parallel to it, but always perpendicular to the current direction.

3. Results and Discussion

3.1. LBMTO-5 Films Structure

Figure 1a,b presents typical 2θ-ω diffractograms of LBMTO-5 thin films deposited on the STO(001). The pattern shows a two sets of LBMTO-5 diffraction peaks 002 and 004 together with those from the substrate, indicating that the films have a strong out-of-plane texture. The visible Pendellosung fringes indicate a smooth film/substrate interface and high crystal quality. The results of coherence length along [002] revealed a well-defined film thickness of 97 nm. The crystalline structure of the target material is rhombohedral, while the film is tetragonal due to substrate influence. The calculated c-axis parameter of the LBMTO-5 thin film is 3.935 Å. Figure 1c shows the corresponding RSMs around the asymmetric (−103) node revealing that the in-plane constant lattice of LBMTO-5 thin film is identical to that of the STO substrate. The epitaxial growth of the LBMTO-5 manganite on the STO substrate was confirmed by performing the XRD azimuth scans (φ-scans) on the {103} planes family of both substrate and thin films (Figure 1d). Each of the LBMTO-5 and STO exhibit four peaks, separated by 90°, suggesting 'cube-on-cube' epitaxial growth (presented in Supplementary Materials, Figure S1).

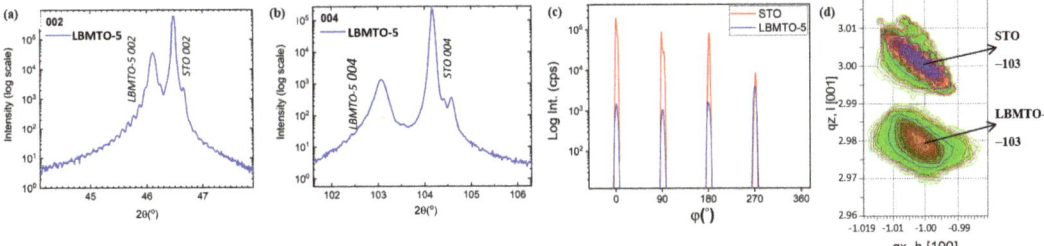

Figure 1. Diffractograms of LBMTO-5 thin film deposited on a STO(001) substrate. (**a,b**) Typical 2θ-ω scans showing the 002 and 004 peaks; (**c**) Azimuth φ-scan on the {103} skew planes of STO substrate and LBMTO-5; (**d**) asymmetric RSMs around the −103 node.

3.2. Magnetic and Magnetocaloric Effect Studies

The magnetic characterization was realized on the 97 nm LBMTO-5/STO epitaxial thin film grown on the STO(001) substrate. Firstly, we measured the temperature dependences of field-cooled magnetization (M-T curves) in a magnetic field H with intensity of 100 Oe (equivalent of 0.01 T induction) applied along in-plane (H//film) and out-of-plane (H⊥film) directions, respectively. As shown in Figure 2, the in-plane magnetization values are higher than the ones measured in the out-of-plane configuration, with a maximum of about 5 times at 5 K. This observation supports the anisotropy with a preferential in-plane magnetic easy axis of the epitaxial thin film. The ferromagnetic–paramagnetic (FM–PM) transition of the LBMTO-5 film grown on the STO substrate can also be observed from these curves. The value of the Curie temperature T_C is found to be around 234 K. This was estimated with much higher precision in the in-plane geometry at the intersection point of the two tangents to the iso field curve that bounds the transition temperature, compared with the perpendicular geometry (see Figure 2). This experimental value is slightly lower than the 264 K value found in the unstressed powder [18] and the 295 K value found in LBMTO-2/STO epitaxial thin films [15]. Similar to the bulk case, it can be seen that an extremely low doping (5% atomic percentage) with the non-magnetic Ti^{4+} replacing Mn^{4+} causes a sudden break of the ferromagnetic Mn^{3+}-O^{2-}-Mn^{4+} interactions without any ferromagnetic compensation. Furthermore, it is important to emphasize the role of in-plane compressive strain for stretching MnO_6 octahedra in the out-of-plane direction, which induces a T_C shift to lower temperatures [20].

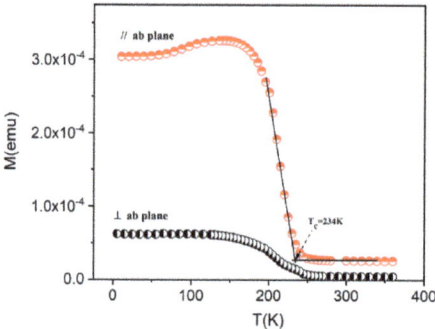

Figure 2. Plots of the field-cooled magnetization versus temperature of a LBMTO-5 thin film, measured under a magnetic field of 0.01 T parallel and perpendicular to the *ab* plane.

To evaluate the MCE characteristics of the present thin films, magnetic field dependent isothermal magnetizations were measured in-plane geometry, under fields up to 5 T induction (5 × 10^4 Oe intensity) in the temperature range from 190 to 260 K with a temperature step of 10 K (see Figure 3a). All the M-H data have been corrected by subtracting the diamagnetic background (dominated by the substrate STO). The shape of the magnetization curves near T_C is typical of a second-order transition as usually observed in titanium-substituted (La,Ba)MnO$_3$ manganites [19].

Based on the Banerjee criteria [21], the positive and negative sign of the slope of Arrott plots near T_C correspond to second-order magnetic phase transition (SOMPT) and first-order magnetic phase transition (FOMPT), respectively. Apparently, the positive slopes of Arrott plots in close proximity to T_C for LBMTO-5 /STO thin films, confirms SOMPT (see Supplementary Materials, Figure S2). On the other hand, maximum values of $(-\Delta S_M)^{max}$ were predicted to show the proportional relationships $\sim (\mu_0 H/T_C)^{2/3}$, which confirm the long-range ferromagnetic order (presented in Supplementary Materials, Figure S3). In a next step, the magnetic entropy change (ΔS_M) was obtained using the Maxwell equation, which can be calculated for magnetization isotherms taken at discrete fields and temperatures [22]. The behavior of $(-\Delta S_M)$ per unit volume is shown in Figure 3b.

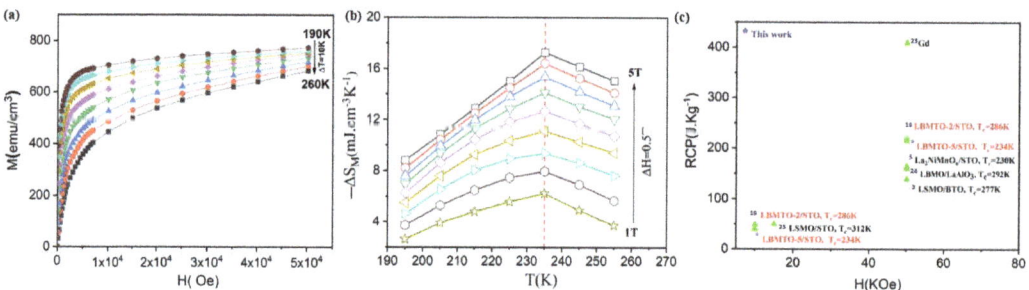

Figure 3. The magnetic properties and magnetocaloric effect of the LBMTO-5/STO(001) film. (a) Isothermal magnetization curves measured at different temperatures around T_C. Inset shows the Arrott plot of $\mu_0 H/M$ versus M^2. (b) Temperature dependence of the magnetic entropy change (ΔS_M) under different applied magnetic fields. (c) Relative cooling power (RCP) as a function of the applied magnetic field compared with several other magnetic refrigerants thin films from the literature. The corresponding references [3,5,19,23–25] are marked in the left side of each material.

The temperature dependence of the isothermal entropy change ΔS_M has a uniform distribution around T_C where the maximum is also reached. This maximum is increasing with the external magnetic field. One can see from Figure 3b that the peak position of

$(-\Delta S_M)^{max}$ remains unchanged, as an additional confirmation of SOMPT. Hence, our highest value of the $(-\Delta S_M)^{max}$ obtained under a field induction change of 5 T, with incremental steps of 0.5 T, is 17.27 mJ/cm^3 K (2.60 J/Kg K) and compares favorably with that obtained in LBMTO-2/STO epitaxial thin film. The maximum relative cooling power (RCP) is determined using the Wood and Potter method [26,27] and it is compared in Figure 3c with several other magnetic refrigerants' thin films with T_C in the range of 230–321 K, as reported earlier in the literature. Similar to LBMTO-2/STO [19], our results indicate potential applications of the LBMTO-5/STO film for micro-scale magnetic cooling.

Moreover, the maximum values of $(-\Delta S_M)^{max}$ have proven to be proportional to $(\mu_0 H/T_C)^{2/3}$ (see Figure S3 in Supplementary Materials), which confirms a long-range ferromagnetic order in the investigated epitaxial LBMTO-5 thin films below T_C.

3.3. Magnetotransport and Magnetoresistance

Lightly doped manganese oxides with perovskite structures belong to a strong magnetic–electronic coupling system. In order to explore the assessment of the relationship between the structural and physical properties of the films, the magneto-transport properties were exploited. In Figure 4 are shown the resistivity curves for the LBMTO-5/STO film in 0 T and 5 T induction field applied in parallel (H// ab plane of the film) and perpendicular (H⊥ ab plane of the film) directions, respectively. The current always flows in the plane of the film, along the magnetically easy axis that corresponds to the [010] crystallographic direction. For both configurations, the resistivity (ρ-T) displays a ferromagnetic metallic (M) behavior at low temperatures which is transformed into a paramagnetic semiconductor (SC) (i.e., dρ/dT < 0) at high temperatures. The next paragraphs will analyze in detail these two different behaviors.

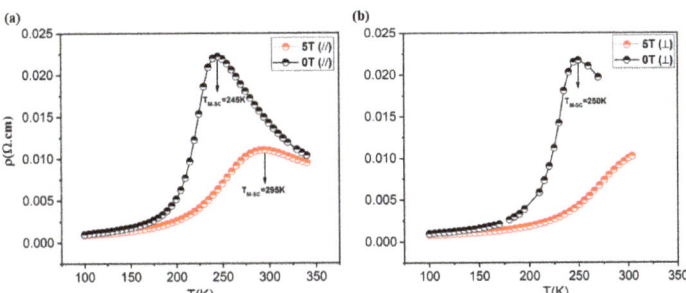

Figure 4. Temperature dependence resistivity of LBMTO-5 thin film in zero field and in a magnetic field of 5 T induction, applied parallel (**a**) and perpendicular (**b**) to the ab plane, respectively.

A low residual resistivity of 9.64×10^{-4} Ω.cm is obtained at 100 K in zero-field that confirms the good quality of the film, in agreement with HRXRD results (see Section 3.1.). The maxima of the curves in Figure 4 clearly define the transition metal–semiconductor transition temperature T_{M-SC}, which has a value close to the T_C. This suggests a strong interplay between the magnetic and transport properties in the LBMTO-5 film. It should be noted that the maximum resistivity, and thus the T_{M-SC}, shifts to a higher temperature with the increasing magnetic field. This behavior is in accordance with a delocalization of polarons that is characteristic of double-exchange ferromagnets. Moreover, a crystallographic strain present in the film is expected to reduce the resistivity and to shift the transition temperature T_{M-SC} toward higher temperatures with respect to their bulk counterpart [12].

In order to understand the charge transport mechanisms responsible for the conduction along the LBMTO-5/STO structure, the Zener double exchange (ZDE) polynomial law [18,28] is employed in the low-temperature ferromagnetic metallic state corresponding to the temperature range of 98–175 K. The ZDE polynomial low has the form:

$$\rho(T) = \rho_0 + \rho_2 T^2 + \rho_{4.5} T^{4.5} \tag{1}$$

Here, ρ_0 is the resistivity due to point-defect scattering; ρ_2 represents the electrical resistivity due to the electron–electron scattering; $\rho_{4.5}$ is the resistivity contributions due to electron–electron, electron–magnon and electron–phonon scattering processes; T^2 and $T^{4.5}$ are the corresponding temperature values.

The experimental data of Figure 4 are fitted according to the ZDE Equation (1) and the results for the parallel configuration are displayed in Figure 5 (for the perpendicular configuration see Supplementary Materials, Figure S4). The consistency of the fits is evident from the values of 0.999 of the correlation coefficient (R^2). Moreover, the expected increase in the ρ_0, ρ_2 and $\rho_{4.5}$ parameters in the bulk form [18] is mainly due to the scattering of charge carriers by the grain boundaries. In the paramagnetic semiconductor region, the $(\rho\text{-}T)_{//}$ curve is usually described by the small polarons model [29,30]. This model has the form:

$$\rho(T) = BT \exp\frac{E_a}{k_B T} \tag{2}$$

where E_a is the activation energy for hopping conduction and B is the residual resistivity. The activation energy E_a and the model fitting is displayed in Figure 6. The obtained value of the E_a in the absence of an external magnetic field is lower than E_a deduced elsewhere in the bulk counterpart. A lower value of the E_a suggests that the polaron hopping becomes easier in the epitaxial thin film than in the bulk form.

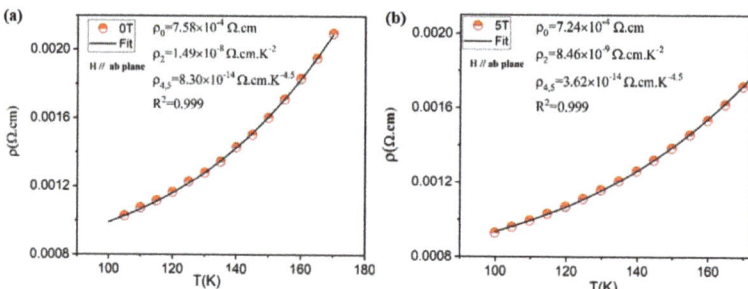

Figure 5. Fitting plots of the resistivity data of LBMTO-5 thin film in the low temperature ferromagnetic metallic state by using Equation (1) for a zero-field (**a**) and under 5 T induction (**b**), in the parallel geometry.

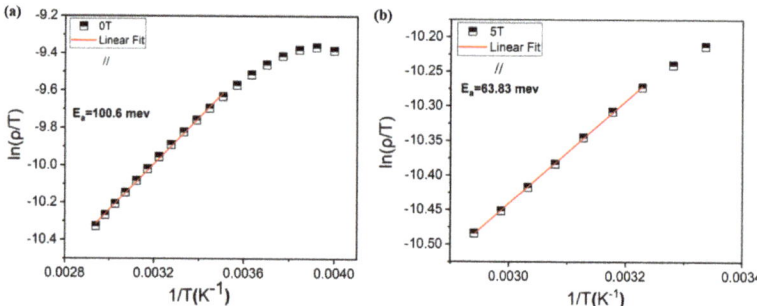

Figure 6. Fitting plots of the resistivity data of $\ln(\rho/T)$ as a function of $1/T$ of LBMTO-5 thin film in the paramagnetic semiconductor region by using Equation (2) for a zero-field (**a**) and under 5 T (**b**), in the parallel geometry.

The magnetoresistance of the LBMTO-5/STO structure is determined according to the formula: $MR = \left|\frac{R(H=0)-R(H)}{R(H)}\right|$, as shown in Figure 7. It can be observed that the higher values of MR are obtained when the magnetic field is applied in-plane parallel to the film (parallel geometry). As shown in Figure 7, the $MR_{//}$ (for H = 5 T) reaches a maximum value of 82% at a temperature close to the T_C. The LBMTO-5/STO epitaxial thin film has a larger MR in comparison with LBMTO-2/STO (MR (5 T) = 60% at 300 K), LCMO-STO (MR (6.8 T) = 32% at 272 K) and LCMO-LAO (MR (6.8 T) = 22.5% at 274 K) thin films, as reported by Egilmez et al. [12].

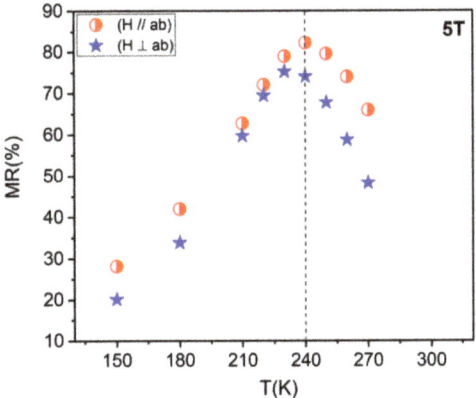

Figure 7. Temperature dependence of magnetoresistance MR of the LBMTO-5 thin film, measured under a magnetic field of 5 T along in-plane and out-of-plane geometry.

To gain more insight into the physical mechanism of the magnetoresistance anisotropy observed for our LBMTO-5/STO epitaxial films, the normalized resistivity (ρ/ρ_{max}) is measured as a function of the angle θ between the applied field H and the *ab* plane. Note that here, the θ = 0° corresponds to the configuration where the sample is perpendicular to the direction of H, being equivalent to the out-of-plane or perpendicular geometry. Figure 8 shows a typical angular dependence of (ρ/ρ_{max}) for LBMTO-5/STO structure at a temperature near the metal–semiconductor transition (250 K) and in a field of 5 T induction.

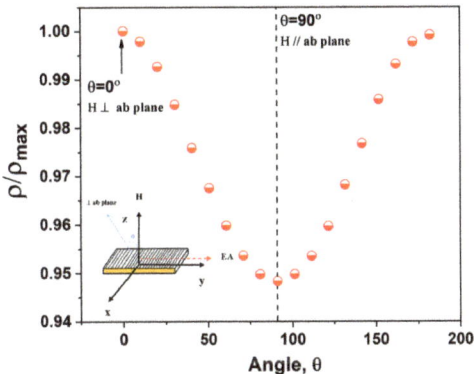

Figure 8. Angular dependence of the (ρ/ρ_{max}) of LBMTO-5/STO system in a magnetic field of 5 T and at a temperature near the metal–semiconductor transition (250 K). Inset: schematic view of the measurement with the sample rotating in the in-plane (parallel) and the out-of-plane (perpendicular) geometry.

In agreement with the initial expectations that were also confirmed by magnetic measurements (see Figure 2, Section 3.2.), a low resistivity value is obtained in the parallel configuration at θ = 90°, equivalent to the in-plane geometry that matches with a maximum of MR (see Figure 7), as compared to perpendicular configuration. Obviously, there is an in-plane anisotropy with the $(\rho/\rho_{max})_\perp - (\rho/\rho_{max})_{//} \cong 9\%$. This can be easily understood because for the existing in-plane anisotropy, a more complete saturation of the magnetization can be achieved in the in-plane geometry (parallel configuration).

4. Conclusions

To summarize, an epitaxial LBMTO-5 epilayer was successfully grown on a STO(001) substrate. The magnetic, magnetocaloric effect and magnetoresistance of the LBMTO-5 film with a thickness of 97 nm have been studied at different directions of the applied magnetic field with respect to the sample plane. Herein, the epilayer exhibits a second-order FM–PM phase transition around 234 K and an in-plane magnetic uniaxial easy axis (along the direction of the longer edge of the sample). Further, at increasing temperature, a gradual metal to semiconductor transition finishing at 245 K is observed. The carriers transport belongs to a small polaron hopping model in paramagnetic semiconductors. Under 5 T magnetic field applied parallel to the film surface, the maximum of the $(-\Delta S_M)$ and of the relative cooling power RCP are 17.27 mJ/cm^3 K and 1400 mJ/cm^3 K, respectively. These results highlight the potential applications of the present films for micro-scale magnetic cooling. Another important finding is that the LBMTO-5/STO epitaxial thin film has a giant magnetoresistance as high as 82% at a temperature close to the T_C, which may be interesting for electro-magnetic applications. A magnetoresistance anisotropy reaching a maximum value around T_C is also characteristic of such films.

Supplementary Materials: The following supporting information can be downloaded at: https://www.mdpi.com/article/10.3390/ma15228003/s1, Figure S1: Schematic diagrams of the in-plane lattice arrangement for LBMTO-5/STO(001); Figure S2: The Arrott plot of $\mu_0 H/M$ versus M^2; Figure S3: Magnetic entropy change versus the $(\mu_0 H/T_C)^{2/3}$; Figure S4: Fitting plots of the resistivity data in the low temperature ferromagnetic metallic state in the perpendicular geometry.

Author Contributions: Conceptualization, M.O. and V.K.; methodology, C.F.C., I.P., A.C.G. and A.L.; software, M.O. and I.P.; validation, M.O., A.C.G. and V.K.; formal analysis, M.O. and V.K.; investigation, C.F.C., I.P., A.C.G., A.L. and V.K.; resources, M.O., V.K. and B.B.; data curation, M.O., I.P. and V.K.; writing—original draft preparation, M.O.; writing—review and editing, M.O., B.B. and V.K.; visualization, M.O., V.K. and B.B.; supervision, M.O., V.K. and A.C.G.; project administration, M.O., B.B. and V.K.; funding acquisition, M.O., V.K. and B.B. All authors have read and agreed to the published version of the manuscript.

Funding: M.O. acknowledge the Tunisian Ministry of Higher Education and Scientific Research (Core program: 03-19PEJC03 project) for the financial support. NIMP authors acknowledge the Romanian Ministry of Research, Innovation and Digitalization in the framework of Core Program PN19-03 (Contract no. 21 N/08.02.2019) as well as UEFISCDI through the project EEA-NO-179/2020, (contract no. 39/2021) and PN-III-P2-2.1-PED-2021-0378 (contract no. 575PED/2022).

Institutional Review Board Statement: Not applicable.

Informed Consent Statement: Not applicable.

Data Availability Statement: Not applicable.

Conflicts of Interest: The authors declare no conflict of interest.

References

1. Weiße, A.; Fehske, H. Interplay of charge, spin, orbital and lattice correlations in colossal magnetoresistance manganites. *Eur. Phys. J. B* **2002**, *30*, 487. [CrossRef]
2. Millis, A.J. Lattice effects in magnetoresistive manganese perovskites. *Nature* **1998**, *392*, 147. [CrossRef]
3. Giri, S.K.; MacManus-Driscoll, J.L.; Li, W.; Wu, R.; Nath, T.K.; Maity, T.S. Strain induced extrinsic magnetocaloric effects in La$_{0.67}$Sr$_{0.33}$MnO$_3$ thin films, controlled by magnetic field. *J. Phys. D: Appl. Phys.* **2019**, *52*, 165302. [CrossRef]

4. Wang, Y.; Shao, J.; Yu, Y.; Shi, Q.; Zhu, Y.; Miao, T.; Lin, H.; Xiang, L.; Li, Q.; Cai, P.; et al. Enhanced magnetocaloric effect in manganite nanodisks. *Phys. Rev. Mater.* **2019**, *3*, 084411. [CrossRef]
5. Matte, D.; de Lafontaine, M.; Ouellet, A.; Balli, M.; Fournier, P. Tailoring the Magnetocaloric Effect in La2NiMnO6 Thin Films. *Phys. Rev. Appl.* **2018**, *9*, 054042. [CrossRef]
6. Belo, J.H.; Pires, A.L.; Araújo, J.P.; Pereira, A.M. Magnetocaloric materials: From micro- to nanoscale. *Journal of Materials Research* **2019**, *34*, 134. [CrossRef]
7. Zhang, H.; Wang, Y.; Wang, H.; Huo, D.; Tan, W. Room-temperature magnetoresistive and magnetocaloric effect in $La_{1-x}Ba_x MnO_3$ compounds: Role of Griffiths phase with ferromagnetic metal cluster above Curie temperature. *Journal of Applied Physics* **2022**, *131*, 043901. [CrossRef]
8. Thompson, S.M. The discovery, development and future of GMR: The Nobel Prize 2007. *J. Phys. D* **2008**, *41*, 093001. [CrossRef]
9. Markovich, V.; Jung, G.; Yuzhelevski, Y.; Gorodetsky, G.; Mukovskii, Y.M. Anisotropic magnetoresistance in low-doped $La_{0.78}Ca_{0.22}MnO_3$ crystals. *J. Appl. Phys.* **2011**, *109*, 07D702. [CrossRef]
10. Perna, P.; Maccariello, D.; Ajejas, F.; Guerrero, R.; Méchin, L.; Flament, S.; Santamaria, J.; Miranda, R.; Camarero, J. Engineering large anisotropic magnetoresistance in $La_{0.7}Sr_{0.3}MnO_3$ films at room temperature. *Adv. Funct. Mater.* **2017**, *27*, 1700664. [CrossRef]
11. Fan, J.; Xie, Y.; Qian, F.; Ji, Y.; Hu, D.; Tang, R.; Liu, W.; Zhang, L.; Tong, W.; Ma, C.; et al. Isotropic magnetoresistance and enhancement of ferromagnetism through repetitious bending moments in flexible perovskite manganite thin film. *Alloy. Compd.* **2019**, *806*, 753. [CrossRef]
12. Egilmez, M.; Ma, R.; Chow, K.H.; Jung, J. The anisotropic magnetoresistance in epitaxial thin films and polycrystalline samples of $La_{0.65}Ca_{0.35}MnO_3$. *J. Appl. Phys.* **2009**, *105*, 7061. [CrossRef]
13. Zener, C. Interaction between the d-shells in the transition metals. II. Ferromagnetic compounds of manganese with perovskite structure. *Phys. Rev.* **1951**, *82*, 403. [CrossRef]
14. Anderson, P.W.; Hasegawa, H. Considerations on double exchange. *Phys. Rev.* **1955**, *100*, 675. [CrossRef]
15. Galca, A.C.; Oumezzine, M.; Leca, A.; Chirila, C.F.; Kuncser, V.; Kuncser, A.; Ghica, C.; Pasuk, I.; Oumezzine, M. Structure, transition temperature, and magnetoresistance of titanium-doped lanthanum barium manganite epilayers onto STO 001 substrates. *Appl. Phys. Lett.* **2017**, *111*, 182409. [CrossRef]
16. Kim, Y.J.; Kumar, S.; Lee, C.G.; Koo, B.H. Study on structural, magnetic and transport properties of $La_{0.7}Ca_{0.3}Mn_{1-x}Co_xO_3$ (x = 0.01-0.05) thin films. *J. Ceram. Soc. JAPAN* **2009**, *117*, 612. [CrossRef]
17. Yoshimatsu, K.; Wadati, H.; Sakai, E.; Harada, T.; Takahashi, Y.; Harano, T.; Shibata, G.; Ishigami, K.; Kadono, T.; Koide, T.; et al. Spectroscopic studies on the electronic and magnetic states of Co-doped perovskite manganite $Pr_{0.8}Ca_{0.2}Mn_{1-y}Co_yO_3$ thin films. *Phys. Rev. B* **2013**, *88*, 1. [CrossRef]
18. Oumezzine, M.; Peña, O.; Kallel, S.; Kallel, N.; Guizouarn, T.; Gouttefangeas, F.; Oumezzine, M. Electrical and magnetic properties of $La_{0.67}Ba_{0.33}Mn_{1-x}(Me)_xO_3$ perovskite manganites: Case of manganese substituted by trivalent (Me = Cr) and tetravalent (Me = Ti) elements. *Appl. Phys. A* **2014**, *114*, 819. [CrossRef]
19. Oumezzine, M.; Galca, A.C.; Pasuk, I.; Chirila, C.F.; Leca, A.; Kuncser, V.; Tanase, L.C.; Kuncser, A.; Ghica, C.; Oumezzine, M. Structural, magnetic and magnetocaloric effects in epitaxial $La_{0.67}Ba_{0.33}Ti_{0.02}Mn_{0.98}O_3$ ferromagnetic thin films grown on 001-oriented $SrTiO_3$ substrates. *Dalton Trans.* **2016**, *45*, 15034. [CrossRef]
20. Kim, J.; Ryu, S.; Jeen, H. Strain-effected physical properties of ferromagnetic insulating $La_{0.88}Sr_{0.12}MnO_3$ thin films. *RSC Adv.* **2019**, *9*, 2645. [CrossRef]
21. Banerjee, B.K. On a generalised approach to first and second order magnetic transitions. *Phys. Lett.* **1964**, *12*, 16. [CrossRef]
22. Moya, X.; Hueso, L.; Maccherozzi, F.; Tovstolytkin, A.; Podyalovskii, D.; Ducati, C.; Phillips, L.; Ghidini, M.; Hovorka, O.; Berger, A.; et al. Giant and reversible extrinsic magnetocaloric effects in $La_{0.7}Ca_{0.3}MnO_3$ films due to strain. *Nat. Mater.* **2013**, *12*, 52. [CrossRef] [PubMed]
23. Pecharsky, V.K.; Gschneidner, J.K.A. Giant magnetocaloric effect in $Gd_5(Si_2Ge_2)$. *Phys. Rev. Lett.* **1997**, *78*, 4494. [CrossRef]
24. Morelli, D.T.; Mance, A.M.; Mantese, J.V.; Micheli, A.L. Magnetocaloric properties of doped lanthanum manganite films. *J. Appl. Phys.* **1996**, *79*, 373. [CrossRef]
25. Kumar, S.V.; Chukka, R.; Chen, Z.; Yang, P.; Chen, L. Strain dependent magnetocaloric effect in $La_{0.67}Sr_{0.33}MnO_3$ thin-films. *AIP Advances.* **2013**, *3*, 052127. [CrossRef]
26. Wood, M.E.; Potter, W.H. General analysis of magnetic refrigeration and its optimization using a new concept: Maximization of refrigerant capacity. *Cryogenics* **1985**, *25*, 667. [CrossRef]
27. Gschneidner, K.A., Jr.; Pecharsky, V.K. Magnetocaloric Materials. *Annu. Rev. Mater. Sci.* **2000**, *30*, 387–429. [CrossRef]
28. Snyder, G.J.; Hiskers, R.; DiCarolis, S.; Beasley, M.R.; Geballe, T.H. Intrinsic electrical transport and magnetic properties of $La_{0.67}Ca_{0.33}MnO_3$ and $La_{0.67}Sr_{0.33}MnO_3$ MOCVD thin films and bulk material. *Phys. Rev. B* **1996**, *53*, 14434. [CrossRef]
29. Emin, D.; Holstein, T. Adiabatic Theory of an Electron in a Deformable Continuum. *Phys. Rev. Lett.* **1976**, *36*, 323. [CrossRef]
30. Chaikin, P.M.; Beni, G. Thermopower in the correlated hopping regime. *Phys. Rev. B* **1976**, *13*, 647. [CrossRef]

Article

The New Reliable pH Sensor Based on Hydrous Iridium Dioxide and Its Composites

Nikola Lenar, Robert Piech and Beata Paczosa-Bator *

Faculty of Materials Science and Ceramics, AGH University of Science and Technology, Mickiewicza 30, PL-30059 Krakow, Poland
* Correspondence: paczosa@agh.edu.pl; Tel.: +48-01-2617-5021; Fax: +48-01-2634-1201

Abstract: The new reliable sensor for pH determination was designed with the use of hydrous iridium dioxide and its composites. Three different hIrO$_2$-based materials were prepared and applied as solid-contact layers in pH-selective electrodes with polymeric membrane. The material choice included standalone hydrous iridium oxide; composite material of hydrous iridium oxide, carbon nanotubes, and triple composite material composed of hydrous iridium oxide; carbon nanotubes; and poly(3-octylthiophene-2,5-diyl). The paper depicts that the addition of functional material to standalone metal oxide is beneficial for the performance of solid-state ion-selective electrodes and presents the universal approach to designing this type of sensors. Each component contributed differently to the sensors' performance—the addition of carbon nanotubes increased the electrical capacitance of sensor (up to 400 μF) while the addition of conducting polymer allowed it to increase the contact angle of material changing its wetting properties and enhancing the stability of potentiometric response. Hydrous iridium oxide contacted electrodes exhibit linear response in wide linear range of pH (2–11) and stable potentiometric response (the lowest potential drift of 0.036 mV/h is attributed to the electrode with triple composite material).

Keywords: pH sensing; solid-contact electrodes; iridium dioxide; composite materials; carbon nanotubes; poly(3-octylthiophene-2,5-diyl)

Citation: Lenar, N.; Piech, R.; Paczosa-Bator, B. The New Reliable pH Sensor Based on Hydrous Iridium Dioxide and Its Composites. *Materials* 2023, 16, 192. https://doi.org/10.3390/ma16010192

Academic Editor: Thomas Dippong

Received: 2 December 2022
Revised: 21 December 2022
Accepted: 22 December 2022
Published: 25 December 2022

Copyright: © 2022 by the authors. Licensee MDPI, Basel, Switzerland. This article is an open access article distributed under the terms and conditions of the Creative Commons Attribution (CC BY) license (https://creativecommons.org/licenses/by/4.0/).

1. Introduction

The quantity pH is intended to be a measurement of the activity of hydrogen ions in a solution [1]. According to IUPAC recommendations [2], potentiometry with the use of hydrogen-sensitive electrodes is the only accurate method of pH measurement. Conventionally, pH measurements are performed with the use of a glass electrode [2–6].

The glass electrode was the first electrode acknowledged as an ion-selective electrode designed by Klemensiewicz and Haber [7], who recognized the practical application coming from the research on glass membranes performed by Cremer [8]. Since invented in the 1900s, the glass electrode has been widely used for pH detection and was unlikely to be replaced because of its great analytical parameters [5,6,9]. The glass electrode belongs to the oldest group of potentiometric sensors named conventional electrodes, characterized with the presence of the internal solution. This type of an electrode requires vertical position and stable temperature and pressure to avoid the phase change of an inner solution [10]. Moreover, the presence of glass bubble, acting as a pH-selective membrane, makes the glass electrode brittle and vulnerable to mechanical damage [5]. Because of those disadvantages, coming from the electrode's construction rather than the poor analytical performance, the glass electrode is likely to find its substitute in all-solid-state, ion-selective electrodes.

The first electrode representing this new group was designed by Cattrall and Freiser in 1971 [11] and was called the coated wire electrode as the ion-selective membrane was coated around the metallic wire and later also on the other types of electrodes, such as disc electrodes. This type of an electrode allowed it to overcome all disadvantages related to the presence of an inner solution and the problem of brittleness; however, the analytical performance was unfortunately deteriorated. It later turned out that the indirect contact of an electrode and ion-selective membrane, which is electronic and ionic conductor, causes the blockage of the charge transfer resulting in high resistance and deterioration of potential stability. The stability is reportedly limited by the small (double-layer) capacitance formed at the interface between the electronic conductor and the ion-selective membrane [12].

The potential stability of the all solid state electrode can be improved by applying an intermediate layer between the electronic and ionic conductor with suitable ion-to-electron transduction properties [13]. This group of electrodes was named solid-contact electrodes as the solid-contact layer is placed in-between the electronic conductor and ion-selective membrane. This group of electrodes developed rapidly over the last 30 years, and nowadays, solid-contact electrodes are characterized by analytical parameters nearly as excellent as those of the glass electrode [3,14].

Various materials have been applied over the years as solid-contact layers in pH-sensitive ion-selective electrodes including conducting polymers poly(3,4-ethylenedioxythiophene)/poly(styrenesulfonate) PEDOT/PSS [15] and derivative of poly(3,4-ethylenedioxythiophene) PEDOT-C_{14} [16], carbon nanomaterials such as multi-walled carbon nanotubes MWCNTs [17], and hydrous metal oxides—ruthenium dioxide $hRuO_2$ [14].

This study focuses on introducing the new materials based on hydrous iridium dioxide into pH-selective solid-contact electrodes with polymeric membrane. Three different materials were introduced in the scope of this paper including standalone hydrous iridium dioxide ($hIrO_2$), double composite of iridium dioxide with carbon nanotubes ($hIrO_2$-NTs) and triple composite of iridium dioxide, carbon nanotubes and poly(3-octylthiophene-2,5-diyl) ($hIrO_2$-NTs-POT). The last material is a unique combination of three significantly different materials: metal oxide, carbon nanomaterial, and conducting polymer into one composite material applied for the first time as solid-contact layer in pH-selective sensors.

2. Materials and Methods

2.1. Materials

Aqueous solutions were prepared by dissolving salts and acids in distilled and deionized water. Standard solutions of fixed pH value were prepared by dissolving citric acid (POCH, Gliwice, Poland) and boric acid (POCH). The buffer solution of 1 mM citric acid and 1 mM boric acid was titrated with 1M sodium hydroxide and 1M hydrochloric acid. NaOH (POCH) and HCl (POCH) were added to meet the desired pH value that is 4-12 and 2-3, respectively. All chemicals used for solution preparation were of analytical grade and were used as received without any further purification.

Designed pH sensors are ion-selective electrodes characterized by a sandwich structure with ion-selective membrane placed on the solid-contact layer. The solid-contact layer consisted of hydrous iridium dioxide ($hIrO_2$) (Alfa Aesar, Haverhill, MA, USA), multi-walled carbon nanotubes (NTs) (Nanostructured & Amorphous Materials, Inc., Houston, TX, USA), and Poly(3- octylthiophene-2,5-diyl) (POT) (Sigma Aldrich, St. Louis, MO, USA). The ion-selective membrane consisted of hydrogen ionophore V (Calix[4]-aza-crown), sodium tetrakis(4-fluorophenyl)borate dihydrate, 2-Nitrophenyl octyl ether (NPOE), and poly(vinyl chloride) (PVC) of high molecular weight were purchased from Sigma-Aldrich. Dimethylformamide (DMF) and Tetrahydrofuran (THF) used as solvents were also purchased from Sigma-Aldrich.

2.2. Sensor's Preparation

Designed pH sensors are solid-contact ion selective electrodes with pH-selective membrane responsible for selective recognition of hydrogen ions and solid-contact layer between the membrane and electrode's surface.

The preparation procedure starts with preparation of glassy carbon disc (GCD) electrodes' surface by polishing the disc using aluminum oxide paste of descending grain size and rinsing it with deionized water and methanol.

For the purpose of this work 12 items of pH-selective electrodes were prepared: 3 items with hydrous iridium dioxide layer ($hIrO_2$); 3 items with double composite layer of hydrous iridium dioxide and carbon nanotubes ($hIrO_2$ – NTs); 3 items with triple composite layer of hydrous iridium dioxide, carbon nanotubes, and Poly(3- octylthiophene-2,5-diyl) ($hIrO_2$ – NTs – POT); and 3 items without solid-contact layer used as a control group.

Material for hydrous iridium dioxide layer was prepared by ultrasonically dispersing (for 30 min) 7 mg of $hIrO_2$ in 1 mL of DMF. Double composite material was prepared by ultrasonically dispersing (for 30 min) 5 mg of $hIrO_2$ and 10 mg of carbon nanotubes (NTs) in 1 mL of DMF. The triple composite material was prepared by ultrasonically dispersing (for 15 min) 5 mg of $hIrO_2$, 5 mg of NTs, and 10 mg of POT in 1mL of THF. The solution was then centrifuged for 15 min (10 000 RPM), and then, the sediment of solid particles was separated. After centrifugation, the portion of the POT dissolved in THF was removed, and the residue after centrifugation ($hIrO_2$, NTs and undissolved POT) was dispersed again ultrasonically (15 min) in a new amount of THF (1 mL).

Electrodes representing each group were casted with a pH-selective membrane. The membrane was prepared by dispersing 252 mg of the membrane components in 2 mL of THF. The composition of pH-selective membrane was as follows: 0.90% (w/w) hydrogen ionophore V, 66% (w/w) o-NPOE, 32.85% (w/w) PVC, and 0.25% (w/w) sodium tetrakis(4-fluorophenyl)borate dihydrate.

All electrodes were prepared with the use of drop-casting method following the procedure for each group listed below:

1—GC/$hIrO_2$/H^+-ISM sensors were prepared by casting 15 µL of $hIrO_2$ solution onto disc electrode surface, drying the solution in 80 degrees until the solvent (DMF) evaporation and then casting the obtained solid-contact $hIrO_2$ layer with 60 µL of membrane solution.

2—GC/$hIrO_2$+NTs/H^+-ISM sensors were prepared by casting 15 µL of $hIrO_2$-NTs solution onto disc electrode surface, drying the solution in 80 degrees until the solvent (DMF) evaporation and then casting the obtained solid-contact $hIrO_2$-NTs layer with 60 µL of membrane solution.

3—GC/$hIrO_2$+NTs+POT/H^+-ISM sensors were prepared by casting 15 µL of $hIrO_2$-NTs-POT solution onto disc electrode surface, drying the solution in room temperature until the solvent (THF) evaporation, and then casting the obtained solid-contact $hIrO_2$-NTs-POT layer with 60 µL of membrane solution.

Drop-casting method is simple and fast, and no additional binder is required when using this technique as the membrane's solid particles adhere to the electrode (or mediation layer, in case of solid-contact electrodes) after the solvent (THF) evaporation.

All prepared sensors were conditioned in the buffer solution of pH 3 prior to every measurement.

2.3. Conducted Measurements

Potentiometric measurements were performed using a 16- channel mV-meter (Lawson Labs, Inc., Malvern, PA, USA). The potentiometric response towards H^+ ions was examined in the standard solutions buffered with 10 mM citric acid and 10 mM boric acid titrated with sodium hydroxide or hydrochloric acid. The pH values of buffer solutions were fixed on a value from 2 to 12. Potentiometric measurements were conducted versus the reference electrode-Ag/AgCl electrode with 3 M KCl solution (6.0733.100 Metrohm, Herisau, Switzerland) and in the presence of auxiliary electrode-platinum wire.

The chronopotentiometric measurements were carried out with the use of an Autolab General Purpose Electrochemical System (AUT302N.FRA2-AUTOLAB, Metrohm Autolab, Barendrecht, The Netherlands) with NOVA 2.1. software. Designed pH-selective sensors ion-selective electrodes were tested as working electrodes in three-electrode cell with a reference electrode Ag/AgCl electrode with 3 M KCl solution (6.0733.100 Metrohm, Herisau,

Switzerland) and in the presence of auxiliary electrode–platinum wire. Chronopotentiometric tests were conducted in the standard buffer solution of pH 3. A constant current of +100 nA was applied to the working electrode for 60 s, followed by a −100 nA current for another 60 s according to the procedure proposed by Bobacka [12].

The examination of the wetting properties of the materials for solid-contact layers was performed with the use of the Theta Lite contact angle microscope with One Attension software by Biolin Scientific, Frölunda, Sweden.

The microstructure of solid-contact materials was examined using Scanning Electron Microscope and Transmission Electron Microscope. SEM scans were collected using Scanning Electron Microscope-LEO 1530 (Carl Zeiss, Germany) and TEM scans using Transmission Electron Microscope—Tecnai 20 X-TWIN (FEI, Hillsboro, OR, USA).

3. Results

3.1. Materials' Microstructure Characteristics

The microstructures of materials for solid-contact layers were examined using Scanning and Transmission Electron Microscope. SEM scan of iridium dioxide (Figure 1a) depict nanometric particles of IrO_2. The minute size of the oxide's grains indicate that the iridium dioxide itself obtained layer is characterized by quite a high surface area.

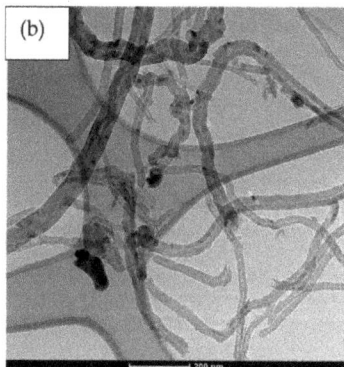
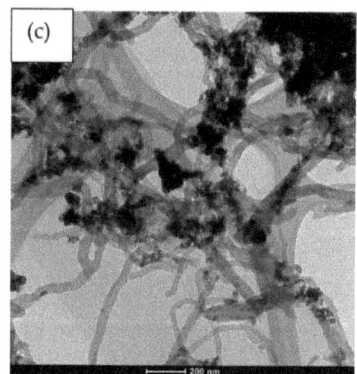

Figure 1. Microstructure of solid-contact materials: (**a**)—SEM scan of IrO_2, (**b**)—TEM scan of IrO_2-NT, (**c**)—TEM scan of IrO_2-NT-POT.

In the TEM scan of IrO_2-NT material (Figure 1b), contrast black spots visible against the carbon nanotubes are the iridium dioxide nanoparticles. As can be seen, oxide particles effectively adhere to carbon nanomaterial, elevating the surface area of the material in comparison with standalone IrO_2.

The TEM scan of IrO$_2$-NT-POT (Figure 1c) revealed carbon nanotubes covered with single particles of iridium dioxide and poly(3-octylthiophene-2,5-diyl). It can be seen that agglomerates of iridium dioxide and poly(3-octylthiophene-2,5-diyl) cover the nanotubes of carbon material, creating a complicated structure of high surface area (Figure 1c). The designed triple composite material is characterized by a high surface area due to the complicated structure. The combination of three different substrates allowed the creation of one complex structure composite material and consequently to achieve the highest surface area of all designed solid-contact materials.

3.2. Potentiometric Response towards Hydrogen Ions

The ionic response of designed solid-contact hIrO$_2$-electrodes was tested during potentiometric measurements in standard solutions of pH values from 2 to 12 in the presence of the group of control (coated-disc) electrodes. The electromotive force (EMF) was recorded during the time of three days of electrodes' conditioning in pH 3 (after 24, 48, and 72 h). The exemplary potentiometric response is presented in the Figure 2 for one electrode of each group after 24 h of the electrodes' conditioning. Average analytical parameters of designed electrodes calculated after 24, 48, and 72 h—the slope of the calibration curve, the standard potential, and the linear range, together with standard deviation values, are presented in the Table 1.

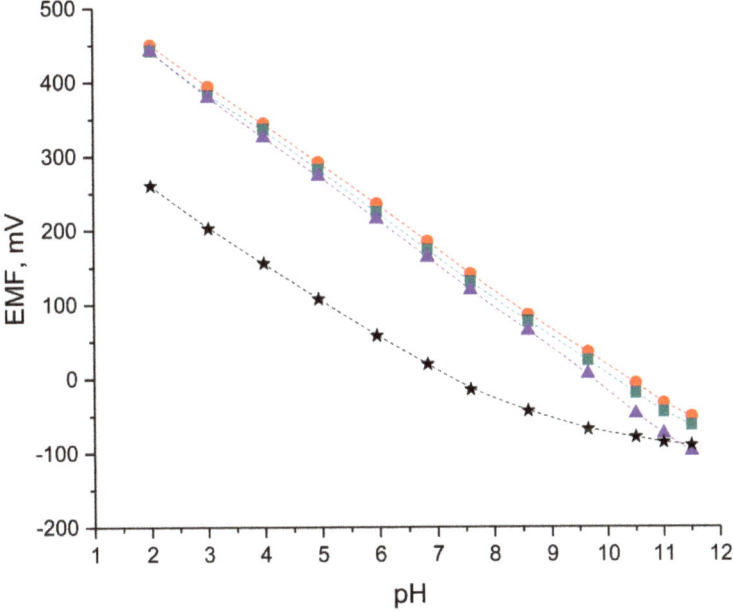

Figure 2. Exemplary potentiometric response ●-GC/hIrO$_2$/H$^+$-ISM, ■-GC/hIrO$_2$+NTs/H$^+$-ISM, ▲-GC/hIrO$_2$+NTs+POT/H$^+$-ISM and ★-GC/pH-ISM electrode after 24 h of electrodes' conditioning in pH 3.

Table 1. Calibration curves parameters of pH-selective electrodes after 24, 48 and 72 h of conditioning (n = 3).

Electrode	Linear Range [pH]	Slope S ± SD [mV/pH]	Standard Potential E^0 [mV]
GC/hIrO$_2$/H$^+$-ISM	2–11	54.12 ± 0.16	557 ± 2
GC/hIrO$_2$-NTs/H$^+$-ISM	2–11	54.40 ± 0.19	548 ± 1
GC/hIrO$_2$-NTs-POT/H$^+$-ISM	2–11.5	57.18 ± 0.07	554 ± 1
GC/H$^+$-ISM	2–7	48.93 ± 0.99	352 ± 8

As expected, for solid-contact pH-selective electrodes the linear range—the range of pH in which the near-Nernstian response towards hydrogen ions is observed, was wider in contrast to coated-disc electrode. The slope of the calibration curve for designed solid-contact electrodes based on hydrous iridium dioxide was closer to the theoretical value (approx. 54 to 57 mV/pH) than for the electrode without the solid-contact layer (only 49 mV/pH).

In addition, the repeatability of the electrodes' response was better for electrodes with $hIrO_2$-based materials since, for the GC/H^+-ISM group, the standard deviation values calculated over three days were of higher values.

Within the groups of solid-contact electrodes with hydrous iridium dioxide ($GC/hIrO_2/H^+$-ISM) and double composite of hydrous iridium dioxide and carbon nanotubes ($GC/hIrO_2$-NTs/H^+-ISM), the potentiometric response towards hydrogen ions was comparable with similar parameters obtained in the same linear range—from pH 2 to 11.

For the third group of solid-contact electrodes with triple composite material, the wider linear range was recognized (from pH 2 to 11.5). In addition, the smallest standard deviation values, describing the repeatability of electrodes' response over three days of conditioning, can be attributed to the $GC/hIrO_2$-NTs-POT/H^+-ISM group. Implementing the triple composite material into the electrodes' construction allowed us to obtain sensors of highly repeatable potentiometric response and the value of the slope of the calibration curve closest to the theoretical value in the pH range from 2 to 11.5.

Potential reversibility was tested in the set of standard solutions of pH values from 3 to 6. The measurement lasted 3 min in the solution of a certain pH, and the stability of potentiometric response was examined when changing the pH of the standard solution.

As can be seen in the Figure 3, the potentiometric response is reversible for all tested groups of electrodes, and the EMF stabilizes immediately after the change in the pH value.

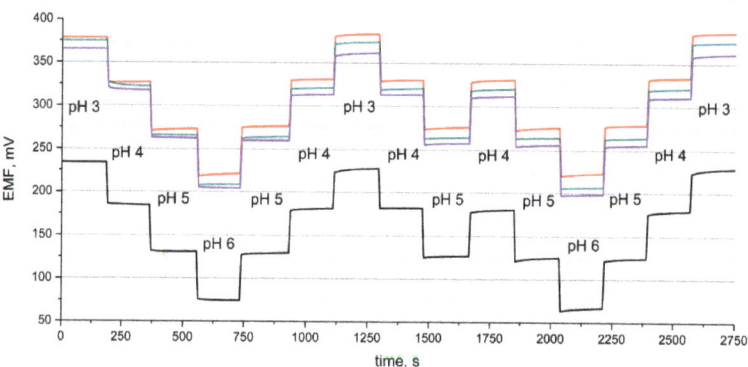

Figure 3. Potential reversibility of $GC/hIrO_2/H^+$-ISM (red line), $GC/hIrO_2$+NTs/H^+-ISM (green line), $GC/hIrO_2$+NTs+POT/H^+-ISM (purple line) and GC/H^+-ISM (black line) electrode recorded in the buffer solution of pH 3, 4, 5 and 6, alternately.

3.3. Potential Stability

The stability of potentiometric response of designed electrodes was tested for 19 hours in the standard buffer solution of pH 3. The course of the measurement with time is presented in the Figure 4. The measurement was conducted versus the single junction potential reference electrode—Ag/AgCl electrode with 3 M KCl solution. The potential stability was characterized by the potential drift parameter, which was calculated as $\Delta EMF/t$ ratio (ΔEMF—electromotive force change, t—time of measurement). The stability of potentiometric response was compared for each group of solid-contact $hIrO_2$-based electrodes and the results were juxtaposed with the potential drift received for coated-disc electrode.

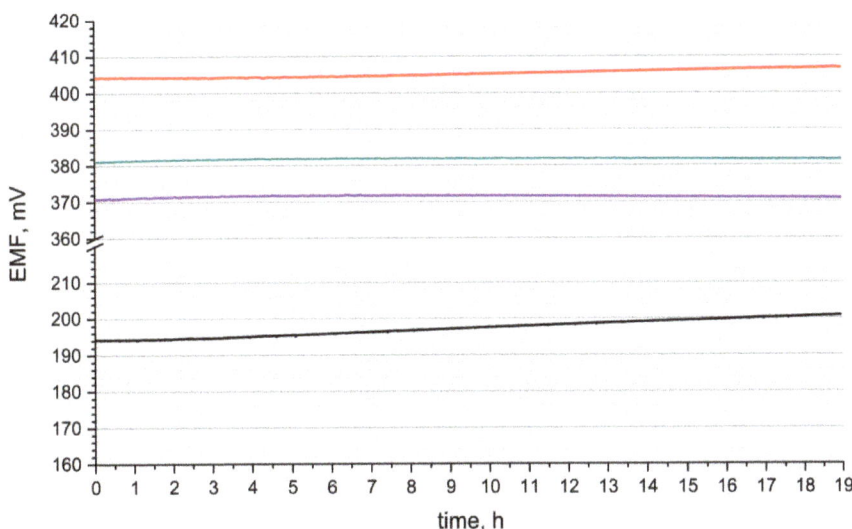

Figure 4. Potential stability of GC/hIrO$_2$/H$^+$-ISM (red line), GC/hIrO$_2$+NTs/H$^+$-ISM (green line), GC/hIrO$_2$+NTs+POT/H$^+$-ISM (purple line) and GC/H$^+$-ISM (black line) electrode recorded in the buffer solution of pH 3.

For the solid-contact electrodes the results of potential drift values were as follows: 0.1 mV/h, 0.077 mV/h and 0.036 mV/h for GC/hIrO$_2$/H$^+$-ISM, GC/hIrO$_2$-NTs/H$^+$-ISM, and GC/hIrO$_2$-NTs-POT/H$^+$-ISM electrode, respectively. As noticed, the addition of carbon nanomaterial—and later on, conducting polymer—allows us to obtain electrodes of lower potential drifts (that is, higher potential stability). This phenomena was observed in our previous work on potassium-selective electrodes as for electrodes with standalone metal oxide (ruthenium dioxide) equaled to 0.085 mV/h [18] and 0.028 mV/h [19] and 0.077 mV/h [20] after implementing composite materials of this oxide with poly(3-octylthiophene-2,5-diyl) and Poly(3,4-ethylenedioxythiophene) Polystyrene Sulfonate, respectively.

In the literature for solid-contact pH-selective electrodes, the following values can be found: 2.4 mV/h [15] for poly(3,4-ethylenedioxythiophene)/poly(styrenesulfonate)—contacted electrode, 0.5 mV/h [17] for multiwalled carbon nanotubes—contacted electrode, and 0.15 mV/h [14] for hydrous ruthenium dioxide—contacted electrode.

The significantly lower potential stability in contrast to the solid-contact electrodes, of 0.42 mV/h of coated-disc electrode GC/H$^+$-ISM-ISM, is due to the blocked interface between the electronic and ionic conductor and the lack of the ion-to-electron transducer between the electrode and ion-selective membrane.

3.4. Redox Test

The EMF response for one electrode representing each group (three solid-contact hIrO$_2$-based electrodes and one coated-disc electrode) was recorded in the set of solutions containing constant amount of a FeCl$_2$ and FeCl$_3$ redox couple (1 mM) with the logarithm of Fe^{2+}/Fe^{3+} ratio equal to −1, −0.5, 0, 0.5, and 1.

As presented in the Figure 5, there was no redox response detected in all tested electrodes. The slight change of potential response was caused by various pH values in examined solutions rather than by a redox signal.

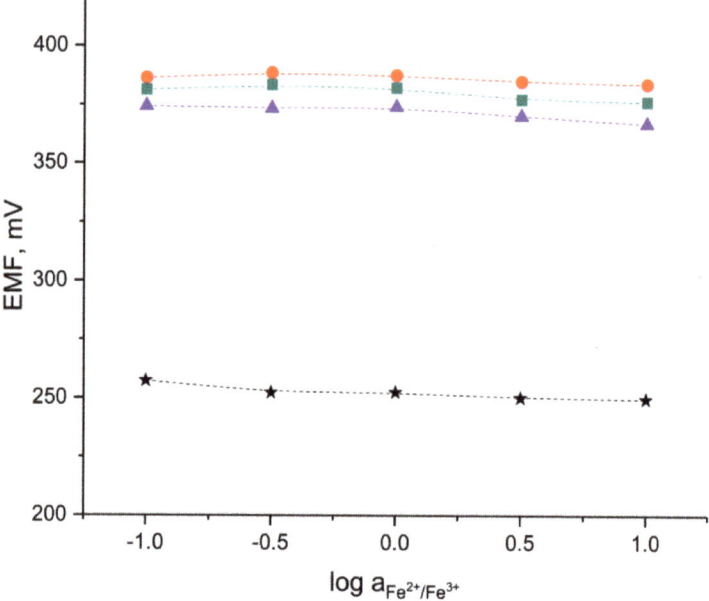

Figure 5. Redox test conducted in the Fe^{2+}/Fe^{3+} redox couple solutions for ●-GC/hIrO$_2$/H$^+$-ISM, ■-GC/hIrO$_2$-NTs/H$^+$-ISM, ▲-GC/hIrO$_2$-NTs-POT/H$^+$-ISM and ★-GC/H$^+$-ISM electrode.

Although the hydrous iridium dioxide-based materials are electronic conductors [21] and should exhibit clear redox response, the polymer-based ion-selective membrane is an electronic insulator and prevents the redox sensitivity of electrodes. Therefore, the examined sensors do not exhibit redox response. This also implicates the proper coverage of tested electrodes with polymeric membrane during the electrodes' preparation with the drop-casting method.

3.5. Light Test

The sensitivity to varying light conditions determines the stability of the sensors during the measurement of different or changing light intensity. This test was performed because of the presence of POT in the solid-contact layer, as this polymer was characterized in the literature as light sensitive [3].

The light test was performed in the standard buffer solution of pH 3 during potentiometric measurement. The EMF was recorded while changing the intensity of light conditions from bright light to darkness. The procedure of the test is presented in the Figure 6. Sensors representing each group of designed electrodes were examined, and a stable potentiometric response was observed for all tested hIrO$_2$-based electrodes. The potentiometric response was stable with time what depict that the light does not influence the performance of pH-selective sensors.

The test proved that despite the presence of a light-sensitive conducting polymer, the designed sensors, with carbon nanotubes-poly(3-octylthiophene-2,5-diyl)-hydrous iridium dioxide triple composite material as a solid-contact, layer are characterized as light-insensitive.

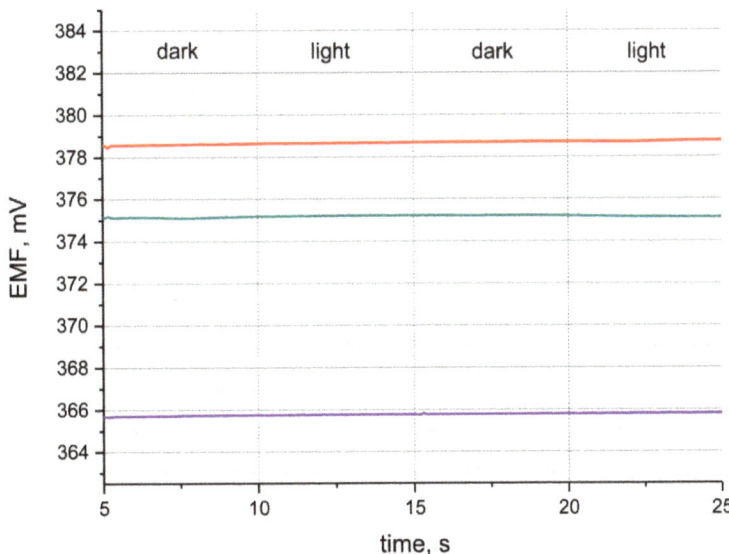

Figure 6. Light test conducted in the buffer solution of pH 2.5 for GC/hIrO$_2$/H$^+$-ISM (red line), GC/hIrO$_2$+NTs/H$^+$-ISM (green line), GC/hIrO$_2$+NTs+POT/H$^+$-ISM (purple line) and GC/H$^+$-ISM (black line) electrode.

3.6. Water Layer Test

The water layer test was performed to evaluate the presence and the influence of the water thin film formed under the ion-selective membrane in the IrO$_2$-based, pH-sensitive electrodes [22].

The experiment was performed according to the procedure proposed by Guzinski [16] in three steps. First, EMF was recorded in the primary ions (H$^+$) solution of pH 2.5. Second, EMF was recorded in the secondary ions solution (0.1 M NaCl), and finally, it was recorded back in the standard H$^+$ solution of pH 2.5. The same experiment was conducted for a coated-disc electrode.

As expected, in the EMF–time chart (presented in the Figure 6) of the GC/H$^+$-ISM electrode, after the second step (contacting the interfering ion), the significant potential drift was observed. It took much longer for the coated-disc electrode to reach the equilibrium potential than it was observed for IrO$_2$-based solid-contact electrodes. For all three groups of designed electrodes (GC/hIrO$_2$/H$^+$-ISM, GC/hIrO$_2$-NTs/H$^+$-ISM, GC/hIrO$_2$-NTs-POT/H$^+$-ISM) no potential drift was observed, and, even after contacting the 0.1 M NaCl solution, the potential response was stable through the time of measurement (as can be seen in the Figure 7), which proves the absence of the undesirable water layer.

The main reason for the absenteeism of the water film under the pH-selective membrane in the solid-contact electrodes is the presence of the IrO$_2$-based materials. The stability of the electrodes' response during the water-layer test depends on the wetting properties of the solid-contact layer. The comparison of the contact angle values for all tested materials for solid-contact layers is presented in the Figure 8.

Standalone IrO$_2$ is characterized by the low contact angle value (18 ± 1°), yet the addition of carbon nanotubes to the metal oxide allowed it to enhance the hydrophobicity of material by elevating the contact angle value (up to 89 ± 2°). The best performance during the experiment with primary and interfering ions (the best stabilization of potentiometric response) can be attributed to the GC/hIrO$_2$-NTs-POT/H$^+$-ISM group of electrodes characterized by the highest contact angle value (equal to 177 ± 3°). The addition of conducting polymer to the composite material changed the wetting properties of the material, making it highly hydrophobic, which, in consequence, contributed to the stability of the potentiometric response during the water layer test.

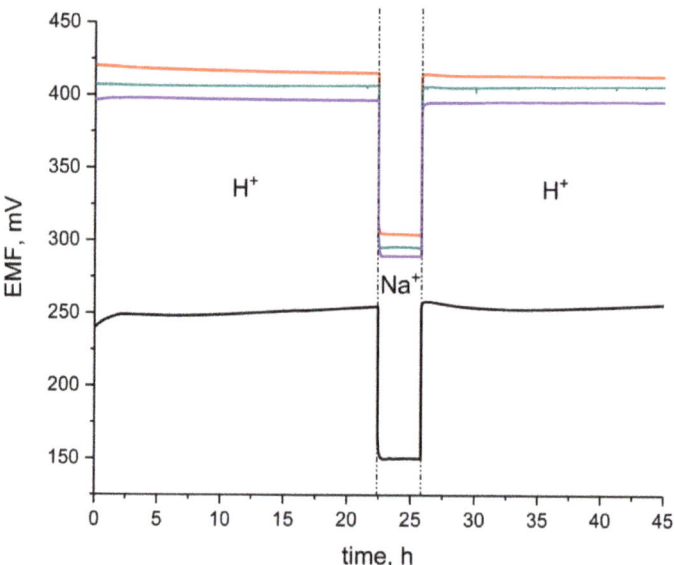

Figure 7. Water-layer test for GC/hIrO$_2$/H$^+$-ISM (red line), GC/hIrO$_2$+NTs/H$^+$-ISM (green line), GC/hIrO$_2$+NTs+POT/H$^+$-ISM (purple line) and GC/H$^+$-ISM (black line) electrode. The test was conducted in the buffer solution of pH 3 (primary ion: H$^+$) and 0.1 M NaCl (secondary ion: Na$^+$).

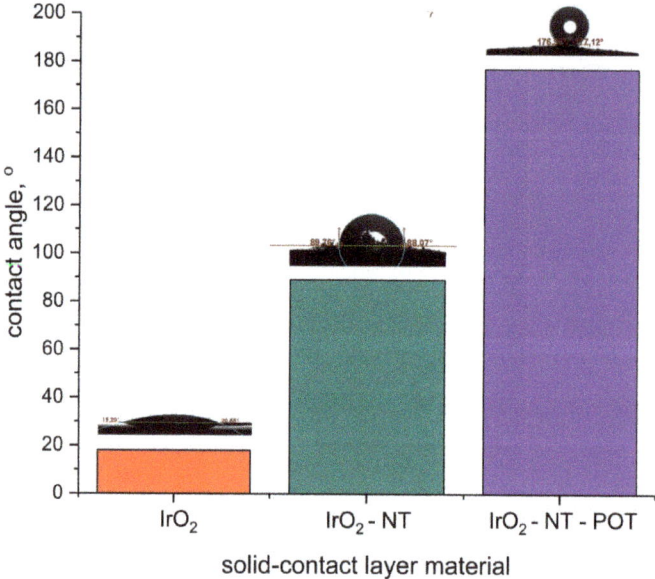

Figure 8. Contact angle values of materials for solid-contact layers for pH-selective electrodes (from left: standalone IrO$_2$, double composite IrO$_2$-NT and triple composite IrO$_2$-NT-POT).

3.7. Electrical Parameters of PH-Sensors

Electrical capacitance, resistance, and potential drift in the forced current condition were calculated based on the results obtained using chronopotentiometry technique using the procedure proposed by Bobacka [12]. The results were presented in Table 2 for each group of pH-selective electrodes and compared. Chronopotentiograms were recorded while the current of 100 nA was forced to flow through the measurement cell. The measurement was conducted in six steps (three steps of recording potential response during +100 nA current flow and three steps of −100 nA current flow, alternately). The electrical parameters were calculated for each step, and average values together with standard deviation values are presented in the Table 2.

Table 2. Electrical parameters calculated for the linear part of the recorded chronopotentiograms with standard deviation values (n = 6 steps).

Electrode	Potential Drift dE/dt ± SD [µV/s]	Resistance R ± SD [kΩ]	Capacitance C ± SD [µF]
GC/hIrO$_2$/H$^+$-ISM	1531 ± 19	807 ± 8	66 ± 8
GC/hIrO$_2$-NTs/H$^+$-ISM	253 ± 3	577 ± 4	174 ± 12
GC/hIrO$_2$-NTs-POT/H$^+$-ISM	309 ± 14	299 ± 3	387 ± 17

As presented in the Table 2, with the increasement of components in the solid-contact layer, the electrical capacitance increases and the resistance decreases. The best electrical parameters that are the highest electrical capacity and the lowest resistance and potential drift can be attributed to the group of electrodes with the triple composite layer.

4. Discussion

Analytical and electrical parameters of designed pH-selective electrodes based on hydrous iridium dioxide and its composites were compared with other electrodes of the same type presented so far in the literature. The compilation is presented in Table 3.

Table 3. Electrical and potentiometric parameters compared for a group of hydrous iridium dioxide-contacted electrodes with other pH-selective solid-contacted electrodes.

Solid Contact Material	pH Linear Range	Slope [mV/pH]	Potential Drift [mV/h]	Capacitance [µF]	Reference
Polypyrrole doped with hexacyanoferrate(II) (PPy-Fe(CN))	2–12	56.9 ± 4.3	0.005	-	[23]
Multi-walled carbon nanotube (MWCNT)	2.89–9.90	58.8 ± 0.4	0.5	30	[17]
Polydopamine-carbon nano-onion (CNO-PDA)	1.50–10.50	60.1 ± 0.3	-	-	[24]
Polyaniline (PANI)	2–9	52.7 ± 1.1	-	-	[25]
Derivative of poly(3,4-ethylenedioxythiophene) (PEDOT-C$_{14}$)	3–11	57.7 ± 0.2	-	-	[16]
poly(3,4-ethylenedioxythiophene)–poly(styrenesulfonate) (PEDOT-PSS)	5–10.3	55.7± 0.5	2.4	-	[15]
Hydrous ruthenium dioxide (RuO$_2$)	2–12	59.31 ± 0.15	0.15	1120	[14]
Hydrous iridium dioxide (IrO$_2$)	2–11	54.12 ± 0.16	0.1	66	this work
Iridium dioxide-carbon nanotubes (IrO$_2$-NT)	2–11	54.40 ± 0.19	0.077	174	this work
Iridium dioxide-carbon nanotubes-poly(3-octylthiophene-2,5-diyl) (IrO$_2$-NT-POT)	2–11.5	57.18 ± 0.07	0.036	387	this work

As presented in the table, electrodes with standalone IrO$_2$, double composite IrO$_2$-NT, and triple composite IrO$_2$-NT-POT as the solid-contact layer are characterized by the linear range complementary to the previous solutions presented in the literature. Applying iridium dioxide-based triple composite material allowed it to receive the calibration plot with the slope value equal to 57.18, which is in agreement with the theoretical Nernstian value. What should be emphasized here is that the obtained IrO_2-NT-POT-contacted electrode is characterized with the outstanding repeatability represented with the standard deviation values from the averaged values of the slope.

Although the capacitance values are not the highest of all the presented solutions, the potential drift values are considerably lower, which depicts the remarkable stability of the potentiometric response of the designed iridium dioxide-based electrodes.

5. Conclusions

It was reportedly noticed that the increase in the number of components of a solid-contact layer is beneficial for the performance of pH-selective electrodes. Each component contributed differently to the final performance of the designed sensors.

The addition of carbon nanotubes to iridium dioxide allowed us to increase the value of electrical capacitance and enhance the potential stability of sensor.

The addition of a conducting polymer to the composite material changed the wetting properties of the material, making it highly hydrophobic, which in consequence contributed to the stability of the potentiometric response during the water-layer test.

All designed groups of pH-selective $hIrO_2$-based sensors exhibit near-Nernstian response in the pH range between 2 and 11. The response towards hydrogen ions turned out to be repeatable and reversible within this range of pH values, and neither redox nor light sensitivity were detected. No presence of a water layer was detected in the solid-contact electrodes with IrO_2- based materials.

Author Contributions: Conceptualization, B.P.-B.; data curation, N.L.; formal analysis, N.L.; funding acquisition, R.P. and B.P.-B.; methodology, N.L. and B.P.-B.; project administration, B.P.-B. and R.P.; resources, R.P. and B.P.-B.; supervision, B.P.-B.; validation, N.L.; visualization, N.L.; writing—original draft, N.L.; writing—review and editing, N.L. All authors have read and agreed to the published version of the manuscript.

Funding: Research project supported by program "Excellence initiative-research university" for the University of Science and Technology.

Institutional Review Board Statement: Not applicable.

Informed Consent Statement: Not applicable.

Data Availability Statement: Not applicable.

Conflicts of Interest: The authors declare no conflict of interest.

References

1. Sørensen, S.P.L.; Linderstrøm-Lang, K.L.C. On the ionisation of proteins. *Trav. Lab. Carlsb.* **1924**, *15*, 6.
2. Spitzer, P.; Wilson, G.S.; Rondinini, S.; Naumann, R.; Covington, A.K.; Camoes, M.F.; Mussini, T.; Milton, M.J.T.; Buck, R.P.; Brett, C.M.A.; et al. Measurement of pH. Definition, standards, and procedures (IUPAC Recommendations 2002). *Pure Appl. Chem.* **2007**, *74*, 2169–2200.
3. Shao, Y.; Ying, Y.; Ping, J. Recent advances in solid-contact ion-selective electrodes: Functional materials, transduction mechanisms, and development trends. *Chem. Soc. Rev.* **2020**, *49*, 4405–4465. [CrossRef] [PubMed]
4. Zdrachek, E.; Bakker, E. Potentiometric Sensing. *Anal. Chem.* **2019**, *91*, 2–26. [CrossRef]
5. Bakker, E.; Bühlmann, P.; Pretsch, E. Polymer membrane ion-selective electrodes-what are the limits? *Electroanalysis* **1999**, *11*, 915–933. [CrossRef]
6. Bakker, E. Ion-Selective Electrodes. In *Encyclopedia of Analytical Science*, 3rd ed.; Elsevier: Amsterdam, The Netherlands, 2019; pp. 231–251.
7. Haber, F.; Klemensiewicz, Z. über elektrische Phasengrenzkräfte. *Z. F. Phys. Chem.* **1909**, *67*, 385. [CrossRef]
8. Cremer, M. *Über die Ursache der Elektromotorischen Eigenschaften der Gewebe, Zugleich ein Beitrag zur Lehre von den Polyphasischen Elektrolytketten*; Oldenbourg: Munich, Germany, 1906.
9. Bakker, E.; Telting-Diaz, E.B. Electrochemical Sensors. *Anal. Chem.* **2002**, *74*, 2781–2800. [CrossRef]

10. Hu, J.; Stein, A.; Bühlmann, P. Rational design of all-solid-state ion-selective electrodes and reference electrodes. *TrAC—Trends Anal. Chem.* **2016**, *76*, 102–114. [CrossRef]
11. Cattrall, R.W.; Freiser, H. Coated wire ion-selective electrodes. *Anal. Chem.* **1971**, *43*, 1905–1906. [CrossRef]
12. Bobacka, J. Potential Stability of All-Solid-State Ion-Selective Electrodes Using Conducting Polymers as Ion-to-Electron Transducers. *Anal. Chem.* **1999**, *71*, 4932–4937. [CrossRef] [PubMed]
13. Nikolskii, B.P.; Materova, E. Solid contact in membrane ion-selective electrodes. In *Ion-Selective Electrode Reviews*; Elsevier: Amsterdam, The Netherlands, 1985; Volume 7, pp. 3–39.
14. Lenar, N.; Paczosa-Bator, B.; Piech, R. Ruthenium dioxide nanoparticles as a high-capacity transducer in solid-contact polymer membrane-based pH-selective electrodes. *Microchim. Acta* **2019**, *186*, 777–788. [CrossRef] [PubMed]
15. Zhang, J.; Guo, Y.; Li, S.; Xu, H. A solid-contact pH-selective electrode based on tridodecylamine as hydrogen neutral ionophore. *Meas. Sci. Technol.* **2016**, *27*, 105101. [CrossRef]
16. Guzinski, M.; Jarvis, J.M.; D'Orazio, P.; Izadyar, A.; Pendley, B.D.; Lindner, E. Solid-contact pH sensor without CO_2 Interference with a Superhydrophobic PEDOT-C14 as solid contact: The ultimate "water layer" test. *Anal. Chem.* **2017**, *89*, 8468–8475. [CrossRef] [PubMed]
17. Crespo, G.A.; Gugsa, D.; MacHo, S.; Rius, F.X. Solid-contact pH-selective electrode using multi-walled carbon nanotubes. *Anal. Bioanal. Chem.* **2009**, *395*, 2371–2376. [CrossRef] [PubMed]
18. Lenar, N.; Paczosa-Bator, B.; Piech, R. Ruthenium dioxide as high-capacitance solid-contact layer in K+-selective electrodes based on polymer membrane. *J. Electrochem. Soc.* **2019**, *166*, B1470–B1476. [CrossRef]
19. Lenar, N.; Paczosa-Bator, B.; Piech, R.; Królicka, A. Poly(3-octylthiophene-2,5-diyl)—Nanosized ruthenium dioxide composite material as solid-contact layer in polymer membrane-based K+-selective electrodes. *Electrochim. Acta* **2019**, *322*, 134718. [CrossRef]
20. Lenar, N.; Piech, R.; Paczosa-Bator, B. Potentiometric sensor with high capacity composite composed of ruthenium dioxide and poly(3,4-ethylenedioxythiophene) polystyrene sulfonate. *Materials* **2021**, *14*, 1891. [CrossRef]
21. Fog, A.; Buck, R.P. Electronic semiconducting oxides as pH sensors. *Sens. Actuators* **1984**, *5*, 137–146. [CrossRef]
22. Fibbioli, M.; Morf, W.E.; Badertscher, M.; De Rooij, N.F.; Pretsch, E. Potential drifts of solid-contacted ion-selective electrodes due to zero-current ion fluxes through the sensor membrane. *Electroanalysis* **2000**, *12*, 1286–1292. [CrossRef]
23. Michalska, A.; Hulanicki, A.; Lewenstam, A. All solid-state hydrogen ion-selective electrode based on a conducting poly(pyrrole) solid contact. *Analyst* **1994**, *119*, 2417–2420. [CrossRef]
24. Zuaznabar-Gardona, J.C.; Fragoso, A. A wide-range sol- id state potentiometric pH sensor based on poly-dopamine T coated carbon nano-onion electrodes. *Sens. Actuators B* **2018**, *273*, 664–671. [CrossRef]
25. Lindfors, T.; Ervelä, S.; Ivaska, A. Polyaniline as pH-sensitive component in plasticized PVC membranes. *J. Electroanal. Chem.* **2003**, *560*, 69–78. [CrossRef]

Disclaimer/Publisher's Note: The statements, opinions and data contained in all publications are solely those of the individual author(s) and contributor(s) and not of MDPI and/or the editor(s). MDPI and/or the editor(s) disclaim responsibility for any injury to people or property resulting from any ideas, methods, instructions or products referred to in the content.

Article

Modification of Electrospun CeO_2 Nanofibers with $CuCrO_2$ Particles Applied to Hydrogen Harvest from Steam Reforming of Methanol

Kai-Chun Hsu [1,2], Chung-Lun Yu [1,2], Heng-Jyun Lei [1,2], Subramanian Sakthinathan [1,2,*], Po-Chou Chen [3,4], Chia-Cheng Lin [1], Te-Wei Chiu [1,2,*], Karuppiah Nagaraj [5], Liangdong Fan [6,*] and Yi-Hsuan Lee [7]

[1] Department of Materials and Mineral Resources Engineering, National Taipei University of Technology, No. 1, Section 3, Zhongxiao East Road, Taipei 106, Taiwan
[2] Institute of Materials Science and Engineering, National Taipei University of Technology, No. 1, Section 3, Chung-Hsiao East Road, Taipei 106, Taiwan
[3] Graduate Institute of Organic and Polymeric Materials, National Taipei University of Technology, No. 1, Section 3, Zhongxiao East Road, Taipei 106, Taiwan
[4] E-Current Co., Ltd., 10F.-5, 50, Section 4, Nanjing East Road, Taipei 10533, Taiwan
[5] SRICT-Institute of Science and Research, UPL University of Sustainable Technology, Vataria, Ankleshwar 393135, Gujarat, India
[6] Department of New Energy Science and Technology, College of Chemistry and Environmental Engineering, Shenzhen University, Shenzhen 518060, China
[7] Department of Mechanical Engineering, National Taipei University of Technology, No. 1, Section 3, Zhongxiao East Road, Taipei 106, Taiwan
* Correspondence: sakthinathan1988@gmail.com (S.S.); tewei@ntut.edu.tw (T.-W.C.); fanld@szu.edu.cn (L.F.); Tel.: +886-963910794 (S.S.); +886-2-2771-2171 (ext. 2742) (T.-W.C.); +86-1062333931 (L.F.)

Abstract: Hydrogen is the alternative renewable energy source for addressing the energy crisis, global warming, and climate change. Hydrogen is mostly obtained in the industrial process by steam reforming of natural gas. In the present work, $CuCrO_2$ particles were attached to the surfaces of electrospun CeO_2 nanofibers to form CeO_2-$CuCrO_2$ nanofibers. However, the $CuCrO_2$ particles did not readily adhere to the surfaces of the CeO_2 nanofibers, so a trace amount of SiO_2 was added to the surfaces to make them hydrophilic. After the SiO_2 modification, the CeO_2 nanofibers were immersed in Cu-Cr-O precursor and annealed in a vacuum atmosphere to form CeO_2-$CuCrO_2$ nanofibers. The $CuCrO_2$, CeO_2, and CeO_2-$CuCrO_2$ nanofibers were examined by X-ray diffraction analysis, transmission electron microscopy, field emission scanning electron microscopy, scanning transmission electron microscope, thermogravimetric analysis, and Brunauer–Emmett–Teller studies (BET). The BET surface area of the CeO_2-$CuCrO_2$ nanofibers was 15.06 m^2/g. The CeO_2-$CuCrO_2$ nanofibers exhibited hydrogen generation rates of up to 1335.16 mL min^{-1} g-cat^{-1} at 773 K. Furthermore, the CeO_2-$CuCrO_2$ nanofibers produced more hydrogen at lower temperatures. The hydrogen generation performance of these CeO_2-$CuCrO_2$ nanofibers could be of great importance in industry and have an economic impact.

Keywords: delafossite; $CuCrO_2$-CeO_2; electrospinning; catalyst; methanol; hydrogen production

Citation: Hsu, K.-C.; Yu, C.-L.; Lei, H.-J.; Sakthinathan, S.; Chen, P.-C.; Lin, C.-C.; Chiu, T.-W.; Nagaraj, K.; Fan, L.; Lee, Y.-H. Modification of Electrospun CeO_2 Nanofibers with $CuCrO_2$ Particles Applied to Hydrogen Harvest from Steam Reforming of Methanol. *Materials* 2022, *15*, 8770. https://doi.org/10.3390/ma15248770

Academic Editor: Dippong Thomas

Received: 9 November 2022
Accepted: 7 December 2022
Published: 8 December 2022

Publisher's Note: MDPI stays neutral with regard to jurisdictional claims in published maps and institutional affiliations.

Copyright: © 2022 by the authors. Licensee MDPI, Basel, Switzerland. This article is an open access article distributed under the terms and conditions of the Creative Commons Attribution (CC BY) license (https://creativecommons.org/licenses/by/4.0/).

1. Introduction

Hydrogen is a viable renewable energy source for addressing the threat of global warming and the decline of fossil fuels. Fuel cells are one of the latest technologies that may effectively convert chemicals into electrical energy to reduce these pollutants [1,2]. Proton-exchange membrane fuel cells (PEMFCs) in particular are systems with zero pollution emissions because they convert chemical energy into electrical energy during the electrochemical reaction of hydrogen and oxygen. Because the anodic Pt-based catalyst can only take less than 10 ppm of CO, PEMFCs typically require a supply of high-purity hydrogen [3,4]. Several methods can be used to produce hydrogen, but the main one is the steam reforming of natural gas. However, the use of hydrogen poses challenges in terms of production and storage [5,6]. The use of steam reforming of methanol (SRM) to produce hydrogen will effectively solve the above problems. SRM has attracted attention because of its low-temperature need for reaction, water-solubility, endothermic process, and high hydrogen yields, which make it ideal for fuel cell applications [7,8]. Decomposition, steam reforming, and partial oxidation are the three primary methods utilized to produce hydrogen from CH_3OH [9,10].

$$CH_3OH \rightarrow CO + 2H_2 \quad \ldots \ldots \ldots \ldots \ldots \ldots \ldots \ldots \ldots \ldots \Delta H^0 = 128 \text{ kJ/mol} \quad (1)$$

$$CH_3OH + H_2O \rightarrow CO_2 + 3H_2 \quad \ldots \ldots \ldots \ldots \ldots \ldots \ldots \Delta H^0 = 131 \text{ kJ/mol} \quad (2)$$

$$CH_3OH + \frac{1}{2}O_2 \rightarrow CO_2 + 2H_2 \quad \ldots \ldots \ldots \ldots \ldots \ldots \ldots \Delta H^0 = 155 \text{ kJ/mol} \quad (3)$$

At a high CO ratio, the decomposition process is an extremely endothermic breakdown process. As a result, this method is ineffective for fuel cells [11]. The partial oxidation reaction process is a highly exothermic reaction with 66% hydrogen output. This process uses pure oxygen instead of air. Finally, the endothermic steam reforming process produces a high rate of hydrogen, up to 75% on a dry basis, with CO as a byproduct. As a result, the steam reforming process is more advantageous for hydrogen production [12,13]. These traditional methods for converting SRM into hydrogen comprise processes like CH_3OH decomposition, water gas shift, and the SRM process [14].

The SRM process is appropriate for producing hydrogen due to its low reaction temperature, adequate water miscibility, high hydrogen concentration ratio, and low CO level. It is also a direct and cost-effective method of hydrogen production. In addition, it is endothermic and can produce a large amount of hydrogen, which is favorable for fuel cell usage [15,16].

The performance of steam re-forming at the reactor is highly impacted by the reaction conditions and catalyst preparation. In the SRM reaction, Ru, Zn, Pd, Ni, Cu, and a combination of these metal-based catalysts are frequently used. Cu-based catalysts are particularly suitable for hydrogen production in the SRM process. Copper and copper-based catalyst materials have a high operating temperature of 573 K, although deactivation occurs at 573–623 K due to the thermal frittage of Cu particles. A further key issue with the SRM process is the deactivation of the catalyst due to the deposition of carbon particles on the Cu catalyst surface [17,18].

To alleviate these issues, a metallic oxide such as ZrO_2, Fe_2O, ZnO, and CeO_2 can be combined with copper and copper-based catalysts to enhance the catalytic performance. With the metallic oxide, the enhanced Cu catalyst will have a fine dispersion with high efficiency and thermal stability. Catalysts have a big impact on the formation of hydrogen in the SRM reaction and the final products. Hence, metal oxide catalysts like Al_2O_3, ZnO/Al_2O_3, ZrO_2/Al_2O, Cr_2O_3/Al_2O_3, $CuO/ZnO/Al_2O_3$, and CeO_2/ZrO_2 are used as a catalyst for the SRM reaction [19,20]. However, the primary issue with the SRM process is the deposition of carbon particles on the surface of the Cu-related catalyst, which lowers the catalyst's effectiveness. To improve the efficiency of hydrogen production, many studies have focused on the use of delafossite materials in the SRM process [21,22].

The chemical formula of delafossite is ABO_2, where A is a cation with linear coordination to two oxygen ions that are often occupied by a cation of a noble metal with a univalent oxidation state, such as Pt^{1+}, Cu^{1+}, or Ag^{1+}. The central metal of the distorted edge-shared BO_6 octahedron is cation B, which has a trivalent charge, such as B^{3+}, Al^{3+}, Ga^{3+}, Cr^{3+}, or Fe^{3+}. Delafossite is a translucent conductive oxide applied in optoelectronic technology [23–25]. However, research regarding its application to catalysis has been scant. Previous studies have applied it to methanol synthesis, N_2O decomposition [26], methanol steam reforming [27], HCl oxidation [28], photocatalytic hydrogen processing, and NO_3 elimination, among other things [29].

Cerium oxide has numerous applications, some being catalysis, ceramics, gas sensors, fuel cell, biomaterials, and solid electrolytes [30]. The most important characteristic of CeO_2 is that it transfers oxygen via the redox potential transfer between Ce^{4+} and Ce^{3+} under oxidation and reduction conditions [31]. Li et al. have reported that adding CeO_2 can reduce the catalytic temperature of Cu-based catalysts and promote catalytic efficiency [32]. Electrospinning was first patented in the United States in 1902 by John Francis Cooley [33]. This method can produce one-dimensional fibers in the micrometer to nanometer diameter ranges with large active surface areas and high porosity. Oxide nanofibers have already been used in energy and environmental applications, such as sensors, catalysis, biotechnology, solar cells, hydrogen energy, and super-capacitors [34].

The CeO_2-$CuCrO_2$ nanofiber catalyst was synthesized by the self-combustion glycine–nitrate process (GNP) and applied for SRM in this study. The CeO_2-$CuCrO_2$ nanofibers had a nanosized, spherical shape with a crystalline delafossite structure. Furthermore, the hydrogen production rate of CeO_2-$CuCrO_2$ nanofibers was compared with those of $CuCrO_2$, CeO_2, and commercial Cu/Al/Zn catalysts. Based on the comparison, the CeO_2-$CuCrO_2$ nanofibers exhibited higher hydrogen yields with lower coke formation during the SRM process as compared with $CuCrO_2$, CeO_2, and commercial Cu/Al/Zn catalysts.

2. Materials and Methods

2.1. Instrumentation

The starting reagents, namely copper nitrate hexahydrate ($[Cu(NO_3)_3 \cdot 6H_2O]$), chromium nitrate nonahydrate ($[Cr(NO_3)_3 \cdot 9H_2O]$), cerium nitrate hexahydrate ($[Ce(NO_3)_3 \cdot 6H_2O]$), N, N-dimethylformamide, Triton X100 polyvinylpyrrolidone (PVP) (M. W = 1,300,000 g·mol^{-1}), and tetraethyl orthosilicate (TEOS), were obtained from SHOWA and Sigma-Aldrich. In this study, the CeO_2 nanofibers and CeO_2-$CuCrO_2$ nanofibers were examined using the appropriate instrumentation techniques. By using an X-ray diffractometer (D2 Phaser, Bruker) with a working voltage of 30 kV and CuK radiation, the crystalline structures of the nanofibers were examined. Field emission scanning electron microscopy (FESEM) (Regulus-8100, HITACHI, Tokyo, Japan) and transmission electron microscopy studies (FE-2100TEM, JEOL, Tokyo, Japan) were used in this work to examine the morphology and particle size of the catalyst. A thermogravimetric analysis/differential scanning calorimeter (TGA/DSC, STA 449 F5, NETZSCH, Selb, Germany) was used to investigate the thermal degradation behavior of electrospun fibers. The specific surface area was calculated using the Brunauer–Emmett–Teller (BET) method using a Micromeritics TriStar II 3030 specification. Before a BET measurement was performed, an appropriate quantity of the produced catalyst was de-gassed at 473 K for 24 h to eliminate the absorbed water. At different relative pressures (P/P_0) ranging from 0 to 0.3, N_2 adsorption isotherms were observed and examined while the catalyst absorbed N_2.

2.2. Preparation of CeO_2 Nanofibers

The precursor solution was synthesized by dissolving 0.625 g of cerium nitrate in 14.4 mL of N, N-dimethylformamide. Following that, 2.4 g of PVP was dissolved in the above precursor solution. After 6 h of stirring, a bright yellow viscous gel-like reaction precursor solution was obtained. This solution was electrospun with a working distance of 15 cm, a voltage of 18 kV, a flow rate of 0.02 mL/h, temperature controlled at 313 K, and humidity of less than 20%. The as-spun fiber was annealed at 873 K with a heating rate of 274 K per minute, and the resulting CeO_2 nanofibers were analyzed by XRD and SEM studies.

2.3. Surface Modification of CeO_2 Nanofibers

The surfaces of the CeO_2 nanofibers were modified as follows. Because $CuCrO_2$ would not easily adhere directly to the surfaces of the CeO_2 nanofibers, the surfaces were coated with SiO_2. Hence, the CeO_2 fibers were dipped into tetraethyl orthosilicate (TEOS) and then annealed in air at 873 K. The modified CeO$_2$ nanofibers were analyzed by XRD, SEM, and TEM studies.

2.4. Preparation of CeO$_2$ Nanofibers Decorated with CuCrO$_2$ Nanoparticles (CeO$_2$-CuCrO$_2$)

CeO$_2$ nanofibers decorated with CuCrO$_2$ nanoparticles (CeO$_2$-CuCrO$_2$) were prepared with the following procedure. The CeO$_2$ fibers were dipped in the precursor, which was a mixture of methanol, chromium nitrate, copper nitrate, and Triton X100, and then dried at 353 K for 2 min before being annealed at 1073 K in a vacuum with a heating rate of 283 K per minute. The prepared CeO$_2$-CuCrO$_2$ nanofibers were analyzed by XRD, SEM, and TEM.

2.5. Steam Reforming of Methanol Process over Electrospun CeO$_2$-CuCrO$_2$ Nanofibers Catalyst

The steam reforming of methanol was performed in a tubular flow reactor using a 25 cm quartz tube with a 1.2 cm inner diameter, nitrogen as the carrier gas with a flow rate of 30 sccm, and 20 mg of catalyst per SRM reaction. The system was connected to a gas chromatograph for analysis. A methanol–water mixture was prepared in a 3:1 molar ratio and heated to 353 K on a hot plate to evaporate methanol–water vapor. A gas tube was inserted into the Erlenmeyer flask beneath the level of the methanol aqueous solution, and then the methanol vapor was carried by nitrogen to the catalyst for the reaction. The nanofibers were sandwiched between quartz cotton in the middle of the quartz tube and then heated to 523, 573, 623, 673, 723, and 773 K, respectively. A gas chromatograph (GC 1000 China Chromatography TCD) was used to analyze each temperature and identify the average values (Figure 1).

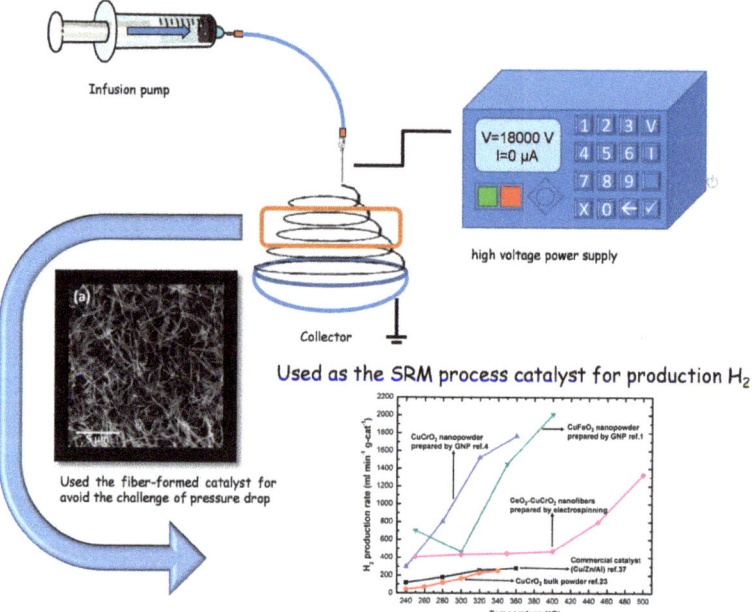

Figure 1. Schematic diagram of the methanol steam reforming process on the electrospun CeO$_2$-CuCrO$_2$ nanofibers catalyst.

3. Results and Discussion

3.1. XRD Analysis

The prepared nanofiber diffraction patterns and crystal structures were studied by XRD studies and analyzed with powder X-ray diffractometric MDI JADE5.0 software tools. Figure 2a presents the XRD pattern of CeO_2 nanofibers, showing the pure CeO_2 cubic phase (JCPDS card PDF#34-0394.) After the electrospun CeO_2 nanofibers were annealed at high temperature, the cerium nitrate decomposed and CeO_2 remained.

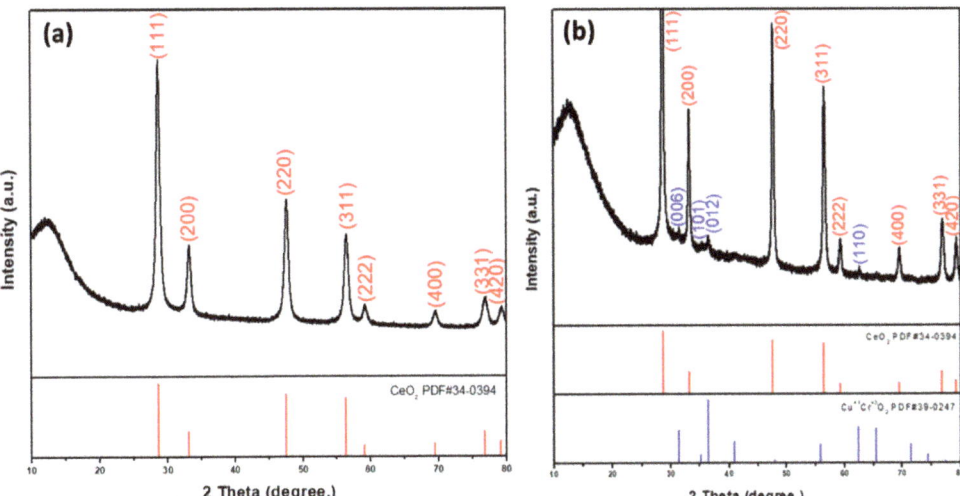

Figure 2. XRD patterns of (a) CeO_2 and (b) CeO_2-$CuCrO_2$ nanofibers.

Figure 2b shows the XRD pattern of $CuCrO_2$-coated CeO_2 nanofibers. The XRD spectra of CeO_2-$CuCrO_2$ nanofibers exhibited that $CuCrO_2$ nanoparticles were attached to the CeO_2 nanofibers. The XRD pattern reveals CeO_2 (PDF#34-0394) and $CuCrO_2$ (PDF#39-0247) phases on the CeO_2-$CuCrO_2$ nanofibers. The XRD pattern of CeO_2 nanofibers after SiO_2 surface modification only reveals CeO_2 (PDF#34-0394) due to annealing at 873 K and its SiO_2 content being too low.

Figure 3 shows the XRD pattern of CeO_2-$CuCrO_2$ after SRM at different temperatures. From the XRD pattern, it can be observed that when the catalytic temperature increases, the peak of $CuCrO_2$(101) at $2\theta = 35.178°$ (PDF#39-0247) gradually disappears. At 773 K, due to the precipitation of $CuCrO_2$ after catalysis, the peaks of the copper (111) and (200) planes can be observed at $2\theta = 43.297°$ and $50.433°$ (PDF#04-0836), respectively.

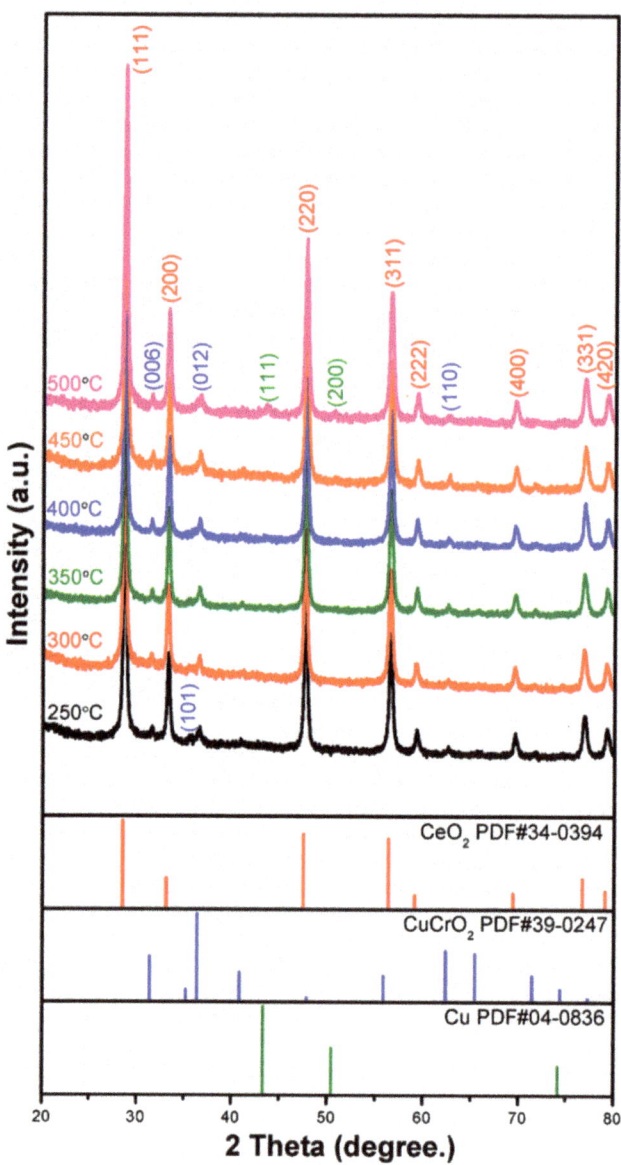

Figure 3. XRD pattern of CeO$_2$-CuCrO$_2$ nanofibers after SRM at different temperatures.

3.2. FESEM Analysis

The morphologies of the CeO$_2$ nanofibers and CeO$_2$-CuCrO$_2$ nanofibers were identified by FESEM and TEM studies. Figure 4 shows the FESEM image of electrospun CeO$_2$ and CeO$_2$-CuCrO$_2$ nanofibers. Figure 4a,b show the CeO$_2$ nanofibers after annealing at a rate of 274 K/min to 873 K in an air atmosphere. The CeO$_2$ nanofibers decreased in size by about 110 nm and disappear from PVP due to annealing. Figure 4c,d show FESEM images of CeO$_2$-CuCrO$_2$ nanofibers. It was found that CuCrO$_2$ is difficult to directly attach to the surface of CeO$_2$; therefore, the surface was modified with SiO$_2$ to improve the CuCrO$_2$

adherence. The FESEM images show that the CeO_2-$CuCrO_2$ nanofibers were very thin, with diameters similar to those of the CeO_2 nanofibers.

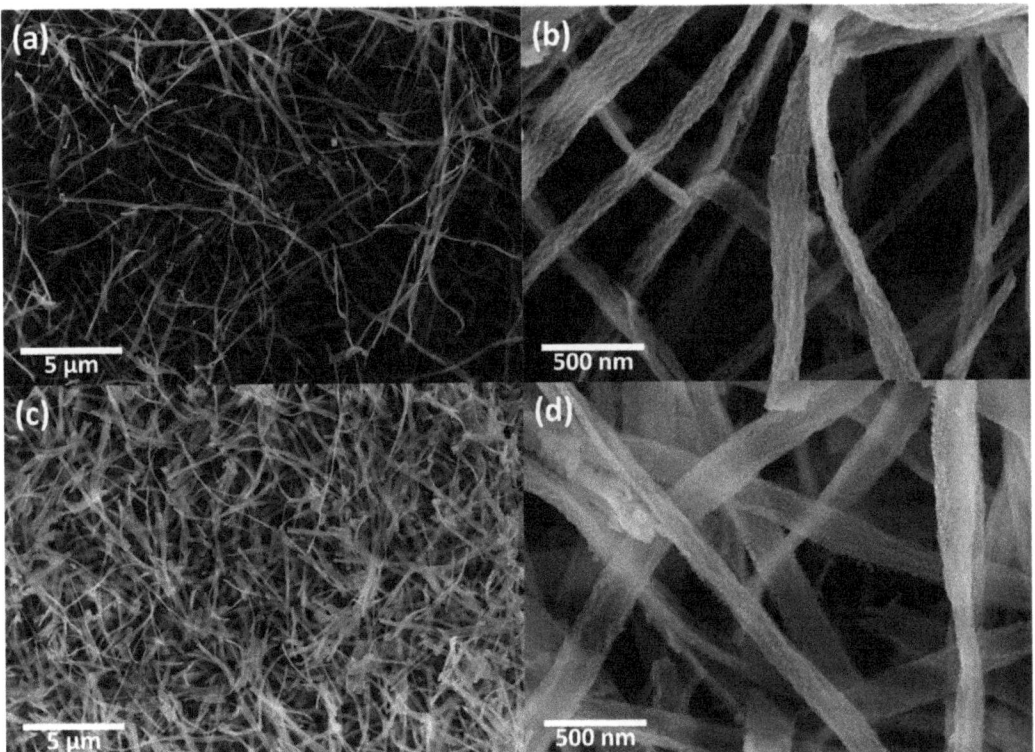

Figure 4. FESEM image of (**a**,**b**) CeO_2 nanofibers and (**c**,**d**) CeO_2-$CuCrO_2$ nanofibers.

Figure 5 shows the FESEM images of CeO_2-$CuCrO_2$ after SRM at different temperatures. The morphologies of CeO_2-$CuCrO_2$ nanofibers after SRM at 523–623 K exhibited the fiber structure, revealing that the CeO_2-$CuCrO_2$ nanofibers had better stability at lower reaction temperatures, as can be seen in Figure 5a–c. Figure 5d shows that the morphology of the nanofibers becomes more fragmented when the catalytic temperature reaches 673 K. From Figure 5e,f, it can be observed that when the catalytic temperature reaches 723 K, copper-precipitated particles begin to appear on the surface of the nanofibers. When the catalytic temperature reaches 773 K, the precipitated particles are scattered on the surface.

Figure 5. FESEM images of CeO$_2$-CuCrO$_2$ nanofibers after SRM at different temperatures (**a**) 523 K, (**b**) 573 K, (**c**) 623 K, (**d**) 673 K, (**e**) 723 K, and (**f**) 773 K.

3.3. TEM and STEM-EDS Analysis

Figure 6a,b shows TEM images of CeO$_2$-CuCrO$_2$ nanofibers. These images confirmed that the CuCrO$_2$ particles were arranged and attached to the surface of CeO$_2$ nanofibers. The STEM image in Figure 7a confirmed that the CeO$_2$ nanofibers had particles arranged on the surface and attached to them. The particle attached to the fiber were investigated by STEM-EDS mapping. The STEM-EDS mapping confirmed the presence of (b) Ce, (c) Si, (d) O, (e) Cu, and (f) Cr in the CeO$_2$-CuCrO$_2$ nanofibers. Figure 7g shows the overall STEM-EDS mapping, which confirmed the presence of Ce, Si, O, Cu, and Cr in the CeO$_2$-CuCrO$_2$ nanofibers. Figure 8 shows the STEM-EDS results of modified CeO$_2$ nanofibers coated with CuCrO$_2$. Figure 8a confirms the presence of Ce, Si, O, Cu, and Cr in the CeO$_2$-CuCrO$_2$ nanofibers. The STEM-EDS spectra of CeO$_2$-CuCrO$_2$ nanofibers after SiO$_2$ surface modification show that (b) Ce, (c) Si, (d) O, (e) Cu, and (f) Cr were present in the CeO$_2$-CuCrO$_2$ nanofibers. Figure 8g shows the STEM-EDS overall spectra confirming that the layer on the fiber surface was mainly composed of CeO$_2$-CuCrO$_2$. Hence, TEM and STEM-EDS studies confirmed the formation of CeO$_2$-CuCrO$_2$ nanofibers.

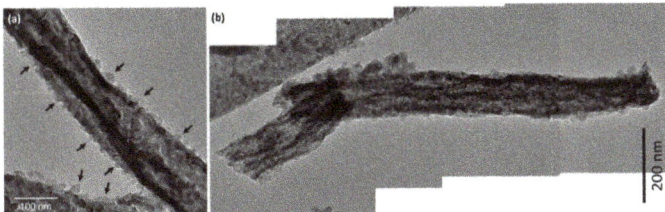

Figure 6. (a,b) TEM image of CeO$_2$-CuCrO$_2$ nanofiber.

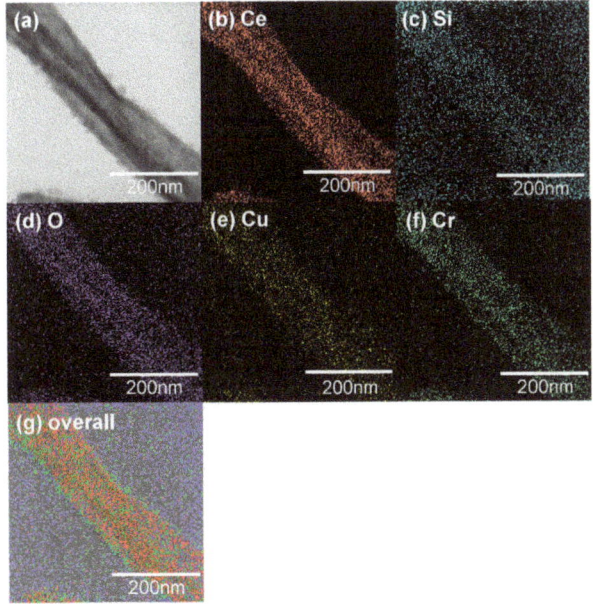

Figure 7. (a) STEM image of CeO$_2$-CuCrO$_2$ nanofiber, (b) Ce, (c) Si, (d) O, (e) Cu, (f) Cr. (g) Overall STEM-EDS mapping of Ce, Si, O, Cu and Cr present in CeO$_2$-CuCrO$_2$ nanofiber.

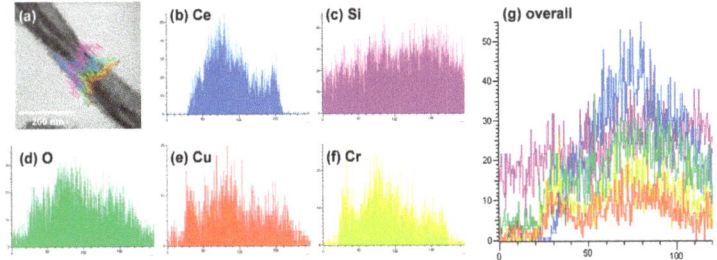

Figure 8. STEM image of CeO$_2$-CuCrO$_2$ nanofibers. (a) Overall STEM spectrum of Ce, O, Si, Cu, and Cr overlap; Elemental mapping of (b) Ce, (c) Si, (d) O, (e) Cu, (f) Cr. (g) Overall STEM-EDX spectrum of Ce, O, Si, Cu, and Cr overlap.

3.4. TGA Analysis

To observe the decomposition mechanism of the as-spun nanofibers at high temperatures, a simultaneous thermogravimetric analyzer was used to observe the TGA/DSC curve, and the temperature was increased to 873 K at a rate of 283 K per minute in air.

Figure 9 shows the TGA/DSC curve of electrospun CeO_2-$CuCrO_2$ fibers. According to the TGA/DSC studies, the weight loss before 373 K is due to the volatilization of the remaining water in the CeO_2-$CuCrO_2$ fibers. The slight weight loss at approximately 423 K and the exothermic slope are due to DMF decomposition, and the endothermic peak at around 493 K is due to cerium nitrate decomposition. After that, a massive and continuous weight loss at about 523 K indicates the significant decomposition of PVP. Moreover, at 573 K to 673 K, an endothermic peak signals the formation of CeO_2-$CuCrO_2$ nanofibers.

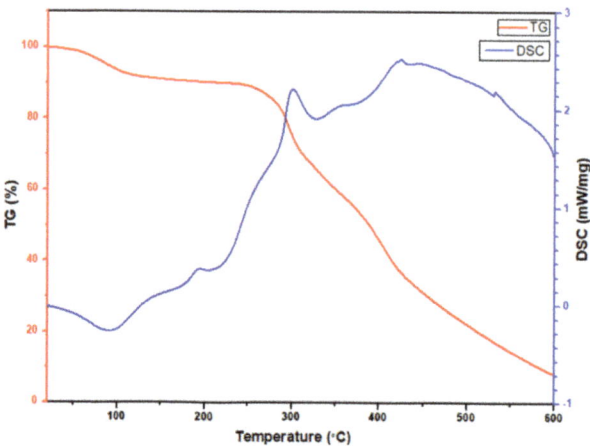

Figure 9. TGA/DSC studies of electrospun CeO_2-$CuCrO_2$ nanofibers.

3.5. Specific Surface Area Analysis

The specific surface area of the CeO_2-$CuCrO_2$ nanofibers is listed in Table 1. Table 1 shows the BET-specific surface areas of delafossite materials produced by solid-state reaction, glycine combustion, and electrospinning. The specific surface area of the CeO_2-$CuCrO_2$ nanofibers produced in this experiment is 15.06 m^2/g, which is larger than the specific surface area of solid-state reactions and other electrospun products.

Table 1. Specific surface area of the different delafossite materials prepared by various processes.

Processes	Specific Surface Area (m^2/g)	Reference
GNP method ($CuFeO_2$)	11.38	[16]
Solid-state ($CuCrO_2$)	1.94	[17]
Solid-state ($CuFeO_2$)	0.57	[17]
Electrospinning ($CuCrO_2$)	7.85	[18]
Electrospinning ($CuFeO_2$)	4.33	[19]
Electrospinning (CeO_2-$CuCrO_2$)	15.06	This study

3.6. Steam Reforming of Methanol Performance

A gas chromatograph was attached to the thermal conductivity detector and used for measuring the rate of hydrogen production. At a flow velocity of 30 sccm and temperatures between 523 K and 773 K, the hydrogen generation was measured using 0.04 g of catalyst, and the hydrogen production rate was converted into mL min^{-1} g-cat^{-1}. To obtain high catalytic performance, the prepared CeO_2-$CuCrO_2$ catalyst was heated without contact with methanol vapor. Additionally, the carrier gas was altered so that the system was filled with methanol steam, and the gas coming from the exit tube was located. As shown in Table 2, the SRM process was carried out over the CeO_2-$CuCrO_2$ catalysts at 523–673 K at a flow rate of 30 sccm. Additionally, when the reaction temperature was raised from

523–673 K, the rate of hydrogen generation increased. The results are shown in Figure 10. In Figure 10, the CeO_2-$CuCrO_2$ nanofiber exhibited an excellent hydrogen production performance compared with $CuCrO_2$ (solid-state method) and commercial Cu/Zn/Al catalysts [24,35]. Table 2 shows the hydrogen production rate of CeO_2-$CuCrO_2$ nanofibers at different temperatures—the hydrogen production rises with an increase in temperature. However, the CeO_2-$CuCrO_2$ nanofibers lose their activity at the reaction temperature is too high, therefore the experiment has not been continued to a high temperature. The CeO_2-$CuCrO_2$ nanofibers are extremely stable in air, in contrast to an H_2-activated catalyst, which is typically harmful when exposed to air due to its high activity and potential for ignition and explosion. Therefore, there is no need to activate the CeO_2-$CuCrO_2$ nanofiber catalyst at high temperatures for the SRM process. This study implies that greater efficiency can be attained than with traditional catalysts if CeO_2-$CuCrO_2$ nanofibers are installed in a fuel cell vehicle. Future research will examine the stability of the catalyst, SRM conditions, and reactor condition optimization.

Table 2. Hydrogen production rate of CeO_2-$CuCrO_2$ nanofibers at different temperatures.

Temperature (Kelvin)	H_2 Production Rate (mL min^{-1} g-cat^{-1})
523	410.66
573	438.48
623	451.52
673	474.30
723	798.28
773	1335.16

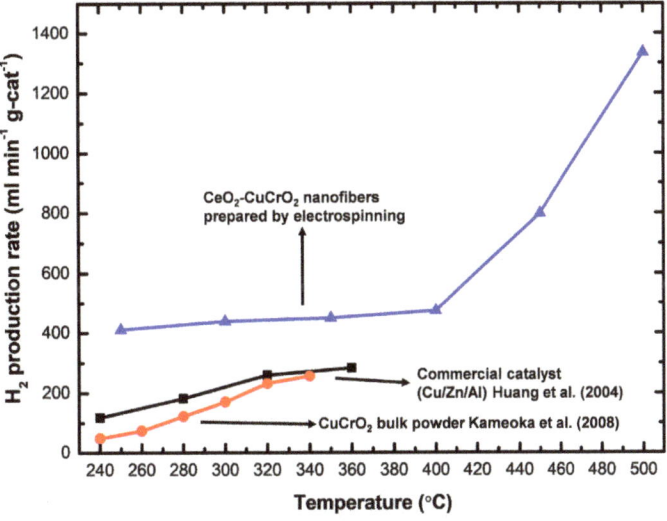

Figure 10. Hydrogen production of electrospinning prepared CeO_2-$CuCrO_2$, compared with $CuCrO_2$ (solid-state method) [24] and commercial Cu/Zn/Al catalysts [35].

4. Conclusions

In this study, the electrospun CeO_2-$CuCrO_2$ nanofiber catalyst was effectively created and used for steam reforming of methanol (SRM). The prepared nanofiber catalyst was evaluated by the field emission scanning electron microscope, transmission electron microscope, X-ray diffractometer energy-dispersive X-ray spectroscopy, thermogravimetric analyzer (STA), and Brunauer-Emmett-Teller analysis. The specific surface area of the CeO_2-$CuCrO_2$ nanofibers is 15.06 m^2/g. The best hydrogen production rate of the CeO_2-$CuCrO_2$

nanofibers, 1335.16 mL min^{-1} g-cat^{-1}, was achieved at a flow rate of 30 sccm and reaction temperature of 773 K. Furthermore, the optimization of reduction conditions and catalyst stability were studied. According to the findings, the increased hydrogen production rate can be ascribed to the stronger catalytic activity, larger surface area, lower reactor temperature, and higher methanol flow rate of the CeO$_2$-CuCrO$_2$ nanofiber catalyst. According to the H$_2$ production performance, the CeO$_2$-CuCrO$_2$ nanofibers can be employed as a better catalyst for commercial H$_2$ production and are suited for fuel cell vehicles without high-temperature activation.

Author Contributions: Conceptualization, Investigation, writing—original draft preparation, K.-C.H.; writing—review and editing, Investigation, Software, C.-L.Y.; writing—review and editing, H.-J.L.; writing—review and editing, S.S.; Validation, P.-C.C.; Methodology, C.-C.L.; Validation, Supervision, Methodology, T.-W.C.; writing—review and editing, L.F.; writing—review and editing, K.N. and Y.-H.L. All authors have read and agreed to the published version of the manuscript.

Funding: This work was supported by the Ministry of Science and Technology of Taiwan (MOST 108-2221-E-027-056, MOST 109-2221-E-027-068, MOST 109-2221-E-027-059, MOST 110-2221-E-027-041, and MOST 109-2113-M-027-001-MY3). This work was supported by the National Science and Technology Council of Taiwan (NSTC 111-2221-E-027-104). The authors appreciate the Precision Research and Analysis Centre of the National Taipei University of Technology (NTUT) for providing the measurement facilities.

Institutional Review Board Statement: Not applicable.

Informed Consent Statement: Not applicable.

Data Availability Statement: Not applicable.

Acknowledgments: This work was supported by the Ministry of Science and Technology of Taiwan (MOST 108-2221-E-027-056, MOST 109-2221-E-027-059, MOST 109-2221-E-027-068, and MOST 109-2113-M-027-001-MY3) and the National Science and Technology Council of Taiwan (NSTC 111-2221-E-027-104). The thanks to Precision Research and Analysis Centre of the National Taipei University of Technology (NTUT) for the measurement support.

Conflicts of Interest: The authors declare no conflict of interest.

References

1. Yu, C.L.; Sakthinathan, S.; Hwang, B.Y.; Lin, S.Y.; Chiu, T.W.; Yu, B.S.; Fan, Y.J.; Chuang, C. CuFeO$_2$–CeO$_2$ Nanopowder Catalyst Prepared by Self-Combustion Glycine Nitrate Process and Applied for Hydrogen Production from Methanol Steam Reforming. *Int. J. Hydrogen Energy* **2020**, *45*, 15752–15762. [CrossRef]
2. Huang, R.J.; Sakthinathan, S.; Chiu, T.W.; Dong, C. Hydrothermal Synthesis of High Surface Area CuCrO$_2$ for H$_2$ production by Methanol Steam Reforming. *RSC Adv.* **2021**, *11*, 12607–12613. [CrossRef] [PubMed]
3. Yu, C.L.; Sakthinathan, S.; Chen, S.Y.; Yu, B.S.; Chiu, T.W.; Dong, C. Hydrogen Generation by Methanol Steam Reforming Process by Delafossite-Type CuYO$_2$ Nanopowder Catalyst. *Microporous Mesoporous Mater.* **2021**, *324*, 111305. [CrossRef]
4. Chiu, T.W.; Hong, R.T.; Yu, B.S.; Huang, Y.H.; Kameoka, S.; Tsai, A.P. Improving Steam-Reforming Performance by Nanopowdering CuCrO$_2$. *Int. J. Hydrogen Energy* **2014**, *39*, 14222–14226. [CrossRef]
5. Abbasi, M.; Farniaei, M.; Rahimpour, M.R.; Shariati, A. Enhancement of Hydrogen Production and Carbon Dioxide Capturing in a Novel Methane Steam Reformer Coupled with Chemical Looping Combustion and Assisted by Hydrogen Perm-Selective Membranes. *Energy Fuels* **2013**, *27*, 5359–5372. [CrossRef]
6. Wiese, W.; Emonts, B.; Peters, R. Methanol Steam Reforming in a Fuel Cell Drive System. *J. Power Sources* **1999**, *84*, 187–193. [CrossRef]
7. Palo, D.R.; Dagle, R.A.; Holladay, J.D. Methanol Steam Reforming for Hydrogen Production. *Chem. Rev.* **2007**, *107*, 3992–4021. [CrossRef]
8. Itoh, N.; Kaneko, Y.; Igarashi, A. Efficient Hydrogen Production via Methanol Steam Reforming by Preventing Back-Permeation of Hydrogen in a Palladium Membrane Reactor. *Ind. Eng. Chem. Res.* **2002**, *41*, 4702–4706. [CrossRef]
9. Pajaie, H.S. Hydrogen Production from Methanol Steam Reforming over Cu/ZnO/Al$_2$O$_3$/CeO$_2$/ZrO$_2$ Nanocatalyst in an Adiabatic Fixed-Bed Reactor. *Iran. J. Energy Environ.* **2012**, *3*, 307–313. [CrossRef]
10. Papavasiliou, J.; Avgouropoulos, G.; Ioannides, T. Production of Hydrogen via Combined Steam Reforming of Methanol over CuO-CeO$_2$ Catalysts. *Catal. Commun.* **2004**, *5*, 231–235. [CrossRef]
11. Chen, W.H.; Lin, B.J. Effect of Microwave Double Absorption on Hydrogen Generation from Methanol Steam Reforming. *Int. J. Hydrogen Energy* **2010**, *35*, 1987–1997. [CrossRef]

12. Agrell, J.; Birgersson, H.; Boutonnet, M. Steam Reforming of Methanol over a Cu/ZnO/Al$_2$O$_3$ Catalyst: A Kinetic Analysis and Strategies for Suppression of CO Formation. *J. Power Sources* **2002**, *106*, 249–257. [CrossRef]
13. Kameoka, S.; Tanabe, T.; Tsai, A.P. Self-Assembled Porous Nano-Composite with High Catalytic Performance by Reduction of Tetragonal Spinel CuFe$_2$O$_4$. *Appl. Catal. A Gen.* **2010**, *375*, 163–171. [CrossRef]
14. Christopher, J.; Swamy, C.S. Catalytic Activity and XPS Investigation of Dalofossite Oxides, CuMO$_2$ (M=Al, Cr or Fe). *J. Mater. Sci.* **1992**, *27*, 1353–1356. [CrossRef]
15. Saadi, S.; Bouguelia, A.; Trari, M. Photocatalytic Hydrogen Evolution over CuCrO$_2$. *Sol. Energy* **2006**, *80*, 272–280. [CrossRef]
16. Shen, J.P.; Song, C. Influence of Preparation Method on Performance of Cu/Zn-Based Catalysts for Low-Temperature Steam Reforming and Oxidative Steam Reforming of Methanol for H$_2$ Production for Fuel Cells. *Catal. Today* **2002**, *77*, 89–98. [CrossRef]
17. Abrokwah, R.Y.; Deshmane, V.G.; Owen, S.L.; Kuila, D. Cu-Ni Nanocatalysts in Mesoporous MCM-41 and TiO$_2$ to Produce Hydrogen for Fuel Cells via Steam Reforming Reactions. *Adv. Mater. Res.* **2015**, *1096*, 161–168. [CrossRef]
18. Navarro, R.M.; Peña, M.A.; Fierro, J.L.G. Production of Hydrogen by Partial Oxidation of Methanol over a Cu/ZnO/Al$_2$O$_3$ Catalyst: Influence of the Initial State of the Catalyst on the Start-up Behaviour of the Reformer. *J. Catal.* **2002**, *212*, 112–118. [CrossRef]
19. Lindström, B.; Pettersson, L.J.; Menon, G. Activity and Characterization of Cu/Zn, Cu/Cr and Cu/Zr on γ-Alumina for Methanol Reforming for Fuel Cell Vehicles. *Appl. Catal. A Gen.* **2002**, *234*, 111–125. [CrossRef]
20. Shokrani, R.; Haghighi, M.; Ajamein, H.; Abdollahifar, M. Hybrid Sonochemic Urea-Nitrate Combustion Preparation of CuO/ZnO/Al$_2$O$_3$ Nanocatalyst Used in Fuel Cell-Grade Hydrogen Production from Methanol: Effect of Sonication and Fuel/Nitrate Ratio. *Part. Sci. Technol.* **2018**, *36*, 217–225. [CrossRef]
21. Basile, A.; Parmaliana, A.; Tosti, S.; Iulianelli, A.; Gallucci, F.; Espro, C.; Spooren, J. Hydrogen Production by Methanol Steam Reforming Carried out in Membrane Reactor on Cu/Zn/Mg-Based Catalyst. *Catal. Today* **2008**, *137*, 17–22. [CrossRef]
22. Huang, Y.H.; Wang, S.F.; Tsai, A.P.; Kameoka, S. Reduction Behaviors and Catalytic Properties for Methanol Steam Reforming of Cu-based Spinel Compounds CuX$_2$O$_4$ (X=Fe, Mn, Al, La). *Ceram. Int.* **2014**, *40*, 4541–4551. [CrossRef]
23. Bichon, P.; Asheim, M.; Sperle, A.J.T.; Fathi, M.; Holmen, A.; Blekkan, E.A. Hydrogen from Methanol Steam-Reforming over Cu-based Catalysts with and without Pd Promotion. *Int. J. Hydrog. Energy* **2007**, *32*, 1799–1805. [CrossRef]
24. Kameoka, S.; Okada, M.; Tsai, A.P. Preparation of a Novel Copper Catalyst in Terms of the Immiscible Interaction between Copper and Chromium. *Catal. Lett.* **2008**, *120*, 252–256. [CrossRef]
25. Sato, S.; Takahashi, R.; Sodesawa, T.; Yuma, K.I.; Obata, Y. Distinction between Surface and Bulk Oxidation of Cu through N$_2$O Decomposition. *J. Catal.* **2000**, *196*, 195–199. [CrossRef]
26. Hwang, B.Y.; Sakthinathan, S.; Chiu, T.W. Production of Hydrogen from Steam Reforming of Methanol Carried out by Self-Combusted CuCr$_{1-x}$Fe$_x$O$_2$ (x=0–1) Nanopowders Catalyst. *Int. J. Hydrog. Energy* **2019**, *44*, 2848–2856. [CrossRef]
27. Amrute, A.P.; Larrazábal, G.O.; Mondelli, C.; Pérez-Ramírez, J. CuCrO$_2$ delafossite: A Stable Copper Catalyst for Chlorine Production. *Angew. Chem. Int. Ed.* **2013**, *52*, 9772–9775. [CrossRef]
28. Ketir, W.; Bouguelia, A.; Trari, M. NO$_3^-$ Removal with a New Delafossite CuCrO$_2$ Photocatalyst. *Desalination* **2009**, *244*, 144–152. [CrossRef]
29. Singh, S. Cerium Oxide Based Nanozymes: Redox Phenomenon at Biointerfaces. *Biointerphases* **2016**, *11*, 04B202. [CrossRef]
30. Hornés, A.; Hungría, A.B.; Bera, P.; López Cámara, A.; Fernández-García, M.; Martínez-Arias, A.; Barrio, L.; Estrella, M.; Zhou, G.; Fonseca, J.J.; et al. Inverse CeO$_2$/CuO Catalyst as an Alternative to Classical Direct Configurations for Preferential Oxidation of CO in Hydrogen-Rich Stream. *J. Am. Chem. Soc.* **2010**, *132*, 34–35. [CrossRef]
31. Li, Y.; Cai, Y.; Xing, X.; Chen, N.; Deng, D.; Wang, Y. Catalytic Activity for CO Oxidation of Cu-CeO$_2$ Composite Nanoparticles Synthesized by a Hydrothermal Method. *Anal. Methods* **2015**, *7*, 3238–3245. [CrossRef]
32. Tucker, N.; Stanger, J.J.; Staiger, M.P.; Razzaq, H.; Hofman, K. The History of the Science and Technology of Electrospinning from 1600 to 1995. *J. Eng. Fibers Fabr.* **2012**, *7*, 63–73. [CrossRef]
33. Thavasi, V.; Singh, G.; Ramakrishna, S. Electrospun Nanofibers in Energy and Environmental Applications. *Energy Environ. Sci.* **2008**, *1*, 205–221. [CrossRef]
34. Chao, T.C.; Chiu, T.W.; Fu, Y. Fabrication and Characteristic of Delafossite-Type CuFeO$_2$ Nanofibers by Electrospinning Method. *Ceram. Int.* **2018**, *44*, S80–S83. [CrossRef]
35. Huang, X.; Ma, L.; Wainwright, M.S. The influence of Cr, Zn and Co additives on the performance of skeletal copper catalysts for methanol synthesis and related reactions. *Appl. Catal. A Gen.* **2004**, *257*, 235–243. [CrossRef]

Article

Magnetic and Magnetocaloric Properties of Nano- and Polycrystalline Manganites La$_{(0.7-x)}$Eu$_x$Ba$_{0.3}$MnO$_3$

Roman Atanasov [1], Rares Bortnic [1], Razvan Hirian [1], Eniko Covaci [2], Tiberiu Frentiu [2], Florin Popa [3] and Iosif Grigore Deac [1,*]

1. Faculty of Physics, Babes-Bolyai University, Str. Kogalniceanu 1, 400084 Cluj-Napoca, Romania
2. Faculty of Chemistry and Chemical Engineering, Babes-Bolyai University, Str. Arany Janos 11, 400028 Cluj-Napoca, Romania
3. Materials Science and Engineering Department, Technical University of Cluj-Napoca, Blvd. Muncii 103-105, 400641 Cluj-Napoca, Romania
* Correspondence: iosif.deac@phys.ubbcluj.ro

Citation: Atanasov, R.; Bortnic, R.; Hirian, R.; Covaci, E.; Frentiu, T.; Popa, F.; Deac, I.G. Magnetic and Magnetocaloric Properties of Nano- and Polycrystalline Manganites La$_{(0.7-x)}$Eu$_x$Ba$_{0.3}$MnO$_3$. *Materials* 2022, 15, 7645. https://doi.org/10.3390/ma15217645

Academic Editor: Daniela Iannazzo

Received: 7 October 2022
Accepted: 19 October 2022
Published: 31 October 2022

Publisher's Note: MDPI stays neutral with regard to jurisdictional claims in published maps and institutional affiliations.

Copyright: © 2022 by the authors. Licensee MDPI, Basel, Switzerland. This article is an open access article distributed under the terms and conditions of the Creative Commons Attribution (CC BY) license (https://creativecommons.org/licenses/by/4.0/).

Abstract: Here, we report synthesis and investigations of bulk and nano-sized La$_{(0.7-x)}$Eu$_x$Ba$_{0.3}$MnO$_3$ (x ≤ 0.4) compounds. The study presents a comparison between the structural and magnetic properties of the nano- and polycrystalline manganites La$_{(0.7-x)}$Eu$_x$Ba$_{0.3}$MnO$_3$, which are potential magnetocaloric materials to be used in domestic magnetic refrigeration close to room temperature. The parent compound, La$_{0.7}$Ba$_{0.3}$MnO$_3$, has Curie temperature T_C = 340 K. The magnetocaloric effect is at its maximum around T_C. To reduce this temperature below 300 K, we partially replaced the La ions with Eu ions. A solid-state reaction was used to prepare bulk polycrystalline materials, and a sol-gel method was used for the nanoparticles. X-ray diffraction was used for the structural characterization of the compounds. Transmission electron spectroscopy (TEM) evidenced nanoparticle sizes in the range of 40–80 nm. Iodometry and inductively coupled plasma optical emission spectrometry (ICP-OES) was used to investigate the oxygen content of the studied compounds. Critical exponents were calculated for all samples, with bulk samples being governed by tricritical mean field model and nanocrystalline samples governed by the 3D Heisenberg model. The bulk sample with x = 0.05 shows room temperature phase transition T_C = 297 K, which decreases with increasing x for the other samples. All nano-sized compounds show lower T_C values compared to the same bulk samples. The magnetocaloric effect in bulk samples revealed a greater magnetic entropy change in a relatively narrow temperature range, while nanoparticles show lower values, but in a temperature range several times larger. The relative cooling power for bulk and nano-sized samples exhibit approximately equal values for the same substitution level, and this fact can substantially contribute to applications in magnetic refrigeration near room temperature. By combining the magnetic properties of the nano- and polycrystalline manganites, better magnetocaloric materials can be obtained.

Keywords: manganites; nanoparticle perovskites; crystallography; magnetic behavior; phase transition; critical behavior; magnetocaloric effect

1. Introduction

The search for more efficient refrigeration methods has been ongoing ever since humans looked at the snowy peaks on the mountains from under the blistering sun [1]. Although the desire for a cooled environment and long-lasting food was ever present, no significant progress was made until the advent of electricity [2]. Then, vapor compression cooling systems became dominant as refrigeration became prevalent. However, it has been proven that such refrigeration is harmful to the environment; hence, new ways must be found and researched [2].

A good candidate for such a new method is the use of the magnetocaloric effect in, for example, intermetallic compounds of Gadolinium (Gd) [3], where the efficiency of the Carnot cycle can reach 60% [3], whereas in the conventional gas compression method (CGC), it is only about 5–10% [4]. However, since Curie temperatures of Gd alloys are lower (276 K) than that of Gd (294 K) [3], several other candidates have been investigated [5].

In addition, manganites of the type $A_{1-x}B_xMnO_3$ (where A is a trivalent rare earth cation and B is a divalent alkaline earth cation [6] are known for their colossal magnetoresistance (CMR) effect, which is at a maximum close to the Curie temperature T_C [7]. This effect refers to when a transition from insulator to metal occurs at a temperature denoted by T_p. The sharp transition from ferromagnetic to paramagnetic phase at T_C is important for a high magnetic entropy change. The two temperatures T_C and T_P are close to each other depending on the size of the domain walls; the larger the wall, the bigger the distance between them, which requires more energy to "flip" the orientation of the neighboring domain [8]. In nanoparticles, the sizes of the particles vary and the disconnection between them causes the change in magnetic entropy to be more gradual and smaller in magnitude [9].

Besides the CMR effect, the most important property of such materials is the large magnetic entropy change which occurs when the external magnetic field varies in some compounds. In recent times, rare earth manganites such as $La_{1-x}Sr_xMnO_3$ and $Pr_{1-x}Ba_xMnO_3$ [10,11] have been of increasing interest for exhibiting such large entropy changes.

The optimal ratio of doping in samples (such as the parent sample for this study: $La_{1-x}Ba_xMnO_3$) is x = 0.3, where the ratio of Mn^{3+} and Mn^{4+} allows for the optimal double exchange process [8].

Magnetic and electrical behavior of these types of compounds depend on the preparation method (which influences the domain wall size), the ratio of Mn^{3+}/Mn^{4+} ions, and the size difference between the rare earth element and the alkali metal [4,7,9]. In the case where the mismatch is great, as in the case with La and Ba, a separation between T_C and T_P is observed, and in addition, the magnetic entropy change close to T_C is sharp because of strong spin-lattice coupling, which is a good sign for a high magnetocaloric effect [10]. In cases where the difference is relatively small, as with La and Ca, the grain boundary is smaller and the distance between T_C and T_p is also smaller [8]. The number of La^{3+} ions affects the critical exponents, and in the case of $(La,Ba)MnO_3$, they correspond to the short-range Heisenberg model [11].

In this paper, the critical and magnetocaloric behaviors of $La_{(0.7-x)}Eu_xBa_{0.3}MnO_3$ (where x = 0.05, 0.1, 0.2, 0.3, 0.4) in bulk material and nano-sized particles are discussed. $La_{0.7}Ba_{0.3}MnO_3$ has a large magnetic entropy change at 340 K. It has been shown that substitution of Eu in place of La atoms in $La_{0.7}Sr_{0.3}MnO_3$ samples leads to lowering of T_C [12,13] below room the temperature, where the magnetocaloric effect could be important for domestic cooling applications. As a result, the smaller ionic radius of an Eu atom was chosen for this study as a substitute for La in order to promote higher disorder and to manipulate the values of T_C. Bulk compounds were prepared by solid-state reaction method, and nanoparticles were made with a modified sol-gel method. All samples have crystal structures belonging to the Rhombohedral (R-3c) symmetry group. Magnetic critical behavior analysis revealed that bulk samples are governed by the tricritical mean field model, while the nano-samples are governed by the 3D Heisenberg model. It was found that the relative cooling power increased with the level of doping, while the magnetic entropy change vs. temperature graphs were the sharpest in the sample with x = 0.05.

The paper is organized as follows. In Section 2, we describe the preparation routes for the bulk polycrystalline and for nano-sized samples, as well as the methods we used to characterize them from structural, morphological, oxygen stoichiometric, electrical, and magnetic perspectives. In Section 3, we present the results of our investigations and the analyses of the obtained data. We also discuss the critical magnetic behavior and the magnetocaloric effect of the samples. Finally, Section 4 summarizes the conclusions resulting from this study.

2. Materials and Methods

The bulk samples were prepared by solid-state reaction. Precursors, consisting of oxides La_2O_3 (99.9%), Eu_2O_3 (99.99%), MnO_2 (99.9%), and carbonate $BaCO_3$ (99.9%) from Alfa Aesar, were mixed by hand in an agate mortar using a pestle for approximately 3 h each. The mixed powder was then calcinated at 1100 °C for 24 h in air. After that, the samples were pressed at 3 tons into pellets and sintered at 1350 °C for 30 h in air to produce 2 g samples.

The nano-sized samples were prepared with the sol-gel method. Nitrates of La (99.9%), Eu (99.9%), Ba (99%), and Mn (98%) from Alfa Aesar were used as precursors. They were dissolved in pure water at 60 °C for 45–60 min, after which 10 g of sucrose of 99% purity was added. The mixture was stirred for another 45 min to allow for positive ions to attach to the sucrose chain. The temperature was then reduced and pectin was added 20 min before the end of mixing in order to expand the xero-gel. The mixture was dried in a sand bath for 24 h and then placed in a high-air flow oven at 1000 °C for 2 h.

Both systems were structurally categorized using X-ray diffraction (XRD), and the data were analyzed using the FULLPROF Rietveld refinement technique. Scanning electron microscopy (SEM) was used to determine grain sizes along with the Williamson–Hall method for analyzing XRD data. EDX was used to confirm sample stoichiometry. Transmission Electron Microscopy (TEM) was used for determining the average size of nanoparticles.

Oxygen stoichiometry was determined by iodometric analytical titration and with inductively coupled plasma atomic emission spectroscopy (ICP-OES). In iodometry, an amount of the sample is placed in hydrochloric (HCl) acid, in which only Mn positive ions react with negative ions of Cl to produce Cl_2. The gas is then pushed by nitrogen into another vessel containing potassium iodide. Iodine molecules are formed as a result, and the solution is titrated with sodium thiosulfate. Then, the ratio of Mn^{3+} and Mn^{4+} is calculated.

Electrical properties were measured using the four-point technique in a cryogen-free superconducting setup. Four-point chips, measuring voltage and current separately, were placed in a range of temperature between 10 K and 300 K in applied magnetic fields of up to 7 T.

Magnetic measurements were made using a Vibrating Sample Magnetometer (VSM) in the range of 4–300 K and in magnetic fields of up to 4 T.

3. Results

3.1. Structural Analysis

X-ray diffraction patterns show that all of the samples are single phase. The amount of impurities was smaller than 5% in all of the samples. A shift in 2θ to the right with increasing substitution of Eu indicates smaller cell dimensions. The patterns for nanocrystalline samples exhibit wider peaks due to their sizes. Figure 1 shows stacked XRD patterns for bulk and nanocrystalline samples, respectively.

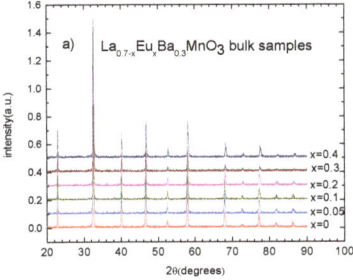
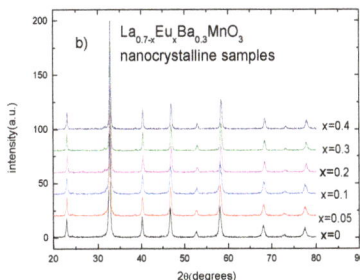

Figure 1. X-ray diffraction patterns for (**a**) $La_{0.7-x}Eu_xMnO_3$ polycrystalline bulk samples and (**b**) $La_{0.7-x}Eu_xMnO_3$ nano-sized samples.

Rietveld refinement analysis confirms the rhombohedral structure of the parent compound for each studied sample with an R-3c space group. Figure 2 presents the values for lattice parameter (*a*) and cell volume (V) for all of the samples as a function of Eu content (x). As observed, with increasing substitution level, the lattice parameter becomes smaller. This is due to Eu^{3+} ions having a smaller ionic radius (1.206 Å) than La^{3+} ions (1.5 Å) [14,15]. In turn, this changes the Mn-O-Mn angle, creating distortion in the Mn-O octahedral [14].

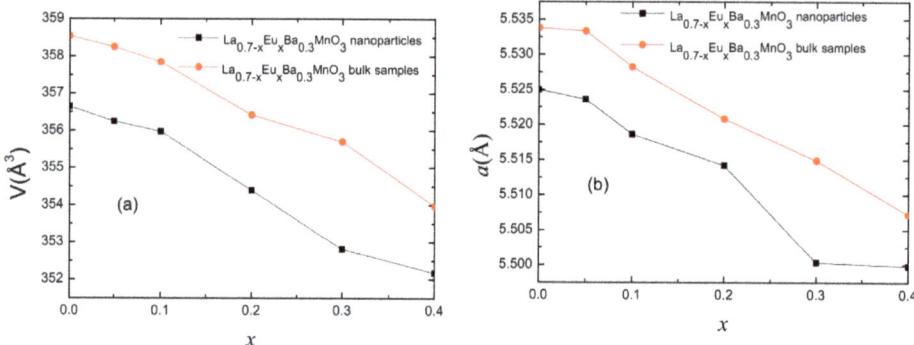

Figure 2. (**a**) Plot of cell volume for polycrystalline bulk and nano-sized samples. (**b**) Plot of cell parameter "*a*" for polycrystalline bulk and nano-sized samples.

In order to understand the stability of these structures, the Goldschmidt tolerance factor was calculated using following relation [7,14]:

$$t = \frac{R_A + R_0}{\sqrt{2}(R_B + R_0)}, \quad (1)$$

where R_A is the radius of A cation, R_B is the radius of B cation, and R_0 is the radius of the anion.

It should be noted that with an increasing Eu ion content in the samples, the tolerance factor decreases slightly but remains consistent with keeping an orthorhombic/rhombohedral structure. Eu ions cause a decrease in R_A and an increase in disorder [14]. In turn, this will decrease orbital overlap and the band gap. The values of the tolerance factor are within the values for an orthorhombic/rhombohedral structure [15]. The angle of Mn-O-Mn bonds increased in the parent samples for both bulk and nanocrystalline samples from 167° to 169° for x = 0.05. Furthermore, as shown in Tables 1 and 2, the bond length of Mn-O diminishes with each additional substitution.

Table 1. Calculated tolerance factors, Mn-O lengths, and crystallite sizes for nanocrystalline samples using the Williamson–Hall and Rietveld methods, including strain values.

Eu Content (Nano)	*t* (Tolerance Factor)	Mn-O (Å)	Williamson–Hall Size (nm)	Average Rietveld Size (nm)	Strain
x = 0	0.997	1.962	36.23	18.14	0.0023
x = 0.05	0.992	1.959	41.46	20.62	0.0024
x = 0.1	0.987	1.958	29.15	21.27	0.0019
x = 0.2	0.976	1.956	54.61	29.03	0.0023
x = 0.3	0.966	1.953	49.95	31.57	0.0019
x = 0.4	0.956	1.952	45.74	34.37	0.0017

Table 2. Calculated Mn-O lengths and crystallite sizes for polycrystalline bulk samples using the Williamson–Hall and Rietveld methods, including strain values.

Eu Content (Bulk)	Williamson–Hall Size (nm)	Mn-O (Å)	Average Rietveld Size (nm)	Strain
x = 0	110.05	1.966	111.1	0.0017
x = 0.05	110.04	1.963	435.97	0.0018
x = 0.1	172.03	1.962	238.22	0.0017
x = 0.2	146.36	1.959	324.41	0.002
x = 0.3	128.38	1.958	246.71	0.0016
x = 0.4	106.05	1.955	144.42	0.0016

The Williamson–Hall (W-H) method [16] for determining crystallite sizes was used for both systems. Table 1 shows the calculation for nanocrystalline samples with an average size range of 30–55 nm for the crystallites. This agrees with the results of TEM investigation. Pictured in Figure 3, TEM shows that the average size of the particles is about 50 nm, varying between 30 nm and 70 nm. Widely used Scherrer size calculations do not take into account the strain between the grains, and thus, they tend to be lower in value. According to Williamson–Hall calculations, the size varies from 106 nm to 172 nm, while scanning electron microscopy (SEM) shows the grain size to be 3–10 µm. The same can be observed for Rietveld crystallite size results; although they are bigger than from the WH method, they are still smaller than the SEM results. This can be attributed to the fact that a single grain contains several crystallites.

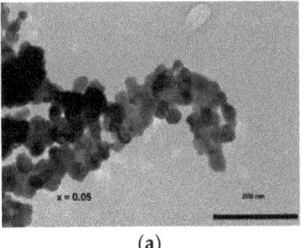
(a)
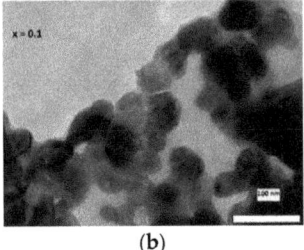
(b)
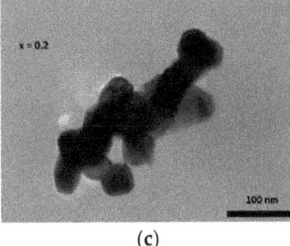
(c)

Figure 3. Selected TEM pictures for $La_{1-x}Eu_xMnO_3$ nano-sized samples for x = 0.05 (a), x = 0.1 (b), x = 0.2 (c).

Energy dispersive X-ray spectroscopy (EDX) was also carried out for the bulk samples. As seen in Figure 4b, where a typical example is presented, the stoichiometry of heavy elements (including lanthanum, europium, barium, and manganese) is in very good agreement with the theoretical values. The distribution of the elements on the surface is considerably uniform. It is good to note here that oxygen is a much smaller atom and does not interact with X-rays nor heavier atoms [17] (pp. 279–307). For that reason, iodometry was implemented as a reliable way to calculate oxygen content.

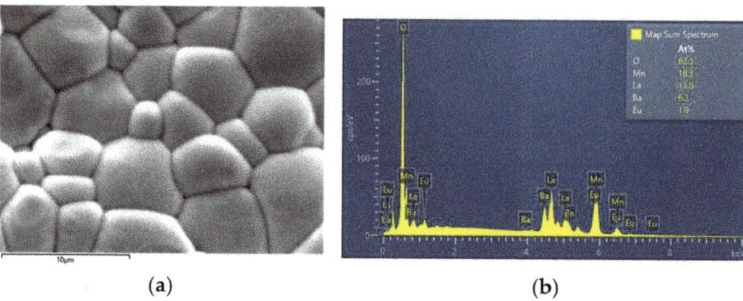

Figure 4. (a) SEM picture of the $La_{0.65}Eu_{0.05}Ba_{0.3}MnO_3$ sample. (b) EDX for the $La_{0.65}Eu_{0.05}Ba_{0.3}MnO_3$ sample.

3.2. Oxygen Content

Iodometric titration and inductively coupled plasma atomic emission spectroscopy were used to study the oxygen content of the bulk samples.

Iodometry is a reliable and popular method of determining oxygen content in manganites [18], as it involves direct measurement of the fraction of Mn^{3+} vs. Mn^{4+} ions. In this study, all of the samples exhibited an excess of Mn^{3+} content. This could be attributed to the oxygen deficiency during calcination and sintering. An average of 75% Mn^{3+} would result in an average oxygen content of $O_{2.97}$ in the range of 2.96–2.99 [19], which would affect its electrical and magnetic properties [19]. The experiment showed acceptable dispersion and error. Standard deviation is a measure of dispersion of data values, or how close they are to the "mean" value, while relative standard deviation is the percentage value of the standard deviation around the "mean". In our study, the relative standard deviation did not exceed 2.7%, showing close proximity to the mean. Results are shown in Table 3.

Table 3. Average oxygen content calculated using iodometry and inductively coupled plasma optical emission spectrometry (ICP-OES).

Eu Content	Average Mn^{3+} Ratio	Standard Deviation	Relative Standard Deviation (%)	Average Oxygen Content	ICP-OES
x = 0	0.7306	0.0159	2.18	$O_{2.98\pm0.02}$	$O_{2.94\pm0.14}$
x = 0.05	0.7257	0.0083	1.14	$O_{2.99\pm0.01}$	$O_{2.93\pm0.13}$
x = 0.1	0.7032	0.0189	2.69	$O_{2.99\pm0.02}$	$O_{2.99\pm0.13}$
x = 0.2	0.7282	0.0136	1.87	$O_{2.98\pm0.02}$	$O_{3.13\pm0.15}$
x = 0.3	0.7565	0.0076	0.99	$O_{2.97\pm0.01}$	$O_{3.03\pm0.15}$
x = 0.4	0.7612	0.0138	1.81	$O_{2.97\pm0.02}$	$O_{3.13\pm0.17}$

The results of iodometry were confirmed with inductively coupled plasma optical emission spectroscopy (ICP-OES) [20]. The process involves passing of the elements through a plasma of argon, which causes excitation and emission of specific wavelengths of light. The results are presented in Table 3. The average oxygen content according to ICP-OES falls well within the error limit of the iodometry results.

3.3. Electrical Measurements

An investigation of electrical resistivity at 0, 1, 2 T was carried out using the four-point probe method. The graphs are shown in Figure 5. As can be seen, the sample with x = 0.3 exhibits an expected behavior typical of a ferromagnetic manganite, with a maximum at T_p. The samples with x = 0–0.2 have similar behaviors, while the sample with x = 0.4 exhibits semi-conducting behavior. The results for Curie temperatures (T_C, as obtained from magnetic measurements), T_p, the values of the resistivities in the absence of a magnetic field (ϱ_{peak}, at 0 T), and the values of magnetoresistance (MR) at T_p are presented in Table 4. Magnetoresistance was calculated using the following formula [16]:

$$MR\% = [(\varrho(H) - \varrho(0))/\varrho(0)] \times 100, \qquad (2)$$

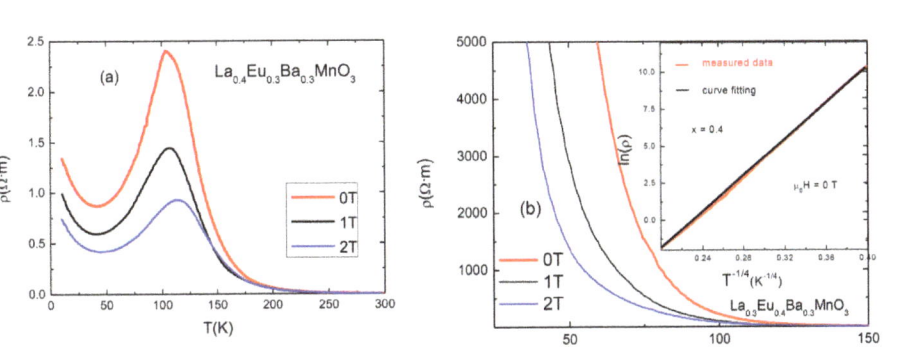

Figure 5. Resistivity vs. temperature graphs for (**a**) x = 0.3 and (**b**) x = 0.4. The inset shows the fitting of $\ln(\varrho)$ as a function of $T^{-1/4}$ for $\mu_0 H = 0$ T.

Table 4. Experimental values for $La_{0.7-x}Eu_xBa_{0.3}MnO_3$ bulk materials: electrical properties.

Compound (Bulk)	T_C (K)	T_P (K)	ϱ_{peak} (Ωcm) in 0 T	MR_{Max} (%) (1 T)	MR_{Max} (%) (2 T)
$La_{0.7}Ba_{0.3}MnO_3$	340	295	0.693	5.8	12.9
$La_{0.65}Eu_{0.05}Ba_{0.3}MnO_3$	297	256	0.812	4.2	11.8
$La_{0.6}Eu_{0.1}Ba_{0.3}MnO_3$	270	220	0.084	32.9	52.6
$La_{0.5}Eu_{0.2}Ba_{0.3}MnO_3$	198	165	21.753	22.7	42.1
$La_{0.4}Eu_{0.3}Ba_{0.3}MnO_3$	142	103	240.455	40.4	63.6
$La_{0.3}Eu_{0.4}Ba_{0.3}MnO_3$	99	-	100×10^9	-	-

The first observation to be made here is that the T_p metallic–insulator transition temperature for each sample at 0 T magnetic field is lower than its T_C; for example, T_C (x = 0.05) = 297 K, and T_p (x = 0.05) = 256 K. This is due to the effect of grain boundaries which act as a semiconductor (or insulator) pushing the inter-grain coupling to lower temperatures [21]. When an external magnetic field is applied, the peak shifts to the right, increasing the conductive properties of the sample. This can be attributed to the lowering of spin fluctuations and to the delocalization of charge carriers caused by the applied magnetic field, which improves the double exchange interaction [22].

Each addition of a smaller-sized Eu^{3+} ion in place of an La^{3+} ion causes disorder [7]. It also changes the angle between Mn-O-Mn by "pulling" oxygen towards the A-site [23]. A decrease in the angle changes the overlap of the electron orbital, which reduces the hopping amplitude of the electrons and causes them to be more localized. This can be observed in the systematic lowering of the T_p of samples with an increasing level of substitution. For high Eu content (x = 0.4), the disorder and the reduced value of the Mn-O-Mn angle resulted in a semiconductor-like behavior of the electrical conductivity of the compound.

From the values for resistivity (ϱ_{peak}) and magnetoresistance (MR_{Max}) in Table 4, it is eviden, that both of them tend to increase with increasing substitution level of Eu. The maximum observed resistivity for the x = 0.3 sample is 240 Ω·cm and (MR_{Max}) (2 T) = 63.6%, while the resistivity increases to several orders of MΩ·m for x = 0.4.

The samples with x < 0.4 have typical CMR resistivity behavior as a function of temperature and applied magnetic field, as can be seen in Figure 5, except in the low-temperature region where an upturn of resistivity occurs. The analysis of these temperature dependences is usually made both for the metallic regime before metal–insulator transition (MIT) and for the semiconducting behavior in the range of higher temperatures [24]. Within the metallic region behavior, the dominant scattering phenomena are electron-electron and electron–magnon [24] (pp. 21–32) with $\varrho = \varrho_0 + \varrho_2 T^2 + \varrho_{4.5} T^{4.5}$. At higher temperatures, after MIT, the resistivity shows semiconductor behavior and its temperature dependence can be described by using the variable range hopping (VRH) and small polaron hopping (SPH) models [25].

For the semiconducting sample, with x = 0.4 (Figure 5b), the best fit for the resistivity behavior is the expression corresponding to the VRH model for a three-dimensional system: $\varrho(T) = \varrho_0 \exp(T_0/T)^{0.25}$ (as shown in the inset of Figure 5b, for $\mu_0 H = 0$ T), where ϱ_0 is the prefactor and T_0 is a characteristic temperature which is related to the density of states at the Fermi level and to the localization length.

It is interesting that in spite of the semiconducting behavior, this sample shows CMR properties. This behavior suggests an electrical conduction mechanism which takes place (by tunneling) between isolated manganite grains which have negative magnetoresistance.

The behavior of resistivity at temperatures below T_p is of interest in this study. A minimum in resistivity can be observed at around 30–50 K before resistivity increases again. This behavior is exhibited by all samples except the one with the highest amount of Eu, where resistivity increases drastically below T_p. In the literature, the minimum in resistivity was partially attributed to Kondo-like effects [14]. These minima are caused by small magnetic impurities which localize electrons of opposite spin, thus increasing the scattering of conduction electrons. However, this scenario is quite different from that of polycrystalline manganites with different size grains separated by (disordered matter) grain boundaries; this rules out the hypothesis of the Kondo effect [26].

The upturn can be better explained by the combined effect of electron–electron interactions, electron–phonon scattering, and weak localization [24,27]. In addition, the disorder and strain from the grain boundaries can also act as supplementary localization factors of charge carriers. Besides these, the electrical conductivity of the grain boundaries depreciates with decreasing temperature, leading to increased resistivity. The upturn in the thermal dependence of resistivity is a consequence of both intrinsic (intragrain) effects and extrinsic grain boundary scattering/tunneling effects [28], as was also found in some other polycrystalline complex transition metal oxides. The sample with the highest Eu content exhibits a continual increase in resistivity below T_p, which can be explained by the size and quality of the grain boundaries. It is evident that grain boundaries play a dominant role in the electrical behavior of the samples.

3.4. Magnetic Properties

All of the samples exhibit ferromagnetic–paramagnetic transitions. Typical and selected magnetization vs. temperature ($M(T)$) plots are presented in Figure 6. Curie temperatures can be calculated from the derivative of the magnetization with respect the temperature, with the inflection point corresponding to T_c, which can be seen in Tables 5–7 [29]. With increasing Eu content x, the values of T_c gradually and systematically become lower as a result of the induced disorder. The sample with x = 0.05 has Curie temperature $T_C = 297$ K.

Nano-scale particles also show ferromagnetic behavior, but their T_C values are lower compared to the equivalent bulk material, as shown in Figure 6. This is due to the size of particles and their "surface effects" which occur as a result of a large surface-to-volume ratio [30], i.e., disorder effects in the surface layer of the particles which contain an increased number of broken chemical bond and other defects, resulting in spin canting and reducing the magnetic moment of the particles [31]. An average difference in T_C values is 70–90 K for the samples with the same substitution level: T_C (x = 0.1 bulk) = 270 K and

T_C (x = 0.1 nano) = 200 K. It is also interesting that the slope of the magnetization change of the $M(T)$ curves in nanoparticles is much lower than in bulk, which can be attributed to the distribution of the particle sizes within the samples. In general, bulk material has higher values of magnetization and a narrower temperature range of the magnetic phase transition, as can be seen in Figure 6.

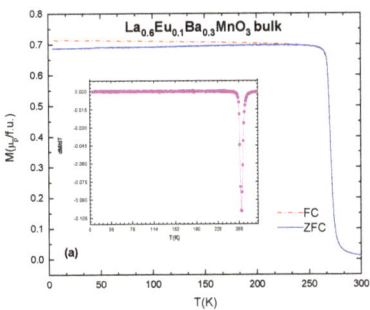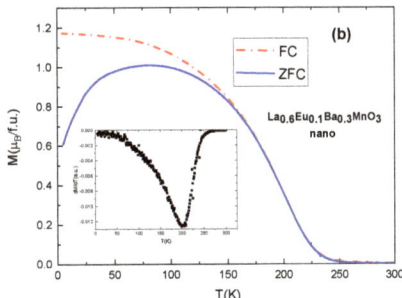

Figure 6. (**a**) ZFC-FC curves and derivative of magnetization (in inset) for the bulk sample with x = 0.1; (**b**) ZFC-FC curves and derivative (in inset) for nanocrystalline sample with x = 0.1.

Arrott plots (M^2 vs. H/M) allow the determination of the magnetic phase transition order [32]. According to the Banerjee criterion, positive and negative slopes of the curves correspond to second- and first-order magnetic phase transitions, respectively [28]. All of the samples exhibit positive slopes for these curves, indicating second-order magnetic phase transitions. A selected sample of Arrott plots are presented in Figure 7.

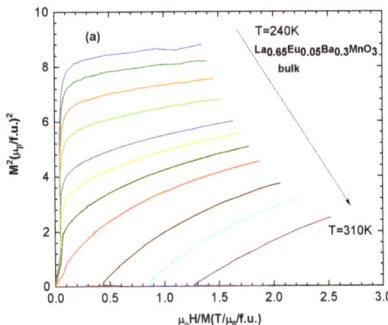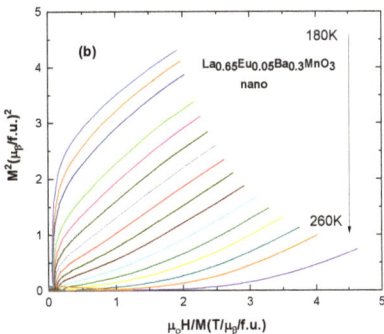

Figure 7. Arrott plot (M^2 vs. H/M) for (**a**) the bulk sample with x = 0.05 and for (**b**) the nanocrystalline sample with x = 0.05.

Arrott plots are based on Landau's mean field theory, [33]. The Gibbs free energy around a critical point is defined as

$$G(T,M) = G_O + MH + aM^2 + bM^4 + \ldots, \tag{3}$$

where a and b are coefficients which depend on temperature. Minimizing the Gibbs free energy with respect to magnetization, we obtain

$$H/M = 2a + 4bM^2, \tag{4}$$

According to Gibbs free energy equations, the isotherm lines must be parallel and straight, but that is not observed in Arrott plots. The problem lies in inexactness of the critical exponents [34]. β relates to spontaneous magnetization and α is related to the inverse of susceptibility χ [34,35].

These equations can be generalized as follows [34]:

$$M_S(T) = M_0 \, (-\varepsilon)^\beta, \; T < T_C, \quad (5)$$

$$\chi^{-1}(T) = \left(\frac{h_0}{M_0}\right)\varepsilon^\gamma, \; T > T_C, \quad (6)$$

$$M = D \, (\mu_0 H^{1/\delta}), \; T = T_C, \quad (7)$$

where ε is the reduced temperature $(T - T_c)/T_c$ and M_0, h_0/M_0, and D are critical amplitudes.

For mean field theory, we have $\gamma = 1$, $\beta = 0.5$, and $\delta = 3$. For the 3D Heisenberg model, we have $\gamma = 1.366$, $\beta = 0.355$, and $\delta = 4.8$. For the Ising model, we have $\gamma = 1.24$, $\beta = 0.325$, and $\delta = 4.82$. For the tricritical mean field model, we have $\gamma = 1$, $\beta = 0.25$, and $\delta = 5$ [24,36].

The modified Arrott plot method [35] is an iterative method. It begins with an Arrott–Noakes plot ($M^{1/\beta}$ vs. $\mu_0 H / M^{1/\gamma}$), which involves finding proper exponents which will make the lines parallel and straight to determine the β and γ critical exponents [10]. Spontaneous magnetization $M_S(T,0)$ is found from the intercepts of isotherms with the ordinate of the plot. The inverse of the susceptibility $\chi_0^{-1}(T)$ is taken from the intercept with the abscissa. Further fitting of these values into Equations (6)–(8) refines the values to those very close to real ones. The Widom scaling relation $\beta + \gamma = \beta \delta$ [10] gives the value of δ. Selected graphs for modified Arrott plots are presented in Figure 8; the data are presented in full in Table 5.

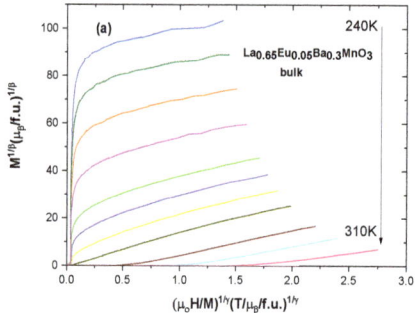
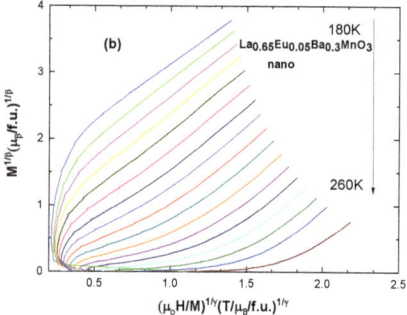

Figure 8. Modified Arrott plots for (**a**) the bulk sample with x = 0.05 and for (**b**) the nanocrystalline sample with x = 0.05.

Figure 9 shows the values for critical exponents for two of the samples: one of bulk and one of nano-scale. It is evident that the bulk samples are more closely governed by the tricritical mean field model (β = 0.234, γ = 0.915, δ = 4.9) for ferromagnets and the nanoparticle samples are governed by the 3D Heisenberg model (β = 0.548, γ = 1.968, δ = 4.589), rather than mean field theory model.

Usually, the 3D Heisenberg model can describe the critical properties of the short-range interactions in doped manganites, together with other theoretical models, such as the mean field and tricritical mean field models. The critical exponent values are related to the range of exchange interaction J(r), spin, and system dimensionality. Within renormalization group theory [37], $J(r) = 1/r^{d+\sigma}$ (d—dimensionality of the system; σ—range of interaction). For σ greater than 2, the 3D Heisenberg model is valid. For σ < 3/2, the mean field theory of long-range interaction is valid. For an intermediate range, a different universality class occurs. For the tricritical point, the critical exponents are universal: β = 0.25, γ = 1, and δ = 5. The tricritical point sets a boundary between two different ranges of order phase transitions. In this study, we focused on the general different magnetic properties of nano- and bulk polycrystalline manganites, not on their high-detail magnetic critical behavior. The modified Arrott plot (MAP) method is very accurate, and its agreement with the Kouvel–Fisher (KF) method and critical isotherm (CI) plots is usually quite remarkable [24,38]; indeed, some authors only analyze these data [39]. This is why we report here the results obtained by using MAP analysis.

Table 5. Critical exponent values for all samples.

Compound		γ	β	δ	T_c (K)
x = 0	bulk	1.065	0.288	4.69	340
x = 0.05	bulk	0.915	0.234	4.9	297
x = 0.1	bulk	1.07	0.24	5.45	270
x = 0.2	bulk	0.976	0.246	4.967	198
x = 0.3	bulk	0.933	0.255	4.659	142
x = 0.4	bulk	1.022	0.249	5.104	99
x = 0	nano	1.823	0.493	4.698	263
x = 0.05	nano	1.968	0.548	4.589	220
x = 0.1	nano	1.867	0.521	4.584	200
x = 0.2	nano	1.755	0.477	4.679	136
x = 0.3	nano	1.931	0.537	4.596	90
x = 0.4	nano	1.789	0.512	4.49	64
Mean field model		1	0.5	3	
3D Heisenberg model		1.366	0.355	4.8	
Ising model		1.24	0.325	4.82	
Tricritical mean field model		1	0.25	5	

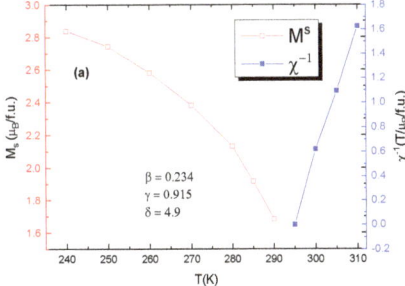
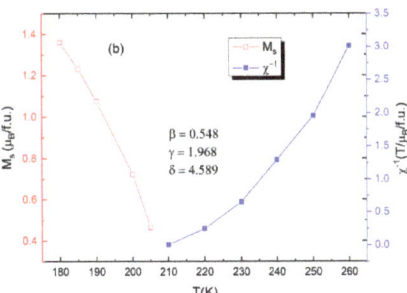

Figure 9. Calculated values for critical exponents for (**a**) the bulk sample with x = 0.05 (**b**) for the nanocrystalline sample with x = 0.05.

All of the samples exhibit a very small coercive field as evident from hysteresis curves, with the largest values of 170 Oe for x = 0.05 bulk and 960 Oe for x = 0.4 nano-sample. This can be largely attributed to low anisotropy and lack of pinning sites [40]. Nanoparticles show greater coercivity than their bulk counterparts, as shown in Tables 6 and 7. It has been found, in previous works, that nanoparticle coercivity tends to increase with a decreasing size in the multi-domain range, and then it decreases in the single-domain range until it reaches a superparamagnetic state, when it becomes zero [41]. Low coercivity is of utmost importance for the magnetocaloric effect.

Suitability for cooling via the magnetocaloric effect is determined by the value of magnetic entropy change in the material [40]. An appropriate ratio between Mn^{3+} and Mn^{4+} is needed for optimal ferromagnetic behavior. About 30% Mn^{4+} gives the best results [8]. When the number of charge carries increases, the risk of entering a charge ordered state occurs, where electron hopping is prohibited by rigid atomic distribution [8]. For second-order phase transitions (which were established via Arrott plots earlier), magnetic entropy change (ΔS_M) can be calculated from magnetization M ($\mu_0 H$) isotherm data and is approximated by the following equation [42]:

$$\Delta S_m(T, H_0) = S_m(T, H_0) - S_m(T, 0) = \frac{1}{\Delta T}\int_0^{H_0}[M(T+\Delta T, H) - M(T, H)]dH, \quad (8)$$

To estimate the magnetocaloric effect, we plot $-\Delta S_M$ vs. T (temperature) for values of external magnetic field ($\mu_0 H$) of 1, 2, 3, 4 T in Figure 10. Furthermore, relative cooling power (RCP) is calculated as the product of entropy change (ΔS_M) and temperature change at half maximum (δT_{FWHM})

$$RCP(S) = -\Delta S_m(T, H) \times \delta T_{FWHM} \quad (9)$$

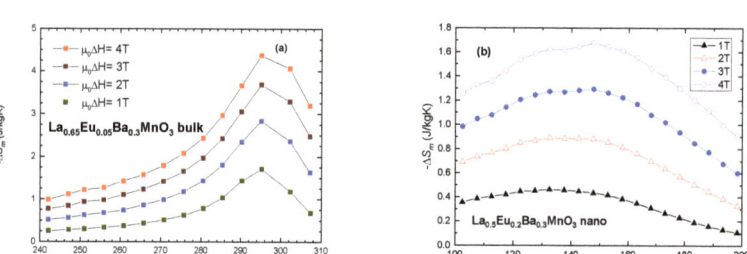

Figure 10. Magnetic entropy change vs. temperature for selected samples: (a) x = 0.05 bulk and (b) x = 0.2 nanocrystalline sample.

Table 6 shows the progression of magnetic entropy change in bulk samples and Table 7 shows the same for nano-sized samples. It is evident that maximum magnetic entropy change occurs at temperatures very close to T_c. Of interest in both systems is the fact that both bulk and nanocrystalline samples with x ≤ 0.3 exhibit RCP values which are in the range of the values recommended for magnetocaloric materials [43]. An interesting note to be made is that the values of Relative Cooling Power (RCP) for similar samples are comparable to each other. RCP (x = 0.2) bulk = 212.4 9 (J/kg) and RCP (x = 0.2) nano = 218.6 (J/kg) (Tables 6 and 7) for $\mu_0 \Delta H$ = 4 T. The values for the maximum magnetic entropy change are very good in comparison with similar compounds presented in the literature as potential magnetocaloric materials [10,44]. Another interesting and useful note is the shape of the curves of magnetic entropy change for bulk and nanocrystalline samples. The bulk material exhibits a sharp curve, while nanoparticles have lower but wider curves. This is due to the size distribution of the nanoparticles and their separation. While the peak entropy change $|\Delta S_M|$ for bulk material is higher (4.2 J/KgK) than for nano-material (1.63 J/kgK), the width of the temperature change (δT_{FWHM}) is reversed, with 37 K for bulk and 95 K for equivalent samples (x = 0.05) when $\mu_0 \Delta H$ = 4 T. This fact can be useful for the cooling industry, where a wider range of applications is preferable.

Table 6. Experimental values for $La_{0.7-x}Eu_xBa_{0.3}MnO_3$ bulk materials: magnetic measurements.

Compound (Bulk)	T_C (K)	M_s (μ_B/f.u.)	H_{ci} (Oe)	$\|\Delta S_M\|$ (J/kgK) $\mu_0\Delta H = 1\,T$	$\|\Delta S_M\|$ (J/kgK) $\mu_0\Delta H = 4\,T$	RCP (S) (J/kgK) $\mu_0\Delta H = 1\,T$	RCP (S) (J/kgK) $\mu_0\Delta H = 4\,T$	Refs
$La_{0.7}Ba_{0.3}MnO_3$	340	4.04	200	1.33	3.5	53.7	158.4	This work
$La_{0.65}Eu_{0.05}Ba_{0.3}MnO_3$	297	3.87	172	1.71	4.2	42.7	155.4	This work
$La_{0.6}Eu_{0.1}Ba_{0.3}MnO_3$	270	3.84	63	1.6	4.1	40	187.7	This work
$La_{0.5}Eu_{0.2}Ba_{0.3}MnO_3$	198	3.7	67	1.41	3.7	38.1	212.6	This work
$La_{0.4}Eu_{0.3}Ba_{0.3}MnO_3$	142	3.78	66	1.7	3.5	42.6	176.4	This work
$La_{0.3}Eu_{0.4}Ba_{0.3}MnO_3$	99	3.46	120	1.02	2.83	25.7	133.3	This work
$La_{0.7}Ca_{0.3}MnO_3$	256			1.38		41		[10]
$La_{0.7}Sr_{0.3}MnO_3$	365			-	4.44 (5 T)		128 (5 T)	[10]
$La_{0.6}Nd_{0.1}Ca_{0.3}MnO_3$	233			1.95		37		[10]
$Gd_5Si_2Ge_2$	276			-	18 (5 T)	-	535 (5 T)	[10]
Gd	293			2.8		35		[10]

Table 7. Experimental values for $La_{0.7-x}Eu_xBa_{0.3}MnO_3$ nano materials.

Compound (Nano)	T_C (K)	M_s (μ_B/f.u.)	H_{ci} (Oe)	$\|\Delta S_M\|$ (J/kgK) $\mu_0\Delta H = 1\,T$	$\|\Delta S_M\|$ (J/kgK) $\mu_0\Delta H = 4\,T$	RCP(S) (J/kgK) $\mu_0\Delta H = 1\,T$	RCP(S) (J/kgK) $\mu_0\Delta H = 1\,T$	Refs
$La_{0.7}Ba_{0.3}MnO_3$	263	2.79	4800	1.04	1.37	105.4	130.1	This work
$La_{0.65}Eu_{0.05}Ba_{0.3}MnO_3$	220	2.95	410	0.43	1.63	43.3	155.6	This work
$La_{0.6}Eu_{0.1}Ba_{0.3}MnO_3$	200	2.6	390	0.93	1.23	93.5	135.3	This work
$La_{0.5}Eu_{0.2}Ba_{0.3}MnO_3$	136	2.96	280	0.46	1.68	47.8	218.4	This work
$La_{0.4}Eu_{0.3}Ba_{0.3}MnO_3$	90	2.3	590	0.39	1.99	38.3	187.7	This work
$La_{0.3}Eu_{0.4}Ba_{0.3}MnO_3$	64	2.09	960	0.25	1.09	23.3	119.9	This work
$La_{0.67}Ca_{0.33}MnO_3$	260				0.97 (5 T)		27 (5 T)	[45]
$Pr_{0.65}(Ca_{0.6}Sr_{0.4})_{0.35}MnO_3$	220			0.75		21.8		[46]
$La_{0.6}Sr_{0.4}MnO_3$	365			1.5		66		[47]

4. Conclusions

The compounds $La_{0.7-x}Eu_xBa_{0.3}MnO_3$ (x = 0, 0.05, 0.1, 0.2, 0.3, 0.4) were synthetized by solid-state reaction to produce a bulk material whose crystalline structure and morphology were investigated by XRD and SEM. The sol-gel method was used to produce nano-scale particles, which were analyzed by XRD and TEM, showing an average size of 30–70 nm. Both systems are single phase, with rhombohedral (R-3c) lattice symmetry. Cell parameters diminish with addition of Eu in the lattice structure. Mn-O bond length tends to shorten with an increasing level of Eu, and the addition of Eu increases the angle Mn-O-Mn. ZFC-FC plots suggest single magnetic phase and low magnetic anisotropy. The sample with x = 0.05 shows a magnetic phase transition at 297 K for bulk compound, while the $La_{0.7}Ba_{0.3}MnO_3$ nanoparticle sample has T_C = 263 K. Both systems show very small coercivity, with nano-sized samples being slightly larger: 63 Oe for the bulk sample with x = 0.1 and 390 Oe for the corresponding nanocrystalline sample. Iodometry was used in order to estimate oxygen content in samples, showing a lower concentration of Mn^{4+} ions, leading to the lowest value of oxygen content to be $O_{2.97\pm0.02}$ for the x = 0.4 sample. General oxygen deficiency was confirmed by inductively coupled plasma optical emission spectrometry (ICP-OES). The samples with x < 0.4 show metallic–insulator transition at a temperature T_p lower than magnetic transition temperatures T_c. The sample with x = 0.4 exhibits increasing resistivity with decreasing temperature below 160 K, suggesting either the ferromagnetic clusters are non-metallic or they are too small to compensate the increase of resistivity induced by disorder. All samples show negative magnetoresistance. The sample with x = 0.3 exhibits magnetoresistance MR (2 T) = 63.6%. Arrott plots confirm second-order magnetic phase transition for all of the samples. Modified Arrott plot analysis revealed the critical exponents for bulk samples to be in the tricritical mean field model range and in the 3D Heisenberg model range for nanocrystalline samples. The maximum magnetic entropy change of 4.2 J/kgK was observed for the x = 0.05 bulk sample for $\mu_0 \Delta H$ = 4 T. Nanocrystalline samples exhibit lower peak magnetic entropy change (1.63 J/kgK for x = 0.05), but T_{fwhm} exceeds 100 K (130K for x = 0.2). Relative cooling power RCP is comparable between equivalent samples for both systems: 212.4 J/kg for x = 0.3 in nanocrystalline samples and 212.6 J/kg for bulk. The value of RCP close to room temperature phase transition T_C = 297 K for the bulk sample $La_{0.65}Eu_{0.05}Ba_{0.3}MnO_3$ is 155.4 J/Kg, possessing the required parameters of a magnetocaloric material. Since the temperature range (δT_{FWHM}) for nano-sized samples $La_{0.7}Ba_{0.3}MnO_3$ and $La_{0.65}Eu_{0.05}Ba_{0.3}MnO_3$ covers a wide range (95 K) including room temperature, they may be used in multistep refrigeration processes. However, we should be cautious in taking high RCP values for nano-materials at face value [48]; potentially, fewer compounds would be necessary for use in stacks of refrigeration systems. This wide range of effective cooling in nanoparticles together with high entropy change in bulk material can be combined for suitable commercial cooling.

Author Contributions: R.A., conceptualization, investigation, methodology, writing—original draft, writing—review and editing, visualization, supervision; R.H., R.B., E.C., F.P. and T.F., methodology, investigation, writing—review and editing; I.G.D., conceptualization, investigation, methodology, visualization, supervision, writing—review and editing. All authors have read and agreed to the published version of the manuscript.

Funding: This research received no external funding.

Institutional Review Board Statement: Not applicable.

Informed Consent Statement: Not applicable.

Data Availability Statement: Data presented in this study are available in this article.

Acknowledgments: The authors acknowledge Ioan Ursu from the Faculty of Physics, Babes Bolyai University, for assistance.

Conflicts of Interest: The authors declare no conflict of interest.

References

1. Fidler, J.C. A History of Refrigeration throughout the World. *Int. J. Refrig.* **1979**, *2*, 249–250. [CrossRef]
2. Briley, G.C. A History of Refrigeration. *ASHRAE J.* **2004**, *46*, S31–S34.
3. Gschneidner, K.A.; Pecharsky, V.K. Thirty Years of near Room Temperature Magnetic Cooling: Where We Are Today and Future Prospects. *Int. J. Refrig.* **2008**, *31*, 945–961. [CrossRef]
4. Dorin, B.R.; Avsec, J.; Plesca, A. The Efficiency of Magnetic Refrigeration and a Comparison with Compresor Refrigeration Systems. *J. Energy Tecnol.* **2018**, *11*, 59–69.
5. Moya, X.; Kar-Narayan, S.; Mathur, N.D. Caloric Materials near Ferroic Phase Transitions. *Nat. Mater.* **2014**, *13*, 439–450. [CrossRef]
6. Dagotto, E.; Hotta, T.; Moreo, A. Colossal Magnetoresistant Materials: The Key Role of Phase Separation. *Phys. Rep.* **2001**, *344*, 1–153. [CrossRef]
7. Dagotto, E. *Nanoscale Phase Separation and Colossal Magnetoresistance*, 1st ed.; Springer Science & Business Media: New York, NY, USA, 2002; pp. 271–284.
8. Coey, J.M.D.; Viret, M.; Von Molnár, S. Mixed-Valence Manganites. *Adv. Phys.* **1999**, *48*, 167–293. [CrossRef]
9. Rostamnejadi, A.; Venkatesan, M.; Alaria, J.; Boese, M.; Kameli, P.; Salamati, H.; Coey, J.M.D. Conventional and Inverse Magnetocaloric Effects in $La_{0.45}Sr_{0.55}MnO_3$ Nanoparticles. *J. Appl. Phys.* **2011**, *110*, 043905. [CrossRef]
10. Varvescu, A.; Deac, I.G. Critical Magnetic Behavior and Large Magnetocaloric Effect in $Pr_{0.67}Ba_{0.33}MnO_3$ Perovskite Manganite. *Phys. B Condens. Matter* **2015**, *470–471*, 96–101. [CrossRef]

11. Weng, S.; Zhang, C.; Han, H. 3D-Heisenberg Ferromagnetic Characteristics in a La$_{0.67}$Ba$_{0.33}$MnO$_3$ Film on SrTiO$_3$. *Eur. Phys. J. B* **2021**, *94*, 91. [CrossRef]
12. Liedienov, N.A.; Pashchenko, A.V.; Pashchenko, V.P.; Prokopenko, V.K.; Revenko, Y.F. Structure defects, phase transitions, magnetic resonance and magneto-transport properties of La$_{0.6-x}$Eu$_x$Sr$_{0.3}$Mn$_{1.1}$O$_{3-\delta}$ ceramics. *Low Temp. Phys.* **2016**, *42*, 1102–1111. [CrossRef]
13. Amaral, J.S.; Reis, M.S.; Amaral, V.S.; Mendonca, T.M.; Araujo, J.P.; Sa, M.A.; Tavares, P.B.; Vieira, J.M. Magnetocaloric effect in Er- and Eu-substituted ferromagneticLa-Sr manganites. *J Magn. Magn. Mat.* **2009**, *290*, 686–689.
14. Raju, K.; Manjunathrao, S.; Venugopal Reddy, P. Correlation between Charge, Spin and Lattice in La-Eu-Sr Manganites. *J. Low Temp. Phys.* **2012**, *168*, 334–349. [CrossRef]
15. Rao, C.N.R. Perovskites. In *Encyclopedia of Physical Science and Technology*; Elsevier: Amsterdam, The Netherlands, 2003; pp. 707–714.
16. Nath, D.; Singh, F.; Das, R. X-Ray Diffraction Analysis by Williamson-Hall, Halder-Wagner and Size-Strain Plot Methods of CdSe Nanoparticles—A Comparative Study. *Mater. Chem. Phys.* **2020**, *239*, 2764–2772. [CrossRef]
17. Wolfgong, W.J. Chemical Analysis Techniques for Failure Analysis. In *Handbook of Materials Failure Analysis with Case Studies from the Aerospace and Automotive Industries*; Butterworth-Heinemann: Oxford, UK, 2016; pp. 279–307.
18. Licci, F.; Turilli, G.; Ferro, P. Determination of Manganese Valence in Complex La-Mn Perovskites. *J. Magn. Magn. Mater.* **1996**, *164*, L268–L272. [CrossRef]
19. Tali, R. Determination of Average Oxidation State of Mn in ScMnO$_3$ and CaMnO$_3$ by Using Iodometric Titration. *Damascus Univ. J. Basic Sci.* **2007**, *23*, 9–19.
20. Covaci, E.; Senila, M.; Ponta, M.; Frentiu, T. Analitical performance and validation of optical emission and atomic absorption spectrometry methods for multielemental determination in vegetables and fruits. *Rev. Roum. Chim.* **2020**, *65*, 735–745. [CrossRef]
21. Deac, I.G.; Tetean, R.; Burzo, E. Phase Separation, Transport and Magnetic Properties of La$_{2/3}$A$_{1/3}$Mn$_{1-x}$Co$_x$O$_3$, A = Ca, Sr ($0.5 \leq x \leq 1$). *Phys. B Condens. Matter* **2008**, *403*, 1622–1624. [CrossRef]
22. Gross, R.; Alff, L.; Büchner, B.; Freitag, B.H.; Höfener, C.; Klein, J.; Lu, Y.; Mader, W.; Philipp, J.B.; Rao, M.S.R.; et al. Physics of Grain Boundaries in the Colossal Magnetoresistance Manganites. *J. Magn. Magn. Mater.* **2000**, *211*, 150–159. [CrossRef]
23. Raju, K.; Pavan Kumar, N.; Venugopal Reddy, P.; Yoon, D.H. Influence of Eu Doping on Magnetocaloric Behavior of La0.67Sr0.33MnO$_3$. *Phys. Lett. Sect. A Gen. At. Solid State Phys.* **2015**, *379*, 1178–1182. [CrossRef]
24. Kubo, K.; Ohata, N. A Quantum Theory of Double Exchange. *J. Phys. Soc. Jpn.* **1972**, *33*, 21–32. [CrossRef]
25. Liedienov, N.A.; Wei, Z.; Kalita, V.M.; Pashchenko, A.V.; Li, Q.; Fesych, I.V.; Turchenko, V.A.; Hou, C.; Wei, X.; Liu, B.; et al. Spin-dependent magnetism and superparamagnetic contribution to the magnetocaloric effect of non-stoichiometric manganite nanoparticles. *Appl. Mater. Today* **2022**, *26*, 101340. [CrossRef]
26. Panwar, N.; Pandya, D.K.; Agarwal, S.K. Magneto-Transport and Magnetization Studies of Pr$_{2/3}$Ba$_{1/3}$MnO$_3$:Ag$_2$O Composite Manganites. *J. Phys. Condens. Matter* **2007**, *19*, 456224. [CrossRef]
27. Panwar, N.; Pandya, D.K.; Rao, A.; Wu, K.K.; Kaurav, N.; Kuo, Y.K.; Agarwal, S.K. Electrical and Thermal Properties of Pr$_{2/3}$(Ba$_{1-x}$Cs$_x$)$_{1/3}$MnO$_3$ Manganites. *Eur. Phys. J. B* **2008**, *65*, 179–186. [CrossRef]
28. Banerjee, B.K. On a Generalised Approach to First and Second Order Magnetic Transitions. *Phys. Lett.* **1964**, *12*, 16–17. [CrossRef]
29. Joy, P.A.; Anil Kumar, P.S.; Date, S.K. The Relationship between Field-Cooled and Zero-Field-Cooled Susceptibilities of Some Ordered Magnetic Systems. *J. Phys. Condens. Matter* **1998**, *10*, 11049–11054. [CrossRef]
30. Arun, B.; Suneesh, M.V.; Vasundhara, M. Comparative Study of Magnetic Ordering and Electrical Transport in Bulk and Nano-Grained Nd$_{0.67}$Sr$_{0.33}$MnO$_3$ Manganites. *J. Magn. Magn. Mater.* **2016**, *418*, 265–272. [CrossRef]
31. Peters, J.A. Relaxivity of Manganese Ferrite Nanoparticles. *Prog. Nucl. Magn. Reson. Spectrosc.* **2020**, *120*, 72–94. [CrossRef]
32. Arrott, A. Criterion for Ferromagnetism from Observations of Magnetic Isotherms. *Phys.Rev.* **1957**, *108*, 1394–1396. [CrossRef]
33. Jeddi, M.; Gharsallah, H.; Bejar, M.; Bekri, M.; Dhahri, E.; Hlil, E.K. Magnetocaloric Study, Critical Behavior and Spontaneous Magnetization Estimation in La$_{0.6}$Ca$_{0.3}$Sr$_{0.1}$MnO$_3$ Perovskite. *RSC Adv.* **2018**, *8*, 9430–9439. [CrossRef]
34. Stanley, H.E. *Introduction to Phase Transitions and Critical Phenomena*; Oxford University Press: Oxford, UK, 1987; pp. 7–10.
35. Arrott, A.; Noakes, J.E. Approximate Equation of State for Nickel Near Its Critical Temperature. *Phys. Rev. Lett.* **1967**, *19*, 786–789. [CrossRef]
36. Pathria, R.K.; Beale, P.D. Phase Transitions: Criticality, Universality, and Scaling. In *Statistical Mechanics*; Elsevier: Amsterdam, The Netherlands, 2022; pp. 417–486.
37. Fisher, M.E.; Ma, S.K.; Nickel, B.G. Critical Exponents for Long-Range Interactions. *Phys. Rev. Lett.* **1972**, *29*, 917. [CrossRef]
38. Smari, M.; Walha, I.; Omri, A.; Rousseau, J.J.; Dhari, E.; Hlil, E.K. Critical parameters near the ferromagnetic–paramagnetic phase transition in La$_{0.5}$Ca$_{0.5-x}$Ag$_x$MnO$_3$ compounds ($0.1 \leq x \leq 0.2$). *Ceram. Int.* **2014**, *40*, 8945–8951. [CrossRef]
39. Kim, D.; Revaz, B.; Zink, B.L.; Hellman, F.; Rhyne, J.J.; Mitchell, J.F. Tri-critical Point and the Doping Dependence of the Order of the Ferromagnetic Phase Transition of La$_{1-x}$Ca$_x$MnO$_3$. *Phys. Rev. Lett.* **2002**, *89*, 227202. [CrossRef] [PubMed]
40. Zverev, V.; Tishin, A.M. Magnetocaloric Effect: From Theory to Practice. In *Reference Module in Materials Science and Materials Engineering*; Elsevier: Amsterdam, The Nederlands, 2016; pp. 5035–5041. [CrossRef]
41. Majumder, D.D.; Majumder, D.D.; Karan, S. Magnetic Properties of Ceramic Nanocomposites. In *Ceramic Nanocomposites*; Woodhead Publishing Series in Composites Science and Engineering; Banerjee, R., Manna, I., Eds.; Elsevier B.V.: Amsterdam, The Netherlands, 2013; pp. 51–91. [CrossRef]

42. Souca, G.; Iamandi, S.; Mazilu, C.; Dudric, R.; Tetean, R. Magnetocaloric Effect and Magnetic Properties of $Pr_{1-x}Ce_xCo_3$ Compounds. *Stud. Univ. Babeș-Bolyai Phys.* **2018**, *63*, 9–18. [CrossRef]
43. Naik, V.B.; Barik, S.K.; Mahendiran, R.; Raveau, B. Magnetic and Calorimetric Investigations of Inverse Magnetocaloric Effect in $Pr_{0.46}Sr_{0.54}MnO_3$. *Appl. Phys. Lett.* **2011**, *98*, 112506. [CrossRef]
44. Gong, Z.; Xu, W.; Liedienov, N.A.; Butenko, D.S.; Zatovsky, I.V.; Gural'skiy, I.A.; Wei, Z.; Li, Q.; Liu, B.; Batman, Y.A.; et al. Expansion of the multifunctionality in off-stoichiometric manganites using post-annealing and high pressure: Physical and electrochemical studies. *Phys Chem. Chem. Phys.* **2002**, *24*, 21872–21885. [CrossRef]
45. Hueso, L.E.; Sande, P.; Miguéns, D.R.; Rivas, J.; Rivadulla, F.; López-Quintela, M.A. Tuning of the magnetocaloric effect in nanoparticles synthesized by sol–gel techniques. *J. Appl. Phys.* **2002**, *91*, 9943–9947. [CrossRef]
46. Anis, B.; Tapas, S.; Banerjee, S.; Das, I. Magnetocaloric properties of nanocrystalline $Pr_{0.65}(Ca_{0.6}Sr_{0.4})_{0.35}MnO_3$. *J. Appl. Phys.* **2008**, *103*, 013912. [CrossRef]
47. Ehsani, M.H.; Kameli, P.; Ghazi, M.E.; Razavi, F.S.; Taheri, M. Tunable magnetic and magnetocaloric properties of $La_{0.6}Sr_{0.4}MnO_3$ nanoparticles. *J. Appl. Phys.* **2013**, *114*, 223907. [CrossRef]
48. Griffith, L.D.; Mudryk, Y.; Slaughter, J.; Pecharsky, V.K. Material-based figure of merit for caloric materials. *J. Appl. Phys.* **2018**, *123*, 034902. [CrossRef]

Article

Modification of Some Structural and Functional Parameters of Living Culture of *Arthrospira platensis* as the Result of Selenium Nanoparticle Biosynthesis

Liliana Cepoi [1], Inga Zinicovscaia [2,3,4,*], Tatiana Chiriac [1], Ludmila Rudi [1], Nikita Yushin [2,5], Dmitrii Grozdov [2], Ion Tasca [1], Elena Kravchenko [2] and Kirill Tarasov [2]

1. Institute of Microbiology and Biotechnology, Technical University of Moldova, 1 Academiei Str., 2028 Chisinau, Moldova
2. Joint Institute for Nuclear Research, 6 Joliot-Curie Str., 141980 Dubna, Russia
3. Horia Hulubei National Institute for R&D in Physics and Nuclear Engineering, 30 Reactorului Str. MG-6, 077125 Bucharest, Romania
4. Institute of Chemistry, 3 Academiei Str., 2028 Chisinau, Moldova
5. Doctoral School of Biological, Geonomic, Chemical and Technological Science, State University of Moldova, 2009 Chisinau, Moldova
* Correspondence: zinikovskaia@mail.ru; Tel.: +7-49-6216-5609

Abstract: Selenium nanoparticles are attracting the attention of researchers due to their multiple applications, including medicine. The biosynthesis of selenium nanoparticles has become particularly important due to the environmentally friendly character of the process and special properties of the obtained particles. The possibility of performing the biosynthesis of selenium nanoparticles via the living culture of *Arthrospira platensis* starting from sodium selenite was studied. The bioaccumulation capacity of the culture, along with changes in the main biochemical parameters of the biomass, the ultrastructural changes in the cells during biosynthesis and the change in the expression of some genes involved in stress response reactions were determined. Protein, lipid and polysaccharide fractions were obtained from the biomass grown in the presence of sodium selenite. The formation of selenium nanoparticles in the protein fraction was demonstrated. Thus, *Arthrospira platensis* culture can be considered a suitable matrix for the biosynthesis of selenium nanoparticles.

Keywords: selenium nanoparticles; spirulina; bioaccumulation; proteins; genes

1. Introduction

Selenium is an indispensable microelement for life, being a component in antioxidant enzymes. Its deficiency can lead to the development of cancer or cardiovascular diseases, whilst an excess of selenium is extremely toxic to living cells. Thus, the use of selenium in different treatments requires a very careful approach, due to the very narrow range between therapeutic and toxic doses [1].

Nanoparticles are particles with diverse properties and a high level of activity compared to bulk elements. In this regard, selenium nanoparticles (SeNPs) are very attractive as alternative remedies for various applications. Thus, SeNPs produced in different ways can be considered a good alternative to antibiotics or preparations designed for cancer treatment [2–5]. For example, SeNPs obtained by pulsed laser ablation in a liquid method showed that antibacterial action against standard and antibiotic-resistant phenotypes of Gram-negative and Gram-positive bacteria was not toxic to healthy skin tissues, but had anticancer effects on human melanoma and glioblastoma cells at a concentration of 1 ppm [3]. In addition, SeNPs have antioxidant properties, as well as anti-inflammatory and anti-diabetic action [1]. In recent years, attention has also been drawn to the theranostic potential of SeNPs [6].

In contrast to other types of nanoparticles, SeNPs are of great prospects for the treatment of cardiovascular diseases. SeNPs, in addition to replacement of the selenium-deficiency characteristic for heart patients, also have the ability to transfer electrical signals in seeded scaffolds [7]. A recent study showed that SeNPs modify chitosan used as a biofilm-forming material in cardiac surgery so that it acquires electrical conductivity that ensures rapid coupling between cardiomyocytes, which significantly improves the mechanical properties of the myocardium [8]. Other nanoparticles with such properties, for example gold ones, are characterized by pronounced cardiotoxicity [9]. SeNPs inhibit the formation of atherosclerosis in apolipoprotein-E-deficient mice by inhibiting hyperlipidemia through suppressing hepatic cholesterol and fatty acid metabolism and alleviate oxidative stress, thereby enhancing antioxidant activity [10]. These properties in SeNPs are explained by their higher biocompatibility compared to other type of nanoparticles, which is ensured by the fact that they can be formed in the cytosol of cells, where elemental selenium resulting from the metabolism of selenites is accumulated [11].

Currently, SeNPs are produced using physical, chemical and biological methods. Biological methods of nanoparticle synthesis are becoming more and more attractive due to the low toxic effects on the environment and low production price [12]. SeNP biosynthesis can be achieved in two ways: using living cells or using different extracts obtained from cells. In both cases, formation of nanoparticles is based on the reduction potential of the matrix used for the synthesis. In the case of living organisms, the reduction of selenite and selenate ions to SeNPs is seen as a strategy to reduce the toxicity of inorganic selenium.

The most frequently used living matrices for the biosynthesis of SeNPs are fungal cultures [13–16] and bacteria [12,17–19]. Cyanobacteria also present great interest for the biosynthesis of SeNPs. Thus, the biosynthesis of SeNPs by *Hapalosiphon* sp. [20], *Synechocystis* sp. [21], *Microcystis aeruginosa* [22], *Anabaena* sp. [23], *Anabaena variabilis*, *Arthrospira indica*, *Gloeocapsa gelatinosa*, *Oscillatoria* sp. and *Phormidium sp* [24] was reported.

Spirulina (*Arthrospira platensis*), due to its unique therapeutic properties, has also been studied as a matrix for the biosynthesis of SeNPs. However, often, the conditions under which these nanoparticles are produced are incompatible with the normal vital processes for cyanobacteria, which are subject to pronounced biodegradation processes. For example, in a previously performed study, to obtain SeNPs, spirulina separated from cultivation medium was transferred to sodium selenite solution, which does not contain the necessary elements to ensure the vital processes of spirulina, as well as the optimal level of salinity and pH not respected. As a result, the process of SeNP biosynthesis was associated with severe biomass biodegradation [25]. A successful synthesis of SeNPs by spirulina (*Spirulina platensis*) was achieved at pH 7.0 when applying a 12:12 photoperiodic regime, as was reported by Alipour and co-authors [26]. The authors obtained homogeneous SeNPs with antioxidant properties, significantly exceeding those of the sodium selenite solution. A current review reporting the research in the field of nanoparticle biosynthesis highlighted their biosynthesis by different cyanobacteria, in particular, extracellularly or on the surface of the cell wall [27]. The intracellular synthesis of SeNPs takes place due to the activity of two main enzymes—selenate reductase and selenite reductase—while extracellular synthesis occurs under the action of different organic compounds with reducing potential [27].

At the same time, different changes in the cyanobacteria cells subjected to selenium ions were mentioned. They include inhibition or reduction in the culture growth rate at certain element concentrations, modification of biochemical parameters, antioxidant properties, the level of reactive oxygen species in cells, structural changes in cells and modification of the expression of some genes associated with the photosynthesis process [20–23,28,29].

Obtaining spirulina (*Arthrospira platensis*) biomass, known for its multiple therapeutic benefits, supplemented with biogenic SeNPs seems to be a very interesting direction, both in terms of fundamental and applied research. The main goal of the present study was to obtain selenium nanoparticles using *Arthrospira platensis*, by adding selenium salt during biomass growth, maintaining high biomass productivity.

2. Materials and Methods

2.1. Experiment Design

As the object of the study, cyanobacteria *Arthospira platensis* (*A. platensis*, spirulina) CNMN-CB-02 from collection of non-pathogenic microorganisms (Institute Microbiology and Biotechnology, TUM, Chisinau, Moldova) was used. The composition of the medium used for biomass growth is presented in [30]. The amount of inoculum was 0.4–0.45 g/L. The culture was cultivated at temperature 28–32 °C, pH—8–10 and continuous illumination with an intensity of 55–85 µM photons/m^2/s. The culture was shaken daily for 2 h. The duration of the cultivation cycle was 6 days. On the third day of the cultivation cycle, Na_2SeO_3 in concentrations 25–200 mg/L, with a step of 25 mg/L, was added to the spirulina medium. At the end of the cultivation cycle, biomass was separated from the medium by filtration. Selenium concentration in initial and experimental solutions was determined using an inductively coupled plasma-optical emission spectrometer, PlasmaQuant 9000 Elite (Analytik Jena, Jena, Germany). Calibration solutions and standards for measurements were prepared from IV-STOCK-27 (Inorganic Ventures, Christiansburg, VA, USA) standard solution. All control standards were analyzed after every 5 samples. For the biochemical tests, the biomass was subjected to a repeated freezing–thawing procedure.

2.2. Determination of Selenium Content in Biomass

The determination of selenium content in spirulina biomass and its fractions was carried out according to GOST R 51309-99 "Drinking water. Determination of elements content by atomic spectrometry methods", which is based on the vaporization of the solution to be analyzed containing selenium in an air flame with acetylene and the measurement of the absorbance of the flame (vapors containing selenium) at wavelengths 196–207.5 nm. The determinations were made based on the calibration curve. Biomass mineralization was primarily produced by using concentrated H_2O_2 and HNO_3. The biomass was mineralized for 2–3 h in a sand bath until a colorless solution was obtained.

2.3. Determination of the Amount of Spirulina Biomass

The amount of biomass was determined spectrophotometrically by recording the absorbance of the suspension at a wavelength of 750 nm and the subsequent calculation in g/L.

2.4. Determination of the Biochemical Composition of Spirulina Biomass

The content of protein was determined based on the principle of formation of copper complex with peptide bonds and its subsequent reduction under alkaline conditions. The formed complex reduces the Folin-Ciocalteu reagent (Lowry Method). The content of carbohydrates was determined spectrophotometrically based on the formation of hydroxymethylfurfural under interaction of carbohydrates with the Anthon reagent ($C_{14}H_{10}O$) in acid medium. For determination of phycobiliprotein content, we used the water extract obtained by repeated freezing–thawing of 10 mg of biomass in 1 mL of distilled water. The absorbance of the supernatant was measured at 620 nm for c-phycocyanin and 650 nm for allophycocyanin. The lipid content was determined spectrophotometrically using phosphovanilinic reagent. Malondialdehyde (MDA) was determined based on reactive products of thiobarbituric acid. The content of chlorophyll a and β carotene was determined spectrophotometrically based on the determination of absorbance of ethanolic extract obtained from spirulina biomass. More detail about each parameter determination can be found in [30].

2.5. RNA Extraction and Quantitative RT-PCR (RT-qPCR)

Total RNA was extracted from the pellet obtained after centrifugation of 25 mL of algal culture in three biological repeats using TRIzol™ Reagent (Invitrogen™, Waltham, MA, USA) according to manufacturer's protocol. The integrity of isolated RNA was evaluated using an QIAxcel Advanced system (Qiagen, Hilden, Germany). cDNA was obtained with Maxima First Strand cDNA Synthesis Kit for RT-qPCR, with dsDNase (Thermo Scientific™, Waltham, MA, USA) following the manufacturer's instructions. The RT-qPCR was performed using iTaq Universal SYBR Green Supermix (BioRad, Hercules, CA, USA) on a CFX96 Touch Real-Time PCR Detection System (BioRad, USA) with primers for the following genes: FeSOD, GOGAT, hsp90, rbcL, POD and 16S rRNA (reference) (Table S1). The relative transcript levels of the selected genes were analyzed using the $\Delta\Delta Ct$ method.

2.6. Transmission Electron Microscopy (TEM) Analysis and Energy Dispersive X-ray Analysis

The morphology of spirulina cells (control and exposed to selenium) was described using a JEM-1400 transmission electron microscope (Jeol, Akishima, Japan) at an accelerating voltage of 100 kW. Microprobe analysis of SeNPs was conducted with an energy-dispersive X-ray analysis spectrometer (EDAX, Pleasanton, CA, USA). The acquisition time ranged from 60 to 100 s, and the accelerating voltage was 20 kV.

3. Results and Discussion

3.1. Uptake of Se and Its Accumulation in the Arthrospira platensis Biomass

The removal of selenium by *A. platensis* from the nutrient medium varied between 12.8 and 26.6% (Figure 1). The lowest Se removal from medium was attained at a Na_2SeO_3 concentration of 75 mg/L (the selenium concentration being 47.03 mg/L), while the highest at a Na_2SeO_3 concentration of 125 mg/L (the selenium concentration being 78.38 mg/L).

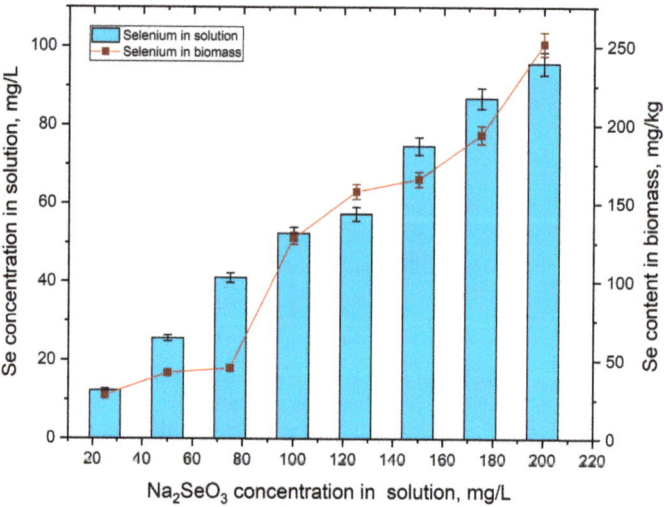

Figure 1. Selenium accumulation in *Arthrospira platensis* biomass (n = 3, the error bars represent the standard deviation of the measurements).

Selenium content in biomass increased with the rise in its concentration in the nutrient medium. At a Na_2SeO_3 concentration range of 25–75 mg/L, the content of selenium in biomass varied between 28 and 45 µg/g. At a Na_2SeO_3 concentration of 100 mg/L, the content of selenium in the biomass was 128 µg/g, which is 2.9-times higher compared to the value obtained at a salt concentration of 75 mg/L. A further increase in the Na_2SeO_3 concentration in the spirulina cultivation medium resulted in a slow continuous accumulation of selenium in biomass, reaching a value of 250 µg/g at a salt concentration of 200 mg/L.

It is known that spirulina efficiently accumulates selenium from various salts, including sodium selenite, in a dose- and time-dependent manner [31,32]. The tolerance of cyanobacteria to selenite is relatively high. For example, different strains of spirulina growth were not inhibited by sodium selenite present in the medium in concentrations up to 400 mg/L [32]. Selenium accumulation in biomass can also be influenced by other factors, such as sulfur concentration in the medium or the presence of various organic compounds [22,31]. In the present study, spirulina was grown at optimal conditions, ensuring the accumulation of a large amount of high-quality biomass; thus, a balance can be obtained between all the valuable components in the biomass, including the amount of accumulated selenium. Since high concentrations of Na_2SeO_3 led to unfavorable changes in biomass composition, the effect of concentrations higher than 200 mg/L was not analyzed, even though they may ensure a higher level of accumulation of selenium in biomass.

3.2. The Content of Selenium in Arthrospira platensis Biomass Fractions

In order to identify the cellular components mainly responsible for selenium accumulation, the biomass grown in the presence of Na_2SeO_3 at a concentration of 200 mg/L was subjected to fractionation, after which the selenium content in each fraction was determined. The results are shown in Figure 2.

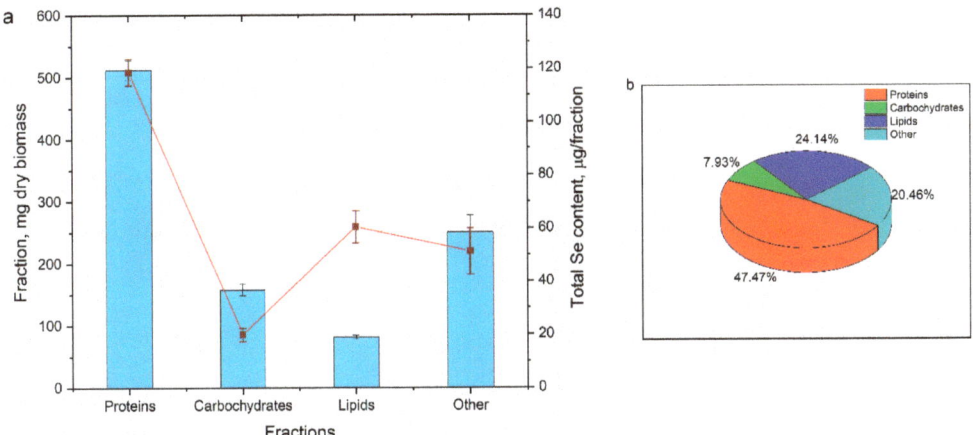

Figure 2. Amount of Se accumulated in biomass fractions: (**a**) total content of Se, µg Se/g per fraction; (**b**) distribution of Se accumulated by fractions, % of the accumulated amount (n = 3, the error bars represent the standard deviation of the measurements).

The highest content of selenium was determined in the protein fraction, constituting 118.7 µg relative to the amount of the extracted protein, so an accumulation capacity of 0.23 µgSe/mg of proteins was obtained. The amount of lipids extracted from spirulina biomass amounted to 81 mg/g of dry biomass. In this fraction, 60.34 µg of Se was accumulated, which is the second-highest amount of accumulated selenium in terms of absolute values. The capacity of selenium accumulation by lipids was significantly higher compared to proteins and constituted 0.74 µgSe per mg of extracted lipids. Carbohydrates showed the lowest selenium accumulation capacity, in absolute numbers, accumulating 19.8 µg in the integral fraction of carbohydrates extracted from one gram of biomass, and in relative values, 0.126 µgSe per mg of extracted carbohydrates. Thus, almost half of the selenium (47.47%) was accumulated in the protein fraction, 24.14% in the lipid fraction and 7.93 in the carbohydrate fraction.

A similar distribution of selenium accumulation by spirulina was reported in other studies. For example, the same pattern was noticed when spirulina accumulated selenium from other selenium compounds, such as iron selenite hexahydrate or germanium selenide [29]. Li and co-authors demonstrated that over 85% of the selenium accumulated in spirulina biomass from inorganic salts is converted to organic selenium. Approximately one-quarter of the total selenium was accumulated in the polypeptide fraction, more than

10% was accumulated in lipids and approximately 2% in carbohydrates. It was assumed that approximately half of the accumulated selenium bound to free amino acids, oligopeptides or other similar compounds [32].

3.3. Formation of Selenium Nanoparticles

The obtained fractions of biomass were subjected to microscopic examination (TEM) in order to identify SeNPs, in case of their formation. Nanostructures were identified only in the total protein fraction (Figure 3). From the obtained images, it can be seen that SeNPs formed in the protein fractions were scattered and spherical, with a size in a range of 2–8 nm (Figure 3). The size of nanoparticles, including selenium, determines both their ability to reach target tissues as well as their retention capacity. Small-sized nanoparticles exhibit significant tumor targeting, with minimum to no nonspecific uptakes, but are also characterized by faster elimination [33]. A final hydrodynamic diameter of nanoparticles smaller than 5.5 nm resulted in their rapid and efficient renal excretion and elimination from the body [34,35]. However, recently, this statement has been questioned, since even nanoparticles of larger sizes but with specific properties are easily eliminated through the kidney [36]. Nevertheless, small nanoparticles are quickly eliminated from the bloodstream through the kidneys. This is of major importance when SeNPs are applied in cancer treatment—nanoparticles that have not been accumulated by tumor tissues are quickly removed from the bloodstream, thereby preventing undesirable effects on other tissues and organs.

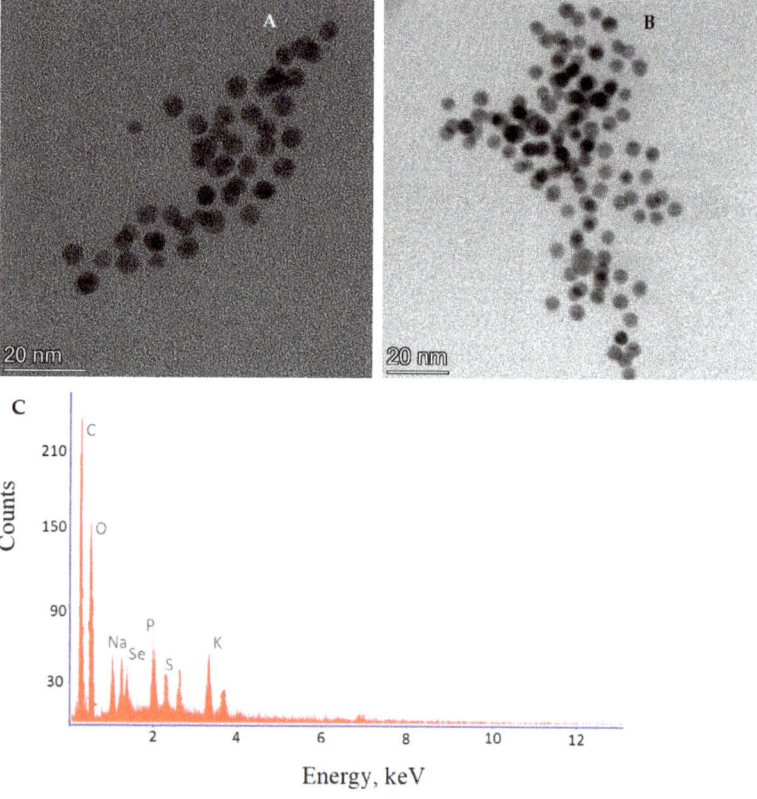

Figure 3. SeNPs in protein fraction ((**A**)—camera imaging 634.1 kx Ceta; (**B**)—STEM diffraction 802.3 kx BF, (**C**)—EDX spectra of selenium nanoparticles).

According to the previously published results, the formation of nanoparticles by spirulina culture takes place preferentially extracellularly [27]. At the same time, there is evidence that prokaryotic organisms can produce SeNPs intracellularly, this being an active process involving the specific enzymes selenate reductase and selenite reductase. Selenite can be also reduced to elemental Se in a non-enzymatic way with the participation of glutathione (GSH) or enzymatically using nonspecific enzymes, such as respiratory and/or detoxifying enzymes, for example, periplasmic nitrite reductase and sulfite reductase [37].

It is known that at high concentrations of selenium, in the present study, 200 mg/L, *A. platensis* is able to resist the action of selenite ions through their reduction to elemental selenium. Selenium is poorly soluble and, therefore, less toxic than selenite ions [32]. Thus, the cyanobacteria apply, in this case, the tactics of reducing toxicity, ensuring a fairly high degree of tolerance to this element.

Depending on the biosynthesis conditions, as, for example, the precursor used and the microorganism culture used as a matrix, SeNPs with a wide range of sizes (from 5 to 530 nm) can be obtained [4,17–19,26]. Nanoparticles obtained intracellularly are smaller compared to those produced extracellularly. In the present study, the size of the nanoparticles was closer to the minimum size limit reported in the literature.

The presented results were obtained under laboratory conditions, when small volumes were used, and their extrapolation to large-scale production requires further appropriate experiments. However, starting from the fact that performing a biosynthesis process at conditions optimal for spirulina growth allows for maintaining a high level of biomass productivity, it can be assumed that the success of this method transfers to the industrial level.

3.4. Content of Biomass and Its Biochemical Composition

The accumulation of *A. platensis* biomass at its cultivation in standard conditions and in the presence of different concentrations of sodium selenite can be seen in Figure 4.

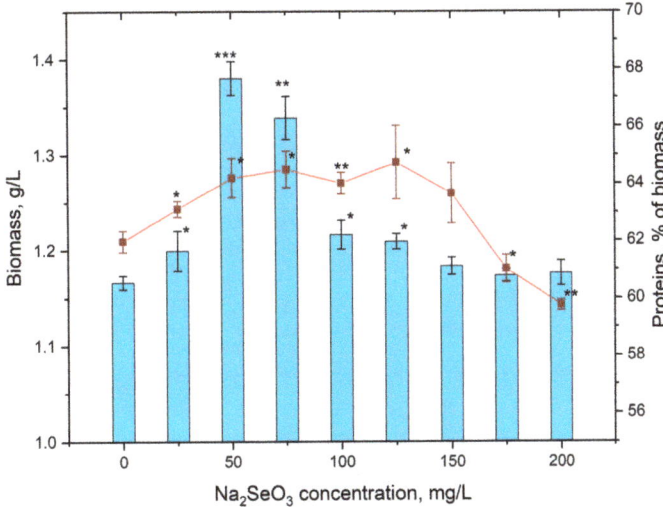

Figure 4. Accumulation of *Arthrospira platensis* biomass and content of protein in biomass under conditions of supplementing the nutrient medium with sodium selenite (* $p < 0.05$, ** $p < 0.005$, *** $p < 0.0005$ for differences between control and experimental sample; n = 3). The error bars represent the standard deviation of the measurements.

The applied concentrations of Na_2SeO_3 did not have a pronounced inhibiting effect on the biomass accumulation during spirulina growth. At salt concentrations of 150,

175 and 200 mg/L, the differences in biomass accumulation in experimental and control samples were not statistically significant. At other Na$_2$SeO$_3$ concentrations, a statistically significant increase in the biomass took place. However, in absolute values, the increase in the amount of biomass was important only at Na$_2$SeO$_3$ concentrations of 50 and 75 mg/L, when the amount of biomass increased by 18.4 and 14.8%, respectively, compared to the control. The stimulatory effects of low concentrations of selenium present in the medium on spirulina growth were obtained when other selenium compounds were used or other experimental conditions were applied [23,24,28]. It is considered that an external supply of Se (by element) up to 150 mg/L has beneficial effects on the accumulation of spirulina biomass [38]; however, it should be mentioned that there is a pronounced dependence between the level of spirulina tolerance to selenium and the experimental conditions.

The stimulating effects of selenium on biomass productivity were observed not only for spirulina, but for other cyanobacteria as well. For example, the growth of the cyanobacterium *Hapalosiphon* sp. was stimulated by Se at concentrations up to 200 ppm, while higher concentrations inhibited biomass growth [20]. In the cyanobacteria *Synechocystis* sp., selenium concentrations up to 200 ppm of selenite did not provide an increase in the amount of biomass, but neither did they inhibit it [21].

The protein content in the biomass of spirulina grown in the presence of different concentrations of sodium selenite varied within very narrow limits and constituted 59.76–64.74% of the dry biomass (Figure 4). The control biomass contained 61.98% of proteins. Although the limits of variation were small, the observed differences have statistical significance, and it should be mentioned that in the case of salt concentrations of 25–125 mg/L, a slight increase in the content of proteins was observed. At a Na$_2$SeO$_3$ concentration of 150 mg/L, the content of protein in biomass did not differ from the control, while at concentrations of 175 and 200 mg/L, a slight decrease was observed.

Carbohydrates and phycobiliptroteins are two components in spirulina biomass, which are extremely labile and react to different external factors (Figure 5). In conditions of toxicity of external effects, an increase in the amount of carbohydrates and a concomitant decrease in the content of phycobilins are observed. In the present study, however, a pronounced increase in the content of both groups of compounds, which can be more associated with the effect of stimulating certain biosynthetic processes in the spirulina culture, was observed. The highest increase in carbohydrate content—by 10.6–17.6% compared to the control—was observed at sodium selenite concentrations of 150–200 mg/L. Similarly, at these concentrations, the level of significance of the differences between experimental samples and control was high ($p < 0.005$). In the case of the other concentrations, the differences between control and samples were lower or not observed.

The total content of phycobiliproteins in the spirulina biomass obtained in the experimental variants was significantly higher compared to the control (by 5.1–30.0%). The lowest difference between the control and the experiment samples was observed at a sodium selenite concentration of 25 mg/L. With the increase in the salt concentration, the content of total phycobiliproteins in the biomass also increased. Thus, at a concentration of sodium selenite of 50 mg/L, the amount of phycobiliproteins increased by 16.1% compared to the control, at a concentration of 75 mg/L—by 22.5% and at concentration of 100 mg/L—by 27.5%. An increase in the Na$_2$SeO$_3$ concentration from 100 to 175 mg/L did not result in a further increase in the content of phycobilins, the amount of these compounds remaining at the plateau level reached at a Na$_2$SeO$_3$ concentration of 100 mg/L. The Na$_2$SeO$_3$ concentration of 200 mg/L was characterized by a reduction in the level of phycobiliproteins by 14.6% compared to the concentration of 175 mg/L. Although it is an important decrease, the total content of phycobiliproteins in the spirulina biomass still remained quite high compared to the control, exceeding it by 15.5%.

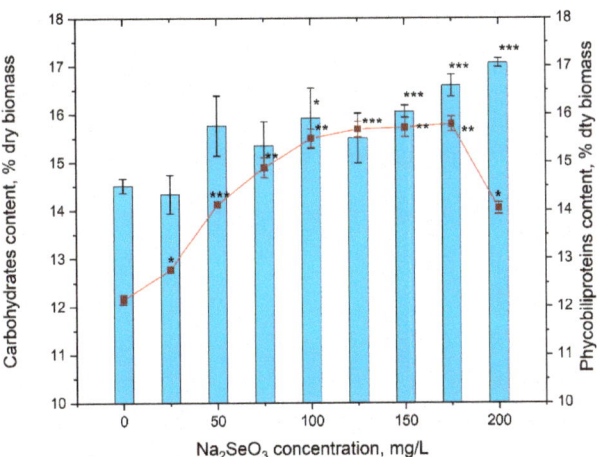

Figure 5. The content of carbohydrates and phycobiliptoteins in *Arthrospira platensis* biomass under conditions of supplementing the nutrient medium with sodium selenite (* $p < 0.05$, ** $p < 0.005$, *** $p < 0.0005$ for differences between control and experimental sample; n = 3). The error bars represent the standard deviation of the measurements.

Primary photosynthetic pigments are most often studied as a response reaction of photosynthetic organisms to the action of various external factors. In this study, the change in the content of chlorophyll α and β carotene in the control and experimental biomass was monitored (Figure 6).

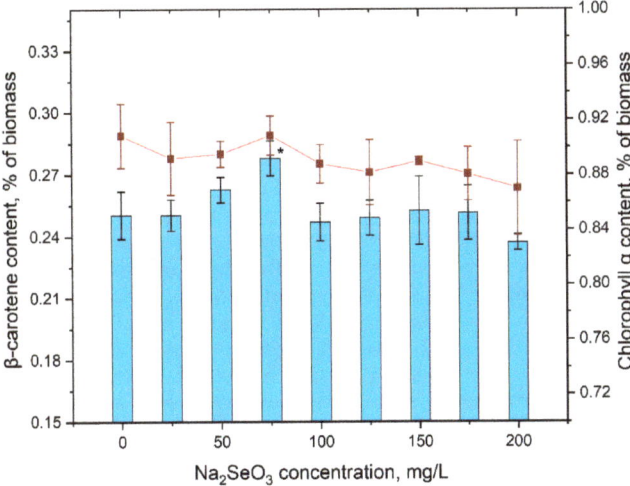

Figure 6. The content of β carotene and chlorophyll a in *Arthrospira platensis* biomass under conditions of supplementing the nutrient medium with sodium selenite (* $p < 0.05$ for differences between control and experimental sample; n = 3). The error bars represent the standard deviation of the measurements.

As can be seen from the presented data, the content of chlorophyll α and β carotene did not change, or changed insignificantly in the biomass of spirulina grown on the medium with the addition of sodium selenite at concentrations up to 200 mg/L. Only in the case of a concentration of 75 mg/L, an increase by 11.2% (statistically veridical at a level of significance $p < 0.05$) in the carotene content in the spirulina biomass was observed. For other Na$_2$SeO$_3$ concentrations, the level of these two extremely important pigments was comparable with control values. To a large extent, the quantitative stability of these cellular components ensures an adequate level of assimilation processes and, therefore, the growth of cyanobacteria under conditions of the presence of selenite ions in the nutrient medium.

The greatest quantitative changes in biochemical parameters were observed at the level of lipids and end products of their peroxidation (Figure 7).

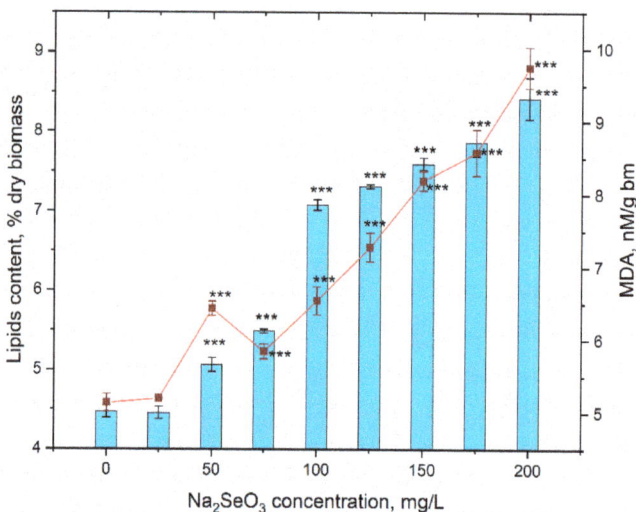

Figure 7. The content of lipids and MDA in *Arthrospira platensis* biomass under conditions of supplementing the nutrient medium with sodium selenite (*** $p < 0.0005$ for differences between control and experimental sample; n = 3). The error bars represent the standard deviation of the measurements.

Starting from a sodium selenite concentration of 50 mg/L, a significant increase in the lipid content in the spirulina biomass was observed. This increase was in a dose-dependent manner, amplified by a rise in salt concentration in the medium. Thus, at a Na$_2$SeO$_3$ concentration of 50 mg/L, the content of lipids increased by 13.3% compared to the control; then, at a concentration of 200 mg/L, this increase was 88.5%. Practically the same pattern was observed for the malonic dialdehyde and its content in biomass increased by 13.7–89.7% compared to control. The content of MDA increased significantly in spirulina biomass grown in the medium with the addition of sodium selenite, but this is associated with a proportional increase in the content of total lipids in the biomass. Thus, in relative terms, the amount of lipid degradation products related to the amount of lipids remained a stable parameter.

The quantitative increase in lipid peroxidation products in spirulina cells under the influence of sodium selenite was also attested by other researchers [28], but in the mentioned study, the content of total lipids in the biomass was not monitored. In general, however, an increase in the amount of lipid oxidative degradation products is an indicator of toxic effects and should be treated with caution.

3.5. Gene Expression Analysis

In the present study, one of the aims was to highlight the change in the expression of some genes associated with the generalized response to stress conditions. Figure 8 shows the results with reference to the change in the expression of *FeSOD*, *hsp90*, *rbcL*, *GOGAT* and *POD* genes in *A. platensis* cells that were grown in the presence of sodium selenite applied at a concentration of 200 mg/L.

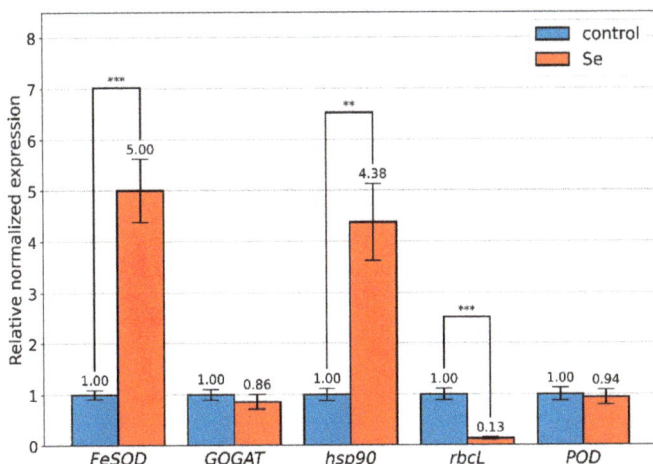

Figure 8. Relative expression difference in selected genes between control *A. platensis* and after Se treatment (200 mg/L of Na$_2$SeO$_3$), ** $p < 0.01$, *** $p < 0.001$; n = 3. The error bars represent the standard deviation of the measurements.

To estimate the possible response to selenium treatment, the transcriptional abundance of heat-shock protein (*HSP90*), glutamate synthase (*GOGAT*), iron-superoxide dismutase (*FeSOD*), peroxidase (*POD*) and the large subunit of Rubisco (*rbcL*) genes was measured in the control and *A. platensis* biomass treated with 200 mg/L of sodium selenite. The results showed the increased expression level for *FeSOD* and *hsp90* genes and decreased transcription level for *rbcL* (Figure 8). The expression levels of *GOGAT* and *POD* genes were not significantly different between control cells and cells treated with selenium.

Previous research has shown that cyanobacterial cells can change the expression of a number of genes in response to stressful conditions, such as genes encoding heat-shock proteins or genes involved in metabolic activity and the production of antioxidant enzymes [39–41]. In the present study, the expression level of five genes—*FeSOD*, *GOGAT*, *hsp90*, *rbcL* and *POD*—was assessed.

The *FeSOD* and *POD* genes encode two of the most important first-line antioxidant enzymes of spirulina—Fe-component superoxide dismutase and peroxidase. The first enzyme catalyzes the dismutation reaction of the superoxide radical into hydrogen peroxide, and the peroxidase—the decomposition reaction of peroxide into water and oxygen. In addition to the production of antioxidant enzymes, the defense mechanisms of cyanobacteria in relation to stress factors include the inclusion of other genes associated with the stress response, such as the genes of heat-shock proteins (HSPs), ribulose bisophosphate carboxylase/oxygenase (Rubisco), glutamate synthase and GOGAT [42].

The increase in the relative expression of the *FeSOD* gene is predictable under stress conditions. In the present study, stress conditions were not confirmed by means of monitoring changes in biomass production or changes in biochemical parameters. In contrast, a 5-fold increase in *FeSOD* expression compared to the control can be considered as evidence of unfavorable oxidative stress in *A. platensis* cells grown in the presence of 200 mg/L of sodium selenite. The situation may become more complicated over time because the expression level of POD, which is encoding peroxidase, remained at the control level, this enzyme being responsible for removal of the reaction product (H$_2$O$_2$) formed as a result of superoxide dismutase activity.

HSPs are involved not only in the case of thermal stress, but also in other forms of stress, because they ensure the stability of proteins and the avoidance of their misfolding [42–45]. In the case of the influence of selenium on *A. platensis*, an increase in the transcriptional abundance of *hsp90* (4.4-times compared to the control) was observed, which points to a need to maintain the stability of cellular proteins. The same upregulation of *hsp90* was observed, for example, when growing spirulina at temperatures lower or higher than the optimum one [42]. The role of HSPs in the adaptation to oxidative stress conditions caused by physical (radiation) and chemical factors (action of methyl viologen) was demonstrated in *Synechococcu* sp. [46].

Gene *rbcL* (the large subunit of ribulose bisphosphate carboxylase/oxygenase; Rubisco) is involved in ensuring the growth of photosynthetic organisms and the metabolic activity necessary to maintain life, as the *GOGAT* gene, which codes the glutamate synthase enzyme [42–46]. *rbcL* is directly involved in carbon assimilation, thus, ensuring the process of photosynthesis and organism growth, and GOGAT catalyzes the reaction of glutamate formation from glutamines and 2-oxoglutarate. No change in the expression level of *GOGAT* was observed, but the reduction in the expression level of *rbcl* was very significant (7.7-times). Downregulation of *rbcl* is observed in cyanobacteria under different stress conditions, for example, under conditions of low temperature or osmotic stress [47].

Thus, the change in the expression of three genes associated with stress from the five studied confirms that the culture of *A. platensis* was exposed to stress caused by the presence of selenium in the nutrient medium, although, apparently, the cyanobacterial culture developed normally.

3.6. Ultrastructural Changes in Arthrospira platensis under the Influence of Selenium

The changes that take place in the ultrastructure of *A. platensis* cells can be seen in Figure 9. In Figure 9A–C, the typical spirulina ultrastructure can be seen. The most significant structures characteristic of the cell cytoplasm are compactly arranged thylakoids, well-structured thylakoids with well-visualized membranes. The cytoplasm is dense and adheres tightly to the cytoplasmic membrane. The cell wall is tightly attached to the cytoplasmic membrane; it is dense and well visible.

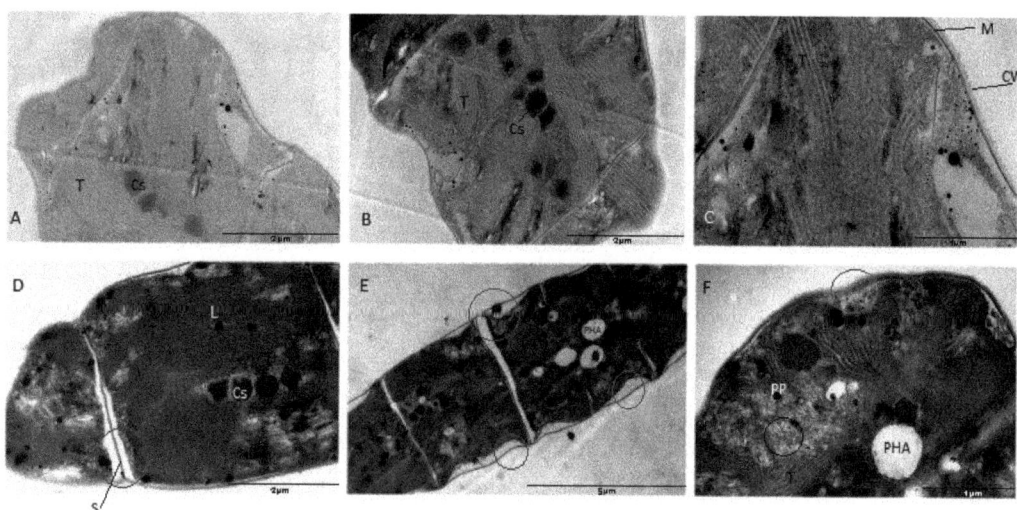

Figure 9. The ultrastructure of *A. platensis* cells (**A–C**)—control, (**D–F**)—cells grown in the medium containing sodium selenite in concentration of 200 mg/L. Cs—carboxisomes; T—thylacoids; M—membrane; CW—cell wall; L—lipid inclusions; PHA—polyhydroxyalkanoates; PP—polyphosphate bodies; S—septum; black circles—disintegration and disorganization of thylakoids, red circles—space between cell membrane and cell wall/intercellular septum, blue circle—modification of the cell wall/exopolysaccharides density. Scale bar in (**A**) 2 µm, (**B**) 2 µm, (**C**) 1 µm, (**D**) 2 µm, (**E**) 5 µm, (**F**) 1 µm.

Under the action of selenium, changes in the structure of spirulina occurred, the main ones being a decrease in the degree of compaction of thylakoids, appearance of translucent spaces between the cytoplasmic membrane and the cell wall, appearance of large polyphosphate bodies, polyhydroxyalkanoates (which are a form of storage of reserves of carbon) and a decompaction of the cell wall.

Such changes are typical for *A. platensis* cells exposed to various external actions and stress factors. For example, in the case of the action of heavy metals (Cd), first of all, the disintegration and disorganization of thylakoid membranes, as well as the appearance of polyphosphate bodies [48] are observed. Destruction of photosynthetic membranes and storage of carbon reserves in the form of polyhydroxyalkanoates was also observed in spirulina under conditions of insufficient nutrients, especially nitrogen [49].

4. Conclusions

Arthrospira platensis easily tolerates the presence of the high concentrations of selenium (up to 125 mg/L) in the medium, growth and biomass accumulation, being within the limits of the values characteristic for control biomass. Moreover, at selenium concentrations up to 50 mg/L, the amount of biomass accumulated during the cultivation cycle increased by up to 18% compared to the control. Under these conditions, most of the biochemical parameters remained within the normal limits for the species, although the values obtained in the experimental variants may differ significantly compared to the control. The content of the primary photosynthetic pigment did not change significantly, regardless of the selenium concentration applied. The content of lipids and carbohydrates in biomass increased with increasing sodium selenite concentration added to the nutrient medium. The content of protein and phycobilin also increased, and the dose-dependent character of this relationship was maintained up to a concentration of sodium selenite of 175 mg/L. With an increase in the content of lipids, the level of malonic dialdehyde in the cells also increased.

During the cultivation of spirulina in the nutrient media containing sodium selenite, the culture actively accumulates selenium in a dose-dependent manner. Most of the bioaccumulated selenium was determined in the protein (47.5% of the accumulated selenium) and the lipid (24.1%) fractions. In the protein fraction, selenium nanoparticles with a size of 2–8 nm were formed.

As a result of the action of selenium on *Arthrospira platensis*, the expression of genes associated with stress changed. Thus, the expression level of *FeSOD* and *Hsp90* increased, which may be associated with the need to manage the increased flow of reactive oxygen species and to stabilize the proteins subjected to the action of the xenobiotics. The significant decrease in the level of *rbcl* expression could cause a decrease in the efficiency of carbon fixation from carbon dioxide.

Selenium also caused ultrastructural changes in *Arthrospira platensis* expressed in the damage and disorder of thylakoids, the detachment of the cytoplasmic membrane from the cell wall, the change in the density of the cell wall and the formation of carbon reserves in the cells, indicating negative effects of selenium ions.

Thus, *Arthrospira platensis* tolerates high concentrations of selenium, accumulates important amounts of this element and carries out the biosynthesis of selenium nanoparticles, which are mainly located in the protein and lipid fractions. The process is accompanied by biochemical, ultrastructural and gene expression changes associated with the response of spirulina to stress conditions.

Supplementary Materials: The following supporting information can be downloaded at: https://www.mdpi.com/article/10.3390/ma16020852/s1, Table S1: List of primers.

Author Contributions: Conceptualization, L.C. and I.Z.; methodology, L.C., N.Y., T.C., L.R., I.T. and E.K.; software, D.G.; formal analysis, D.G.; investigation, N.Y., I.Z., L.C., T.C., K.T. and L.R.; data curation, L.R., L.C., K.T., and I.Z.; writing—original draft preparation, L.C. and I.Z.; writing—review and editing, all authors; visualization, L.C. and I.Z. All authors have read and agreed to the published version of the manuscript.

Funding: The results were obtained within the project 20.80009.5007.05 funded by the NARD, Republic of Moldova.

Institutional Review Board Statement: Not applicable.

Informed Consent Statement: Not applicable.

Data Availability Statement: Not applicable.

Conflicts of Interest: The authors declare no conflict of interest.

References

1. Khurana, A.; Tekula, S.; Saifi, M.A.; Venkatesh, P.; Godugu, C. Therapeutic Applications of Selenium Nanoparticles. *Biomed. Pharmacother.* **2019**, *111*, 802–812. [CrossRef] [PubMed]
2. Ferro, C.; Florindo, H.F.; Santos, H.A. Selenium Nanoparticles for Biomedical Applications: From Development and Characterization to Therapeutics. *Adv. Healthc. Mater.* **2021**, *10*, 2100598. [CrossRef] [PubMed]
3. Geoffrion, L.D.; Hesabizadeh, T.; Medina-Cruz, D.; Kusper, M.; Taylor, P.; Vernet-Crua, A.; Chen, J.; Ajo, A.; Webster, T.J.; Guisbiers, G. Naked Selenium Nanoparticles for Antibacterial and Anticancer Treatments. *ACS Omega* **2020**, *5*, 2660–2669. [CrossRef] [PubMed]
4. Shoeibi, S.; Mashreghi, M. Biosynthesis of Selenium Nanoparticles Using Enterococcus Faecalis and Evaluation of Their Antibacterial Activities. *J. Trace Elem. Med. Biol.* **2017**, *39*, 135–139. [CrossRef] [PubMed]
5. Vinković Vrček, I. Selenium Nanoparticles: Biomedical Applications. In *Selenium Molecular and Integrative Toxicology*; Michalke, B., Ed.; Springer International Publishing: Cham, Switzerland, 2018; pp. 393–412. ISBN 978-3-319-95389-2.
6. Nayak, V.; Singh, K.R.; Singh, A.K.; Singh, R.P. Potentialities of Selenium Nanoparticles in Biomedical Science. *New J. Chem.* **2021**, *45*, 2849–2878. [CrossRef]
7. Cristallini, C.; Vitale, E.; Giachino, C.; Rastaldo, R. Nanoengineering in Cardiac Regeneration: Looking Back and Going Forward. *Nanomaterials* **2020**, *10*, 1587. [CrossRef]
8. Kalishwaralal, K.; Jeyabharathi, S.; Sundar, K.; Selvamani, S.; Prasanna, M.; Muthukumaran, A. A Novel Biocompatible Chitosan–Selenium Nanoparticles (SeNPs) Film with Electrical Conductivity for Cardiac Tissue Engineering Application. *Mater. Sci. Eng. C* **2018**, *92*, 151–160. [CrossRef]
9. Yang, C.; Tian, A.; Li, Z. Reversible Cardiac Hypertrophy Induced by PEG-Coated Gold Nanoparticles in Mice. *Sci. Rep.* **2016**, *6*, 20203. [CrossRef]
10. Xiao, S.; Mao, L.; Xiao, J.; Wu, Y.; Liu, H. Selenium Nanoparticles Inhibit the Formation of Atherosclerosis in Apolipoprotein E Deficient Mice by Alleviating Hyperlipidemia and Oxidative Stress. *Eur. J. Pharmacol.* **2021**, *902*, 174120. [CrossRef]
11. Weekley, C.M.; Harris, H.H. Which Form Is That? The Importance of Selenium Speciation and Metabolism in the Prevention and Treatment of Disease. *Chem. Soc. Rev.* **2013**, *42*, 8870. [CrossRef]
12. Pyrzynska, K.; Sentkowska, A. Biosynthesis of Selenium Nanoparticles Using Plant Extracts. *J. Nanostruct. Chem.* **2022**, *12*, 467–480. [CrossRef]
13. Bafghi, M.H.; Darroudi, M.; Zargar, M.; Zarrinfar, H.; Nazari, R. Biosynthesis of Selenium Nanoparticles by *Aspergillus Flavus* and *Candida Albicans* for Antifungal Applications. *Micro Nano Lett.* **2021**, *16*, 656–669. [CrossRef]
14. Diko, C.S.; Zhang, H.; Lian, S.; Fan, S.; Li, Z.; Qu, Y. Optimal Synthesis Conditions and Characterization of Selenium Nanoparticles in Trichoderma Sp. WL-Go Culture Broth. *Mater. Chem. Phys.* **2020**, *246*, 122583. [CrossRef]
15. Srivastava, N.; Mukhopadhyay, M. Biosynthesis and Structural Characterization of Selenium Nanoparticles Using Gliocladium Roseum. *J. Clust. Sci.* **2015**, *26*, 1473–1482. [CrossRef]
16. Zhang, H.; Zhou, H.; Bai, J.; Li, Y.; Yang, J.; Ma, Q.; Qu, Y. Biosynthesis of Selenium Nanoparticles Mediated by Fungus Mariannaea Sp. HJ and Their Characterization. *Colloids Surf. A: Physicochem. Eng. Asp.* **2019**, *571*, 9–16. [CrossRef]
17. Presentato, A.; Piacenza, E.; Anikovskiy, M.; Cappelletti, M.; Zannoni, D.; Turner, R.J. Biosynthesis of Selenium-Nanoparticles and -Nanorods as a Product of Selenite Bioconversion by the Aerobic Bacterium Rhodococcus Aetherivorans BCP1. *New Biotechnol.* **2018**, *41*, 1–8. [CrossRef]
18. Shakibaie, M.; Khorramizadeh, M.R.; Faramarzi, M.A.; Sabzevari, O.; Shahverdi, A.R. Biosynthesis and Recovery of Selenium Nanoparticles and the Effects on Matrix Metalloproteinase-2 Expression. *Biotechnol. Appl. Biochem.* **2010**, *56*, 7–15. [CrossRef]
19. Ullah, A.; Yin, X.; Wang, F.; Xu, B.; Mirani, Z.A.; Xu, B.; Chan, M.W.H.; Ali, A.; Usman, M.; Ali, N.; et al. Biosynthesis of Selenium Nanoparticles (via Bacillus Subtilis BSN313), and Their Isolation, Characterization, and Bioactivities. *Molecules* **2021**, *26*, 5559. [CrossRef]
20. Chavan, R.B.; Bhattacharjee, M.B. Role of Alginate and Oxalic Acid in Ameliorating Se Toxicity in Hapalosiphon Cyanobacterium. *Int. J. Curr. Microbiol. App. Sci.* **2016**, *5*, 132–139. [CrossRef]
21. Gouget, B.; Avoscan, L.; Sarret, G.; Collins, R.; Carrière, M. Resistance, Accumulation and Transformation of Selenium by the Cyanobacterium *Synechocystis Sp.* PCC 6803 after Exposure to Inorganic Se^{VI} or Se^{IV}. *Radiochim. Acta* **2005**, *93*, 683–689. [CrossRef]
22. Zhou, C.; Huang, J.-C.; Gan, X.; He, S.; Zhou, W. Selenium Uptake, Volatilization, and Transformation by the Cyanobacterium Microcystis Aeruginosa and Post-Treatment of Se-Laden Biomass. *Chemosphere* **2021**, *280*, 130593. [CrossRef] [PubMed]

23. Banerjee, M.; Kalwani, P.; Chakravarty, D.; Singh, B.; Ballal, A. Functional and Mechanistic Insights into the Differential Effect of the Toxicant 'Se(IV)' in the Cyanobacterium Anabaena PCC 7120. *Aquat. Toxicol.* **2021**, *236*, 105839. [CrossRef] [PubMed]
24. Afzal, B.; Yasin, D.; Husain, S.; Zaki, A.; Srivastava, P.; Kumar, R.; Fatma, T. Screening of Cyanobacterial Strains for the Selenium Nanoparticles Synthesis and Their Anti-Oxidant Activity. *Biocatal. Agric. Biotechnol.* **2019**, *21*, 101307. [CrossRef]
25. Zinicovscaia, I.; Chiriac, T.; Cepoi, L.; Rudi, L.; Culicov, O.; Frontasyeva, M.; Rudic, V. Selenium Uptake and Assessment of the Biochemical Changes in *Arthrospira* (*Spirulina*) *Platensis* Biomass during the Synthesis of Selenium Nanoparticles. *Can. J. Microbiol.* **2017**, *63*, 27–34. [CrossRef] [PubMed]
26. Alipour, S.; Kalari, S.; Morowvat, M.H.; Sabahi, Z.; Dehshahri, A. Green Synthesis of Selenium Nanoparticles by Cyanobacterium Spirulina Platensis (Abdf2224): Cultivation Condition Quality Controls. *BioMed Res. Int.* **2021**, *2021*, 1–11. [CrossRef]
27. Afzal, B.; Fatma, T. Selenium Nanoparticles: Green Synthesis and Exploitation. In *Emerging Technologies for Nanoparticle Manufacturing*; Patel, J.K., Pathak, Y.V., Eds.; Springer International Publishing: Cham, Switzerland, 2021; pp. 473–484. ISBN 978-3-030-50702-2.
28. Cepoi, L.; Zinicovscaia, I.; Zosim, L.; Chiriac, T.; Rudic, V.; Rudi, L.; Djur, S.; Elenciuc, D.; Miscu, V.; Ludmila, B.; et al. Metals Removal by Cyanobacteria and Accumulation in Biomass. In *Cyanobacteria for Bioremediation of Wastewaters*; Zinicovscaia, I., Cepoi, L., Eds.; Springer International Publishing: Cham, Switzerland, 2016; pp. 61–111. ISBN 978-3-319-26749-4.
29. Cepoi, L.; Chiriac, T.; Rudi, L.; Djur, S.; Zosim, L.; Bulimaga, V.; Batir, L.; Elenciuc, D.; Rudic, V. Spirulina as a Raw Material for Products Containing Trace Elements. In *Recent Advances in Trace Elements*; Chojnacka, K., Saeid, A., Eds.; John Wiley & Sons, Ltd: Chichester, UK, 2018; pp. 403–420. ISBN 978-1-119-13378-0.
30. Cepoi, L.; Zinicovscaia, I.; Rudi, L.; Chiriac, T.; Rotari, I.; Turchenko, V.; Djur, S. Effects of PEG-Coated Silver and Gold Nanoparticles on Spirulina Platensis Biomass during Its Growth in a Closed System. *Coatings* **2020**, *10*, 717. [CrossRef]
31. Chen, T.; Zheng, W.; Wong, Y.-S.; Yang, F.; Bai, Y. Accumulation of Selenium in Mixotrophic Culture of Spirulina Platensis on Glucose. *Bioresour. Technol.* **2006**, *97*, 2260–2265. [CrossRef]
32. Li, Z.-Y.; Guo, S.-Y.; Li, L. Bioeffects of Selenite on the Growth of Spirulina Platensis and Its Biotransformation. *Bioresour. Technol.* **2003**, *89*, 171–176. [CrossRef]
33. Kang, H.; Rho, S.; Stiles, W.R.; Hu, S.; Baek, Y.; Hwang, D.W.; Kashiwagi, S.; Kim, M.S.; Choi, H.S. Size-Dependent EPR Effect of Polymeric Nanoparticles on Tumor Targeting. *Adv. Healthc. Mater.* **2020**, *9*, 1901223. [CrossRef]
34. Soo Choi, H.; Liu, W.; Misra, P.; Tanaka, E.; Zimmer, J.P.; Itty Ipe, B.; Bawendi, M.G.; Frangioni, J.V. Renal Clearance of Quantum Dots. *Nat. Biotechnol.* **2007**, *25*, 1165–1170. [CrossRef]
35. Choi, H.S.; Liu, W.; Liu, F.; Nasr, K.; Misra, P.; Bawendi, M.G.; Frangioni, J.V. Design Considerations for Tumour-Targeted Nanoparticles. *Nat. Nanotech.* **2010**, *5*, 42–47. [CrossRef] [PubMed]
36. Ferretti, A.M.; Usseglio, S.; Mondini, S.; Drago, C.; La Mattina, R.; Chini, B.; Verderio, C.; Leonzino, M.; Cagnoli, C.; Joshi, P.; et al. Towards Bio-Compatible Magnetic Nanoparticles: Immune-Related Effects, in-Vitro Internalization, and in-Vivo Bio-Distribution of Zwitterionic Ferrite Nanoparticles with Unexpected Renal Clearance. *J. Colloid Interface Sci.* **2021**, *582*, 678–700. [CrossRef] [PubMed]
37. Zhang, Y.; Jin, J.; Huang, B.; Ying, H.; He, J.; Jiang, L. Selenium Metabolism and Selenoproteins in Prokaryotes: A Bioinformatics Perspective. *Biomolecules* **2022**, *12*, 917. [CrossRef]
38. Chen, T.-F.; Zheng, W.-J.; Wong, Y.-S.; Yang, F. Selenium-Induced Changes in Activities of Antioxidant Enzymes and Content of Photosynthetic Pigments in *Spirulina Platensis*. *J. Integr. Plant Biol.* **2008**, *50*, 40–48. [CrossRef] [PubMed]
39. Jeamton, W.; Mungpakdee, S.; Sirijuntarut, M.; Prommeenate, P.; Cheevadhanarak, S.; Tanticharoen, M.; Hongsthong, A. A Combined Stress Response Analysis of Spirulina Platensis in Terms of Global Differentially Expressed Proteins, and MRNA Levels and Stability of Fatty Acid Biosynthesis Genes: Combined Stress Response Analysis of Spirulina Platensis. *FEMS Microbiol. Lett.* **2008**, *281*, 121–131. [CrossRef] [PubMed]
40. Ludwig, M.; Bryant, D.A. Synechococcus Sp. Strain PCC 7002 Transcriptome: Acclimation to Temperature, Salinity, Oxidative Stress, and Mixotrophic Growth Conditions. *Front. Microbio.* **2012**, *3*, 354. [CrossRef] [PubMed]
41. Nakamoto, H.; Suzuki, N.; Roy, S.K. Constitutive Expression of a Small Heat-Shock Protein Confers Cellular Thermotolerance and Thermal Protection to the Photosynthetic Apparatus in Cyanobacteria. *FEBS Lett.* **2000**, *483*, 169–174. [CrossRef]
42. Ismaiel, M.M.S.; Piercey-Normore, M.D. Gene Transcription and Antioxidants Production in *Arthrospira (Spirulina) Platensis* Grown under Temperature Variation. *J. Appl. Microbiol.* **2021**, *130*, 891–900. [CrossRef]
43. Deng, Y.; Zhan, Z.; Tang, X.; Ding, L.; Duan, D. Molecular Cloning and Expression Analysis of RbcL CDNA from the Bloom-Forming Green Alga Chaetomorpha Valida (Cladophorales, Chlorophyta). *J. Appl. Phycol.* **2014**, *26*, 1853–1861. [CrossRef]
44. Liu, L.; Wang, J.; Han, Z.; Sun, X.; Li, H.; Zhang, J.; Lu, Y. Molecular Analyses of Tomato GS, GOGAT and GDH Gene Families and Their Response to Abiotic Stresses. *Acta Physiol. Plant* **2016**, *38*, 229. [CrossRef]
45. Panyakampol, J.; Cheevadhanarak, S.; Sutheeworapong, S.; Chaijaruwanich, J.; Senachak, J.; Siangdung, W.; Jeamton, W.; Tanticharoen, M.; Paithoonrangsarid, K. Physiological and Transcriptional Responses to High Temperature in Arthrospira (*Spirulina*) Platensis C1. *Plant Cell Physiol.* **2015**, *56*, 481–496. [CrossRef] [PubMed]
46. Hossain, M.M.; Nakamoto, H. Role for the Cyanobacterial HtpG in Protection from Oxidative Stress. *Curr. Microbiol.* **2003**, *46*, 70–76. [CrossRef] [PubMed]
47. Mori, S.; Castoreno, A.; Lammers, P.J. Transcript Levels of *RbcR1*, *NtcA*, and *RbcL/S* Genes in Cyanobacterium *Anabaena* Sp. PCC 7120 Are Downregulated in Response to Cold and Osmotic Stress. *FEMS Microbiol. Lett.* **2002**, *213*, 167–173. [CrossRef] [PubMed]

48. Rangsayatorn, N.; Upatham, E.S.; Kruatrachue, M.; Pokethitiyook, P.; Lanza, G.R. Phytoremediation Potential of Spirulina (Arthrospira) Platensis: Biosorption and Toxicity Studies of Cadmium. *Environ. Pollut.* **2002**, *119*, 45–53. [CrossRef] [PubMed]
49. Deschoenmaeker, F.; Facchini, R.; Cabrera Pino, J.C.; Bayon-Vicente, G.; Sachdeva, N.; Flammang, P.; Wattiez, R. Nitrogen Depletion in Arthrospira Sp. PCC 8005, an Ultrastructural Point of View. *J. Struct. Biol.* **2016**, *196*, 385–393. [CrossRef]

Disclaimer/Publisher's Note: The statements, opinions and data contained in all publications are solely those of the individual author(s) and contributor(s) and not of MDPI and/or the editor(s). MDPI and/or the editor(s) disclaim responsibility for any injury to people or property resulting from any ideas, methods, instructions or products referred to in the content.

Article

Effect of Core–Shell Rubber Nanoparticles on the Mechanical Properties of Epoxy and Epoxy-Based CFRP

Tatjana Glaskova-Kuzmina [1,*], Leons Stankevics [1], Sergejs Tarasovs [1], Jevgenijs Sevcenko [1], Vladimir Špaček [2], Anatolijs Sarakovskis [3], Aleksejs Zolotarjovs [3], Krishjanis Shmits [3] and Andrey Aniskevich [1]

[1] Institute for Mechanics of Materials, University of Latvia, Jelgavas 3, LV-1004 Riga, Latvia
[2] Synpo, S. K. Neumanna 1316, 530 02 Pardubice, Czech Republic
[3] Institute of Solid State Physics, Kengaraga 8, LV-1063 Riga, Latvia
* Correspondence: tatjana.glaskova-kuzmina@lu.lv

Abstract: The aim of the research was to estimate the effect of core–shell rubber (CSR) nanoparticles on the tensile properties, fracture toughness, and glass transition temperature of the epoxy and epoxy-based carbon fiber reinforced polymer (CFRP). Three additives containing CSR nanoparticles were used for the research resulting in a filler fraction of 2–6 wt.% in the epoxy resin. It was experimentally confirmed that the effect of the CSR nanoparticles on the tensile properties of the epoxy resin was notable, leading to a reduction of 10–20% in the tensile strength and elastic modulus and an increase of 60–108% in the fracture toughness for the highest filler fraction. The interlaminar fracture toughness of CFRP was maximally improved by 53% for ACE MX 960 at CSR content 4 wt.%. The glass transition temperature of the epoxy was gradually improved by 10–20 °C with the increase of CSR nanoparticles for all of the additives. A combination of rigid and soft particles could simultaneously enhance both the tensile properties and the fracture toughness, which cannot be achieved by the single-phase particles independently.

Keywords: epoxy; CFRP; core–shell rubber nanoparticles; tensile properties; fracture toughness; glass transition temperature

1. Introduction

Epoxy resins having relatively high tensile strength and modulus of elasticity, a low creep, and a good stability at elevated temperatures are extensively used as matrices in composite technology for different applications [1,2]. Nevertheless, due to high crosslinking, they are characterized by a high degree of brittleness and a poor resistance to crack initiation/propagation [3].

Their toughness could be improved by adding core–shell rubber (CSR) nanoparticles that are made of a soft rubbery core and a rigid shell around it which are mainly manufactured by emulsion polymerization and then added to the polymer resins. In comparison with the phase-separating rubbers, this method allows the advantage of controlling the particle size by changing the core and shell diameters [4]. The materials that are usually used for the core are siloxane, butadiene, and acrylate polyurethane, while poly (methyl methacrylate) (PMMA) is preferred to be used as the shell materials due to it having a good compatibility with the epoxy polymers [5,6].

It was determined that the addition of CSR particles led to a significant reduction in the tensile properties of the epoxy resin (DGEBA) and almost no effect on its glass transition temperature (T_g) [3]. For the 15 wt.% content of CSR in the epoxy, the elastic modulus and tensile strength of the epoxy were diminished by 27 and 36%, respectively. However, for the same composition of CSR filler particles, the fracture energy was improved by 550%. Similar results were obtained for the epoxy that was filled with CSR particles from 0 to 38 vol.%, revealing a gradual increase in the T_g and Poisson's ratio and a significant

decrease in the tensile and compressive properties of the CSR-modified composites which were explained by rubber having a lower Young's modulus and a higher Poisson's ratio in comparison with the epoxy [7]. By using SEM of fracture surfaces and analytical models, several toughening mechanisms (shear band yielding, core-to-shell debonding and plastic void growth) were defined [3,7].

In general, the fracture toughness of epoxy was improved by adding both rigid and soft particles [8,9]. The rigid particles toughen the epoxy through crack pinning and crack deflection/bifurcation effects, while the toughening mechanisms of the soft particles are filler-debonding, and the subsequent void grows as well as the matrix shear band.

The research aimed to estimate the effect of core–shell rubber (CSR) particles on the tensile properties, fracture toughness and glass transition temperature of the epoxy and epoxy-based CFRP. The novelty of this work is in the multi-step approach for the evaluation of the toughening effects for both the epoxy and epoxy-based CFRP and considering their mechanical properties. The application of the proposed solution with improved fracture toughness both for the epoxy and epoxy-based CFRP could broaden their use in aerospace, automotive, marine and sporting goods due to them having a longer lifetime and enhanced safety features.

2. Materials and Methods

2.1. Materials

CHS-Epoxy 582 (Spolchemie, Usti nad Labem, Czech Republic) [10] was used as matrix material. It is a diglycidyl ether of bisphenol A (DGEBA) with a reactive diluent that has an epoxide equivalent weight (EEW) of 165–173 g/mol. This epoxy resin is recommended for different applications in composites, adhesives, wind energy, construction, electronics and corrosive coatings. The hardener Telalit 0420 (Spolchemie, Usti nad Labem, Czech Republic) which is a cycloaliphatic amine was mixed with epoxy resin at a ratio of 100:25 [11].

Three additives containing CSR nanoparticles which were dispersed in DGEBA with different particle sizes and core material ACE MX 125, 156 and 960 were supplied by Kaneka (Westerlo, Belgium). The information regarding core material and CSR size are given in Table 1. For all of the additives that were studied, the concentration of CSR nanoparticles in DGEBA was 25 wt.%, the shell material was PMMA, and the density was 1.1 g/cm^3 [12]. Carbon fiber fabric KC (0/90) in plane weave and of a specific surface of 160 g/m^2 was supplied by Havel Composites (Svésedlice, Czech Republic) [13] and used for the manufacturing of CFRP laminates.

Table 1. CSR types dispersed in the epoxy [12].

Additive Name	Core Material	CSR Size, nm
ACE MX-125	Styrene butadiene	100
ACE MX-156	Polybutadiene	100
ACE MX-960	Siloxane	300

2.2. Manufacturing of the Test Samples

For pure epoxy samples, the epoxy resin was manually mixed with the hardener for approx. 10 min and the mixture was further degassed by using the vacuum pump. For CSR-modified epoxy resin, a certain content of CSR nanoparticles (2, 4, and 6 wt.%) was added to the epoxy and manually mixed, degassed, and then mixed with the hardener for approx. 10 min. After degassing, all of the mixtures were poured into silicon molds. The curing and post-curing conditions were chosen based on supplier recommendations [10]: overnight at room temperature (RT), 2 h at 60 °C, 1 h at 80 °C, and 1 h at 120 °C.

The silicon molds were used for the manufacturing of the test samples to determine the tensile properties [14,15] and fracture toughness [16] of the epoxy and epoxy modified with CSR particles. Thus, five dog-bone samples and five tapered double cantilever beam (TDCB) samples were manufactured for each test and CSR particle type and each filler fraction.

Double cantilever beam (DCB) CFRP samples were produced by lay-up technology by using woven carbon fiber fabric $(0/90)_{12}$, which was cured at RT, cut into samples and post-cured as CSR-modified epoxy resin. The CSR nanoparticle fraction of 4 wt.% in the epoxy resin was used for the manufacturing of all of the CFRP plates based on the highest results of fracture toughness obtained for the modified epoxy in TDCB tests. At least five DCB samples were manufactured and tested for each CSR nanoparticle additive.

2.3. Testing Methods

2.3.1. Morphology Analysis

The morphology of the fracture surfaces for CFRP samples was examined by using a high-resolution SEM-FIB electron microscope Helios 5 UX (Thermo Scientific, Walthamm, MA, USA), which was operated at 1 kV and 25 pA with scan interlacing and integration to avoid charging.

2.3.2. Tensile Tests

For the test specimens of epoxy and epoxy that was modified with CSR nanoparticles, quasi-static tensile tests were performed by using Zwick 2.5 universal testing machine with a crosshead speed of 2 mm/min at RT. The tensile strength was defined as the maximal achieved value of stress in the specimen, and the elastic modulus was calculated from the slope of a secant line between 0.05 and 0.25% strain on a stress–strain plot. Five test samples per each CSR type and fraction were tested, and the values that are provided correspond to their arithmetic mean value.

2.3.3. Fracture Toughness Tests

A specimen with a sharp pre-crack is needed for the precise measurement of the stress intensity factor (SIF). TDCB specimens produced in the silicone molds had an initial notch with a 1 mm width and a round end, which may substantially increase the apparent fracture toughness of the material. Therefore, the initial pre-crack of 2–5 mm length was made in the specimen before testing by the sharp knife strike. Moreover, side grooves of a depth of approx. 2 mm were produced to minimize the crack deflection and to keep the crack path along the midplane of the specimens [16]. The tests were conducted on Zwick 2.5 universal testing machine at RT with a constant displacement rate of 1 mm/min. SIF was calculated using Mode I load for a crack length < 20 mm within a constant SIF region.

For the specimen without side grooves, the SIF can be evaluated as follows [16]:

$$K_{ng} = 2P_c \frac{\sqrt{m}}{b}, \qquad (1)$$

where P_c is the critical load, b is the width of the specimen, and m is a geometrical parameter, which for the specimen of the considered geometry equals 0.6 mm^{-1}. For the specimen with side grooves, Equation (2) should be modified as

$$K_g = K_{ng} \left(\frac{b}{b_n} \right)^{0.56}, \qquad (2)$$

where b_n is the reduced width of the specimen at the grooves' location, and the exponent value was determined from a series of 3D finite element simulations with grooves of different depths (see Appendix A).

2.3.4. Interlaminar Fracture Toughness Tests

The Mode I interlaminar fracture toughness tests of carbon 0/90 woven fabric laminates were carried out according to ASTM: D5528 [17] using specimens with dimensions of 25 × 3 × 125 mm^3. Though this standard was specified for unidirectional laminates, it has been successfully applied for laminates with different lay-up configurations [18]. According to this standard, a linear elastic behavior is assumed in the calculation of strain energy release rate, which is reasonable when the zone of damage at the delamination front is small relative to the thickness of the DCB sample. Opening Mode I interlaminar fracture toughness, G_{IC}, was evaluated from the load–deflection curve at the point of deviation from linearity (NL). The NL calculation of G_{IC} considers that the delamination starts to grow from the insert in the interior of the specimen at this point. The tests were performed by using Zwick 2.5 testing machine with a crosshead speed of 1 mm/min at RT and Canon EOS40D to record photos every 3 s for the analysis of the crack propagation until a failure occurred. ImageJ 1.38x software [19] was used to estimate the delamination length in DCB samples. At least five DCB samples per each CSR type at 4 wt.% in the epoxy resin used for the impregnation of cross-ply CFRP laminates were tested.

The Modified Beam Theory [17] method was used for the calculation of Mode I interlaminar fracture toughness assuming the correction for the rotation at the delamination front (Δ):

$$G_I = \frac{3P\delta}{2b(a + |\Delta|)}, \qquad (3)$$

where P is the load, δ is the load point displacement, a is the delamination length, and Δ is determined experimentally by generating the least squares plot of the cube root of compliance as a function of delamination length.

Moreover, for the specimens with loading blocks, two correction parameters—a parameter F accounting for the shortening of the moment arm and the tilting of the end blocks and a displacement parameter N accounting for the stiffening of the specimen by the blocks—are recommended [17]:

$$F = 1 - \frac{3}{10}\left(\frac{\delta}{a}\right)^2 - \frac{3}{2}\left(\frac{\delta t}{a^2}\right), \qquad (4)$$

$$N = 1 - \left(\frac{L'}{a}\right)^3 - \frac{9}{8}\left[1 - \left(\frac{L'}{a}\right)^2\right]\left(\frac{\delta t}{a^2}\right) - \frac{9}{35}\left(\frac{\delta t}{a^2}\right)^2, \qquad (5)$$

where L' and t are the geometrical parameters of the blocks.

Then, the corrected formula for interlaminar fracture toughness by using the Modified Beam Theory method takes the form:

$$G_I = \frac{3P\delta}{2b(a + |\Delta|)} \cdot \left(\frac{F}{N}\right). \qquad (6)$$

2.3.5. Density Measurements

The density of the epoxy and epoxy that was modified with CSR particles was defined at RT by using hydrostatic weighing in isopropyl alcohol and Mettler Toledo XS205DU balance with a precision of ±0.05 mg. First, the density of isopropyl alcohol was determined by using a sinker of a known volume of 10 cm^3. Then, the mass of the samples was registered in the air (m_a) and the liquid of known density (m_l). The density of the samples was determined by the formula:

$$\rho = \frac{m_a}{m_a - m_l}(\rho_l - \rho_a) + \rho_a, \qquad (7)$$

where ρ_l and ρ_a are the densities of the liquid (0.785 g/cm^3 for isopropyl alcohol) and air (0.0012 g/cm^3), respectively.

2.3.6. Thermal Mechanical Analysis

The glass transition temperature (T_g) of the epoxy and epoxy modified by CSR particles was estimated by conducting thermomechanical analysis (TMA) tests using TMA/SDTA841e (Mettler Toledo, Greifensee, Switzerland). The samples were heated from 30 to 150 °C at a heating rate of 3 °C/min and a force of 0.02 N, and then, they were subsequently cooled. According to ASTM standard E1545 [20], the glass transition corresponds to the inflection in the dimensional change when plotted against the temperature upon which the material changes from a hard (brittle) state into a soft (rubbery) state. The glass transition temperature was evaluated as the extrapolated onset of the kink in the experimental TMA curve, which was displayed as a function of temperature. At least three tests were conducted for each CSR type and fraction, and the values that are provided correspond to their arithmetic mean value.

3. Results and Discussion

3.1. Morphology of the Fracture Surface

The microscopy analysis of the fracture surfaces of the pure epoxy-based CFRP shown in Figure 1a revealed smooth and glassy surfaces with straight and sharp crack paths, which are characteristic of a brittle damage property and a weak resistance to crack initiation and propagation [8]. No delamination on the interface between the carbon fibers and the epoxy resin was noticed. The fracture surfaces of all four wt.% CSR-modified compositions that are provided in Figure 1b–d proved that the dispersion of CSR nanoparticles was good, and no significant agglomeration of CSR nanoparticles was found. The diameter of the CSR nanoparticles which were evaluated using ImageJ software was slightly higher than the data that are provided in Table 1 by the manufacturer. For the additives ACE MX-125 and ACE-MX-156, the diameter was very similar, 126 ± 28 nm and 126 ± 26 nm, accordingly. In comparison with these two additives, the ACE MX-960 CSR particles were much larger and had a wide diameter scatter—440 ± 248 nm. It could be an indication that most of the CSR nanoparticles were debonded as particles' debonding and subsequent plastic void growth is considered one of the most important toughening mechanisms for CSR/epoxy composites [3,8,9,21].

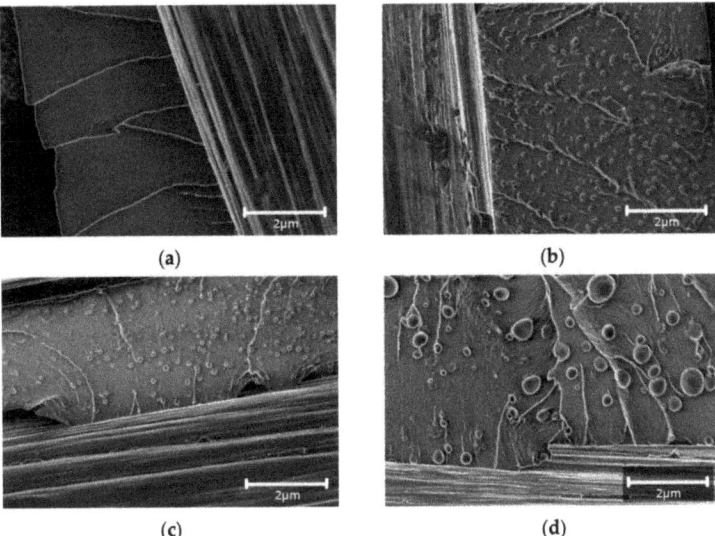

Figure 1. SEM images of fracture surface for the CFRP impregnated with the neat epoxy (**a**) and epoxy/CSR particles (4 wt.%) for the additives: ACE MX-125 (**b**), ACE MX-156 (**c**), and ACE MX-960 (**d**) (scale—2 μm, magnification—×25000).

For all of the CSR nanoparticle additives (Figure 1b,c), it can be noticed that the fracture surface was much rougher, and the crack paths became more curved following the CSR-particle circular shape. It could be also observed that in comparison with the undamaged CSR nanoparticles, the ones on the crack path were not perfectly spherical, thereby revealing their valuable contribution to the crack propagation process [8].

3.2. Density and Porosity

The results that were obtained for the density of epoxy/CSR nanoparticle composites are shown in Figure 2. According to Figure 2, the addition of all of the additives containing CSR nanoparticles led to a decrease in the density of CSR-modified epoxy. By using the mixture rule, the density of the composite material could be estimated:

$$\rho_c = \rho_f \times v_f + \rho_m \times (1 - v_f), \quad (8)$$

where ρ_f and ρ_m are the density of the filler (CSR nanoparticles) and polymer matrix (epoxy), respectively, and while v_f is the volume fraction of the filler, accordingly. The density of the epoxy was experimentally found to be 1.159 ± 0.002 g/cm^3. Considering the known density of the 25%-CSR-modified epoxy of 1.1 g/cm^3 [12], the density of the CSR particles was found to be 0.91 g/cm^3 [3]. Therefore, the addition of the filler particles of a lower density to the epoxy resin has resulted in a slight decrease (by approx. 2%) of the density for the composite. The higher the filler content was, then the lower that the density of the composite was.

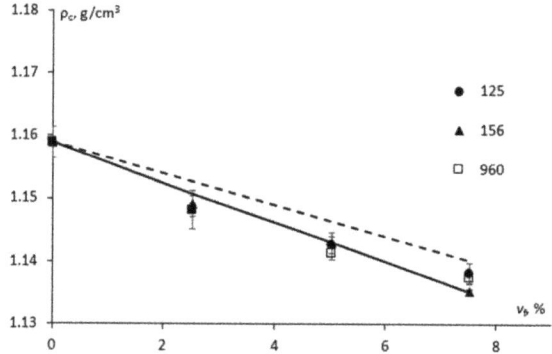

Figure 2. The density of the epoxy modified with different additives containing CSR nanoparticles (indicated on the graph) as a function of filler volume fraction (symbols—experimental data, dashed and solid lines—estimation by Equations (8) and (10), respectively.

The volume fraction of filler could be evaluated as follows [22]:

$$v_f = \frac{\rho_m \times c_f}{\rho_m \times c_f + \rho_f \times (1 - c_f)}, \quad (9)$$

where c_f is the weight fraction of the filler.

As seen in Figure 2, by using the mixture rule (Equation (8)), an overestimated value for the density of all of the CSR-modified epoxy materials was obtained. Therefore, efforts were made to evaluate the density of the composites having additional phase, air-filled pores, which could exist in the composites, and as a result, could lead to them having a lower density:

$$\rho_c = \rho_f \times v_f + \rho_p \times v_p + \rho_m \times (1 - v_f - v_p), \quad (10)$$

where ρ_p is the density of the air and v_p is the volume fraction of pores in the composites, respectively.

The volume fraction of the pores can be derived from Equation (10):

$$v_p = \frac{v_f \times \left(\rho_f - \rho_m\right) + \rho_m - \rho_c}{\rho_m - \rho_p}. \qquad (11)$$

According to Figure 2, it is obvious that though the estimated volume fraction of the pores was only 0.8–2% by using the modified mixture rule (Equation (10)), a better correlation with the experimental data was obtained. It was used in the calculation of the elastic modulus of the epoxy that was filled with the CSR nanoparticles.

3.3. Tensile Properties and Glass Transition Temperature

The stress–strain curves for the epoxy and epoxy that was filled with the ACE MX-156 CSR particles are given in Figure 3a. Analogous results were obtained for the other additives. According to Figure 3b, the elastic modulus of all of the studied materials significantly decreased with the increasing CSR content. The elastic modulus of 1.99 ± 0.04 GPa was found for the unmodified epoxy. For the modified epoxy, it had the lowest value for ACE MX-960 at all of the filler fractions, which could be attributed to the lower effective stiffness of the particles due to the highest CSR size in comparison with the other additives (see Table 1) [2]. The tensile strength of the epoxy (73 ± 3 MPa) decreased by approx. 10–20% with the addition of the CSR particles. Again, slightly lower tensile strengths were found for ACE MX-960 in comparison to the other CSR nanoparticles. Moreover, it could be noted from Figure 3a that the maximal deformation increased (from 4.9 ± 0.6% to 7.2 ± 0.5%) with the increase of the CSR content, thereby revealing the plasticization/softening effect resulting from the inclusion of softer filler particles in a brittle matrix.

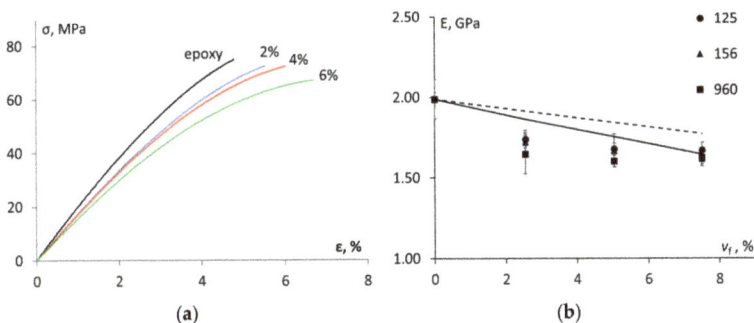

Figure 3. Stress–strain curves for the epoxy and epoxy filled with ACE MX-156 at different filler fractions indicated on the graph (**a**) and elastic modulus vs. filler volume fraction (dots -experimental results for different CSR nanoparticles, dashed and solid lines—evaluation by Equation (14) and by Equation (12), respectively (**b**).

Several analytical models, e.g., the Halpin–Tsai [3], Lewis–Nielsen [3,7] and Mori–Tanaka ones [7], were used to predict the significant reduction of the elastic modulus for the epoxy that was filled with the CSR nanoparticles. Most models epitomize an ideal composite by making several assumptions, e.g., that the polymer matrix and the filler particles are linear-elastic and isotropic, thereby having a perfect bond between them [23–25]. Moreover, the porosity and agglomeration of the filler particles negatively affecting the mechanical properties are usually neglected. In this work, the Hansen model [26,27] considering the spherical particles that are embedded in spherical shells of the matrix was used. It was applied in two steps: 1. to estimate the elastic modulus of the epoxy matrix containing a certain volume fraction of the pores from Equation (8), and 2. to determine the elastic modulus of the epoxy (with the pores) that was filled with the CSR nanoparticles.

According to the Hansen model, the elastic modulus of the matrix that was filled with spherical particles was estimated by using the following formula:

$$E_c = \frac{\left(1-v_f\right) + \left(1+v_f\right) E_f/E_m}{\left(1+v_f\right) + \left(1-v_f\right) E_f/E_m} \times E_m, \quad (12)$$

where E_f and E_m are the elastic moduli of the filler and the matrix, respectively.

For the first step considering the epoxy matrix that was filled with the pores (air bubbles), Equation (12) becomes simplified since $E_f/E_m \ll 1$, and it takes the form

$$E_c^I = \frac{(1-v_p)}{(1+v_p)} \times E_m. \quad (13)$$

For the second step considering the epoxy matrix (with pores) that was filled with the CSR nanoparticles, Equation (12) was modified to include both pores and CSR nanoparticles

$$E_c^{II} = \frac{\left(1-v_f\right) + \left(1+v_f\right) E_f/E_c^I}{\left(1+v_f\right) + \left(1-v_f\right) E_f/E_c^I} \times E_c^I, \quad (14)$$

where the elastic modulus of the CSR particles E_f = 4 MPa [2], and volume fraction of the filler and pores, v_f and v_p, were evaluated from Equations (9) and (11), respectively.

The results of the evaluation by Equations (12) and (14) are shown in Figure 3b. Generally, it could be concluded that at the higher filler contents, the Hansen model allowed us to predict the reduction of the elastic modulus by approx. 20% due to the addition of the soft CSR particles in the epoxy resin. It could be either noticed that the consideration of the pores (0.8–2.0 vol.%) improves the description of the experimental results. In general, the addition of ACE MX-960 to the epoxy resin led to marginally lower values of elastic modulus than those of the two other CSR-containing additives. It could indicate a higher volume of the softcore when it is compared to the total particle (core plus shell) volume since the size of these particles is the greatest when it is compared to the other ones (see Table 1).

The results that were obtained for the glass transition temperature as evaluated using the TMA diagrams are provided in Figure 4. The glass transition temperature of the epoxy was approx. 78.1 ± 2.2 °C which was within the range (70–140 °C) of the reported values of T_g for DGEBA type epoxy [3,8,28]. Contradictory results are provided in the literature revealing the occurrence of an improvement [7], a reduction [8] or almost no effect [3,4,21,28] on T_g for the epoxy with the addition of the CSR nanoparticles. According to Figure 4, a gradual increase of 10–20 °C was obtained for the epoxy that was filled with all of the additives containing the CSR nanoparticles, which could be attributed to the high crosslink density and toughening effect of rubber modifiers, thereby testifying to their dissolution in the epoxy continuous phase.

3.4. Fracture Toughness

The representative load–displacement curves for TDCB tests are provided in Figure 5a. Obviously, the soft CSR nanoparticles were effective as tougheners for the epoxy resin. According to Figure 5a, the critical load of the epoxy was significantly improved with the increase of CSR nanoparticles of ACE MX-960. Similar results were obtained for the rest of the additives containing the CSR nanoparticles.

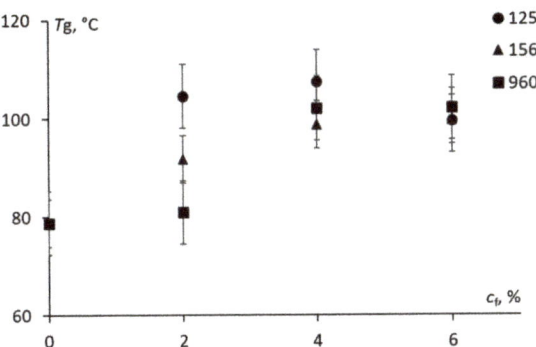

Figure 4. The glass transition temperature of the epoxy modified with different additives containing CSR nanoparticles (indicated on the graph) as a function of filler weight fraction.

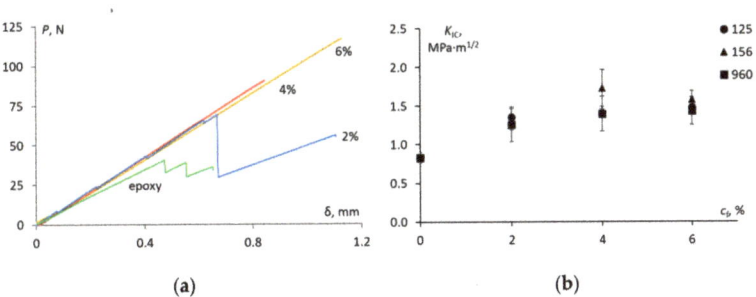

Figure 5. Load–displacement curves for TDCB tests of the epoxy and epoxy filled with ACE MX-960 at different filler weight fractions indicated on the graph (**a**) and fracture toughness for the epoxy and epoxy modified with ACE MX-125, 156 and 960 at different weight filler fractions (indicated on the graph) (**b**).

The fracture toughness of the epoxy which was evaluated by Equation (2) was 0.83 ± 0.07 MPa·m$^{1/2}$ which is slightly lower than the values that are reported in the literature for the epoxy resins [7,8]. As seen in Figure 5b, the addition of the CSR nanoparticles led to a gradual improvement in the fracture toughness for all of the types of additives. No considerable distinction in the fracture toughness among the additives was detected, thereby proving that small (100 nm) and large (300 nm) CSR particles were equally efficient. Though generally, ACE MX-156 showed the greatest enhancement in the fracture toughness value which was approx. 108% at the CSR content of 4 wt.%. The optimum rubber content beyond which the fracture toughness did not improve was reported [7,28]. In this work, according to Figure 5b, the optimum CSR nanoparticle content could be estimated as 4 wt.% for all of the additives. Of course, this result is only relevant for certain dispersion conditions of the CSR particles in the epoxy. Nevertheless, the manual mixing of the CSR nanoparticles in the epoxy resulted in a good dispersion of the CSR nanoparticles as seen by the SEM and a considerable improvement in the fracture toughness. Additionally, a low fraction of pores that was indirectly estimated from the density measurements revealed the sufficient quality of the manufactured samples.

3.5. Interlaminar Fracture Toughness

The typical DCB load vs. displacement curves of the unmodified and CSR-modified CFRP laminate specimens are shown in Figure 6a. The saw-like drops on the load–displacement diagrams after the critical load was achieved were obviously caused by the woven 0/90 lay-up configuration of the carbon fabric that was used to produce the composite laminate. According to Figure 6a, the effect of all of the additives containing the CSR nanoparticles was substantial, thereby leading to the improvement of the critical load of the CFRP by 32–70%. The Mode I interlaminar fracture toughness of CFRP which was evaluated by Equation (6) was enhanced from 390 ± 50 J/m^2 to a maximal value of 599 ± 13 J/m^2 as shown in Figure 6b for the CFRP with 4 wt.% of ACE MX-960.

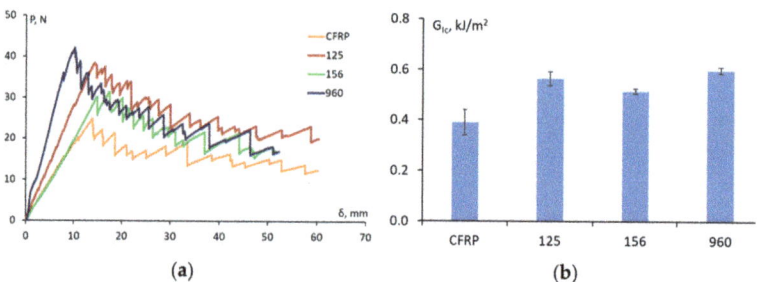

Figure 6. Typical load-crack opening displacement curves for CFRP and CFRP modified with ACE MX (indicated on the graph) at 4 wt.% (**a**) and interlaminar fracture toughness evaluated by Equation (6) for different materials studied (indicated on the graph) (**b**).

However, the toughening effect of the CSR nanoparticles in the epoxy did not fully transfer to the epoxy-based CFRP composite laminates. E.g., the use of an epoxy that was modified with 4 wt.% of ACE-MX 156 having the maximal improvement of fracture toughness by 108% as a matrix for CFRP laminates resulted in the improvement to the interlaminar fracture toughness by only 32%. The interlaminar fracture toughness was maximally improved by 53% for ACE MX-960 at CSR content 4 wt.%. A further increase in the CSR fraction could result in greater improvement of the interlaminar fracture toughness of the CFRP, though, it should be emphasized that rubber toughening has also the side effect of increasing the viscosity of the epoxy resin, thereby negatively contributing to the fabrication of composite laminates [8,9,18,28]. Additionally, at higher values of the filler fraction, a significant agglomeration can occur, thereby causing a local stress concentration and a detrimental effect on the toughening performance of the filler particles [29,30].

4. Conclusions

The epoxy resin was modified by the addition of three types of CSR nanoparticles of different contents. On the one hand, the addition of all of the additives containing the soft CSR nanoparticles resulted in a minor decrease in the density, and a substantial reduction in the elastic modulus and tensile strength of the epoxy resin. The Hansen model was applied to describe the elastic modulus of the epoxy having a certain fraction of the CSR nanoparticles and pores, and a good agreement with the experimental results was found at the high CSR contents.

On the other hand, it was testified that the fracture toughness of the epoxy was significantly improved by the addition of all of the investigated types of CSR. The optimum CSR nanoparticle content was found to be 4 wt.% for all of the CSR nanoparticle types, thereby resulting in the improvement of the fracture toughness of the epoxy by 60–108%. No considerable distinction in the fracture toughness among the additives was detected, thereby proving that the small (100 nm) and large (300 nm) CSR nanoparticles were equally efficient.

Moreover, the effect of all of the additives containing the CSR nanoparticles was substantial, leading to the improvement in the interlaminar fracture toughness of the CFRP by 32–53%. Although, the toughening effect of the CSR nanoparticles in the epoxy was two times higher than it was in the epoxy-based CFRP composite laminates.

Additionally, a gradual increase of the glass transition temperature was obtained for the epoxy that was filled with all of the additives containing CSR nanoparticles, which could be attributed to the high crosslink density and toughening effect of rubber modifiers, thereby testifying to their dissolution in the epoxy continuous phase.

The possible combination of rigid and soft particles could be a compromise to simultaneously improve both the tensile properties and the fracture toughness, which cannot be achieved by the single-phase particles independently.

Author Contributions: The study concept was devised by T.G.-K., A.A. and V.Š.; methodology, validation, and formal analysis were carried out by A.A., S.T. and T.G.-K.; the investigation, resources, and data curation were performed by L.S., T.G.-K., J.S., A.S., K.S. and A.Z.; writing—original draft preparation was performed by T.G.-K., A.A. and S.T.; supervision was performed by A.A.; project administration was performed by A.A. and V.Š. All authors have read and agreed to the published version of the manuscript.

Funding: This research was funded by M-Era.Net project MERF "Matrix for carbon reinforced epoxy laminates with reduced flammability" grant No. 1.1.1.5/ERANET/20/04 from the Latvian State Education Development Agency and M-Era.Net project "EPIC—European Partnership for Improved Composites" funded by grant No. TH06020001. A.S., K.S. and A.Z. are grateful to funding received from the European Union Horizon 2020 Framework program H2020-WIDESPREAD-01-2016-2017-TeamingPhase2 under grant agreement No. 739508, project CAMART2.

Institutional Review Board Statement: Not applicable.

Informed Consent Statement: Not applicable.

Data Availability Statement: Not applicable.

Conflicts of Interest: The authors declare no conflict of interest.

Appendix A

This Appendix presents the results of a three-dimensional finite element analysis of the grooved and flat TDCB specimens to investigate the influence of the geometrical parameters of the grooves on the stress intensity factor at the crack tip. The geometry of the TDCB sample that was used in this work was proposed in [16], and it is presented in Figure A1a. Grooves of different depths and shapes were analyzed by changing the angle γ of the grooves (45 and 90 degrees were used for simulations) and the internal radius R of the grooves, as shown in Figure A1b.

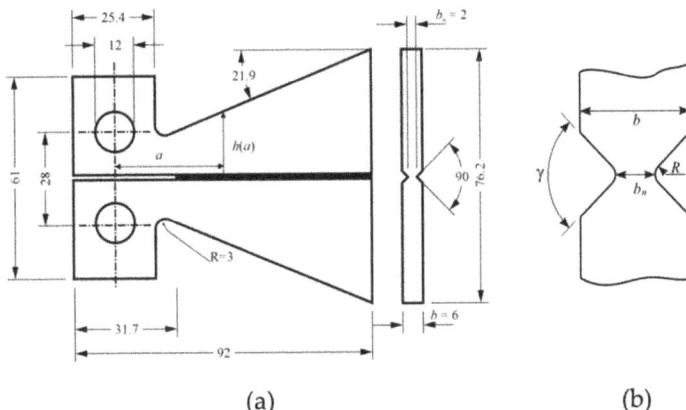

Figure A1. The geometry of the TDCB specimen with dimensions in millimeters (**a**) and a detailed view of the grooved section of the specimen (**b**).

The influence of the depth of the grooves on the SIF can be estimated using simple energy considerations and the known relation between the stress intensity factor and the energy release rate. Assuming, that energy that is stored in the loaded arms of flat and grooved TDCB specimen is the same, the SIF of the grooved specimen can be written as

$$K_g = K_{ng} \left(\frac{b}{b_n} \right)^{0.5}, \qquad (A1)$$

where K_{ng} is the SIF for a flat specimen defined in Equation (1). However, this simple analysis does not take into account the stress concentration at the bottom of the grooves. Freed and Craft [31] suggested an alternative form of Equation (A1):

$$K_g = K_{ng} \left(\frac{b}{b_n} \right)^n, \qquad (A2)$$

where the value of exponent n is in the range $0.5 - 1$ and should be estimated through numerical analysis or by fitting the experimental data. The value $n = 1$ corresponds to the limiting case of a flat specimen with groves angle $\gamma = 180°$.

Lemmens et al. [32] used a 3D finite element simulation of grooved specimens and obtained the value of the exponent n in Equation (A2) to be equal 0.51 and 0.6 for the center and edge of the crack front, respectively. Gómez et al. [33] used a more complex model with a curved crack front and concluded that the best fit n value is close to 0.5. However, both of these works used grooves with an angle equal to 45° and no influence of the groove's sharpness was investigated.

In this work, finite element code ABAQUS was used to calculate the distribution of a stress intensity factor along the front of the initial pre-crack using the standard procedure that is available in ABAQUS. The finite element mesh of the grooved sample near the crack tip is shown in Figure A2a. Quadratic 15-node wedge elements were used for the whole model, except for the zone around the crack tip, where 20-node brick elements were generated in a circular manner with one-side-collapsed quarter point elements for the inner circle, as shown in Figure A2b. The calculations with different mesh densities showed that 15 elements through the specimen's width gave sufficient accuracy in the middle section of the crack, except for the small zones near the crack edges.

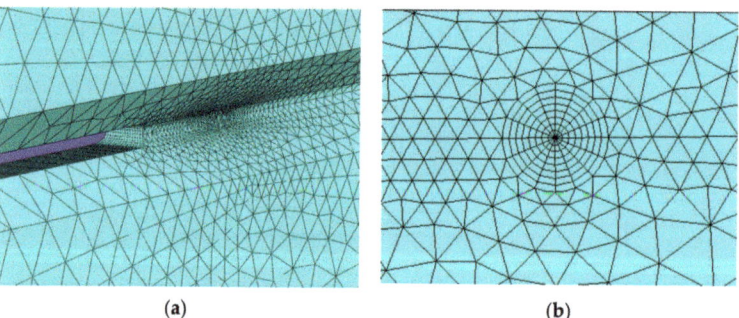

(a) (b)

Figure A2. Finite element mesh of a specimen with grooves (**a**) and arrangement of elements along the crack front for the calculation of stress intensity factors (**b**).

Several finite element models with a width of the specimen b that is equal to 6, 12 and 24 mm and different grooves geometries were analyzed. The SIF distribution along the crack front for a specimen with a width that is equal to 6 mm is presented in Figure A3, where the results are normalized with respect to the 2D plane strain solution in Equation (1). The results show that the SIF values gradually increase as the reduced width of the grooved specimen decreases. The side grooves also influence the SIF distribution along the crack front, making the SIF higher near the surface of the specimen, which is contrary to the flat specimen ($b/b_n = 1$), where the SIF is slightly higher at the center point of the crack front.

Figure A3a presents the results of the calculations for the specimens with different grooves angles, 45 and 90 degrees, respectively. The results show that the SIF is slightly higher for the grooves of 90°, which can be explained by the fact that more material is removed from the specimen in this case, thereby resulting in a higher compliance of the arms of a specimen under the same load. The influence of the sharpness of the grooves on the distribution of the SIF along the crack front is presented in Figure A3b, where the grooves with inner radii of 0.5, 0.25 and 0.15 mm were used for the calculations. As could be expected, the smaller that the inner radius of the grooves was, then the higher the SIF near the surface of the specimen was, however, at the same time, in the middle section of the specimen, the SIF is lower for the sharp grooves, resulting in almost the same average SIF for all three radii of the grooves.

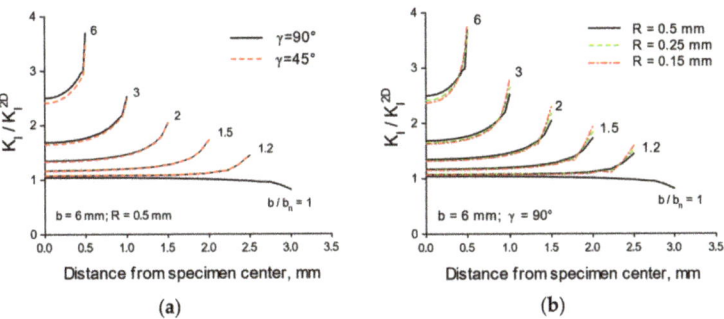

Figure A3. Stress intensity factor distribution along the crack front for grooves of different depths and two angles (**a**) and different radii at the tip of the groove (**b**). Numerical results are normalized by a two-dimensional solution.

The average values of the SIF along the crack front for different geometries of the grooves were approximated by a power function, and the results for the TDCB specimens with the width b that was equal to 6 mm and groove's angle that was equal to 45° and 90° are presented in Figure A4.

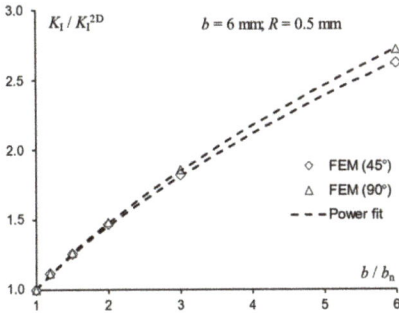

Figure A4. Approximation of the numerical data for the normalized average stress intensity factor by a power function.

The best fit values of the exponent n are listed in Table A1 for different geometries of the TDCB specimen and grooves.

Table A1. Exponent n value determined by fitting of simulations data for different grooves angles and specimen widths.

Grooves Angle γ	Specimen Width b, mm		
	6	12	24
45°	0.54	0.55	0.59
90°	0.56	0.59	0.65

References

1. Aniskevich, A.; Glaskova-Kuzmina, T. Effect of moisture on elastic and viscoelastic properties of fibre reinforced plastics: Retrospective and current trends. In *Creep and Fatigue in Polymer Matrix Composites*, 2nd ed.; Guedes, R.M., Ed.; Woodhead Publishing: Amsterdam, The Netherlands, 2019; pp. 83–120. [CrossRef]
2. Nunes, S.G.; Saseendran, S.; Joffe, R.; Amico, S.C.; Fernberg, P.; Varna, J. On Temperature-Related Shift Factors and Master Curves in Viscoelastic Constitutive Models for Thermoset Polymers. *Mech. Compos. Mater.* **2020**, *56*, 573–590. [CrossRef]
3. Giannakopoulos, G.; Masania, K.; Taylor, A.C. Toughening of epoxy using core-shell particles. *J. Mater. Sci.* **2011**, *46*, 327–338. [CrossRef]
4. Day, R.J.; Lovell, P.A.; Wazzan, A.A. Thermal and mechanical characterization of epoxy resins toughened using preformed particles. *Polym. Int.* **2001**, *50*, 849–857. [CrossRef]
5. Qian, J.Y.; Pearson, R.A.; Dimonie, V.L.; El-Aasser, M.S. Synthesis and application of core–shell particles as toughening agents for epoxies. *J. Appl. Polym. Sci.* **1995**, *58*, 439–448. [CrossRef]
6. Shen, J.; Zhang, Y.; Qiu, J.; Kuang, J. Core-shell particles with an acrylate polyurethane core as tougheners for epoxy resins. *J. Mater. Sci.* **2004**, *39*, 6383–6384. [CrossRef]
7. Quan, D.; Ivankovic, A. Effect of core-shell rubber (CSR) nano-particles on mechanical properties and fracture toughness of an epoxy polymer. *Polymer* **2015**, *66*, 16–28. [CrossRef]
8. Tang, L.C.; Zhang, H.; Sprenger, S.; Ye, L.; Zhang, Z. Fracture mechanisms of epoxy-based ternary composites filled with rigid-soft particles. *Compos. Sci. Technol.* **2012**, *72*, 558–565. [CrossRef]
9. Dadfar, M.R.; Ghadami, F. Effect of rubber modification on fracture toughness properties of glass reinforced hot cured epoxy composites. *Mater. Des.* **2013**, *47*, 16–20. [CrossRef]
10. Material Datasheet for CHS-Epoxy 582 by Spolchemie (Czech Republic). Available online: https://www.spolchemie.cz/en/product.chs-epoxy-582/?msclkid=77435be2bb3011ec852121a4067dc4ca (accessed on 11 August 2022).
11. Material Datasheet for Telalit 0420 by Spolchemie (Czech Republic). Available online: https://www.spolchemie.cz/en/product.telalit-0420/?msclkid=2065552fbb3111ec82d92b19f5b9d09c (accessed on 11 August 2022).
12. Material Datasheet for ACE MX-125, ACE MX-156, and ACE MX-960 by Kaneka (Belgium). Available online: https://www.kaneka.be/sites/default/files/uploads/brochures/MX/Kaneka-leaflet-Kane-Ace-Product-properties.pdf?msclkid=a2acc2d9bb3311ecb184043310b58907 (accessed on 11 August 2022).
13. Material Datasheet for Carbon Fabric by Havel Composites (Czech Republic). Available online: https://havel-composites.com/cs/produkty/uhlikova-tkanina-kc-160g-m2-3k-platno-s-120-cm-3568-4896 (accessed on 20 September 2022).
14. ISO 527-1; Plastics—Determination of Tensile Properties-Part 1: General Principles. ISO: Geneva, Switzerland, 2012.
15. ISO 527-2; Plastics—Determination of Tensile Properties-Part 2: Test Conditions for Moulding and Extrusion Plastics. ISO: Geneva, Switzerland, 2012.
16. Beres, W.; Ashok, K.K.; Thambraj, R. A tapered double-cantilever-beam specimen designed for constant-K testing at elevated temperatures. *J. Test. Eval.* **1997**, *25*, 536–542. [CrossRef]
17. ASTM Standard D 5528-01; Standard Test Method for Mode I Interlaminar Fracture Toughness of Unidirectional Fiber-Reinforced Polymer Matrix Composites. ASTM International: West Conshohocken, PA, USA, 2004.
18. Boon, Y.D.; Joshi, S.C. A review of methods for improving interlaminar interfaces and fracture toughness of laminated composites. *Mater. Today Commun.* **2022**, *22*, 100830. [CrossRef]
19. ImageJ Software. Available online: https://imagej.nih.gov/ij/index.html (accessed on 11 August 2022).
20. ASTM Standard E 1545-01; Standard Test Method for Assignment of the Glass Transition Temperature by Thermomechanical Analysis. ASTM International: West Conshohocken, PA, USA, 2002.
21. Van Velthem, P.; Gabriel, S.; Pardoen, T.; Bailly, C.; Ballout, W. Synergy between phenoxy and CSR tougheners on the fracture toughness of highly cross-linked epoxy-based composites. *Polymers* **2021**, *13*, 2477. [CrossRef] [PubMed]
22. Glaskova-Kuzmina, T.; Zotti, A.; Borriello, A.; Zarrelli, M.; Aniskevich, A. Basalt fibre composite with carbon nanomodified epoxy matrix under hydrothermal ageing. *Polymers* **2021**, *13*, 532. [CrossRef] [PubMed]
23. Glaskova-Kuzmina, T.; Aniskevich, A.; Papanicolaou, G.; Portan, D.; Zotti, A.; Borriello, A.; Zarrelli, M. Hydrothermal aging of an epoxy resin filled with carbon nanofillers. *Polymers* **2020**, *12*, 1153. [CrossRef] [PubMed]
24. Forental', G.A.; Sapozhnikov, S.B. Numerical-experimental estimation of the mechanical properties of an epoxy nanocomposite. *Mech. Compos. Mater.* **2011**, *47*, 521–528. [CrossRef]

25. Lagzdins, A.; Zilaucs, A.; Beverte, I.; Andersons, J.; Cabulis, U. A refined strut model for describing the elastic properties of highly porous cellular polymers reinforced with short fibers. *Mech. Compos. Mater.* **2017**, *53*, 321–334. [CrossRef]
26. Hansen, T.C. Influence of aggregate and voids on modulus of elasticity of concrete, cement mortar, and cement paste. *Int. Concr. Abstr. Portal* **1965**, *62*, 193–216. [CrossRef]
27. Yoshitake, I.; Rajabipour, F.; Mimura, Y.; Scanlon, A. A prediction method of tensile Young's modulus of concrete at early age. *Adv. Civ. Eng.* **2012**, *2012*, 391214. [CrossRef]
28. Becu, L.; Maazouz, A.; Sautereau, H.; Gerard, J.F. Fracture behavior of epoxy polymers modified with core-shell rubber particles. *J. Appl. Polym. Sci.* **1997**, *65*, 2419–2431. [CrossRef]
29. Quan, D.; Mischo, C.; Binsfeld, L.; Ivankovic, A.; Murphy, N. Fracture behaviour of carbon fibre/epoxy composites interleaved by MWCNT- and graphene nanoplatelet-doped thermoplastic veils. *Compos. Struct.* **2020**, *235*, 111767. [CrossRef]
30. Ning, H.; Li, J.; Hu, N.; Yan, C.; Liu, Y.; Wu, L.; Liu, F.; Zhang, J. Interlaminar mechanical properties of carbon fiber reinforced plastic laminates modified with graphene oxide interleaf. *Carbon* **2015**, *91*, 224–233. [CrossRef]
31. Freed, C.; Krafft, J. Effect of side grooving on measurements of plane-strain fracture toughness. *J. Mater.* **1966**, *1*, 770–790.
32. Lemmens, R.J.; Dai, Q.; Meng, D.D. Side-groove influenced parameters for determining fracture toughness of self-healing composites using a tapered double cantilever beam specimen. *Theor. Appl. Fract. Mech.* **2014**, *74*, 23–29. [CrossRef]
33. Garoz Gómez, D.; Gilabert, F.A.; Tsangouri, E.; Van Hemelrijck, D.; Hillewaere, X.K.D.; Du Prez, F.E.; Van Paepegem, W. In-depth numerical analysis of the TDCB specimen for characterization of self-healing polymers. *Int. J. Solids Struct.* **2015**, *64–65*, 145–154. [CrossRef]

Article

A Novel Dual-Emission Fluorescence Probe Based on CDs and Eu³⁺ Functionalized UiO-66-(COOH)₂ Hybrid for Visual Monitoring of Cu²⁺

Jie Che [1], Xin Jiang [1], Yangchun Fan [1], Mingfeng Li [1], Xuejuan Zhang [2], Daojiang Gao [1], Zhanglei Ning [1,*] and Hongda Li [3,*]

[1] College of Chemistry and Materials Science, Sichuan Normal University, Chengdu 610068, China
[2] The Experiment Center, Shandong Police College, Jinan 250014, China
[3] Liuzhou Key Laboratory for New Energy Vehicle Power Lithium Battery, School of Electronic Engineering, Guangxi University of Science and Technology, Liuzhou 545006, China
* Correspondence: zlning@sicnu.edu.cn (Z.N.); hdli@gxust.edu.cn (H.L.); Tel.: +86-28-84760802 (Z.N.)

Abstract: In this work, CDs@Eu-UiO-66(COOH)₂ (denoted as CDs-F2), a fluorescent material made up of carbon dots (CDs) and a Eu³⁺ functionalized metal–organic framework, has been designed and prepared via a post-synthetic modification method. The synthesized CDs-F2 presents dual emissions at 410 nm and 615 nm, which can effectively avoid environmental interference. CDs-F2 exhibits outstanding selectivity, great sensitivity, and good anti-interference for ratiometric sensing Cu²⁺ in water. The linear range is 0–200 μM and the limit of detection is 0.409 μM. Interestingly, the CDs-F2's silicon plate achieves rapid and selective detection of Cu²⁺. The change in fluorescence color can be observed by the naked eye. These results reveal that the CDs-F2 hybrid can be employed as a simple, rapid, and sensitive fluorescent probe to detect Cu²⁺. Moreover, the possible sensing mechanism of this dual-emission fluorescent probe is discussed in detail.

Keywords: metal–organic frameworks; probe; fluorescent; CDs; copper ions

1. Introduction

As one of the essential trace transition metals in organisms, copper ions (Cu²⁺) are widely distributed in the environment for various life activities [1]. For the human body, Cu²⁺ is required for cellular respiration, bone formation, and cardiovascular disease prevention [2]. However, excessive amounts of Cu²⁺ cause drowsiness, elevated blood pressure, liver damage, acute hemolytic anemia, neurotoxicity, and neurodegenerative diseases [3,4]. In recent years, multiple analytical techniques have been applied to the quantitative analysis of Cu²⁺, such as atomic absorption spectrometry [5], colorimetric methods [6], electrochemical techniques [7], and inductively coupled plasma emission spectrometry [8]. However, expensive instruments, tedious operation, long reaction times, and the need for trained operators greatly limit their application [2]. Hence, developing a facile, fast, and reliable method for the detection of Cu²⁺ in aqueous solutions is exceptionally significant.

Compared with traditional analytical methods, fluorescence technology has attracted the attention of researchers because fluorescence measurements are generally low-cost and have fast detection speed, high sensitivity, and easy visualization [9–11]. At present, fluorescent probes have been widely investigated for detecting Cu²⁺ [12,13]. Among them, metal–organic frameworks (MOFs) have aroused considerable interest owing to their flexible and adjustable structure, porosity, and large specific surface area, which is beneficial for concentrating trace amounts and enhancing the contact area between the probe and the analyte [14–17]. Wang et al. designed the fluorescent probe AuNCs/ZIF-8, which exhibited fluorescence turn-off responses to Cu²⁺ in the concentration range of 2–15 μM with a detection limit of 0.9 μM [18]. Jiang et al. successfully synthesized a PPN probe based on a natural β-pinene derivative nopinone, and the color of the solution could be observed to change from colorless to yellow after adding Cu^{2+} [19]. However, these fluorescent probes are based on a single emission, which makes the detection accuracy susceptible to environmental conditions including temperature, viscosity, pH, and operations. In contrast, the dual-emission ratiometric fluorescence probes can achieve good self-calibration in detection processes by measuring the fluorescence intensity ratio at two different wavelengths as a signal parameter [20–22]. Due to the fact that the measured fluorescence ratio signal is not influenced by the instrument and environment, the ratiometric fluorescent probes have high sensitivity, selectivity, and linear range. In addition, the dual-emission ratiometric probes have distinguishable visible color changes, which is greatly helpful for fast and on-site measuring by the naked eye. Thus, rationally designing and developing a MOFs-based ratiometric fluorescent probe with the improvement in detection performances is highly desirable.

In this paper, we report a novel fluorescent probe based on CDs and Eu^{3+} functionalized UiO-66 through a post-synthetic modification method (Scheme S1). The UiOs series material is an octahedral cage conformation MOFs material formed by ligating organic ligands with metallic zirconium as the metal center. UiO-66 is one of the more common materials in the UiOs series. The CDs-F2 hybrid exhibits a typical blue-emitting behavior from CDs and a clear red-emitting of Eu^{3+}. CDs-F2 composite shows outstanding fluorescence properties for sensitive detection of Cu^{2+} in aqueous solutions. The results indicated that CDs-F2 can rapidly and sensitively detect Cu^{2+} in "on-off" mode. Moreover, the CDs-F2 film was made with a silicon plate, realizing precise visible detection by the distinguishable fluorescence color change. The work in this paper may provide an effective and intuitive method for the rapid detection of Cu^{2+} in water.

2. Experimental Details

2.1. Reagents and Instruments

All measurements were performed at room temperature. All chemical reagents and solvents are commercially available and were used directly without further purification.

The instruments and characterization are identical to our reported works [16,23].

2.2. Synthesis of UiO-66-(COOH)$_2$ (Denoted as F1)

F1 was prepared according to the previous literature with some modifications [24]. The reactants, including $ZrCl_4$ (1.0600 g), 1,2,4,5-benzenetetracarboxylic (H_4btec, 0.1260 g), and p-Phthalic acid (PTA, 0.4989 g), were dispersed in a mixture of DMF (50 mL) and acetic acid (5 mL) at room temperature and further sonicated for 30 min. Then, the well-mixed liquid was heated in an oven at 160 °C for 24 h. After cooling, the as-obtained white solid was separated by centrifugation and washed with distilled water and methanol. In order to remove residual DMF from the samples, the solids were suspended in 30 mL of acetone for one week and the acetone solution was changed daily. Finally, the product was recovered in a vacuum at 70 °C.

2.3. One-Pot Synthesis of CDs@Eu-UiO-66(COOH)$_2$ (CDs-F2)

CDs were synthesized on the basis of previous studies [25]. CDs-F2 was prepared by a one-pot post-synthesis modification. The mixture of 200 mg F1 and 0.04 M $Eu(NO_3)_3 \cdot 6H_2O$ in 25 mL of CDs was stirred at room temperature for 24 h. Subsequently, the hybrid product was obtained by centrifugal washing and dried under vacuum at 70 °C.

2.4. Luminescence Sensing Experiments

Fluorescence detection of metal ions in water was carried out at room temperature. A total of 2 mg CDs-F2 is dispersed into distilled water (3 mL) and conduct ultrasound for 30 min, then 1 mL of different metal ions (0.01 M) was added into the dispersion (M^{n+} = K^+, Mg^{2+}, Cd^{2+}, Co^{2+}, Cr^{3+}, Sr^{2+}, Mn^{2+}, Ni^{2+}, Ca^{2+}, Cu^{2+}). Finally, their fluorescence spectra were collected.

2.5. Preparation of Fluorescent Films

The slide with dimensions of 10 mm × 25 mm × 1 mm was washed alternately with ethanol and distilled water and then put at room temperature to dry. The silica gel solution was obtained by adding 100 mg of sodium carboxymethylcellulose to 20 mL of distilled water, stirring until dissolved, and then adding 7.5 g of silica gel with continuous stirring. The silica gel solution was applied evenly on the dry slides, dried naturally, and then heated in an oven at 60 °C for 30 min to obtain silica gel plates. A total of 150 µL of CDs-F2 suspension was disposed of on the silica gel sheets and dried at 60 °C for 3 h to obtain the silica film sample.

3. Result and Discussion

3.1. Optimization of the Fluorescence Properties

The photoluminescent (PL) properties of F1, Eu-UiO-66-(COOH)$_2$ (denoted as F2), CDs, and CDs-F2 have been investigated in detail at room temperature. F1 shows a broad emission band peaked at 410 nm in the blue region (Figure S1a), resulting from π-π* transitions of the ligands [26]. Subsequently, a series of F2 samples were prepared by introducing different starting doping amounts of Eu^{3+} into the F2 compound. The emission spectra were collected as presented in Figure S1b. Except for the wide emission band centered at 410 nm, F2 exhibited several new emission peaks at 592 nm, 615 nm, 652 nm, and 700 nm, which were attributed to the $^5D_0 \rightarrow {}^7F_J$ (J = 1–4) transitions of Eu^{3+} [27]. Both the emission of the ligand and emissions of Eu^{3+} appear simultaneously, indicating that Eu^{3+} has been successfully incorporated. In response to the increase in Eu^{3+} concentration, the emission intensity of Eu^{3+} remains much weaker than the emission intensity of the ligand, which indicates that the energy transfer effect between the organic ligand and Eu^{3+} is ineffective.

Considering the excellent physicochemical stability and abundant surface functional groups of CDs, we introduced CDs as a guest molecule to form an effective energy transfer to enhance the luminescence efficiency of Eu^{3+}. The excitation and emission spectra of the blue-emitting CDs are shown in Figure S2a,b. The excitation spectrum displays two strong bands centered at 250 nm and 360 nm, respectively. Depending on the different excitation wavelengths, the emission spectra of CDs revealed different intensities, while the shape and profile for the emission peak changed little. The maximum emission peak at 430 nm of the prepared CDs was mainly caused by the surface state defects rather than by the eigenstate emission and their synergistic effect [28]. Moreover, different concentrations of Eu^{3+} were added to the CD solution. Figure S2c,d reveal that the emission intensity of CDs gradually decreased with the increase in Eu^{3+} concentration under the excitation at 360 nm or 250 nm UV light, which supported the presence of energy transfer between Eu^{3+} and CDs. However, the characteristic emission of Eu^{3+} was not observed. Such a phenomenon is caused by the high-energy vibrational coupling of Eu^{3+} with -OH in water, leading to the quenching of fluorescence belonging to Eu^{3+} [29]. Therefore, a rigid environment is needed to reduce the energy loss of the Eu^{3+} nonradiative transition.

Afterward, Eu^{3+} and CDs were introduced simultaneously into the matrix material F1. Excitingly, from the emission spectra of the synthesized CDs-F2 samples (Figure S3), both the red characteristic emission peak (615 nm) attributed to the Eu^{3+} and CDs' blue characteristic emission peak (410 nm) can be observed. The strongest emission peak of the CDs is blue-shifted (from 410 nm to 430 nm). This phenomenon may be caused by the transformation of CDs from solutions to composite powders [30]. The characteristic emission intensity of CDs and Eu^{3+} changes with the starting doping amounts of CDs and Eu^{3+}. Eventually, when the CDs and Eu^{3+} are tuned to 25 mL and 0.04 M, the characteristic emission intensity of CDs and Eu^{3+} basically reached a relatively balanced state ($I_{410\,nm}/I_{615\,nm} \approx 1$). This condition was selected for subsequent studies. To obtain the excitation wavelength of the material, we recorded the excitation and emission spectra of the material. The maximum excitation bands of the material appeared at 287 nm and 264 nm with monitoring wavelengths of 410 nm and 615 nm, respectively (Figure S4a). The emission spectra of the materials were measured at different excitation wavelengths in the range of 260–310 nm. It can be observed that with the increasing excitation wavelength, the emission intensity of Eu^{3+} gradually decreases, while that of CDs firstly increases and then decrease (Figure S4b). Given that the emission intensities at 410 nm and 615 nm are similar to the intensity when excited at 280 nm, 280 nm was adopted as the optimal excitation wavelength for the CDs-F2 hybrid. On the basis of those factors mentioned above, under the optimization of addition content (CDs: 25 mL and Eu^{3+}: 0.04 M) and excitation wavelength (280 nm), the fluorescent excitation and emission spectra of the CDs-F2 are presented in Figure 1a, and the hybrid exhibits a reddish-purple color under the UV lamp with a CIE chromaticity coordinate of (0.2934, 01433) (Figure 1b). Therefore, this CDs-F2 with blue and red emission was explored as a dual-emission ratiometric fluorescence probe for the detection of metal ions.

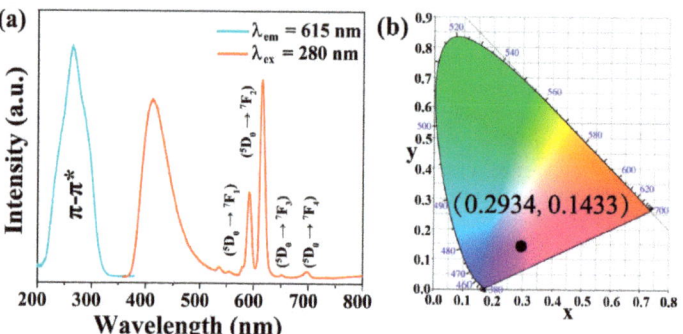

Figure 1. (a) The fluorescence of CDs-F2 and the screening of the excitation wavelength of the sample. (b) The corresponding CIE chromaticity diagram of CDs-F2.

3.2. Characterizations

The composition and crystal structure of the as-obtained products were studied by powder X-ray diffraction (PXRD) (Figure 2a). It can be seen that the XRD pattern of the prepared F1 and CDs-F2 were in good agreement with the simulated results [31]. The morphology of the CDs-F2 sample was studied by scanning electron micrograph (SEM). As shown in Figure S5, the crystal of the CDs-F2 samples exhibits an octahedral structure as reported in UiO-66-based MOFs [24], indicating that the addition of CDs and Eu^{3+} did not change the microstructure of the sample. In addition, with careful observation, the XRD pattern of the CDs-F2 sample appears weak and has broad peaks in the range of 20–40°. This wide peak was speculated to be derived from CDs (Figure 2b), indicating that CDs were successfully incorporated into the F1 material [32]. The FT-IR spectra of the CDs, F1 and CDs-F2 are shown in Figure S6. The FT-IR spectrum of CDs demonstrates that strong peaks related to O-H and N-H appear at 3400 cm^{-1}, while peaks at 1437 cm^{-1} and 1374 cm^{-1} are attributed to the typical stretching vibration band of C-N and C-N=, respectively. Additionally the peak at 761 cm^{-1} was ascribed to N-H oscillation vibration [33,34]. CDs with a large number of functional groups on the surface have the potential to synergize with Eu^{3+} [35]. Moreover, F1 shows an absorption peak at 1710 cm^{-1}, which is derived from the protonated form of -COOH, indicating that F1 contains free carboxyl groups [36,37]. After $Eu(NO_3)_3·6H_2O$ was introduced into F1, the absorption peak disappeared, implying that Eu^{3+} can be encapsulated in the material and coordinated with -COO^-. To further confirm the successful introduction of Eu^{3+} into the F1 material, XPS tests were performed. As shown in Figure 2c, compared with F1, CDs-F2 exhibited a new Eu peak at 1100–1200 eV, which confirms that Eu^{3+} was successfully introduced to the F1 material. Meanwhile, the two binding energies of Eu 3d in the CDs-F2 sample located at 1137.6 eV and 1167.3 eV shift in contrast with that of $Eu(NO_3)_3·6H_2O$ (1137.1 eV and 1166.8 eV, respectively) (Figure S7), further indicating that Eu^{3+} was introduced into the material and coordinated with the free carboxyl group of the ligand [38]. Similarly, the N_2 adsorption measurement was carried out, as shown in Figure 2d. This result reveals the surface area and pore volume of the F1 is 580 m^2/g and 0.24 cm^3/g, respectively. However, after modification, the surface area and pore volume of CDs-F2 decreased to 455 m^2/g and 0.19 cm^3/g, respectively. It was demonstrated again that Eu^{3+} and CDs have been introduced into the pore channel of F1. In addition, the energy-dispersive X-ray analysis (EDX) spectrum is demonstrated in Figure S8. Peaks of elements Zr, Eu, C, O, and N can be detected (besides the element Au and partial C from measurement), which confirmed that the Eu^{3+} and CDs were captured in F1. Based on the above results, CDs-F2 with red and blue double emission has been successfully synthesized.

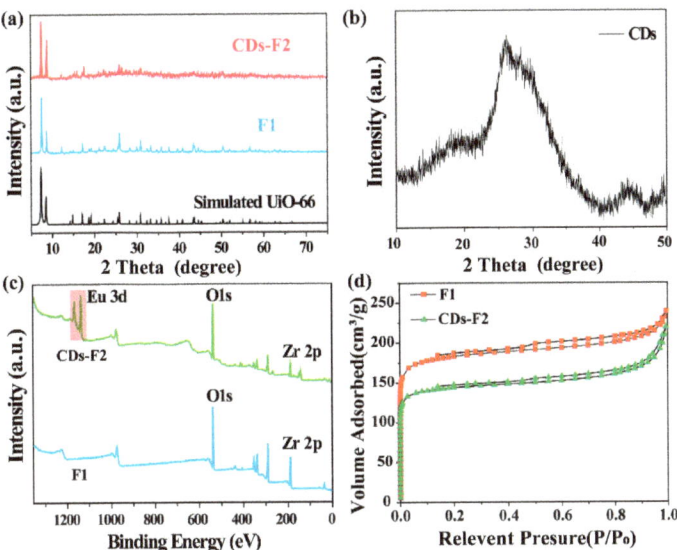

Figure 2. (a) PXRD patterns of simulated UiO-66 and as-prepared F1, CDs-F2, and (b) CDs. (c) XPS spectra of F1 and CDs-F2. (d) The N_2 adsorption isotherms of F1 and CDs-F2.

3.3. Fluorescence Sensing for Cu^{2+}

Considering the remarkable fluorescence properties of CDs-F2, the potential sensing ability of CDs-F2 for metal ions in an aqueous solution was investigated. Selectivity is an essential factor for fluorescent probes. The fluorescence spectra of CDs-F2 in the presence of different metal ions (Ca^{2+}, Mn^{2+}, Ni^{2+}, Sr^{2+}, K^+, Cd^{2+}, Ba^{2+}, Mg^{2+}, Cr^{3+}, Cu^{2+}) are illustrated in Figure 3a. It can be found that after adding different metal ions, the blue emission from CDs at 410 nm is almost unchanged, but the characteristic red emission from Eu^{3+} varies with different ions. The most remarkable one is the solution treated by Cu^{2+}, the intensity ratio of blue and red emission ($I_{410\,nm}/I_{615\,nm}$) is significantly increased (Figure 3b). This result demonstrates that CDs-F2 can act as a ratiometric fluorescence probe for selectively detecting Cu^{2+} among various metal ions, which can effectively avoid environmental interference. In order to obtain more intuitive detection results, silica plates containing CDs-F2 material were prepared. Interestingly, the selective recognition of Cu^{2+} by the fluorescent probe silica plate was clearly distinguishable to the naked eye, resulting in the luminescence transforming from reddish-purple to blue under 254 nm UV light irradiation (Figure 3c). Anti-interference ability is another important aspect of the performance of a fluorescent probe. Competitive experiments were conducted by monitoring the luminescence intensity of CDs-F2 toward coexisting metal ions in the presence and absence of Cu^{2+}. It can be seen that the response of CDs-F2 is not influenced by the coexisting metal ions (Figure S9). When Cu^{2+} is added to the solution, the characteristic emission intensity of CDs (410 nm) in the CDs-F2 materials shows slight variation and that of Eu^{3+} (615 nm) exhibits a significant decrease, whether other metal ions exist or not. It is demonstrated that the CDs-F2 material has excellent anti-interference performance for the recognition of Cu^{2+}.

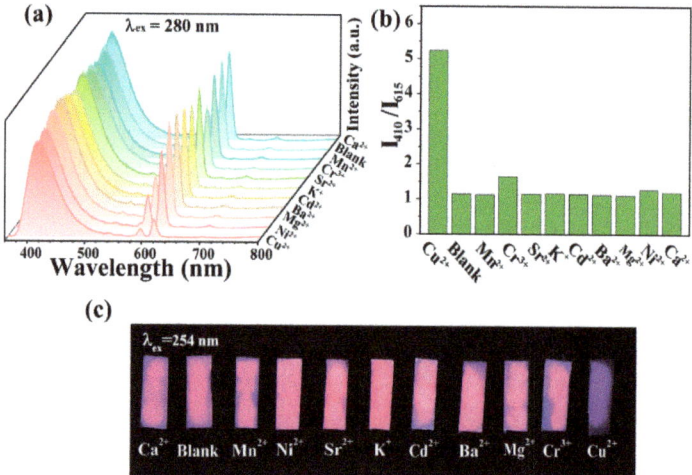

Figure 3. Specific identification performance of CDs-F2 (a) and detailed drawing of the selectivity (b). (c) The photographs of CDs-F2 silica plate containing various metal ions solution under 254 nm UV light irradiation.

Sensitivity is one of the key factors determining the further application of fluorescent probes in practical applications [39]. The sensitivity of this probe to Cu^{2+} was determined by measuring the fluorescence intensity ratio ($I_{410\,nm}/I_{615\,nm}$) of CDs-F2 in aqueous solutions of Cu^{2+} with a concentration range of 1×10^{-6} to 2×10^{-4} M (Figure 4a). The Eu^{3+} characteristic peaks of CDs-F2 weakened sequentially with the increase in Cu^{2+} concentration, while the characteristic peaks of CDs remained stable. As shown in Figure 4b, $I_{410\,nm}/I_{615\,nm}$ exhibited a well-defined linear relationship with the concentration of Cu^{2+}. The correlation equation is $I_{410\,nm}/I_{615\,nm} = 21021[C] + 0.8828$ ($R^2 = 0.9892$). The detection limit for Cu^{2+} was calculated as 0.409 µM according to the 3σ IUPAC standard formula (3σ/K), where σ is the standard deviation of 21 repeated blank tests and K is the slope of the linear equation [16]. This value is much lower than the toxicity level for Cu^{2+} drinking water set by EPA (20 µM) and GB 5749-2022 (15 µM) [40].

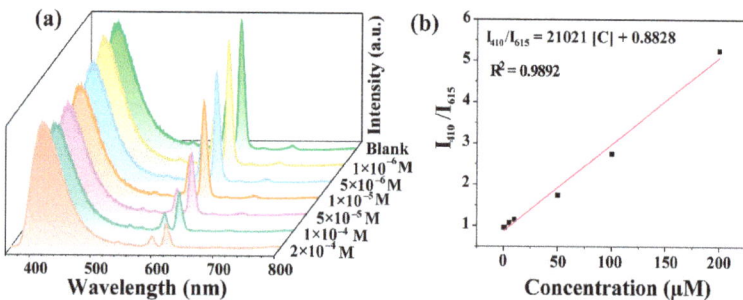

Figure 4. (a) The emission spectrum of CDs-F2 with Cu^{2+} concentration. (b) The linear relationship between fluorescence emission intensity ratio $I_{410\,nm}/I_{615\,nm}$ and Cu^{2+} concentration.

Furthermore, the detection performance of the materials in this study for Cu^{2+} in comparison with other works is listed in Table 1. It could be noticed that the present work exhibits a wide linear relationship and low detection limit response compared to our previous work [41] and the other reported fluorescent probes. All the above evidence indicated that CDs-F2 materials are expected to be applied for rapid and immediate detection of Cu^{2+} in aqueous solutions on site.

Table 1. Comparison of LOD and line range of CDs-F2 with other probes for Cu^{2+} detection.

Material	LOD (μM)	Line Range	Ref.
{[$Mg_3(ndc)_{2.5}(HCO_2)_2(H_2O)$][$NH_2Me_2$] $2H_2O \cdot DMF$}	0.56	10–45 μM	[19]
Eu(FBPT) (H_2O) (DMF)	8.5	0–17 equiv	[42]
2,4,6-trihydroxybenzaldehyde rhodamine B hydrazone	0.48	0–12 μM	[43]
SF@AgNPs	0.333	1–6 μM	[44]
APA-Rh	1.04	0–40 μM	[45]
$Na(Yb,Nd)F_4$@$Na(Yb,Gd)F_4$:Tm@$NaGdF_4$	0.1	0.125–3.125 μM	[46]
MOF-525 NPs	3.5	1.0–250 nM	[47]
FDPP-C8; TDPP-C8	65×10^3 127×10^3	0–4 μM; 0–8 μM	[48]
BOPHY-PTZ	—	0–2 μM	[49]
Tb-MOFs	10	$1\text{–}5 \times 10^3$ μM	[41]
CDs-F2	0.409	0–200 μM	This work

3.4. Possible Sensing Mechanism of CDs-F2 for Cu^{2+}

Furthermore, the possible mechanism of CDs-F2 for Cu^{2+} detection has been studied. In this study, the mechanism may be attributed to the following two reasons [32]: (i) Cu^{2+} induced the framework collapse; (ii) energy transfer between the Cu^{2+} and the composite. A series of experiments were conducted to gain more insight into the possible quenching mechanism. The XRD pattern of CDs-F2 powder after sensing Cu^{2+} (named Cu: CDs-F2) was first collected in sequence to check the crystal structure. As shown in Figure 5a, it can be seen that it is consistent with the XRD diffraction peak of CDs-F2 material, which proves the structure of the CDs-F2 sample remained the same after being treated with Cu^{2+}. Figure 5b shows the excitation spectra of CDs-F2 and the UV absorption spectra of metal ions. The excitation spectra of the probe did not overlap with the excitation spectra of the metal ions, which ruled out the possibility of fluorescence quenching caused by energy transfer between the analyte and the probe [50]. Generally, the fluorescence quenching caused by the formation of non-luminescent intermediates between the fluorophore and the quenching agent is static quenching. In contrast, the fluorescence quenching caused by the collision between the excited fluorophore and the quenching agent is dynamic quenching [51]. To explore whether it is dynamic quenching or static quenching, we studied the fluorescence lifetime of CDs-F2 before and after adding Cu^{2+} (Figure S10). The fluorescence lifetime of CDs-F2 does not change at 410 nm in the presence or absence of Cu^{2+} (0.0104 μs and 0.0103 μs). However, the fluorescence lifetime of the material shortens significantly at 615 nm (260 μs to 19.6 μs). This result suggests that Eu^{3+} and Cu^{2+} occurred dynamic quenching during the sensing process. Subsequently, X-ray photoelectron spectroscopy (XPS) analysis was performed. It can be observed from Figure 6 the binding energies of the Eu 3d orbitals changed from 1137.6 eV and 1167.3 eV to 1136.5 eV and 1166.2 eV, respectively. It may be due to the Cu^{2+} possessing an unsaturated $3d^9$ electron configuration and a lower metal-centered energy level formed by partially filled d orbitals. The d-d transitions between these energy levels are non-emitting and lead to strong reabsorption, which degrades the luminescence of Eu^{3+} [52].

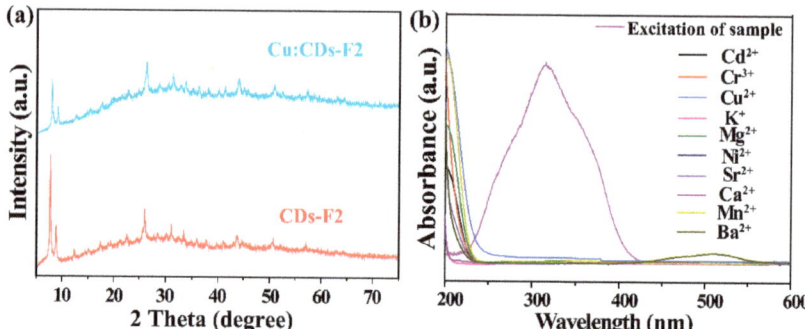

Figure 5. (**a**) PXRD patterns of CDs-F2 before and after sensing the solution of Cu^{2+}. (**b**) UV spectra of heavy metals and excitation spectrum of CDs-F2.

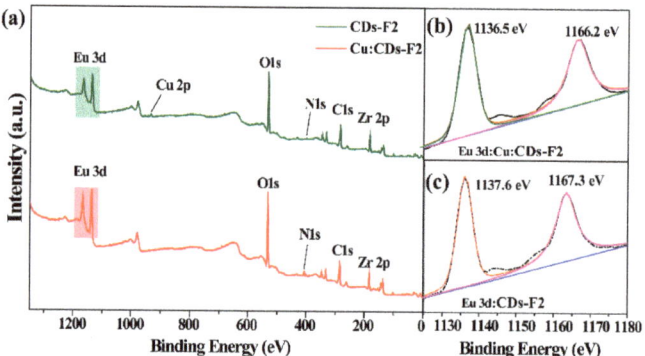

Figure 6. (**a**) XPS spectra of CDs-F2 and Cu: CDs-F2. The binding energy of Eu 3d in Cu: CDs-F2 (**b**) and CDs-F2 (**c**).

4. Conclusions

In summary, the CDs-F2 fluorescent probe with blue and red double emission was successfully prepared through a one-pot post-synthetic method. The CDs-F2 ratiometric fluorescent probe with self-calibration ratio analysis provides more reliable sensing results. The results demonstrate that the developed CDs-F2 can specifically recognize Cu^{2+} and show excellent anti-interference performance when other metal ions coexist. Meanwhile, silica gel plate fluorescent probes were prepared for fast and visual detection of Cu^{2+}. The changed emission color of CDs-F2 in the presence of Cu^{2+} can be easily monitored by the naked eye. Moreover, the sensing mechanism of CDs-F2 for Cu^{2+} detection was systematically investigated. The result reveals that adding Cu^{2+} would affect the energy transfer between the ligand and Eu^{3+}, which would quench the luminescence of Eu^{3+}. This finding indicates that CDs-F2 material can be employed as a fluorescent probe to rapidly and efficiently detect Cu^{2+} in aqueous solutions. At present, the fluorescent probes we prepared have not been put into practical application. In the follow-up study, it is hoped that the practical application of CDs-F2 can be explored.

Supplementary Materials: The following supporting information can be downloaded at: https://www.mdpi.com/article/10.3390/ma15227933/s1, Scheme S1. Schematic diagram of the preparation process and application of CDs-F2. Figure S1. (a) PL spectra of F1. (b) Emission of F2 with different doped content of Eu^{3+}. Figure S2. (a) PL excitation spectra of the CDs. (b) Steady-state emission spectra of CDs at different excitation wavelengths. The steady-state emission spectrum of CDs in the absence and presence of different concentrations of Eu^{3+} (c) λ_{ex} = 360 nm, (d) λ_{ex}= 250 nm. Figure S3.

Emission of synthetic materials in different proportions. Figure S4. (a) Excitation spectra of CDs-F2 monitored at 615 nm and 410 nm, respectively. (b) Emission of CDs-F2 with excitation wavelengths from 260 to 310 nm. Figure S5. SEM image of CDs-F2. Figure S6. FT-IR spectra analysis of F1, CDs-F2, and CDs. Figure S7. Eu 3d XPS spectra of Eu(NO$_3$)$_3$·H$_2$O and CDs-F2. Figure S8. EDX spectra of CDs-F2. Figure S9. The response of CDs-F2 toward coexisting metal ions in the presence and absence of Cu^{2+}: (a) λ_{em} = 410 nm; (b) λ_{em} = 615 nm. Figure S10. Fluorescence lifetimes of CDs-F2 in the absence (a,c) and presence (b,d) of Cu^{2+} in aqueous solution. Table S1. CDs-F2 determined by energy-dispersive analysis by X-rays (EDX).

Author Contributions: Conceptualization, J.C. and X.J.; methodology, J.C.; software, X.J. and Y.F.; investigation, J.C. and Y.F.; resources, Z.N. and D.G.; data curation, M.L.; writing—original draft preparation, J.C. and Y.F.; writing—review and editing, X.Z.; project administration, H.L.; funding acquisition, D.G. and Z.N.; supervision, Z.N. and H.L. All authors have read and agreed to the published version of the manuscript.

Funding: This work was supported by the National Key R&D Program of China (No. 2017YFC0504903), the Open Foundation of Key Laboratory of Special Wastewater Treatment of Sichuan Province Higher Education System (SWWT2020-2), and the project of the Science & Technology Department of Sichuan Province (No. 2021YFG0277).

Institutional Review Board Statement: Not applicable.

Informed Consent Statement: Not applicable.

Data Availability Statement: Not applicable.

Conflicts of Interest: The authors declare no conflict of interest.

References

1. Erdemir, S.; Malkondu, S. Dual-emissive fluorescent probe based on phenolphthalein appended diaminomaleonitrile for Al^{3+} and the colorimetric recognition of Cu^{2+}. *Dye. Pigm.* **2019**, *163*, 330–336. [CrossRef]
2. Liu, H.; Cui, S.; Shi, F.; Pu, S. A diarylethene based multi-functional sensor for fluorescent detection of Cd^{2+} and colorimetric detection of Cu^{2+}. *Dye. Pigm.* **2019**, *161*, 34–43. [CrossRef]
3. Wu, F.; Wang, J.; Pu, C.; Qiao, L.; Jiang, C. Wilson's Disease: A Comprehensive Review of the Molecular Mechanisms. *Int. J. Mol. Sci.* **2015**, *16*, 6419–6431. [CrossRef] [PubMed]
4. Roy, D.; Chakraborty, A.; Ghosh, R. Perimidine based selective colorimetric and fluorescent turn-off chemosensor of aqueous Cu^{2+}: Studies on its antioxidant property along with its interaction with calf thymus-DNA. *RSC Adv.* **2017**, *7*, 40563–40570. [CrossRef]
5. Arain, S.A.; Kazi, T.G.; Afridi, H.I.; Abbasi, A.R.; Panhwar, A.H.; Naeemullah; Shanker, B.; Arain, M.B. Application of dual-cloud point extraction for the trace levels of copper in serum of different viral hepatitis patients by flame atomic absorption spectrometry: A multivariate study. *Spectrochim. Acta A* **2014**, *133*, 651–656. [CrossRef]
6. Li, J.-J.; Ji, C.-H.; Hou, C.-J.; Huo, D.-Q.; Zhang, S.-Y.; Luo, X.-G.; Yang, M.; Fa, H.-B.; Deng, B. High efficient adsorption and colorimetric detection of trace copper ions with a functional filter paper. *Sens. Actuators B* **2016**, *223*, 853–860. [CrossRef]
7. Xie, Y.-L.; Zhao, S.-Q.; Ye, H.-L.; Yuan, J.; Song, P.; Hu, S.-Q. Graphene/CeO$_2$ hybrid materials for the simultaneous electrochemical detection of cadmium(II), lead(II), copper(II), and mercury(II). *J. Electroanal. Chem.* **2015**, *757*, 235–242. [CrossRef]
8. Zhao, L.; Zhong, S.; Fang, K.; Qian, Z.; Chen, J. Determination of cadmium(II), cobalt(II), nickel(II), lead(II), zinc(II), and copper(II) in water samples using dual-cloud point extraction and inductively coupled plasma emission spectrometry. *J. Hazard. Mater.* **2012**, *239–240*, 206–212. [CrossRef]
9. He, L.; Dong, B.; Liu, Y.; Lin, W. Fluorescent chemosensors manipulated by dual/triple interplaying sensing mechanisms. *Chem. Soc. Rev.* **2016**, *45*, 6449–6461. [CrossRef]
10. Yang, J.; Che, J.; Jiang, X.; Fan, Y.; Gao, D.; Bi, J.; Ning, Z. A Novel Turn-On Fluorescence Probe Based on Cu(II) Functionalized Metal–Organic Frameworks for Visual Detection of Uric Acid. *Molecules* **2022**, *27*, 4803. [CrossRef]
11. Zeng, X.; Zhou, Q.; Wang, L.; Zhu, X.; Cui, K.; Peng, X.; Steele, T.W.J.; Chen, H.; Xu, H.; Zhou, Y. A Fluorescence Kinetic-Based Aptasensor Employing Stilbene Isomerization for Detection of Thrombin. *Materials* **2021**, *14*, 6927. [CrossRef] [PubMed]
12. Wang, H.; Pei, Y.; Qian, X.; An, X. Eu-metal organic framework@TEMPO-oxidized cellulose nanofibrils photoluminescence film for detecting copper ions. *Carbohydr. Polym.* **2020**, *236*, 116030. [CrossRef] [PubMed]
13. Qian, X.; Deng, S.; Chen, X.; Gao, Q.; Hou, Y.-L.; Wang, A.; Chen, L. A highly stable, luminescent and layered zinc(II)-MOF: Iron(III)/copper(II) dual sensing and guest-assisted exfoliation. *Chin. Chem. Lett.* **2020**, *31*, 2211–2214. [CrossRef]
14. Yin, Y.; Yang, J.; Pan, Y.; Gao, Y.; Huang, L.; Luan, X.; Lin, Z.; Zhu, W.; Li, Y.; Song, Y. Mesopore to Macropore Transformation of Metal–Organic Framework for Drug Delivery in Inflammatory Bowel Disease. *Adv. Healthc. Mater.* **2021**, *10*, 2000973. [CrossRef] [PubMed]

15. Wang, B.; Zhang, X.; Huang, H.; Zhang, Z.; Yildirim, T.; Zhou, W.; Xiang, S.; Chen, B. A microporous aluminum-based metal-organic framework for high methane, hydrogen, and carbon dioxide storage. *Nano Res.* **2021**, *14*, 507–511. [CrossRef]
16. Feng, L.; Dong, C.; Li, M.; Li, L.; Jiang, X.; Gao, R.; Wang, R.; Zhang, L.; Ning, Z.; Gao, D.; et al. Terbium-based metal-organic frameworks: Highly selective and fast respond sensor for styrene detection and construction of molecular logic gate. *J. Hazard. Mater.* **2020**, *388*, 121816. [CrossRef]
17. Zhou, X.; Liu, L.; Niu, Y.; Song, M.; Feng, Y.; Lu, J.; Tai, X. A Water-Stable Zn-MOF Used as Multiresponsive Luminescent Probe for Sensing Fe^{3+}/Cu^{2+}, Trinitrophenol and Colchicine in Aqueous Medium. *Materials* **2022**, *15*, 7006. [CrossRef]
18. Wang, K.; Qian, M.; Qi, H.; Gao, Q.; Zhang, C. Single Particle-Based Confocal Laser Scanning Microscopy for Visual Detection of Copper Ions in Confined Space. *Chin. J. Chem.* **2021**, *39*, 1804–1810. [CrossRef]
19. Jiang, Q.; Wang, Z.; Li, M.; Song, J.; Yang, Y.; Xu, X.; Xu, H.; Wang, S. A nopinone based multi-functional probe for colorimetric detection of Cu^{2+} and ratiometric detection of Ag^+. *Photochem. Photobiol. Sci.* **2020**, *19*, 49–55. [CrossRef]
20. Li, Y.; Qi, S.; Xia, C.; Xu, Y.; Duan, G.; Ge, Y. A FRET ratiometric fluorescent probe for detection of Hg^{2+} based on an imidazo[1,2-a]pyridine-rhodamine system. *Anal. Chim. Acta* **2019**, *1077*, 243–248. [CrossRef]
21. Lohar, S.; Dhara, K.; Roy, P.; Sinha Babu, S.P.; Chattopadhyay, P. Highly Sensitive Ratiometric Chemosensor and Biomarker for Cyanide Ions in the Aqueous Medium. *ACS Omega* **2018**, *3*, 10145–10153. [CrossRef] [PubMed]
22. Raj, P.; Lee, S.-y.; Lee, T.Y. Carbon Dot/Naphthalimide Based Ratiometric Fluorescence Biosensor for Hyaluronidase Detection. *Materials* **2021**, *14*, 1313. [CrossRef] [PubMed]
23. Li, M.; Dong, C.; Yang, J.; Yang, T.; Bai, F.; Ning, Z.; Gao, D.; Bi, J. Solvothermal synthesis of La-based metal-organic frameworks and their color-tunable photoluminescence properties. *J. Mater. Sci. Mater. Electron.* **2021**, *32*, 9903–9911. [CrossRef]
24. Li, Z.; Sun, W.; Chen, C.; Guo, Q.; Li, X.; Gu, M.; Feng, N.; Ding, J.; Wan, H.; Guan, G. Deep eutectic solvents appended to UiO-66 type metal organic frameworks: Preserved open metal sites and extra adsorption sites for CO_2 capture. *Appl. Surf. Sci.* **2019**, *480*, 770–778. [CrossRef]
25. Huang, J.; Tian, B.; Wang, J.; Wang, Y.; Lu, W.; Li, Q.; Jin, L.; Li, C.; Wang, Z. Controlled synthesis of 3D flower-like $MgWO_4:Eu^{3+}$ hierarchical structures and fluorescence enhancement through introduction of carbon dots. *CrystEngComm* **2018**, *20*, 608–614. [CrossRef]
26. Zhang, Y.; Yan, B. A portable self-calibrating logic detector for gradient detection of formaldehyde based on luminescent metal organic frameworks. *J. Mater. Chem. C* **2019**, *7*, 5652–5657. [CrossRef]
27. Wang, H.; Li, Y.; Ning, Z.; Huang, L.; Zhong, C.; Wang, C.; Liu, M.; Lai, X.; Gao, D.; Bi, J. A novel red phosphor $Li_xNa_{1-x}Eu(WO_4)_2$ solid solution: Influences of Li/Na ratio on the microstructures and luminescence properties. *J. Lumin.* **2018**, *201*, 364–371. [CrossRef]
28. Singhal, P.; Vats, B.G.; Jha, S.K.; Neogy, S. Green, Water-Dispersible Photoluminescent On–Off–On Probe for Selective Detection of Fluoride Ions. *ACS Appl. Mater.* **2017**, *9*, 20536–20544. [CrossRef]
29. Zhang, M.; Zhai, X.; Sun, M.; Ma, T.; Huang, Y.; Huang, B.; Du, Y.; Yan, C. When rare earth meets carbon nanodots: Mechanisms, applications and outlook. *Chem. Soc. Rev.* **2020**, *49*, 9220–9248. [CrossRef]
30. Wang, Y.; Hong, F.; Yu, L.; Xu, H.; Liu, G.; Zhong, D.; Yu, W.; Wang, J. Construction, energy transfer, tunable multicolor and luminescence enhancement of $YF_3:RE^{3+}$(RE = Eu, Tb)/carbon dots nanocomposites. *J. Lumin.* **2020**, *221*, 117072. [CrossRef]
31. Xiaoxiong, Z.; Wenjun, Z.; Cuiliu, L.; Xiaohong, Q.; Chengyu, Z. Eu^{3+}-Postdoped UIO-66-Type Metal–Organic Framework as a Luminescent Sensor for Hg^{2+} Detection in Aqueous Media. *Inorg. Chem.* **2019**, *58*, 3910–3915. [CrossRef] [PubMed]
32. Fu, X.; Lv, R.; Su, J.; Li, H.; Yang, B.; Gu, W.; Liu, X. A dual-emission nano-rod MOF equipped with carbon dots for visual detection of doxycycline and sensitive sensing of MnO_4^-. *Rsc. Adv.* **2018**, *8*, 4766–4772. [CrossRef] [PubMed]
33. Li, X.; Luo, J.; Deng, L.; Ma, F.; Yang, M. In Situ Incorporation of Fluorophores in Zeolitic Imidazolate Framework-8 (ZIF-8) for Ratio-Dependent Detecting a Biomarker of Anthrax Spores. *Anal. Chem.* **2020**, *92*, 7114–7122. [CrossRef]
34. Lan, S.; Wang, X.; Liu, Q.; Bao, J.; Yang, M.; Fa, H.; Hou, C.; Huo, D. Fluorescent sensor for indirect measurement of methyl parathion based on alkaline-induced hydrolysis using N-doped carbon dots. *Talanta* **2019**, *192*, 368–373. [CrossRef]
35. Huo, Q.; Tu, W.; Guo, L. Enhanced photoluminescence property and broad color emission of $ZnGa_2O_4$ phosphor due to the synergistic role of Eu^{3+} and carbon dots. *Opt. Mater.* **2017**, *72*, 305–312. [CrossRef]
36. Xia, C.; Xu, Y.; Cao, M.-M.; Liu, Y.-P.; Xia, J.-F.; Jiang, D.-Y.; Zhou, G.-H.; Xie, R.-J.; Zhang, D.-F.; Li, H.-L. A selective and sensitive fluorescent probe for bilirubin in human serum based on europium(III) post-functionalized Zr(IV)-Based MOFs. *Talanta* **2020**, *212*, 120795. [CrossRef] [PubMed]
37. Peng, X.-X.; Bao, G.-M.; Zhong, Y.-F.; He, J.-X.; Zeng, L.; Yuan, H.-Q. Highly selective detection of Cu^{2+} in aqueous media based on Tb^{3+}-functionalized metal-organic framework. *Spectrochim. Acta A* **2020**, *240*, 118621. [CrossRef] [PubMed]
38. Qu, X.-L.; Yan, B. Stable Tb(III)-Based Metal–Organic Framework: Structure, Photoluminescence, and Chemical Sensing of 2-Thiazolidinethione-4-carboxylic Acid as a Biomarker of CS_2. *Inorg. Chem.* **2019**, *58*, 524–534. [CrossRef]
39. Wang, B.; Yan, B. A turn-on fluorescence probe Eu^{3+} functionalized Ga-MOF integrated with logic gate operation for detecting ppm-level ciprofloxacin (CIP) in urine. *Talanta* **2020**, *208*, 120438. [CrossRef]
40. Fu, Y.; Pang, X.-X.; Wang, Z.-Q.; Chai, Q.; Ye, F. A highly sensitive and selective fluorescent probe for determination of Cu (II) and application in live cell imaging. *Spectrochim. Acta A.* **2019**, *208*, 198–205. [CrossRef]
41. Dong, C.-L.; Li, M.-F.; Yang, T.; Feng, L.; Ai, Y.-W.; Ning, Z.-L.; Liu, M.-J.; Lai, X.; Gao, D.-J. Controllable synthesis of Tb-based metal–organic frameworks as an efficient fluorescent sensor for Cu^{2+} detection. *Rare Met.* **2021**, *40*, 505–512. [CrossRef]

42. Guo, Y.; Wang, L.; Zhuo, J.; Xu, B.; Li, X.; Zhang, J.; Zhang, Z.; Chi, H.; Dong, Y.; Lu, G. A pyrene-based dual chemosensor for colorimetric detection of Cu^{2+} and fluorescent detection of Fe^{3+}. *Tetrahedron Lett.* **2017**, *58*, 3951–3956. [CrossRef]
43. Cheah, P.W.; Heng, M.P.; Saad, H.M.; Sim, K.S.; Tan, K.W. Specific detection of Cu^{2+} by a pH-independent colorimetric rhodamine based chemosensor. *Opt. Mater.* **2021**, *114*, 110990. [CrossRef]
44. Zhou, Y.; Zhang, G.; Xu, T.; Wu, Y.; Dong, C.; Shuang, S. Silk Fibroin-Confined Star-Shaped Decahedral Silver Nanoparticles as Fluorescent Probe for Detection of Cu^{2+} and Pyrophosphate. *Acs Biomater. Sci. Eng.* **2020**, *6*, 2770–2777. [CrossRef] [PubMed]
45. Sağırlı, A.; Bozkurt, E. Rhodamine-Based Arylpropenone Azo Dyes as Dual Chemosensor for Cu^{2+}/Fe^{3+} Detection. *J. Photochem. Photobiol. A Chem.* **2020**, *403*, 112836. [CrossRef]
46. Su, S.; Mo, Z.; Tan, G.; Wen, H.; Chen, X.; Hakeem, D.A. PAA Modified Upconversion Nanoparticles for Highly Selective and Sensitive Detection of Cu^{2+} Ions. *Front. Chem.* **2021**, *8*, 619764. [CrossRef]
47. Zhang, J.; Chen, M.-Y.; Bai, C.-B.; Qiao, R.; Wei, B.; Zhang, L.; Li, R.-Q.; Qu, C.-Q. A Coumarin-Based Fluorescent Probe for Ratiometric Detection of Cu^{2+} and Its Application in Bioimaging. *Front. Chem.* **2020**, *8*, 00800. [CrossRef]
48. Nie, K.; Dong, B.; Shi, H.; Chao, L.; Duan, X.; Jiang, X.-F.; Liu, Z.; Liang, B. N-alkylated diketopyrrolopyrrole-based ratiometric/fluorescent probes for Cu^{2+} detection via radical process. *Dye. Pigm.* **2019**, *160*, 814–822. [CrossRef]
49. He, C.; Zhou, H.; Yang, N.; Niu, N.; Hussain, E.; Li, Y.; Yu, C. A turn-on fluorescent BOPHY probe for Cu^{2+} ion detection. *New J. Chem.* **2018**, *42*, 2520–2525. [CrossRef]
50. Han, Z.; Nan, D.; Yang, H.; Sun, Q.; Pan, S.; Liu, H.; Hu, X. Carbon quantum dots based ratiometric fluorescence probe for sensitive and selective detection of Cu^{2+} and glutathione. *Sens. Actuators B* **2019**, *298*, 126842. [CrossRef]
51. Hao, J.-N.; Xu, X.-Y.; Lian, X.; Zhang, C.; Yan, B. A Luminescent 3d-4f-4d MOF Nanoprobe as a Diagnosis Platform for Human Occupational Exposure to Vinyl Chloride Carcinogen. *Inorg. Chem.* **2017**, *56*, 11176–11183. [CrossRef] [PubMed]
52. Hao, J.-N.; Yan, B. Determination of Urinary 1-Hydroxypyrene for Biomonitoring of Human Exposure to Polycyclic Aromatic Hydrocarbons Carcinogens by a Lanthanide-functionalized Metal-Organic Framework Sensor. *Adv. Funct. Mater.* **2017**, *27*, 1603856. [CrossRef]

Article

WO₃ Nanopores Array Modified by Au Trisoctahedral NPs: Formation, Characterization and SERS Application

Jan Krajczewski [1,*], Robert Ambroziak [2], Sylwia Turczyniak-Surdacka [1,3] and Małgorzata Dziubałtowska [1]

1. Faculty of Chemistry, University of Warsaw, 1 Pasteur St., 02-093 Warsaw, Poland
2. Institute of Physical Chemistry, Polish Academy of Sciences, Kasprzaka 44/52, 01-224 Warsaw, Poland
3. Biological and Chemical Research Centre, Faculty of Chemistry, University of Warsaw, 101 Żwirki i Wigury Street, 20-089 Warsaw, Poland
* Correspondence: jkrajczewski@chem.uw.edu.pl

Abstract: The WO₃ nanopores array was obtained by an anodization method in aqueous solution with addition of F⁻ ions. Several factors affecting the final morphology of the samples were tested such as potential, time, and F⁻ concentrations. The morphology of the formed nanopores arrays was examined by SEM microscopy. It was found that the optimal time of anodization process is in the range of 0.5–1 h. The nanopores size increased with the increasing potential. The XPS measurements do not show any contamination by F⁻ on the surface, which is common for WOₓ samples formed by an anodization method. Such a layer was successfully modified by anisotropic gold trisoctahedral NPs of various sizes. The Au NPs were obtained by seed-mediated growth method. The shape and size of Au NPs was analysed by TEM microscopy and optical properties by UV-VIS spectroscopy. It was found that the WO₃-Au platform has excellent SERS activity. The R6G molecules could be detected even in the range of 10^{-9} M.

Keywords: SERS; plasmonics; WO₃ nanostructures; Au TOH NPs; anodization

Citation: Krajczewski, J.; Ambroziak, R.; Turczyniak-Surdacka, S.; Dziubałtowska, M. WO₃ Nanopores Array Modified by Au Trisoctahedral NPs: Formation, Characterization and SERS Application. *Materials* **2022**, *15*, 8706. https://doi.org/10.3390/ma15238706

Academic Editor: Dippong Thomas

Received: 8 November 2022
Accepted: 5 December 2022
Published: 6 December 2022

Publisher's Note: MDPI stays neutral with regard to jurisdictional claims in published maps and institutional affiliations.

Copyright: © 2022 by the authors. Licensee MDPI, Basel, Switzerland. This article is an open access article distributed under the terms and conditions of the Creative Commons Attribution (CC BY) license (https://creativecommons.org/licenses/by/4.0/).

1. Introduction

SERS spectroscopy (Surface-Enhanced Raman Scattering) is a very useful analytical tool due to its high precision, non-destructive character, and low detection limit [1]. The typical SERS spectrum is a fingerprint of molecules that allows for analyzing even complicated molecules as well as DNA, RNA, or peptide strands [1]. There are some literature reports about even single molecule detection by the SERS method [1].

The development of a novel SERS platform with excellent SERS properties is still desired. The ideal SERS platform should exhibit high EF (Enhancement Factor) and repeatability from point-to-point analysis as well as from sample-to-sample analysis. Time and chemical stability are other important factors.

The ideal SERS platform should consist of repetitively placing nanostructures over the whole sample in a homogenous way. It is possible to form some nanostructures from pure plasmonic metals by lithography [2] and electrochemistry [3], but the simplest way is to form a nanostructured material from another material and then cover its surface with a plasmonic layer. Such a solution was applied for TiO₂ [4] and ZrO₂ NTs (Nanotubes) [5], ZnO nanorods [6], and InP nanowires [7]. However, the main limitation of this method results from the fact that the deposited plasmonic layer flattens the nanostructure's surface. For example, it is proved that in the ZrO₂ NTs, the thickest layer of 50 nm covers the morphology of the nanotubes, which leads to a decrease in EF [8]. Therefore, the other possibility is to deposit the plasmonic nanoparticles on the nanostructured surface [9]. The presence of a nanostructured surface leads to the homogenous distribution of NPs even by simple drop casting and prevents the so-called coffee ring effect [10]. Additionally, the application of NPs with an anisotropic shape instead of a typical semi-spherical shape allows the increase of EF and, in consequence, decreases the LOD (limit of detection) for many analytes.

One of the interesting structures are WO_3 nanopores that exhibit interesting properties. Such media can be used as chemical sensors [11] or in applications where the properties of electrochromism are used [12]. However, the synthesis of highly organized WO_3 nanopores by electrochemical methods is still under development [13–17]. For example, Schmuki et al. proposed a synthesis method without the use of electrolytes containing fluorides [18]. Although nanopores were obtained, they were closed at the top. Another example of a synthesis method leading to well-formed nanopores involves the use of hot, pure orthophosphoric (V) acid, which is dangerous because there is a risk of chemical and thermal burns, and it is also corrosive [19]. However, one of the papers proposed the electrochemical synthesis of WO_3 nanostructures with well-formed nanopores. It is a very promising method and we decided to experimentally determine the influence of the parameters on the obtained morphology. So far, WO_3 nanostructures formed by an anodization method have not been used as a SERS platform, but instead have found an application mainly in the photocatalysis.

In this work, we report about synthesis parameterization of a well-ordered WO_3 nanopores array by an anodization method in aqueous solution containing F^- ions. Parameters like time, F^- concentrations, and applied potential affecting the final morphology of the array were carefully examined. The XPS analysis showed that formed WO_3 nanopores array is not contaminated by F^- ions, which is typical for WO_3 array preparation in organic solvents. After deposition of trisoctahedral Au NPs with various sizes, the nanoarray could be applied as a SERS active platform. It was found that the size of deposited NPs affects the EF.

2. Materials and Methods

2.1. Materials

Tungsten foil (0.25 mm thick and 99.5% purity) was purchased from Alfa Aesar. NaF, NH_4F, HF (40%), Ascorbic Acid (AA), CTAB (Cetyltrimethylammonium bromide), CTAC (Cetyltrimethylammonium chloride), $NaBH_4$ (Sodium borohydride), and NaOH (Sodium hydroxide) were purchased from Sigma-Aldrich (Darmstadt, Germany). Sulfuric acid (H_2SO_4) and isopropanol (99%) were purchased from POCH S.A. A 30% solution of tetrachloroauric(III) acid in diluted HCl was purchased from the Polish State Mint. The water was purified by a Millipore Milli-Q system and had a resistivity of ca. 18 MΩ/cm.

2.2. Synthesis of Flat WO_3 and WO_3 Nanoporous Array

The tungsten foil was cut by guillotine for metals onto pieces with 6 mm × 12 mm dimensions. The W samples were cleaned by immersing in acetone, ethanol, and water, respectively, and sonicated in each solution for 5 min. Then, W foil was electrochemically polished in 0.25 M NaOH aqueous solution in a two-electrode cell where the W foil was used as the anode and the platinum counter electrode as the cathode for 8 min 30 s with a potential of 8 V.

The anodization was carried out in one step in 1 M H_2SO_4 solution containing various amounts of NaF (from 0.02 M to 0.15 M NaF in 100 mL of electrolyte). The potential during the anodization process was constant and was in the 20–60 V range. The anodization was carried out for 60 min. The flat WO_3 sample was formed in the same condition (40 V) in the absence of fluoride ions.

2.3. Synthesis of Au Trisoctahedral Nanoparticles

The trisoctahedral gold nanocrystals were synthesised by modified seed-mediated growth method [20]. In the first step, the small semi-spherical nanoseeds were formed by chemical reduction. For this purpose, 92 µL of 10 mM $HAuCl_4$ was added to 7 mL of 75 mM aqueous solution of CTAB. The solution was vigorously stirred and then 0.42 mL of ice-cold $NaBH_4$ solution (10 mM) was injected at once. The solution changed colour from yellow to brown, indicating the formation of small, gold nanoparticles. The solution was then aged for 2 to 5 h at 30 °C. The trisoctahedrons with an average size of 70 nm were formed by the regrowth of nanoseeds. The growth solution was prepared by mixing 0.25 mL of 10 mM HAuCl$_4$ solution with 9 mL of 22 mM CTAC solution. Then, the freshly prepared ascorbic acid solution (3.06 mL of 38.8 mM) was added. The colour of the solution changed from yellow to colourless. Finally, the 50 μL of 100 times diluted nanoseeds was added and the solution was gently mixed for 10 s. The solution changed colour to intense pink in three minutes. The bigger trisoctahedrons were formed by regrowth of 70 nm trisoctahedrons. All procedures were the same, but instead of adding semi-spherical nanoseeds, the previously formed trisoctahedrons were added.

Finally, the trisoctahedrons were centrifuged to stop the growth process. The supernatant containing nanoparticles was redispersed in the same amount of Millipore water. For SERS measurements, the trisoctahedrons solution was 2 times concentrated by one more centrifugation cycle and redispersed in isopropanol. 60 μL of Au NPs was dropped onto the middle of the WO$_3$ surface.

2.4. Characterization Methods

The SERS platform was prepared by the simple drop casting of the solution containing Au trisoctahedrons of various sizes on the WO$_3$ nanopores surface. In each case, 25 μL of Au trisoctahedrons was dropped four times and then samples were dried at 60 °C in the vacuum.

The SERS measurements were carried out after dropping 10 μL of R6G (Rhodamine 6G) solution of various concentrations in the range of 10 mM to 100 pM.

The morphology of the samples was examined using a scanning electron microscope (SEM, a FEI Nova NanoSEM 450, Brno, Czech Republic). For imaging, low-energy electron detectors, an Everhart–Thornley detector (ETD) and a through-the-lens (TLD) detector, were used in the low- and high- resolution modes, respectively. All modes were performed in the same configuration at a primary beam energy of 10 kV.

Material cross-section was performed using electron-ion (Ga+) scanning (Crossbeam 540X microscope Jena, Germany). Prior to the cut, the sensitive-to-the-ion-beam surface was protected with electron-beam-induced Pt deposition (2 kV, 4 nA) and after that with ion-beam-induced Pt deposition (30 kV, 300 pA). The cross-section was performed with 30 kV and 3 nA, whereas the final polishing was stopped at 30 kV and 700 pA.

Transmission electron microscopy (TEM) analyses were carried out with a Zeiss LIBRA 120 electron microscope working at an accelerating voltage of 120 kV. The microscope was equipped with an in-column OMEGA filter. The sample of nanoparticles obtained was deposited onto a 300-mesh copper grid coated with a Formvar layer.

UV-VIS spectra were recorded using a Thermo Scientific Evolution 201 spectrophotometer.

The Raman measurements were carried out using a Horiba Jobin-Yvon Labram HR800 spectrometer equipped with a Peltier-cooled charge-coupled device detector (1024 × 256 pixels), a 600 groove/mm holographic grating, and an Olympus BX40 microscope with a long-distance 50× objective. An He-Ne laser provided the excitation radiation of a 632.8 nm wavelength.

The chemical states of individual elements were verified by X-ray photoelectron spectroscopy (XPS) using a Microlab 350 (Thermo Electron, East Grinstead, UK) spectrometer. For this purpose, the X-ray excitation source (AlKα anode: power 300 W, voltage 15 kV, beam current 20 mA) was used. The lateral resolution of XPS analysis was about 0.2 cm^2. The high-resolution XPS spectra were recorded using the following parameters: pass energy 40 eV and energy step size 0.1 eV. XPS spectra were reprocessed using the CasaXPS (2.3.18PR1.0) software. Spectra were fitted with GL(30) line shape after Shirley background subtraction and subsequently charge corrected to give a C 1 s at 285 eV.

3. Results and Discussion

3.1. Structural Characterization

The WO_3 nanopores array were formed by a standard anodization process in which nanopores were formed perpendicular to the tungsten substrate. During the anodization, the W foil acts as an anode and hence various tungsten oxides are formed. The platinum counter electrode acts as a cathode on which formation of hydrogen occurs. The whole process can be described by the following equations [21]:

$$W + 2H_2O \leftrightarrow WO_2^{2+} + 4H^+ + 6e^-$$

$$2WO_2^{2+} + H_2O + 2e^- \leftrightarrow W_2O_5 + 2H^+$$

$$W_2O_5 + H_2O \leftrightarrow 2WO_3 + 2H^+ + 2e^-$$

Due to the presence of fluoride ions, WO_3 form soluble a complex with fluoride ions:

$$WO_3 + 2H^+ + 4F^- \leftrightarrow WO_2F_4^{2-} + H_2O$$

The first synthesis showed that the commercially available W foils are not flat enough to produce samples containing large areas homogeneously covered by WO_3 nanopores. Therefore, the tungsten foil was electrochemically polished in an aqueous solution of NaOH. The potential was kept at 8 V and the duration was 8 min 30 s. After that process, SEM (Scanning Electron Microscope) images showed atomically flat W surfaces.

As in a typical anodization method of formation of nanopores metal oxide nanoarray, many parameters could affect size, distribution, and efficiency of nanostructure formation. It was found that the optimal distance between Pt counter electrode and W foil was 1 cm. With greater distance, the formed nanostructures are deformed and are not evenly distributed over the surface of the W foil. The other important factor is the ratio of the surfaces of both electrodes. The samples with well-defined WO_3 nanopores are only formed when the Pt counter electrode surface is not smaller than the surface of the W foil.

For many nanostructure materials formed by the anodization process, it has been reported that the electrolyte solution should be aged. It is a common procedure applied in organic solvents containing EG (ethylene glycol) [22,23], glycerol [24], or others [13,25]. This is caused by the weak solubility of the F^- source in organic media. The solution must be vigorously stirred for the complete dissolution of fluoride salt. This process is accompanied by the formation of an air bubble in the viscous organic solution. The presence of the air bubble disturbs the migration of F^- ions and hence the growth of nanostructures is not homogenous. It was found that there are not any significant changes in the morphology of WO_3 nanoarray formed with freshly prepared H_2SO_4 NaF solution compared to those formed with solution aged for 24 h or for three days.

The quality of the obtained nanopores largely depends on the composition and conditions of anodization [26]. The influence of the type of electrolyte, of voltage and concentration for the selected electrolyte, and of the anodization time were examined.

All electrolytes used were water electrolytes. Anodization was performed at 40 V for 60 min in the presence of 0.1 M NaF and 0.1 M NH_4F in a solution of 1 M H_2SO_4 and with 0.1 M $(NH_4)_2SO_4$ in water. Well-developed nanopores were obtained in electrolytes containing NaF and NH_4F. Only in the $(NH_4)_2SO_4$ solution was the desired structure not obtained. Based on the work of K. Syrek et al., it was found that structures obtained with the use of $(NH_4)_2SO_4$ may arise due to the slow formation of pores in WO_3 [26]. The pore layer growth process is strictly dependent on the presence of factors that can dissolve the WO_3 layer. If no ions complexing with WO_3 such as H^+ or F^- are present in the solution, too few holes are formed in the WO_3 layer to form pores. In the $(NH_4)_2SO_4$ solution, the pH is slightly acidic, but the amount of H^+ ions is insufficient to significantly etch the surface of tungsten (VI) oxide, therefore only single pores are visible in further stages of growth.

The best-formed pores are visible in the case of the sample made in NaF solution (Figure 1). It differs from the NH_4F solution only in the cation. The reason for this is the lower mobility of NH_4^+ ions compared to smaller Na^+ ions. This makes it difficult to dissociate the dissolved substances from the metal surface in the case of samples with NH_4^+ cations (Figure 1a–d). The cross-section image (Figure 1b) allows one to determine the thickness of the obtained oxide layer. It is relatively equal over the entire length imaged and amounts to 270 nm.

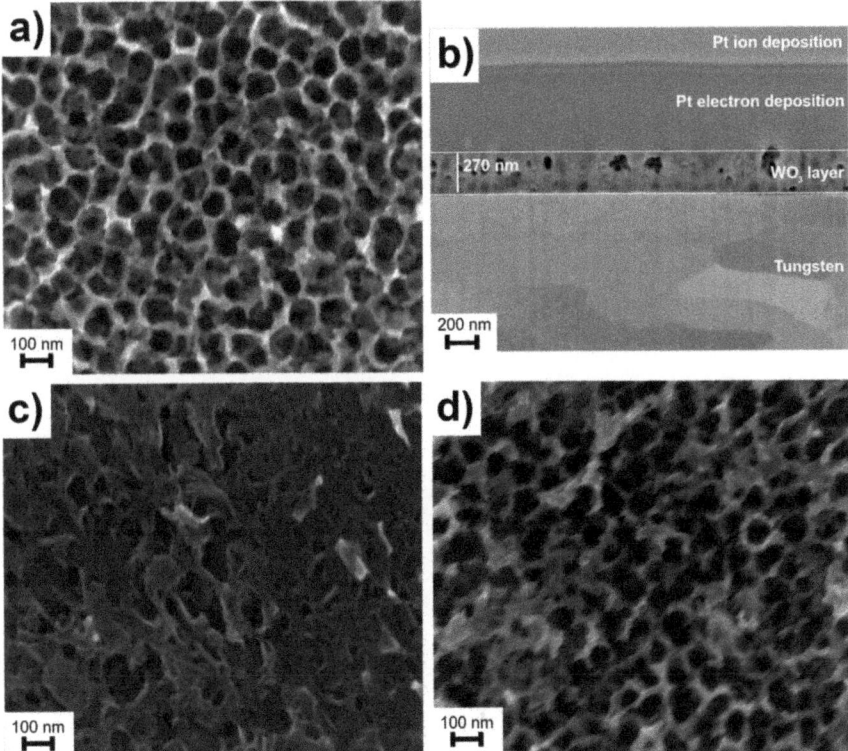

Figure 1. SEM micrographs of WO_3 surfaces obtained using various electrolyte composition: (**a**) 0.1 M NaF in 1 M H_2SO_4, without protective layer; (**b**) FIB cross-section of (**a**) after deposition of Pt protective layer; (**c**) 0.1 M $(NH4)_2SO_4$ in water; and (**d**) 0.1 M NH_4F in 1 M H_2SO_4.

The very important parameter during the formation of semiconducting nanostructures oxide layer by the anodization method is the time of the anodization process. Figure 2 shows steps of the formation of WO_3 nanopores into W foil after anodization in 1 M H_2SO_4 solution of 0.1 M NaF at various times. These SEM images showed that in the first step, the small, randomly distributed pores are formed. Then, increasing processes lead to the formation of more and more WO_3 pores, which finally lead to an assembled layer of WO_3 nanopores. It was found that increasing the anodization time over 1 h does not lead to any significant change in the morphology of the prepared WO_3 nanopores array.

The size of W nanopores could be easily tuned by changing the conditions of the anodic oxidation. The samples used for the SERS activity test were formed in 1 M H_2SO_4 solution 0.1 M NaF and with a potential equal to 40 V. The average porous size was estimated at 71.9 ± 14.6 nm. It was found that the porous size increases when increasing the voltage of the anodization process. When this process was carried out for 1 h with 30 V, the average porous size was 58.3 ± 11.2 nm, while in the case of 50 V for the same amount of time (1 h), the size was 86.2 nm ± 16.2. Figure 3 shows the morphology of the WO_3 nanopores array formed by anodization of W foil at 1 M H_2SO_4 solution 0.1 M NaF for 1 h at various applied potentials. The presented data showed that in low potential, the formation of WO_3 nanopores array cannot be completed, the structures still have defects, and the pore distribution is not homogenous. Samples prepared at higher potentials (40 and 50 V) do not vary between them.

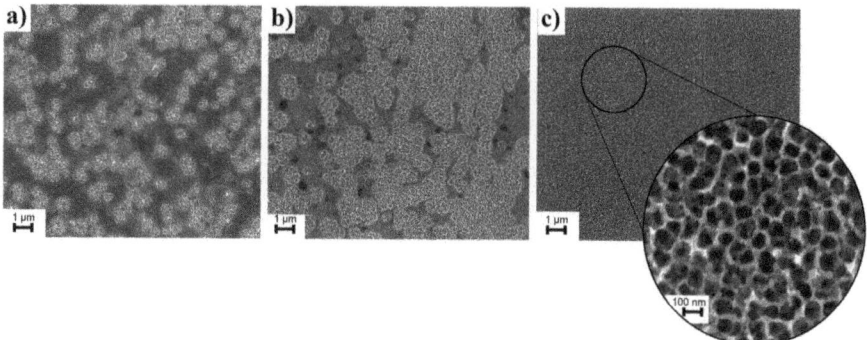

Figure 2. SEM micrographs showing the process of the WO_3 nanopores formation into W foil during anodization in 1 M H_2SO_4 solution of 0.1 M NaF at various times at a constant potential equal to 40 V. (**a**) 15 min; (**b**) 30 min; (**c**) 60 min.

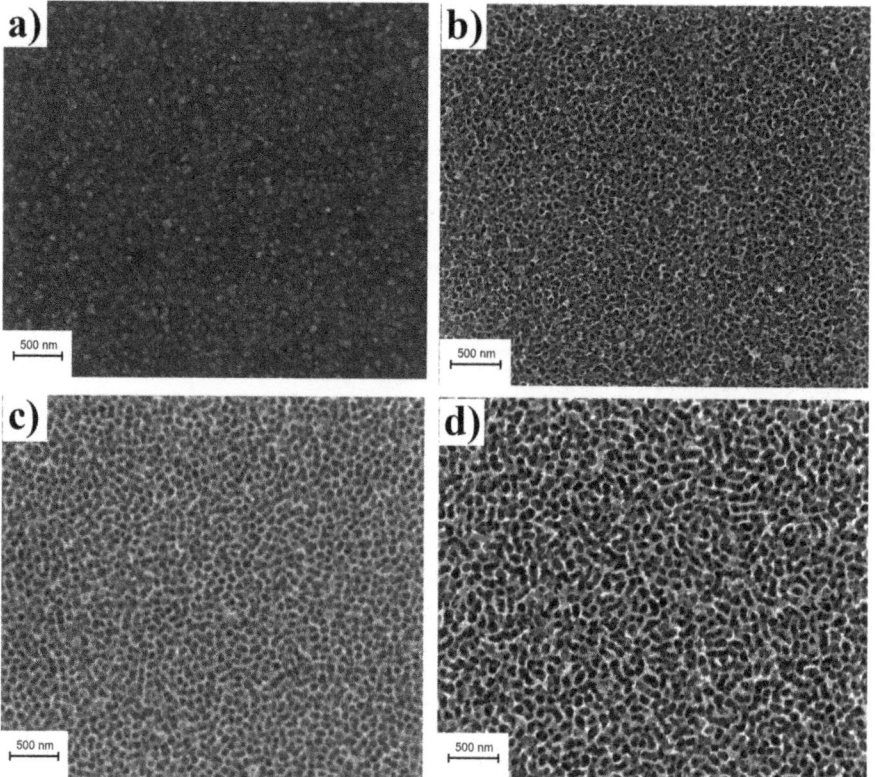

Figure 3. SEM micrographs of WO_3 nanopores array formed by anodization of W foil in 1 M H_2SO_4 solution of 0.1 M NaF for 1 h at various potentials: (**a**) 20 V; (**b**) 30 V; (**c**) 40 V; and (**d**) 50 V.

The composition and concentrations of the electrolyte solution play a crucial role in the formation of WO$_3$ nanopores array by the anodization process. Conducted research showed that NaF ions are necessary for the proper formation of WO$_3$ nanopores. The minimal amount of F$^-$ ions was specified at 0.002 M NaF in 100 mL of electrolyte. The data analysis showed that concentrations in the range of 0.05 to 0.15 M NaF do not lead to any significant change in the size of WO$_3$ pores (Figure 4). All the tested anodization process conditions are shown in Table 1.

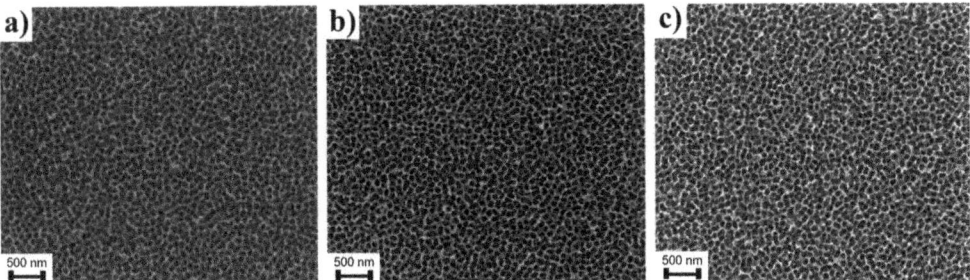

Figure 4. SEM micrographs showing the process of the formation of WO$_3$ nanopores into W foil during anodization in 1 M H$_2$SO$_4$ solution containing various concentrations of NaF: (**a**) 0.05 M; (**b**) 0.1 M; and (**c**) 0.15 M.

Table 1. Detailed list of tested anodization conditions.

H$_2$SO$_4$	NaF	NH$_4$F	(NH$_4$)2SO$_4$	Geometry of WO$_3$
1 M	x	x	x	Flat
1 M	0.05 M	x	x	Nanopores array
1 M	0.1 M	x	x	Nanopores array
1 M	0.15 M	x	x	Nanopores array
1 M	x	0.1 M	x	Nanopores array (some deformations)
1 M	x	x	0.1 M	Rough surface

Results in optimized electrolyte composition: 1 M H$_2$SO$_4$, 0.1M NaF

Time		
15 min	30 min	60 min
Flat with holes	Flat with more holes	Nanopores

Potential (average porous size)			
20 V	30 V	40 V	50 V
No pores	Nanopores (58.3 nm)	Nanopores (71.9 nm)	Nanopores (86.2 nm)

x—absence of reagent.

3.2. The XPS (X-ray Photoelectron Spectroscopy) Surface Characterization

The most intense peaks for W 4f region were assigned to the presence of W6+ (W4f$_{7/2}$ = 35.3 ± 0.1 eV, Δ = 2.1 eV) [27,28] with a characteristic oxide loss structure set 5.5 eV above the W 4f$_{7/2}$ peak (Figure 5). The peak set at lower binding energy (W4f$_{7/2}$ = 33.3 ± 0.2 eV, Δ = 2.1 eV) according to literature might be related to W^{4+} [28–31], since W^{5+} is usually found around 34.3 ± 0.3 eV [28,31–33]. Based on the shape of the spectrum as well as on the fitting function, there is no evidence suggesting the existence of an additional tungsten oxidation state. The signal at O1s spectrum, located at 530.8 eV (O A), is typically assigned to lattice oxygen in WO$_3$, whereas there is no consensus in the literature as to the assignment of the remaining, high-binding energy signals, especially since there is no information on C1s spectrum for some of them. Song Ling Wang et al. [28] ascribe peaks detected with a binding energy at 531.6 eV and 532.4 eV to the presence of adsorbed –OH originated from water and reduced tungsten oxidation states, respectively.

L. Wang et al. [31] suggest that in the case of pristine WO$_3$, the peak at 531.6 eV is presumably related to the surface contamination, whereas the same peak can be assigned to O^{2-} in the vicinity of oxygen vacancies. Therefore, it is suggested that the signals O B (532.4 eV) and O C (533.6 eV) originate from impurities containing groups such as C-O (286.0 eV on C 1 s) and O-C=O (288.8 eV on C 1 s). The C A signal in this case is most probably related to C-C and C-H bonds. Of course, it should be kept in mind that signals in the O1s spectrum may also overlap to some extent, hence the peak marked O B may also be a component of signals derived from O chemisorbed and O C from OH-/H$_2$O. The XPS spectra of the F1s core electrons show no peak originating from fluoride-contacting species, which are commonly found in this kind of samples.

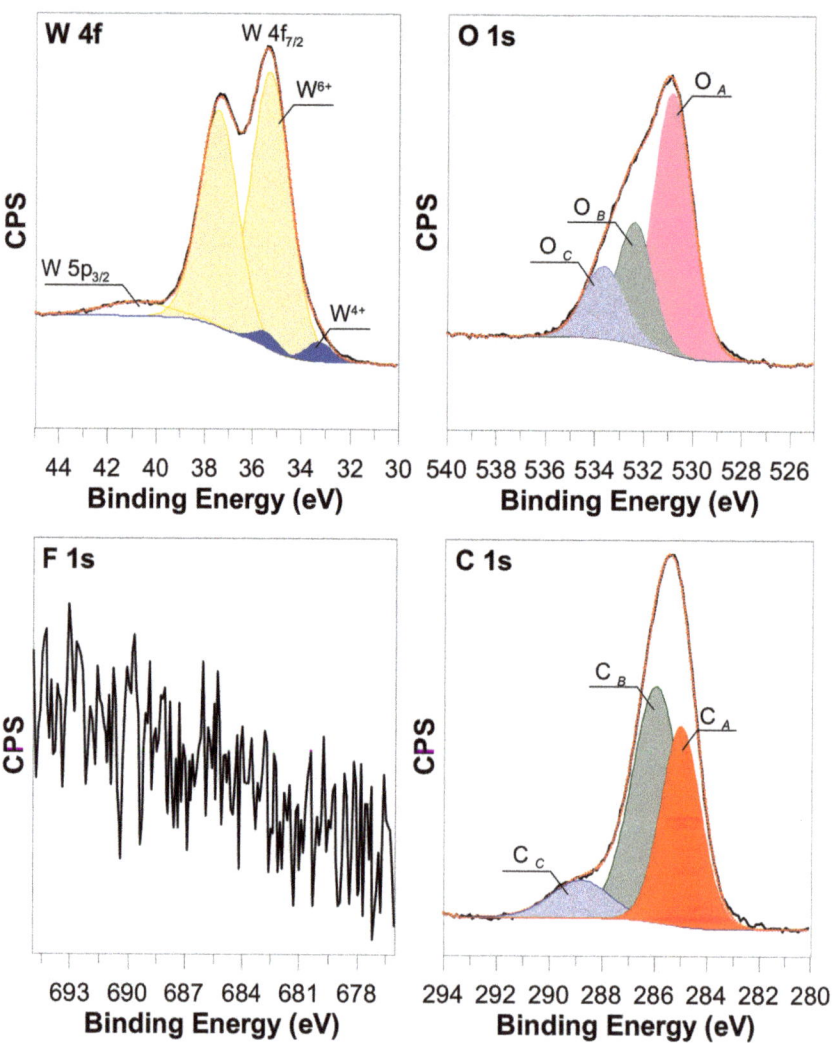

Figure 5. High resolution XPS spectra of WO$_3$ nanopores array formed by an optimized anodization process.

The analyses of the survey spectrum suggest the following about the percentage composition of the surface: O: 27.51%, W: 8.35%, and C: 64.48%.

The common problem during the formation of WO_3 nanostructures by the anodization method is the contamination of the surface by F^- ions which are intercalated into the nanopores structures. This problem is common in the case of organic electrolytes. One of the solutions leading to the elimination of F^- ions is the annealing process. In such cases, the anodized samples were heated in an oven at an elevated temperature in the range of 200–550 °C. However, it was found that temperatures higher than 500 °C led to structural damage to WO_3 nanopores, what could be seen in Figure 6. In many cases, the lower temperature is not sufficient to completely remove F^- ions. However, the XPS analysis of formed samples does not reveal any presence of F^- ions. This effect is probably associated with the aqueous type of the applied electrolyte. The diffusion factor is higher because of the higher viscosity. In consequence, the F^- ions could be diffused from the surface and not intercalated into WO_3 nanostructures.

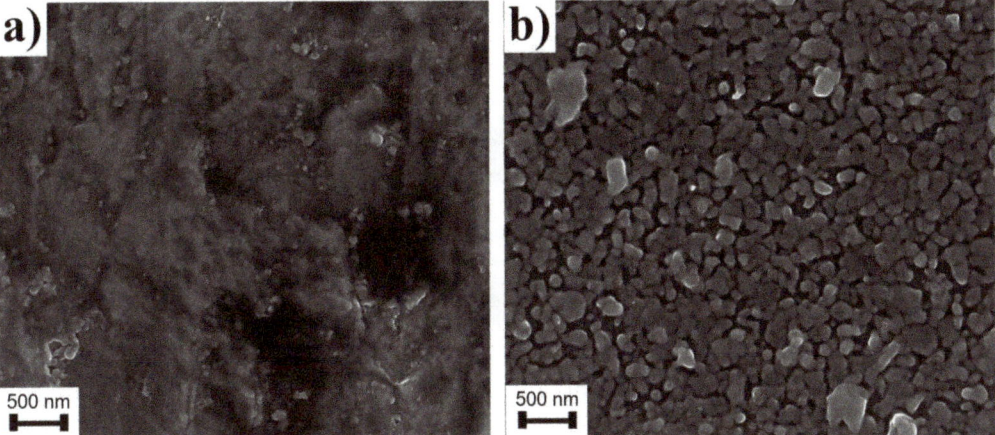

Figure 6. Surface of WO_3 nanopores after annealing at: (**a**) 500 °C; (**b**) 550 °C.

3.3. Au Trisoctahedrons Characterization

The trisoctahedral gold nanoparticles were synthesised by a seed-mediated growth. The Au seeds were stabilized by CTAB and, in consequence, single-crystalline Au seeds were formed, in contrast to citrate-stabilized Au seeds, which are pentatwined. The single-crystalline Au seeds result in more favourable growth into anisotropic shapes. During the growth process, the CTAB was replaced by CTAC. With CTAB, the main product was identified as Au nanocubes. However, the reaction yield and stability are not as high as for Au trisoctahedrons. The Au NPs of bigger sizes were synthesised from the trisoctahedrons of smaller size. The considered shape could be understood as an octahedron with triangular pyramids on each side (Figure 7).

Figure 7. Model of trisoctahedron gold nanoparticles formed by the seed-mediated growth method.

It should be noted that consistent growth of Au NPs does not change the shape and efficiency of the synthesis. Analysis of TEM (transmission electron microscopy) micrographs (Figure 8) showed that the average diameter (measured from edge to edge) is 70 ± 8 nm (62–95 nm) for the smallest NPs, then 93 ± 8 (74–120 nm) for the intermediates, and reach 114 ± 11 (84–135) for the biggest ones. The histograms were based on counting at least 100 nanoparticles for each sample. The small loss in homogeneity of size and shape distribution observed for the biggest Au NPs is caused by the few-steps growth process. Each growth step leads to some spread of size, therefore NPs formed in the third cycle exhibit slightly higher size distribution than others. The optical properties were investigated by UV-Vis spectrophotometry (see Figure 9). The red-shift of SPR (Surface Plasmon Resonance) could be observed associated with the increasing diameter of Au NPs. The 70 nm Au NPs exhibit SPR at 547 nm, the 94 nm Au NPs at 581 nm, while the biggest Au NPs at 610 nm. Moreover, also the half width increase with increasing size of Au NPs. This fact can indicate that the substantial growth of Au NPs leads to a decrease in size homogeneity in the sample.

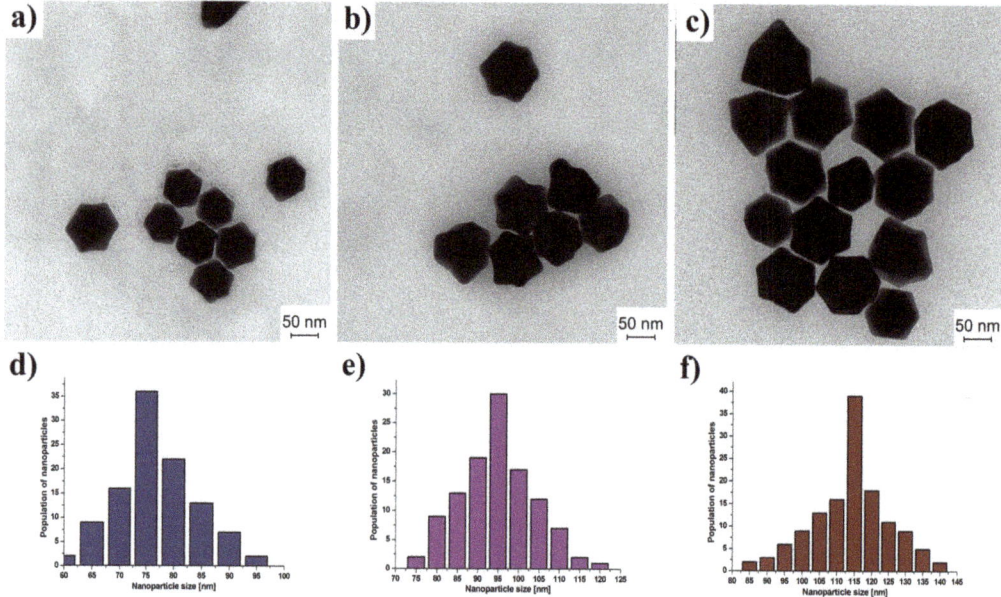

Figure 8. TEM micrographs of trisoctahedral gold nanoparticles of various diameters: (**a**) 70 nm; (**b**) 94 nm; and (**c**) 115 nm. Size distribution of nanoparticles: (**d**) 70 nm; (**e**) 94 nm; and (**f**) 115 nm.

Figure 10 shows substrates with trisoctahedral gold nanoparticles. On the flat surface of WO_3 (Figure 10a), the number of nanoparticles is small. On flat substrates, nanoparticles tend to accumulate at the edge of the droplet, which reduces the possible reinforcements obtained in the centre of the substrate. In the case of nanoparticles on porous substrates, we can observe that particles of sizes 70 and 94 nm can be locked on the pores.

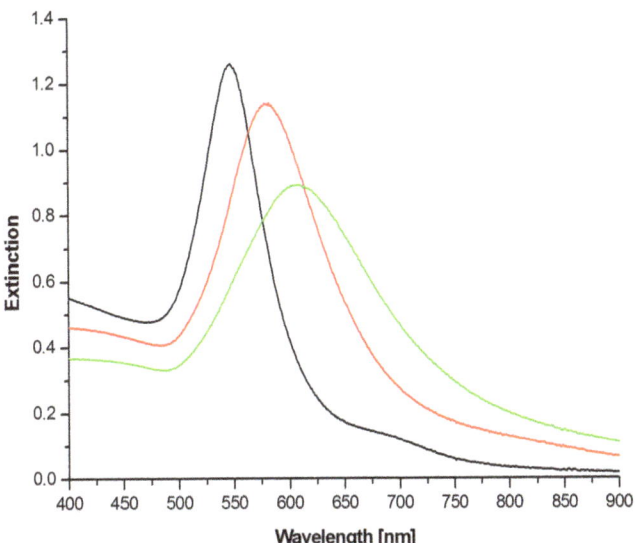

Figure 9. UV-Vis excitation spectra of trisoctahedral gold nanoparticles of various sizes: black—70 nm, red—94 nm, and green—115 nm.

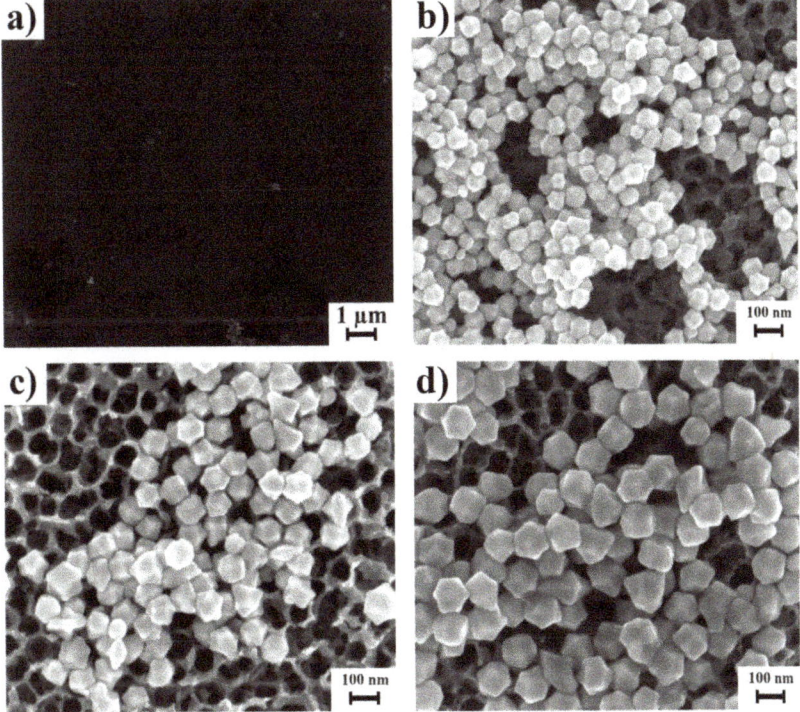

Figure 10. SEM micrographs showing the surface of flat WO$_3$ covered by 94 nm trisoctahedral gold nanoparticles (**a**) and nanopores WO$_3$ covered by trisoctahedral gold nanoparticles of 70 nm (**b**), 94 nm (**c**), and 115 nm (**d**).

3.4. SERS Measurements

R6G was used in the Raman studies. It is a dye often used as a probe molecule in SERS spectroscopy. Spectra of various concentrations of R6G in water were recorded for gold nanoparticles of sizes 71 nm, 94 nm, and 115 nm deposited on a nanostructured WO_3 substrate (Figure 9). Characteristic bands were observed: 613 cm^{-1} corresponding to the C – C – C ring in-plane vibration mode, 772 cm^{-1} – C – H to the out-of-plane bend mode, 1126 cm^{-1} and 1185 cm^{-1} can be attributed to the N – H in-plane bend modes, while 1362 and 1510 cm^{-1} are derived from the N – H in-plane bend modes [34]. The highest intensity of spectra was obtained for nanoparticles with a size of 94 nm. It should be noted that in the case of nanoparticles of a size of 115 nm, there was a change in the band intensity ratio, especially visible for 1362 cm^{-1} and 1510 cm^{-1}. This may be due to the different orientation of the molecule at 115 nm nanoparticles.

As can be seen in Figure 11, the lowest detection limit is reached for the Au NPs of diameter equal to 94 nm. This could result from the overlapping of the SPR band of Au NPs with the applied excitation wavelength. In this way, the generated local electromagnetic field is the highest. The smaller and bigger Au NPs exhibit lower detection limit towards R6G molecules. The SPR peak of 70 nm Au NPs is located at 547 nm. In the SERS experiments, the 632.8 nm laser line was used and hence those values do not overlap with the SPR of those Au NPs. In the case of the biggest Au NPs of 115 nm diameter, the SPR peak is widened, which can indicate a bigger size and shape dispersion, and hence lead to lower SERS enhancement. The reference SERS spectrum was collected on flat WO_3 surfaces covered by 94 nm Au NPs.

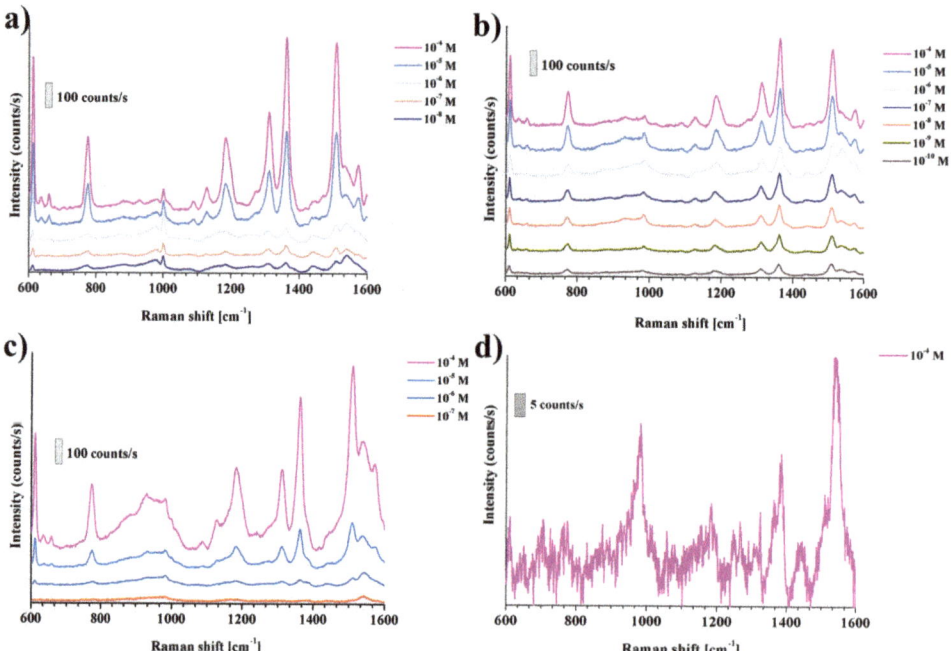

Figure 11. SERS spectra of R6G deposited on WO_3 nanopores surfaces coated with Au trisoctahedral nanoparticles: (**a**) 71 nm; (**b**) 94 nm; and (**c**) 115 nm; (**d**) SERS spectrum of R6G deposited on flat WO_3 coated with Au trisoctahedral nanoparticles (94 nm).

Also, the recovery test was conducted for the proposed WO_3-Au SERS platform. It was found that due to the porous structures of the substrate, simple water rinsing is not sufficient for analyte molecules washout, as the SERS signal from the analyte could be still observed. Also, the sonication procedure in water was applied. However, it was found that not just the analyte molecules were removed, but also some amount of Au NPs are detached, and in consequence, the SERS EF is significantly lower.

4. Conclusions

In this work, we successfully showed an efficient and repetitive method for WO_3 nanopores array formation. Many factors affecting final morphology were carefully examined by SEM microscopy. Trisoctahedron Au NPs were successfully synthesised. It was proved that using smaller trisoctahedron Au NPs as seeds for growth of bigger nanoparticles permits easy tuning of the size of particles while maintaining the well-defined trisoctahedron shape. The UV-Vis spectroscopy showed shifting of SPR into longer wavelengths with successive growth of nanoparticles. The WO_3—Au trisoctahedron modified nanoarray was successfully used as a SERS platform. The highest enhancement was observed for the Au NPs with 94 nm diameter.

Author Contributions: J.K.: Conceptualization, methodology, investigation, formal analysis, writing—review and editing. R.A.: methodology, investigation, formal analysis, writing—review and editing. M.D.: methodology, investigation. S.T.-S.: investigation, formal analysis. All authors have read and agreed to the published version of the manuscript.

Funding: This research received no external funding.

Institutional Review Board Statement: Not applicable.

Informed Consent Statement: Not applicable.

Data Availability Statement: Not applicable.

Acknowledgments: J.K. thanks the University of Warsaw, Faculty of Chemistry for the financial support.

Conflicts of Interest: The authors declare no conflict of interest.

References

1. Cialla, D.; März, A.; Böhme, R.; Theil, F.; Weber, K.; Schmitt, M.; Popp, J. Surface-Enhanced Raman Spectroscopy (SERS): Progress and Trends. *Anal. Bioanal. Chem.* **2012**, *403*, 27–54. [CrossRef] [PubMed]
2. Green, M.; Liu, F.M. SERS Substrates Fabricated by Island Lithography: The Silver/Pyridine System. *J. Phys. Chem. B* **2003**, *107*, 13015–13021. [CrossRef]
3. Fleischmann, M.; Hendra, P.J.; McQuillan, A.J. Raman Spectra of Pyridine Adsorbed at a Silver Electrode. *Chem. Phys. Lett.* **1974**, *26*, 163–166. [CrossRef]
4. Jimenez-Cisneros, J.; Galindo-Lazo, J.P.; Mendez-Rojas, M.A.; Campos-Delgado, J.R.; Cerro-Lopez, M. Plasmonic Spherical Nanoparticles Coupled with Titania Nanotube Arrays Prepared by Anodization as Substrates for Surface-Enhanced Raman Spectroscopy Applications: A Review. *Molecules* **2021**, *26*, 7443. [CrossRef]
5. Pisarek, M.; Roguska, A.; Kudelski, A.; Holdynski, M.; Janik-Czachor, M. Self-Organized TiO_2, Al_2O_3 and ZrO_2 Nanotubular Layers: Properties and Applications. In *Comprehensive Guide for Nanocoatings Technology, Vol 3: Properties and Development*; Aliofkhazraei, E.M., Ed.; Nova Science: New York, NY, USA, 2015; pp. 435–462.
6. Tang, H.; Meng, G.; Huang, Q.; Zhang, Z.; Huang, Z.; Zhu, C. Arrays of Cone-Shaped ZnO Nanorods Decorated with Ag Nanoparticles as 3D Surface-Enhanced Raman Scattering Substrates for Rapid Detection of Trace Polychlorinated Biphenyls. *Adv. Funct. Mater.* **2012**, *22*, 218–224. [CrossRef]
7. Krajczewski, J.; Dumiszewska, E.; Czolak, D.; Turczyniak Surdacka, S.; Kudelski, A. New, Epitaxial Approach to SERS Platform Preparation—InP Nanowires Coated by an Au Layer as a New, Highly Active, and Stable SERS Platform. *Appl. Surf. Sci.* **2023**, *607*, 155096. [CrossRef]
8. Krajczewski, J.; Turczyniak-Surdacka, S.; Dziubałtowska, M.; Ambroziak, R.; Kudelski, A. Ordered Zirconium Dioxide Nanotubes Covered with an Evaporated Gold Layer as Reversible, Chemically Inert and Very Efficient Substrates for Surface-Enhanced Raman Scattering (SERS) Measurement. *Spectrochim. Acta Part A Mol. Biomol. Spectrosc.* **2022**, *275*, 121183. [CrossRef]
9. Ambroziak, R.; Hołdyński, M.; Płociński, T.; Pisarek, M.; Kudelski, A. Cubic Silver Nanoparticles Fixed on TiO_2 Nanotubes as Simple and Efficient Substrates for Surface Enhanced Raman Scattering. *Materials* **2019**, *12*, 3373. [CrossRef]
10. Mampallil, D.; Eral, H.B. A Review on Suppression and Utilization of the Coffee-Ring Effect. *Adv. Colloid Interface Sci.* **2018**, *252*, 38–54. [CrossRef]
11. Li, X.; Bai, J.; Liu, Q.; Li, J.; Zhou, B. WO_3/W Nanopores Sensor for Chemical Oxygen Demand (COD) Determination under Visible Light. *Sensors* **2014**, *14*, 10680–10690. [CrossRef]

12. Nah, Y.-C.; Ghicov, A.; Kim, D.; Schmuki, P. Enhanced Electrochromic Properties of Self-Organized Nanoporous WO_3. *Electrochem. Commun.* **2008**, *10*, 1777–1780. [CrossRef]
13. Zhang, T.; Paulose, M.; Neupane, R.; Schaffer, L.A.; Rana, D.B.; Su, J.; Guo, L.; Varghese, O.K. Nanoporous WO_3 Films Synthesized by Tuning Anodization Conditions for Photoelectrochemical Water Oxidation. *Sol. Energy Mater. Sol. Cells* **2020**, *209*, 110472. [CrossRef]
14. Fernández-Domene, R.M.; Roselló-Márquez, G.; Sánchez-Tovar, R.; Cifre-Herrando, M.; García-Antón, J. Synthesis of WO_3 Nanorods through Anodization in the Presence of Citric Acid: Formation Mechanism, Properties and Photoelectrocatalytic Performance. *Surf. Coat. Technol.* **2021**, *422*, 127489. [CrossRef]
15. Wu, S.; Li, Y.; Chen, X.; Liu, J.; Gao, J.; Li, G. Fabrication of $WO_3 \cdot 2H_2O$ Nanoplatelet Powder by Breakdown Anodization. *Electrochem. Commun.* **2019**, *104*, 106479. [CrossRef]
16. Martins, A.S.; Guaraldo, T.T.; Wenk, J.; Mattia, D.; Boldrin Zanoni, M.V. Nanoporous WO_3 Grown on a 3D Tungsten Mesh by Electrochemical Anodization for Enhanced Photoelectrocatalytic Degradation of Tetracycline in a Continuous Flow Reactor. *J. Electroanal. Chem.* **2022**, *920*, 116617. [CrossRef]
17. Zych, M.; Syrek, K.; Zaraska, L.; Sulka, G.D. Improving Photoelectrochemical Properties of Anodic WO_3 Layers by Optimizing Electrosynthesis Conditions. *Molecules* **2020**, *25*, 2916. [CrossRef] [PubMed]
18. Hahn, R.; Macak, J.M.; Schmuki, P. Rapid Anodic Growth of TiO_2 and WO_3 Nanotubes in Fluoride Free Electrolytes. *Electrochem. Commun.* **2007**, *9*, 947–952. [CrossRef]
19. Altomare, M.; Nguyen, N.T.; Schmuki, P. High Aspect-Ratio WO_3 Nanostructures Grown By Self-Organizing Anodization in Hot Pure O-H_3PO_4. In *ECS Meeting Abstracts*; IOP Publishing: Bristol, UK, 2016; Volume MA2016-01, p. 2123. [CrossRef]
20. Yu, Y.; Zhang, Q.; Lu, X.; Lee, J.Y. Seed-Mediated Synthesis of Monodisperse Concave Trisoctahedral Gold Nanocrystals with Controllable Sizes. *J. Phys. Chem. C* **2010**, *114*, 11119–11126. [CrossRef]
21. Ng, C.; Ye, C.; Ng, Y.H.; Amal, R. Flower-Shaped Tungsten Oxide with Inorganic Fullerene-like Structure: Synthesis and Characterization. *Cryst. Growth Des.* **2010**, *10*, 3794–3801. [CrossRef]
22. Zhan, W.T.; Ni, H.W.; Chen, R.S.; Wang, Z.Y.; Li, Y.; Li, J.H. One-Step Hydrothermal Preparation of TiO_2/WO_3 Nanocomposite Films on Anodized Stainless Steel for Photocatalytic Degradation of Organic Pollutants. *Thin Solid Films* **2013**, *548*, 299–305. [CrossRef]
23. Watcharenwong, A.; Chanmanee, W.; de Tacconi, N.R.; Chenthamarakshan, C.R.; Kajitvichyanukul, P.; Rajeshwar, K. Anodic Growth of Nanoporous WO_3 Films: Morphology, Photoelectrochemical Response and Photocatalytic Activity for Methylene Blue and Hexavalent Chrome Conversion. *J. Electroanal. Chem.* **2008**, *612*, 112–120. [CrossRef]
24. De Tacconi, N.R.; Chenthamarakshan, C.R.; Yogeeswaran, G.; Watcharenwong, A.; de Zoysa, R.S.; Basit, N.A.; Rajeshwar, K. Nanoporous TiO_2 and WO_3 Films by Anodization of Titanium and Tungsten Substrates: Influence of Process Variables on Morphology and Photoelectrochemical Response. *J. Phys. Chem. B* **2006**, *110*, 25347–25355. [CrossRef] [PubMed]
25. Momeni, M.M.; Ghayeb, Y.; Davarzadeh, M. Single-Step Electrochemical Anodization for Synthesis of Hierarchical WO_3–TiO_2 Nanotube Arrays on Titanium Foil as a Good Photoanode for Water Splitting with Visible Light. *J. Electroanal. Chem.* **2015**, *739*, 149–155. [CrossRef]
26. Syrek, K.; Zaraska, L.; Zych, M.; Sulka, G.D. The Effect of Anodization Conditions on the Morphology of Porous Tungsten Oxide Layers Formed in Aqueous Solution. *J. Electroanal. Chem.* **2018**, *829*, 106–115. [CrossRef]
27. Shi, W.; Guo, X.; Cui, C.; Jiang, K.; Li, Z.; Qu, L.; Wang, J.-C. Controllable Synthesis of Cu_2O Decorated WO_3 Nanosheets with Dominant (0 0 1) Facets for Photocatalytic CO_2 Reduction under Visible-Light Irradiation. *Appl. Catal. B Environ.* **2019**, *243*, 236–242. [CrossRef]
28. Wang, S.L.; Mak, Y.L.; Wang, S.; Chai, J.; Pan, F.; Foo, M.L.; Chen, W.; Wu, K.; Xu, G.Q. Visible–Near-Infrared-Light-Driven Oxygen Evolution Reaction with Noble Metal Free WO_2–WO_3 Hybrid Nanorods. *Langmuir* **2016**, *32*, 13016–13053. [CrossRef] [PubMed]
29. Xie, F.Y.; Gong, L.; Liu, X.; Tao, Y.T.; Zhang, W.H.; Chen, S.H.; Meng, H.; Chen, J. XPS Studies on Surface Reduction of Tungsten Oxide Nanowire Film by Ar+ Bombardment. *J. Electron Spectrosc. Relat. Phenom.* **2012**, *185*, 112–118. [CrossRef]
30. Bouvard, O.; Krammer, A.; Schüler, A. In Situ Core-Level and Valence-Band Photoelectron Spectroscopy of Reactively Sputtered Tungsten Oxide Films. *Surf. Interface Anal.* **2016**, *48*, 660–663. [CrossRef]
31. Lu, S.-S.; Zhang, L.-M.; Dong, Y.-W.; Zhang, J.-Q.; Yan, X.-T.; Sun, D.-F.; Shang, X.; Chi, J.-Q.; Chai, Y.-M.; Dong, B. Tungsten-Doped Ni–Co Phosphides with Multiple Catalytic Sites as Efficient Electrocatalysts for Overall Water Splitting. *J. Mater. Chem. A* **2019**, *7*, 16859–16866. [CrossRef]
32. Ciftyürek, E.; Šmíd, B.; Li, Z.; Matolín, V.; Schierbaum, K. Spectroscopic Understanding of SnO_2 and WO_3 Metal Oxide Surfaces with Advanced Synchrotron Based; XPS-UPS and Near Ambient Pressure (NAP) XPS Surface Sensitive Techniques for Gas Sensor Applications under Operational Conditions. *Sensors* **2019**, *19*, 4737. [CrossRef]
33. Lange, M.A.; Krysiak, Y.; Hartmann, J.; Dewald, G.; Cerretti, G.; Tahir, M.N.; Panthöfer, M.; Barton, B.; Reich, T.; Zeier, W.G.; et al. Solid State Fluorination on the Minute Scale: Synthesis of $WO_{3-x}F_x$ with Photocatalytic Activity. *Adv. Funct. Mater.* **2020**, *30*, 1909051. [CrossRef]
34. Li, R.; Li, H.; Pan, S.; Liu, K.; Hu, S.; Pan, L.; Guo, Y.; Wu, S.; Li, X.; Liu, J. Surface-Enhanced Raman Scattering from Rhodamine 6G on Gold-Coated Self-Organized Silicon Nanopyramidal Array. *J. Mater. Res.* **2013**, *28*, 3401–3407. [CrossRef]

Article

One-Step Synthesis of Sulfur-Doped Nanoporous Carbons from Lignin with Ultra-High Surface Area, Sulfur Content and CO_2 Adsorption Capacity

Dipendu Saha *, Gerassimos Orkoulas and Dean Bates

Chemical Engineering Department, Widener University, 1 University Place, Chester, PA 19103, USA
* Correspondence: dsaha@widener.edu; Tel.: +1-610-499-4056; Fax: +1-610-499-4059

Abstract: Lignin is the second-most available biopolymer in nature. In this work, lignin was employed as the carbon precursor for the one-step synthesis of sulfur-doped nanoporous carbons. Sulfur-doped nanoporous carbons have several applications in scientific and technological sectors. In order to synthesize sulfur-doped nanoporous carbons from lignin, sodium thiosulfate was employed as a sulfurizing agent and potassium hydroxide as the activating agent to create porosity. The resultant carbons were characterized by pore textural properties, X-ray photoelectron spectroscopy (XPS), X-ray diffraction (XRD), and scanning electron microscopy (SEM). The nanoporous carbons possess BET surface areas of 741–3626 m^2/g and a total pore volume of 0.5–1.74 cm^3/g. The BET surface area of the carbon was one of the highest that was reported for any carbon-based materials. The sulfur contents of the carbons are 1–12.6 at.%, and the key functionalities include S=C, S-C=O, and SO_x. The adsorption isotherms of three gases, CO_2, CH_4, and N_2, were measured at 298 K, with pressure up to 1 bar. In all the carbons, the adsorbed amount was highest for CO_2, followed by CH_4 and N_2. The equilibrium uptake capacity for CO_2 was as high as ~11 mmol/g at 298 K and 760 torr, which is likely the highest among all the porous carbon-based materials reported so far. Ideally adsorbed solution theory (IAST) was employed to calculate the selectivity for CO_2/N_2, CO_2/CH_4, and CH_4/N_2, and some of the carbons reported a very high selectivity value. The overall results suggest that these carbons can potentially be used for gas separation purposes.

Keywords: porous carbon; sustainability; surface area; CO_2 separation

1. Introduction

Sulfur-doped porous carbon is a unique form of heteroatom doped carbon. Unlike other common types of heteroatoms, such as nitrogen, oxygen, or boron, sulfur atoms are significantly larger than carbon atoms, and therefore, sulfur atoms protrude out of the graphene plant, giving rise to a few unique properties of the parent carbon, such as superconductivity, as revealed in the theoretical studies [1,2]. In addition, the lone pair of electrons in the sulfur atom induces polarizability and interactions with oxygen [3]. There are several specific applications of sulfur-doped porous carbon, including electrocatalysis for fuel cells [4], electrodes for electrochemical capacitors [5], anode material for Li-ion batteries [6], cathodes for Li-S batteries [7], heavy metal removal [8], toxic gas removal [9], H_2 storage [10], CO_2 separation [11], and many others [12].

Most of the time, sulfur-doped carbons are synthesized by carbonizing S-bearing carbon precursors, like thiophenemethanol [13], cysteine [14], algae [15], ionic liquids [16], and others [12]. The detailed list of precursors that have been employed to synthesize sulfur-doped nanoporous carbons are listed in [12]. The porosity within the sulfur-doped carbon is achieved by post-synthesis activation [17] or utilizing templating strategies [13], including hard and soft templates. In our past research, we incorporated sodium thiosulfate ($Na_2S_2O_3$) at elevated temperatures to introduce sulfur functionalities within the porous carbon [8,18–22]. The uniqueness of incorporating $Na_2S_2O_3$ is that it does not require an

S-bearing carbon precursor to synthesize sulfur-doped carbon, as sulfur is contributed by the Na$_2$S$_2$O$_3$.

Lignin is the second most naturally abundant biopolymer present in the environment. It is one of the key constituents of wood along with cellulose and hemicellulose. Although there are three key structural constituents of lignin, including coumaryl, guaiacyl, and sinapyl alcohol, these three components are randomly cross-linked with each other, giving rise to the structural heterogeneity of lignin polymer. The exact structure of lignin polymer depends on the wood (tree) type and processing conditions. Lignin is industrially produced as a by-product in pulp and paper industries and bio-refineries. Although there is much research on the use of lignin, it still lacks prominent value-added utilization. The majority of lignin is used as low-calorie fuel. Historically, lignin was used in several types of specialty carbons, including activated carbon [23,24], mesoporous carbon [25,26], and carbon fibers [27,28]. Synthesis of porous carbon from lignin not may only introduce sustainability in the synthesis but also influences the economy of lignin by increasing its value-added utilization.

In this work, we synthesized sulfur-doped nanoporous carbon from lignin using a one-step approach. We incorporated a varying ratio of sodium thiosulfate (Na$_2$S$_2$O$_3$) and potassium hydroxide (KOH) to simultaneously introduce sulfur functionalities and porosity into the carbon matrix. The resultant carbon was employed for gas separation purposes.

2. Experimental

2.1. Synthesis of Sulfur-Doped Carbons

For all the synthesis purposes, commercially available dealkaline lignin (TCI America) was employed. Typically, the desired components of lignin, sodium thiosulfate (Na$_2$S$_2$O$_3$), and potassium hydroxide (KOH) were mixed in a coffee grinder and then loaded onto an alumina boat. The boat was introduced to the Lindberg-Blue^{TM} (USA) tube furnace. The tube furnace was heated to 800 °C with a ramp rate of 10 °C/min, dwelled at 800 °C for 2 min, and then cooled to room temperature. All the heating and cooling operations were performed under N$_2$ gas. The final products were washed several times with DI water and then filtered and dried. The names of the carbons according to the ratio of lignin, Na$_2$S$_2$O$_3$, and KOH are given in Table 1. The schematic of the synthesis is shown in Figure 1. From the table, it is clear that the total mixture was in the range of 7–12 g, which is the maximum amount of materials that be processed within the porcelain boat onto the tube surface. The ratio of Na$_2$S$_2$O$_3$ and KOH was also adjusted according to the literature and our previous fundings; too low or too high amounts of these materials may result in improper impregnation/activation or breakdown of the entire carbon matrix.

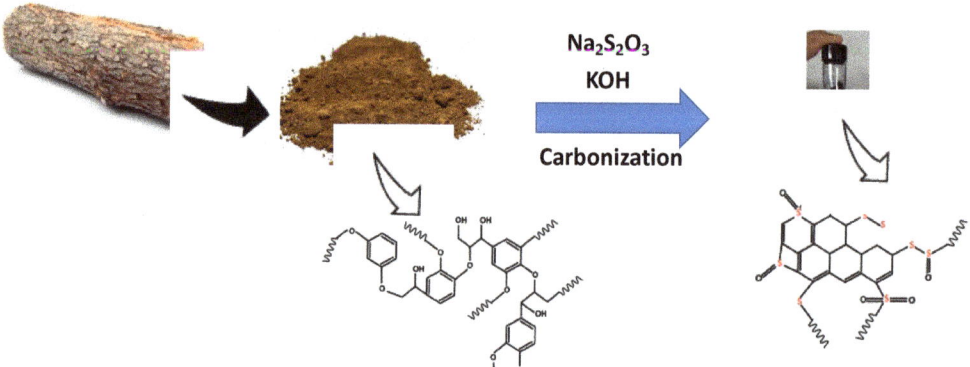

Figure 1. Schematic of one-step synthesis of sulfur-doped nanoporous carbon from lignin.

Table 1. Sample identity.

Sample Identity	Lignin: $Na_2S_2O_3$: KOH
LS-1	3:2:2
LS-2	3:3:1
LS-3	3:1:3
LS-4	3:6:0
LS-5	3:6:3

2.2. Characterization of Sulfur-Doped Carbons

All the carbons were characterized with pore textural properties, x-ray photoelectron spectroscopy (XPS), and scanning electron microscopy (SEM). The pore textural properties, including BET specific surface area (BET SSA) and pore size distribution, were calculated using N_2 adsorption isotherms at 77 K and CO_2 adsorption isotherms at 273 K in Quantachorme's Autosorb iQ-any gas instrument (USA). The non-local density function theory (NLDFT)-based pore size distribution below 12 Å was calculated using CO_2 adsorption isotherm, whereas the larger (>12 Å) pores were calculated using N_2 adsorption isotherm. X-ray photoelectron spectroscopy (XPS) results were obtained in a Thermo-Fisher K-alpha instrument (USA) with a monochromatic Al-Kα as an X-ray anode. The intensity of X-ray energy was set to 1486.6 eV, and the resolution was 0.5 eV. Scanning electron microscopic images (SEM) were obtained in a Carl Zeiss Merlin SEM microscope (USA) operating at 1 kV. X-ray diffraction patterns were obtained in Rigaku miniflex XRD instrument. In order to capture the XRD pattern of the carbon, it was ground to a fine powder in mortar and pestle and introduced within the sample holder.

2.3. Gas Adsorption Studies

Equilibrium adsorption isotherms of pure-component CO_2, CH_4, and N_2 were measured on all the nanoporous carbons at the temperature of 298 K and pressure up to 760 torr in the same Autosorb iQ-any gas instrument. The temperature was maintained by an additional Chiller (Julabo) (USA). All the gases were of ultra-high purity (UHP) grade. About 80 mg of each of the sample was inserted in the sample tube along with filler rod and non-elutriation cap. Each sample was outgassed at 300 °C for 3 h before the adsorption experiment.

3. Results and Discussion

3.1. Material Characteristics

The N_2 adsorption-desorption isotherms at 77 K are shown in Figure 2a. The sharp rise in the low-pressure region suggests the presence of macroporosity. A narrow stretch of hysteresis loop is also observed in all the isotherms, signifying the presence of mesoporosity. The NLDFT-based pore size distribution is shown in Figure 2b. This figure shows that all the carbons have a few pores in the narrow micropore region, including 8.1, 5.5, and 4.7 Å; the pore width around 3.4 Å is attributed to the graphite layer spacing and not a true pore. In the large micropore region, the carbons possess two distinct pores around the 14.7 and 19.3 Å regions. The majority of the carbons also demonstrated a distributed mesoporosity within 20–45 Å, supporting the presence of a hysteresis loop in Figure 2a. The detailed pore textural properties are shown in Table 2. It is observed that LS-3 has the highest BET SSA (3626 m^2/g) and pore volume (1.74 cm^3/g). A porous carbon with a BET surface area more than 3000 m^2/g is very difficult to produce and has been rarely reported in the literature [29–34]; only one work reported the BET surface area higher than that of LS-3 (a MOF-derived porous carbon with BET: 4300 m^2/g) [34]. The lowest porosity belongs to LS-4 (BET: 280 m^2/g; pore volume: 0.157 cm^3/g), synthesized without KOH. It is clear that KOH is the primary agent in creating the porosity, $Na_2S_2O_3$ is primarily used to introduce sulfur functionalities. The influence of $Na_2S_2O_3$ in creating porosity is very small.

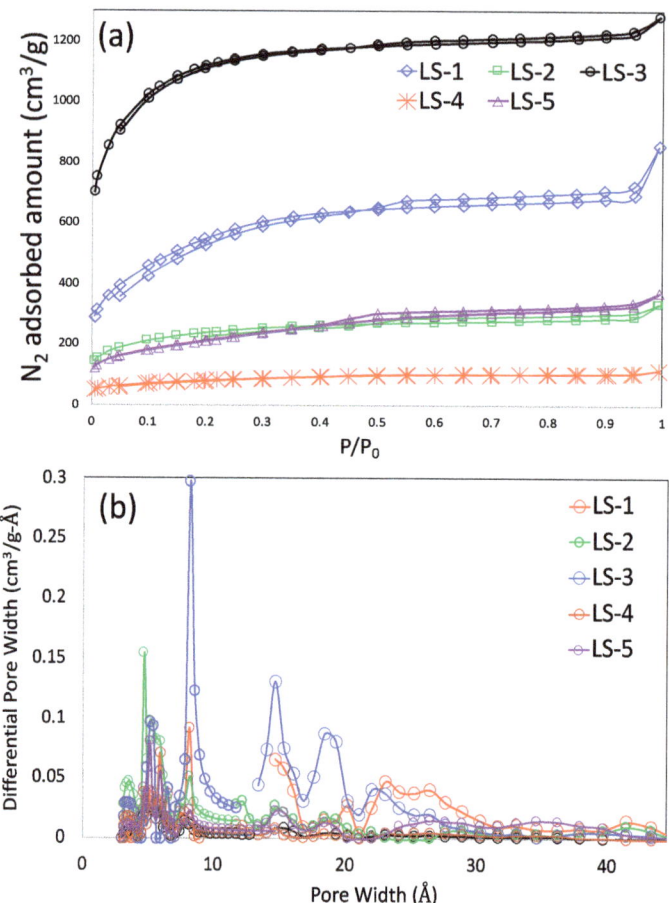

Figure 2. N_2 adsorption–desorption isotherms at 77 K (**a**); NLDFT-based pore size distribution (**b**).

Table 2. Pore textual properties.

Sample Identity	BET SSA (m^2/g)	Total Pore Volume (cm^3/g)	Micropore Volume (cm^3/g)	Mesopore Volume (cm^3/g)
LS-1	1915	1.079	0.532	0.547
LS-2	787	0.443	0.292	0.151
LS-3	3626	1.741	1.44	0.301
LS-4	280	0.157	0.089	0.068
LS-5	741	0.501	0.214	0.287

The quantitative results for XPS are shown in Table 3. The C, S, and O contents were calculated by fitting the C-1s, S-2p, and O-1s peaks, and the representative peak fitting results for LS-3 and LS-5 are shown in Figure 3a–f. As observed in the table, LS-4 has the highest amount of sulfur content (12.6 at.%), followed by LS-5 (8.9 at.%). It is quite intuitive to note that the sulfur content is directly proportional to the addition of $Na_2S_2O_3$ in the course of synthesis; sulfur content decreases in the order of LS-4 > LS-5 > LS-2 > LS-1 > LS-3. Despite LS-5 and LS-4 having the same $Na_2S_2O_3$ contents, a higher KOH in LS-5 causes removal of some of the sulfur contents in the course of activation. It is also interesting to note that LS-4 has about 1 at.% sulfur, which originated from the pristine lignin itself in the course of its industrial production. Within different types of sulfur functionalities, the largest fraction of sulfur is associated with C-S contents in all the porous carbon samples, followed by SO_x and S=C-O. From Table 3, it is obvious that higher sulfur content also caused higher oxygen content (except LS-3), which might have affected sulfur functionalities, resulting in lowering of total carbon content. Within the oxygen-bearing functionalities, the largest group belonged to S=O/C=O/O-H, directly correlating the oxygen contents with sulfur. LS-4, which had the highest sulfur content, had the lowest total carbon content, only 51.3 at.%. According to XPS, C-C sp^2 is the primary carbon structure, suggesting that all the carbons are mostly graphitic in nature. It also needs to be noted that the LS-3, LS-4, and LS-5 have sodium that may have originated from $Na_2S_2O_3$ and/or during the commercial production of lignin.

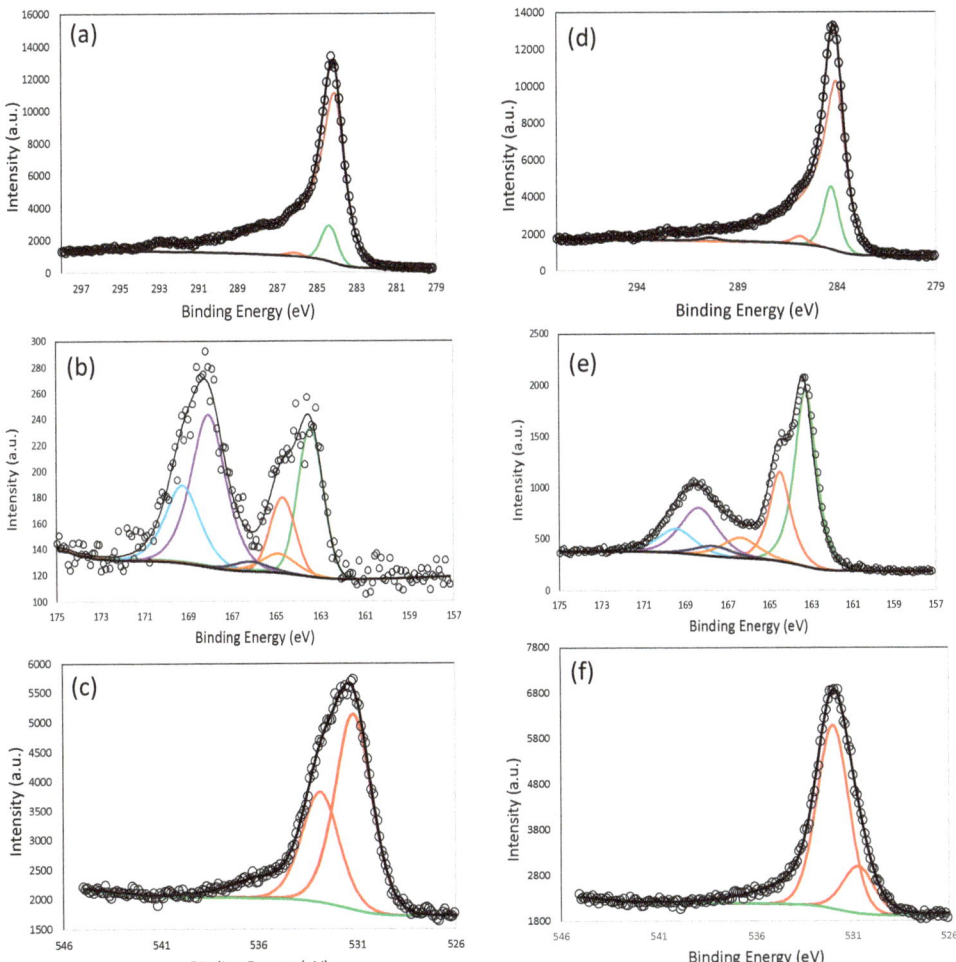

Figure 3. X-ray photoelectron spectroscopy (XPS) analysis; peak deconvolution results of C-1s (**a**), S-2p (**b**), and O-1s (**c**) of LS-3; and C-1s (**d**), S-2p (**e**), and O-1s (**f**) of LS-5.

Table 3. XPS-based quantitative functionalities C, O, and S in LS-1 to -5.

Elements (%)		LS-1		LS-2		LS-3		LS-4		LS-5	
C (%)	C-C sp^2	77.3		69.5		73.0		39.0		58.4	
	C-C sp^3	9.0	87.2	8.8	79.6	6.4	80.1	9.5	51.3	9.5	69.6
	C-O, C-S	0.9		1.2		0.7		2.2		1.3	
	C=O	0		0.0		0.8		0.6		0.5	
S (%)	S-C	1.9		3.5		0.4		6.5		5.5	
	S=C-O	0.2	3.0	0.5	6.5	0.1	1.0	1.1	12.6	1.0	8.9
	SO$_x$	0.9		2.5		0.6		5.0		2.4	
O (%)	S=O, C-O, OH	6.0	8.8	9.8	11.5	8.6	13.5	26.5	30.9	8.9	11.2
	C-O-H	2.8		1.8		4.8		4.5		2.3	

The representative SEM images of LS-3 are shown in Figure 4a–d at different levels of magnification. The carbon particles are irregular in shape, with a size around 20–100 μm. A three-dimensional network of larger pores is observed in the SEM images with macropore size in the range of 1.5–4 μm. Such a pore system along with meso- and micropores may have created a hierarchical porous network in the carbon matrix, which can be highly beneficial for faster diffusion of an adsorbate molecule in the course of diffusion.

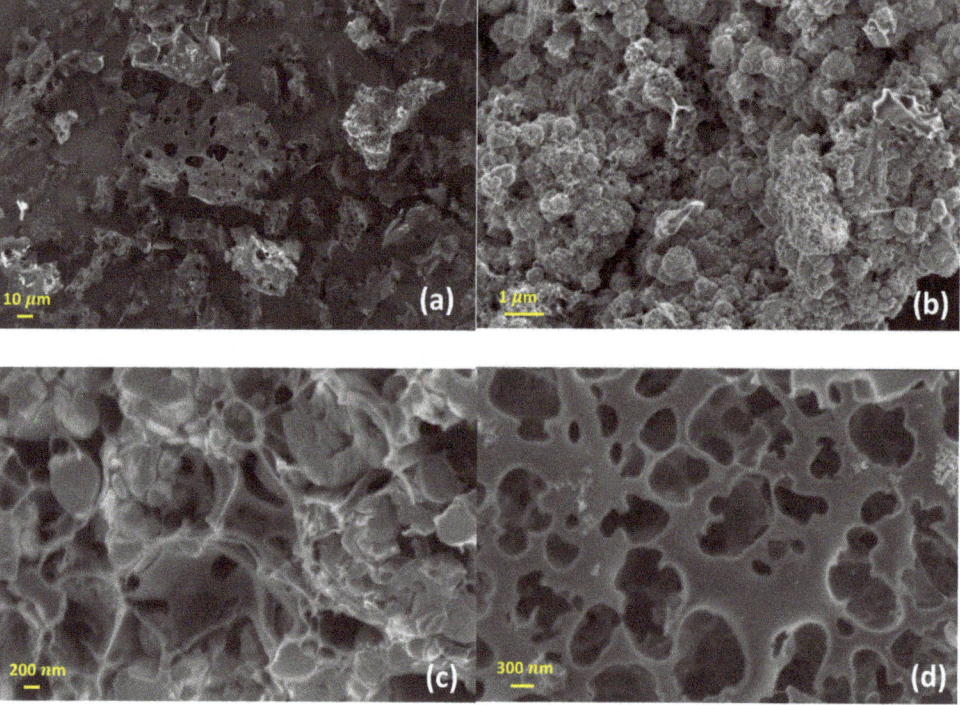

Figure 4. Representative Scanning Electron Imaging (SEM) results of LS-3 (**a–d**).

The X-ray diffraction (XRD) images are shown in Figure 5. Two 'hump'-like and very broad peaks around 23° and 43° are observed for all the carbons. These peaks are remnants of graphitic ordering and are present in almost all sp^2 hybridized carbons [35]. For LS-5, there are a few small peaks observed, which may be associated with salts of Na and/or K, which could originate from $Na_2S_2O_3$, KOH, and the impurities present in pristine lignin itself. Relatively higher amounts of Na also support the XPS observation, and such sodium originates from sodium thiosulfate and/or pristine lignin. We did not pursue any further analysis to reveal details of these salts, as it is beyond the scope of this study.

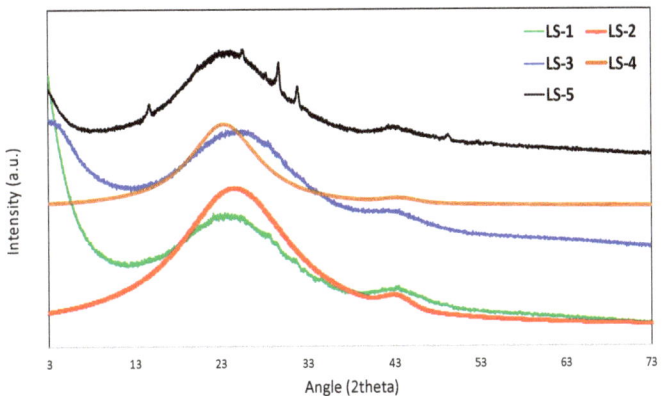

Figure 5. X-ray diffraction (XRD) results.

3.2. Gas Adsorption Studies

The adsorption isotherms CO_2, CH_4, and N_2 are shown in Figure 6a–e for LS-1, -2, -3, -4, and -5, respectively. For all the plots, the CO_2 adsorption amount is higher, followed by CH_4 and N_2. The largest equilibrium adsorption capacity of CO_2 was demonstrated by LS-3 (~10.89 mmol/g at 757 torr pressure), which has the highest BET surface area and micropore volume. Such a high equilibrium uptake of CO_2 is probably the highest CO_2 uptake capacity ever reported for any carbon-based material in the literature. The adsorption of all these gases is influenced by micropore volume. As observed in this study, there is a linear trend of adsorbed amounts of CO_2, CH_4, and N_2, suggesting that the micropore volume played a pivotal role in the adsorption processes. In addition, CO_2 adsorption may also be influenced by the presence of sulfur functionalities. It has been reported that mono- or dioxidized sulfur on the carbon surface causes high enthalpy of CO_2 adsorption of 4–6 kcal/mol, which may be attributed to the negative charge of an oxygen atom, possibly caused by the high positive charge on the sulfur atom [36]. Theoretical calculations also revealed that electron overlap between CO_2 and sulfur functionalities on the carbon surface may enhance the interactions between CO_2 and the carbon substrate [37]. As observed in Figure 5, there is a possible presence of Na and K salts, which might have originated from $Na_2S_2O_3$, KOH, or the pristine lignin itself. To the best of our knowledge, these salts do not have any influence in the adsorption of CO_2, CH_4, or N_2.

Working capacity is generally defined as the difference in the adsorbed amount within the adsorbed pressure of 1 bar (760 torr) and desorbed pressure of 0.1 bar (76 torr). For a suitable adsorbent, consistent working capacity with multiple cycles is required. In this work, we have selected LS-5 as the adsorbent and CO_2 adsorbate gas to study the cyclability of working capacity, and the result is shown in Figure 7. As observed in this figure, the working capacity maintains a constant value within 10 cycles, with a standard deviation of no more than ±0.1.

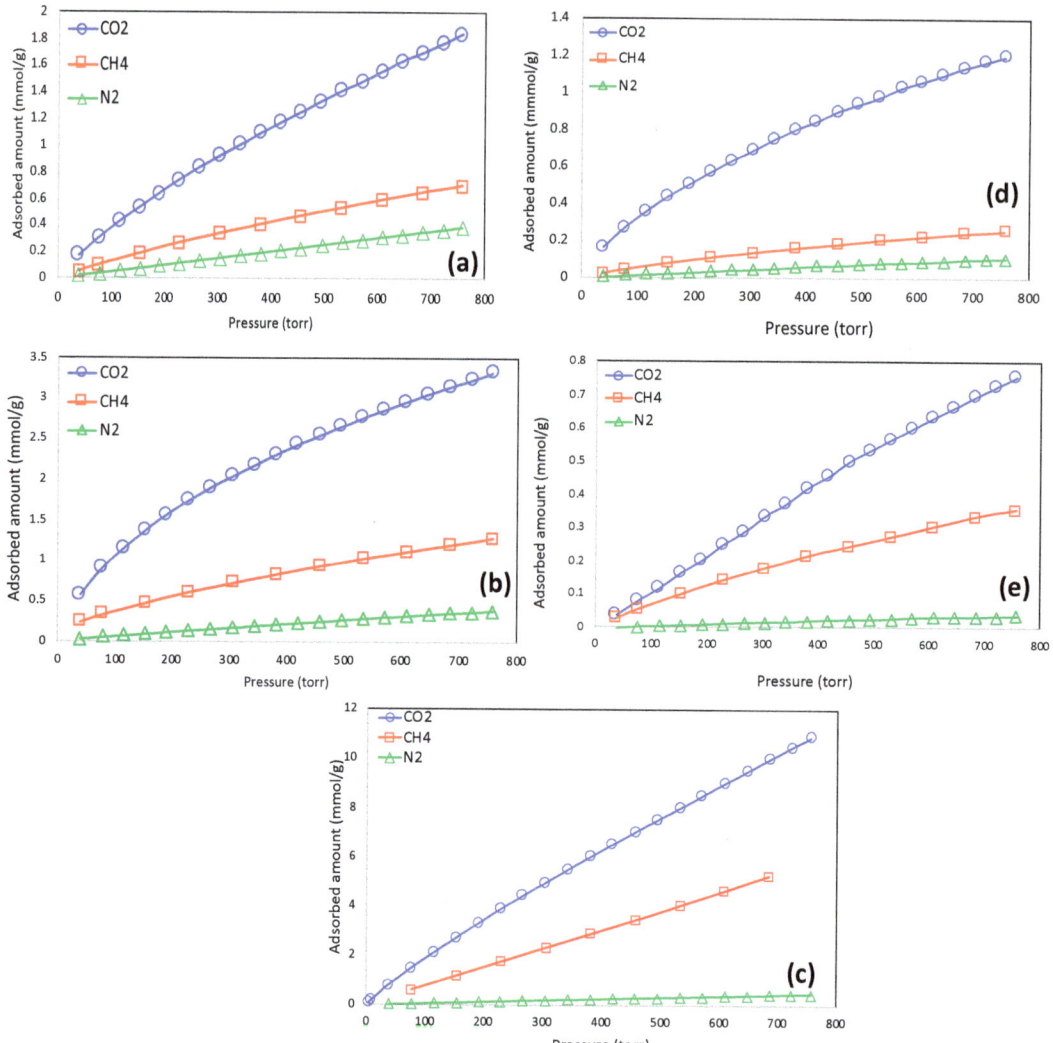

Figure 6. Adsorption isotherms of CO_2, CH_4, and N_2 for LS-1 (**a**), LS-2 (**b**), LS-3 (**c**), LS-4 (**d**), and LS-5 (**e**).

The gas adsorption isotherms were fitted with the Sips isotherm model equation, given below.

$$q = \frac{a_m b p^{1/n}}{1 + b p^{1/n}} \quad (1)$$

where q is the adsorbed amount (mmol/g), p is the pressure (torr), and a_m, b, and n are all Sips constants. The Sips equation is fit employing the solver function of Microsoft Excel, and the values are given in Table S1 of the supporting information.

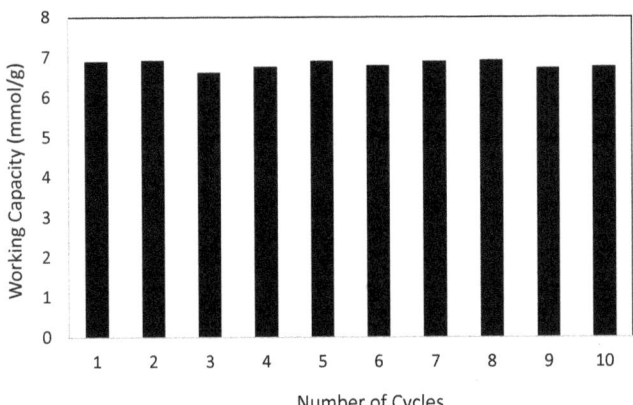

Figure 7. Cyclability of working capacity of CO_2 adsorption in LS-5.

Owing to the experimental difficulty in performing mixed gas adsorption, it is common practice to calculate the selectivity from the pure-component gas adsorption isotherms. Selectivity provides an indication of the preference of the adsorbent materials to prefer one component over another when both species are present in the feed stream. The selectivity ($\alpha_{1/2}$) of component 1 (preferred adsorbate) over component 2 (non-preferred adsorbate) is defined as follows [38]:

$$\alpha_{1/2} = \frac{x_1/y_1}{x_2/y_2} \qquad (2)$$

where x and y are the mole fractions of adsorbate in the adsorbed phase and bulk gas phase, respectively. The most popular way of calculating selectivity from adsorption isotherms is the Ideally Adsorbed Solution Theory (IAST), originally proposed by Myers and Prausnitz [39]. The selectivity values for CO_2/N_2, CO_2/CH_4, and CH_4/N_2 are shown in Figure 8a–c, respectively. From Figure 8a, it is observed that LS-2 and LS-4 have the highest selectivity for CO_2/N_2 (about 180–120) at the lowest pressure, but it decreases significantly at the higher pressure. At the higher pressure, the highest selectivity was demonstrated by LS-3, which is about 80-60. The lowest selectivity was demonstrated by LS-1. For the selectivity of CO_2/CH_4 (Figure 8b), the highest selectivity was demonstrated by LS-4 (20-11), followed by LS-3, LS-2, LS-1, and LS-5. For CH_4/N_2, the highest selectivity was demonstrated by LS-2 and LS-5 (Figure 8c), which was about 80–140 in the lower pressure range and 23-9 in the lower pressure range. The selectivity of CO_2/N_2 is probably one of the highest among other S-doped porous carbons reported in the literature; only the sulfur-doped mesoporous carbon synthesized from resorcinol-formaldehyde in our previous work [22] demonstrated slightly higher selectivity of 190 compared to that of LS-3 and LS-4. For CO_2/CH_4, the selectivity values lie within 3.3–15.7 in the literature [40]. The selectivity of CO_2/CH_4 for LS-4 (21-11) is higher than that reported in the literature. The selectivity of CH_4/N_2 was reported to be as high as 14 in the literature; LS-2 and LS-5 demonstrated a much higher selectivity than this value. It is also important to note that the high equilibrium uptake capacity of a pure preferred component does not always confirm its high selectivity over a non-preferred component; the selectivity also depends on the shape of the pure-component isotherms of both the preferred and non-preferred component. As an example, LS-5 demonstrated very high equilibrium uptake capacity for CO_2 and CH_4; however, it represents the highest selectivity owing to the linear nature of the isotherms.

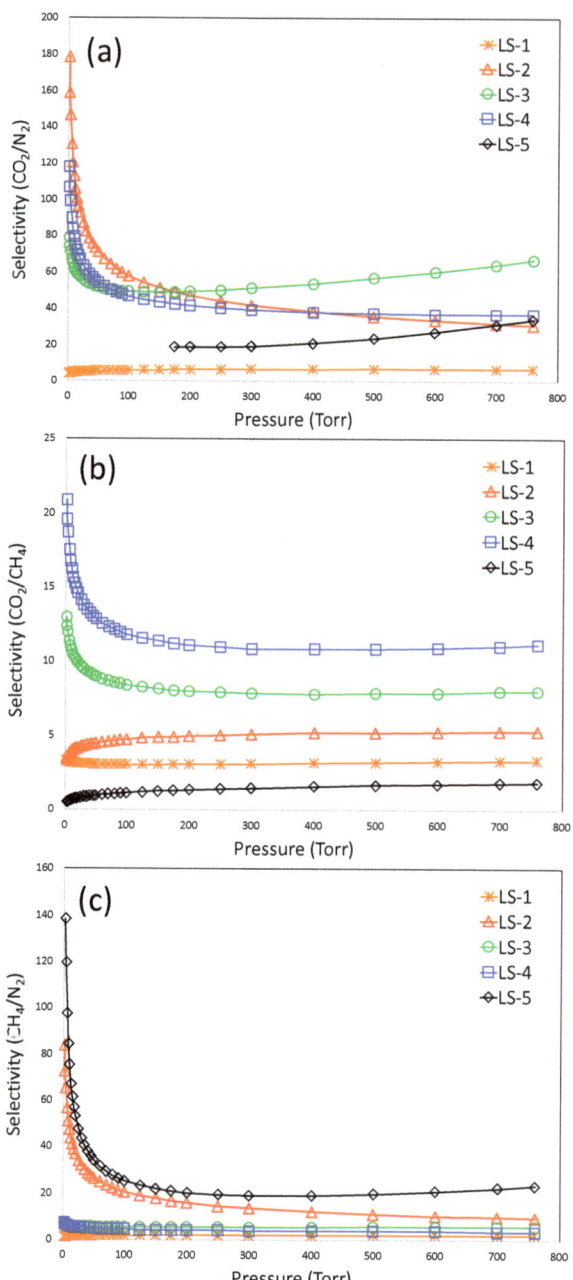

Figure 8. IAST-based selectivity for CO_2/N_2 (**a**), CO_2/CH_4 (**b**), and (**c**) CH_4/N_2.

4. Conclusions

In this work, we successfully synthesized sulfur-doped nanoporous carbon with ultra-high surface area from lignin by one-step carbonization with the help sodium thiosulfate as a sulfurizing agent and potassium hydroxide as an activating agent. The peak deconvolution results of XPS confirmed that the nanoporous carbons possess the sulfur contents of 1 to 12.6 at.%. The porosity analysis revealed that the BET specific surface areas of the carbons are in the range of 741–3626 m^2/g. The surface area of 3626 m^2/g is one of the highest for carbon-based materials reported in literature. Pure-component adsorption isotherms of CO_2, CH_4, and N_2 were measured on all the porous carbons at 298 K, with pressure up to 760 torr. The carbon with the highest BET surface area demonstrated the highest CO_2 uptake of more than 10.89 mmol/g, which is one of the highest for porous carbon-based materials reported in literature. The IAST method was applied to calculate the selectivity of CO_2/N_2, CO_2/CH_4, and CH_4/N_2 from the pure-component isotherm data, and the results demonstrated that these materials can potentially be used for gas separation purposes.

Supplementary Materials: The following supporting information can be downloaded at: https://www.mdpi.com/article/10.3390/ma16010455/s1, Table S1: Sips Constants values.

Author Contributions: Conceptualization, D.S.; Methodology, D.B.; Software, G.O.; Formal analysis, D.S.; Data curation, D.B.; Writing—original draft, D.S.; Supervision, D.S.; Project administration, D.S.; Funding acquisition, D.S. All authors have read and agreed to the published version of the manuscript.

Funding: This work is partly supported by the American Chemical Society sponsored Petroleum Research Fund (ACS-PRF, Grant Number 54205-UNI10). D. Saha also acknowledges the faculty development award from Widener University.

Institutional Review Board Statement: Not applicable.

Informed Consent Statement: Not applicable.

Data Availability Statement: Not applicable.

Acknowledgments: This work is partly supported by the American Chemical Society sponsored Petroleum Research Fund (ACS-PRF, Grant Number 54205-UNI10). D. Saha also acknowledges the faculty development award from Widener University.

Conflicts of Interest: The authors declare no conflict of interest.

References

1. Da Silva, R.R.; Torres, J.H.S.; Kopelevich, Y. Indication of superconductivity at 35 K in graphite-sulfur composites. *Phys. Rev. Lett.* **2001**, *87*, 147001. [CrossRef] [PubMed]
2. Kurmaev, E.; Galakhov, A.; Moewes, A.; Moehlecke, S.; Kopelevich, Y. Inter- layer conduction band states in graphite-sulfur composites. *Phys. Rev. B* **2002**, *66*, 193402. [CrossRef]
3. Paraknowitsch, J.; Thomas, A. Doping carbons beyond nitrogen: An overview of advanced heteroatom doped carbons with boron, sulphur and phosphorus for energy applications. *Energy Environ. Sci.* **2013**, *6*, 2839–2855. [CrossRef]
4. Yang, Z.; Nie, H.; Chen, X.; Chen, X.; Huang, S. Recent progress in doped carbon nanomaterials as effective cathode catalysts for fuel cell oxygen reduction reaction. *J. Power Sources* **2013**, *236*, 238–249. [CrossRef]
5. Hasegawa, G.; Aoki, M.; Kanamori, K.; Nakanishi, K.; Hanada, T.; Tadanaga, K. Monolithic electrode for electric double-layer capacitors based on macro/meso/microporous S-containing activated carbon with high surface area. *J. Mater. Chem.* **2011**, *21*, 2060–2063. [CrossRef]
6. Ito, S.; Murata, T.; Hasegawa, M.; Bito, Y.; Toyoguchi, Y. Study on CXN and CXS with disordered carbon structure as the anode materials for secondary lithium batteries. *J. Power Sources* **1997**, *68*, 245–248. [CrossRef]
7. Zhang, K.; Zhao, Q.; Tao, Z.; Chen, J. Composite of sulfur impregnated in porous hollow carbon spheres as the cathode of Li–S batteries with high performance. *Nano Res.* **2013**, *6*, 38–46. [CrossRef]
8. Saha, D.; Barakat, S.; Van Bramer, S.E.; Nelson, K.A.; Hensley, D.K.; Chen, J. Noncompetitive and competitive adsorption of heavy metals in sulfur-functionalized ordered mesoporous carbon. *ACS Appl. Mater. Interfaces* **2016**, *8*, 34132–34142. [CrossRef]
9. Petit, C.; Kante, K.; Bandosz, T.J. The role of sulfur-containing groups in ammonia retention on activated carbons. *Carbon* **2010**, *48*, 654–667. [CrossRef]
10. Sevilla, M.; Fuertes, A.B.; Mokaya, R. Preparation and hydrogen storage capacity of highly porous activated carbon materials derived from polythiophene. *Int. J. Hydrog. Energy* **2011**, *36*, 15658–15663. [CrossRef]
11. Xia, Y.; Zhu, Y.; Tang, Y. Preparation of sulfur-doped microporous carbons for the storage of hydrogen and carbon dioxide. *Carbon* **2012**, *50*, 5543–5553. [CrossRef]
12. Kicin, W.; Szala, M.; Bystrzejewski, M. Sulfur-doped porous carbons: Synthesis and applications. *Carbon* **2014**, *68*, 1–32. [CrossRef]

13. Shin, Y.; Fryxell, G.E.; Um, W.; Parker, K.; Mattigod, S.V.; Skaggs, R. Sulfur-functionalized mesoporous carbon. *Adv. Funct. Mater.* **2007**, *17*, 2897–2901. [CrossRef]
14. Choi, C.H.; Park, S.H.; Woo, S.I. Heteroatom doped carbons prepared by the pyrolysis of bio-derived amino acids as highly active catalysts for oxygen electro-reduction reactions. *Green Chem.* **2011**, *13*, 406–412. [CrossRef]
15. Saha, D.; Thorpe, R.; Van Bramer, S.; Alexander, N.; Hensley, D.; Orkoulas, G.; Chen, J. Synthesis of Nitrogen and Sulfur Co-Doped Nanoporous Carbons from Algae: Role in CO_2 Separation. *ACS Omega* **2018**, *3*, 18592–18602. [CrossRef] [PubMed]
16. Paraknowitsch, J.P.; Wienert, B.; Zhang, Y.; Thomas, A. Intrinsically sulfur- and nitrogen-co-doped carbons from thiazolium salts. *Chem. Eur. J.* **2012**, *18*, 15416–15423. [CrossRef]
17. Sevilla, M.; Fuertes, A.B. Highly porous S-doped carbons. *Microporous Mesoporous Mater.* **2012**, *158*, 318–323. [CrossRef]
18. Saha, D.; Toof, B.; Krishna, R.; Orkoulas, G.; Gismondi, P.; Thorpe, R.; Comroe, M. Separation of ethane-ethylene and propane-propylene in Ag(I)-doped and sulfurized microporous carbon. *Microporous Mesoporous Mater.* **2020**, *299*, 110099. [CrossRef]
19. Saha, D.; Richards, C.P.; Haines, R.G.; D'Alessandro, N.D.; Kienbaum, M.J.; Griffaton, C.A. Competitive Adsorption of Lead in Sulfur and Iron Dual-Doped Mesoporous Carbons. *Molecules* **2020**, *25*, 403. [CrossRef]
20. DeLuca, G.; Saha, D.; Chakraborty, S. Why Ag(I) grafted porous carbon matrix prefers alkene over alkane? An inside view from ab-initio study. *Microporous Mesoporous Mater.* **2021**, *316*, 110940. [CrossRef]
21. Saha, D.; Comroe, M.L.; Krishna, R. Synthesis of Cu(I)-doped Mesoporous Carbon for Selective Capture of Ethylene from Reaction Products of Oxidative Coupling of Methane. *Microporous Mesoporous Mater.* **2021**, *328*, 111488. [CrossRef]
22. Saha, D.; Orkoulas, G.; Chen, J.; Hensley, D. Adsorptive separation of CO_2 in sulfur-doped nanoporous carbons: Selectivity and breakthrough simulation. *Microporous Mesoporous Mater.* **2017**, *241*, 226–237. [CrossRef]
23. Saha, D.; Van Bramer, S.E.; Orkoulas, G.; Ho, H.-C.; Chen, J.; Henley, D.K. CO_2 capture in lignin-derived and nitrogen-doped hierarchical porous carbons. *Carbon* **2017**, *121*, 257–266. [CrossRef]
24. Saha, D.; Comroe, M.; Krishna, R.; Rascavage, M.; Larwa, J.; You, V.; Standhart, G.; Bingnear, B. Separation of propylene from propane and nitrogen by Ag(I)-doped nanoporous carbons obtained from hydrothermally treated lignin. *Diam. Relat. Mater.* **2022**, *121*, 108750. [CrossRef]
25. Saha, D.; Payzant, E.A.; Kumbhar, A.S.; Naskar, A.K. Sustainable Mesoporous Carbons as Storage and Controlled-Delivery Media for Functional Molecules. *ACS Appl. Mater. Interfaces* **2013**, *5*, 5868–5874. [CrossRef] [PubMed]
26. Saha, D.; Warren, K.E.; Naskar, A.K. Controlled release of antipyrine from mesoporous carbons. *Microporous Mesoporous Mater.* **2014**, *196*, 327. [CrossRef]
27. Kadia, J.F.; Kubo, S.; Gilbert, R.A.V.R.D.; Compere, A.L.; Giriffith, W. Lignin-based carbon fibers for composite fiber applications. *Carbon* **2002**, *40*, 2913–2920.
28. Rong, K.; Wei, J.; Wang, Y.; Liu, J.; Qiao, Z.-A.; Fang, Y.; Dong, S. Deep eutectic solvent assisted zero-waste electrospinning of lignin fiber aerogels. *Green Chem.* **2021**, *23*, 6065–6075. [CrossRef]
29. Srinivas, G.; Krungleviciute, V.; Guo, Z.-X.; Yildirim, T. Exceptional CO_2 capture in a hierarchically porous carbon with simultaneous high surface area and pore volume. *Energy Environ. Sci.* **2014**, *7*, 335–342. [CrossRef]
30. Guo, Y.; Yang, S.; Yu, K.; Zhao, J.; Wang, Z.; Xu, H. The preparation and mechanism studies of rice husk based porous carbon. *Mater. Chem. Phys.* **2002**, *74*, 320–323. [CrossRef]
31. Xue, M.; Chen, C.; Tan, Y.; Ren, Z.; Li, B.; Zhang, C. Mangosteen peel-derived porous carbon: Synthesis and its application in the sulfur cathode for lithium sulfur battery. *J. Mater. Sci.* **2018**, *53*, 11062–11077. [CrossRef]
32. Zhang, C.; Lin, S.; Peng, J.; Hong, Y.; Wang, Z.; Jin, X. Preparation of highly porous carbon through activation of NH_4Cl induced hydrothermal microsphere derivation of glucose. *RSC Adv.* **2017**, *7*, 6486–6491. [CrossRef]
33. Baumann, T.F.; Worslet, M.A.; Han, T.Y.-J.; Satcher, J.H. High surface area carbon aerogel monoliths with hierarchical porosity. *J. Non-Cryst. Solids* **2008**, *354*, 3513–3515. [CrossRef]
34. Kim, H.R.; Yoon, T.-U.; Kim, S.-K.; An, J.; Bae, Y.-S.; Lee, C.Y. Beyond pristine MOFs: Carbon dioxide capture by metal–organic frameworks (MOFs)-derived porous carbon materials. *RSC Adv.* **2017**, *7*, 1266–1270. [CrossRef]
35. Comroe, M.; Kolasinski, K.; Saha, D. Direct Ink 3D Printing of Porous Carbon Monoliths for Gas. *Molecules* **2022**, *27*, 5653. [CrossRef] [PubMed]
36. Seema, H.; Kemp, K.; Le, N.; Park, S.-W.; Chandra, V.; Lee, J.; Kim, K. Highly selective CO_2 capture by S-doped microporous carbon materials. *Carbon* **2014**, *66*, 320–326. [CrossRef]
37. Li, X.; Xue, Q.; Chang, X.; Zhu, L.; Ling, C.; Zheng, H. Effects of Sulfur Doping and Humidity on CO_2 Capture by Graphite Split Pore: A Theoretical Study. *ACS Appl. Mater. Interfaces* **2017**, *9*, 8336–8343. [CrossRef]
38. Yang, R.T. *Gas Separation by Adsorption Processes*; Imperial College Press: London, UK, 1997.
39. Myers, A.L.; Prausnitz, J.M. Thermodynamics of mixed gas adsorption. *AIChE J.* **1965**, *11*, 121–127. [CrossRef]
40. Saha, D.; Grappe, H.; Chakraborty, A.; Orkoulas, G. Post extraction separation, on-board storage and catalytic conversion of methane in natural gas: A review. *Chem. Rev.* **2016**, *116*, 11436–11499. [CrossRef]

Disclaimer/Publisher's Note: The statements, opinions and data contained in all publications are solely those of the individual author(s) and contributor(s) and not of MDPI and/or the editor(s). MDPI and/or the editor(s) disclaim responsibility for any injury to people or property resulting from any ideas, methods, instructions or products referred to in the content.

Article

Silver Depreciation in 3-Polker Coins Issued during 1619–1627 by Sigismund III Vasa King of Poland

Ioan Petean [1,*], Gertrud Alexandra Paltinean [1,*], Emanoil Pripon [2], Gheorghe Borodi [3] and Lucian Barbu Tudoran [3,4]

1. Faculty of Chemistry and Chemical Engineering, Babes-Bolyai University, 11 Arany Janos Street, 400028 Cluj-Napoca, Romania
2. Zalau County Museum of History and Art, 9 Unirii Street, 450042 Zalau, Romania
3. National Institute for Research and Development of Isotopic and Molecular Technologies, 65-103 Donath Street, 400293 Cluj-Napoca, Romania
4. Faculty of Biology and Geology, Babes-Bolyai University, 44 Gheorghe Bilaşcu Street, 400015 Cluj-Napoca, Romania
* Correspondence: ioan.petean@ubbcluj.ro (I.P.); gertrud.paltinean@ubbcluj.ro (G.A.P.)

Citation: Petean, I.; Paltinean, G.A.; Pripon, E.; Borodi, G.; Barbu Tudoran, L. Silver Depreciation in 3-Polker Coins Issued during 1619–1627 by Sigismund III Vasa King of Poland. *Materials* **2022**, *15*, 7514. https://doi.org/10.3390/ma15217514

Academic Editor: Antonio Gil Bravo

Received: 8 October 2022
Accepted: 24 October 2022
Published: 26 October 2022

Publisher's Note: MDPI stays neutral with regard to jurisdictional claims in published maps and institutional affiliations.

Copyright: © 2022 by the authors. Licensee MDPI, Basel, Switzerland. This article is an open access article distributed under the terms and conditions of the Creative Commons Attribution (CC BY) license (https://creativecommons.org/licenses/by/4.0/).

Abstract: The present research is focused on the 3-Polker coins issued during 1619–1627 by Sigismund III Vasa, King of Poland. A major financial crisis took place at that time due to the 30-year War, which started in 1619. There are two theories among historians concerning the silver depreciation of these coins. The most common theory (generally accepted without proof) is that the later years of issue are depreciated below 60% Ag. The second theory is based on the medieval sources that indicate inflation during the years from 1621–1625, but the medieval source only refers to the inflation of the type of coins and does not mention the issuer. Therefore, in this study, we use modern investigation techniques and materials science methods to help historians elucidate the aforementioned aspects regarding the medieval period. The XRD investigation results are in good agreement with the SEM-EDX elemental analysis. The coins from 1619 and 1620 have high silver content, namely, 86.97% and 92.49%, which corresponds to good silver. The amount of Ag found in the coins from 1621–1625 issituated in the range of 63.2–74.6%. The silver titleis suddenly restored in 1626 at about 84.3% and is kept in a good range until the end of this decree under Sigismund III in 1627. In conclusion, the second theory was partly validated by our experimental results, certifying the currency depreciation during 1621–1625, but the silver title was not lower than 54.2%. Notably, even this depreciated silver title assures a good quality of the 3-Polker coins compared to similar coins issued in other countries that were copper–silver-plated. Therefore, the 3-Polker coins were preferably hoarded at that time.Small alterations in the mint mark's design were observed in all the depreciated coins compared to the good ones. This might be a sign for an expert to identify the depreciated coins, a fact which requires supplementary investigations. The silver title's restoration in 1626 also came with a complete change of the mintmark.

Keywords: medieval coins; silver concentration; materials archeometry

1. Introduction

Many modern problems are similar to ones from the medieval and pre-modern ages, especially the financial crises that occur from time to time [1,2]. The medieval minters often hid the financial crises and currency inflation on the inscription MONETA NOVA (a Latin expression meaning new currency) etched on the obverse side along with the minting year [3]. This inscription, combined with the issue year, allows a medieval wise man to identify which coin is good and which is depreciated. Such knowledge assures the success of the wise traders in their business and in their dealings with tax collectors. Inflation occurs by the depreciation of the silver amount in the issued coins. Since all alloys containing at least 40% Ag are white and bright, it would be difficult for medieval people to know whether a certain coin has good title (at least 80% silver) or is depreciated.

The historical background of the 3-Polker coins issued by Sigismund III Vasa, King of Poland and Grand duke of Lithuania, is absolutely necessary for the accuracy of the present study. This is a small silver coin with a 19.8 mm diameter and a thickness of about 0.55 mm, weighting from 1.21 to 1.54 g depending on the minting year. It is also known under the common Polish name as Półtorak, which is equivalent to the Poltura (1.5 Kreutzer) from the Austrian Roman Empire [4]. Its denomination is 1/24 that of the German Thaler and is intended for large circulation. There are some inconstancies and arguments about their silver title. The Thaler was the silver standard for commercial trade, and it was not available in large circulation; rather, it was used in important commercial transactions. Therefore, the 3-Polker was spread widely among the peoples from the Polish–Lithuanian Commonwealth to the vicinity countries [4,5]. The scarce information from the original sources has raised two possible theories about the 3-Polker coins' silver title and particularly about those who generated the inflation during 1620–1625.

The simplest theory is very popular among coin collectors and assumes the general opinion about coins' evolution during the Pre-Modern Age where in the earliest coins are of a good silver quality and the latest are depreciated due to the inflation generated during their circulating period. Thus, the earliest 3-Polker coins issued by Sigismund III in 1614 should be made of good silver, which is expected to constantly decrease to the last year of theirproduction in 1627. Certain clues make this theory generally accepted by the collectors without scientific proofand only based on their own observations.

The more sophisticated theory is represented by the historians' research into the sources ranging as close as possible to the minting time [5,6]. The oldest records regarding the 3-Polker coins issued by Sigismund III mention that in 1614, the initial silver title was about 46.9% purity with a net weight of 1.54 g, corresponding to 0.72 g of pure silver. The records mention that in 1619, the silver title decreased to only 40.6%purity, with a net weight of 1.21 g and a silver content of 0.49 g.

Starting from the second decade of the seventeenth century, the 3-Polker coins appear in the circulation of Transylvanian currency. These coins' flux along with the depreciated issues of the Transylvanian Prince Gabriel Bethlen were considered as the causative agents of the inflation during 1620–1625 [4,6]. The Transylvanian Diet (e.g., legislative organization) held in Cluj in May 1622 settled an exchange course of 4 silver denars for the old Poltura, regardless of the issuing year, and of 2 denars for the issues minted in 1620 and after considering the abovementioned conditions. In the Transylvanian Diet held in Bistrita in the autumn of the same year, the exchange rate was modified to 4 denars for all good Poltura, regardless of the minting year, and the small Poltura were to be exchanged at a rate of 3 denars until the end of 1623 and after at a rate of 2 denars [6].

Nagy Szabó Ferencz of Târgu Mureș (1580–1658) mentions some medieval sources that discuss "copper-rich półtorak" having a silver title of about 12.5%, which came into Transylvania from Silesia [7]. Based on these scripts, some numismatists considered that the "copper rich Polturak" would have been the 3-Polker coins issued by Sigismund III of Poland, while others consider thisto be impossible [4]. The Hungarian historian BuzaJános remarks that the Polturak's name was also used before 1614 for coins with a similar denomination as the 3-Polker such as the imperial coins of the 3 Kreutzer (1/24 of the Thaler's gross value) [8]. Besides all the presented historical aspects regarding the 3-Polker coins' quality, they were found in significant hoards discovered in Salaj and Cluj counties, which are now in the custody of the Zalau Museum and were reported in the following literature: Salajeni Hoard [9]; Verveghiu Hoard [10]; Zalau Hoard No. 1 [11]; Mineu Hoard [12]; Aghires Hoard [13].

It is relatively difficult to determine which is the true supposition between the collector's beliefs and old historical records. Materials science-based research is required to fulfil this purpose. The recent literature reports modern material investigation techniques such as X-ray diffraction (XRD), X-ray photoelectron spectroscopy (XPS), and SEM-EDX, which have beenemployed to reveal the hidden secrets of metallic parts of medieval artefacts such as metallic parts within medieval clothes [14] and metallic fibres within burial robes from medieval crypts [15], and to investigate the surface alteration of metallic artefacts exposed to a closed environment [16].

An optimal match must be found between the artefact and the investigation methods to reveal the artefact's secrets. In our case, regarding the 3-Polker coins, this concerns the Ag–Cu binary system, which is a system with total insolubility and that has an eutectic at the composition of 719 ‰ Ag and 281 ‰ Cu weight percent [17,18]. Metallographic microscopy is one of the most powerful tools used to investigate alloys' microstructures [17,19,20], but it requires local grinding and polishing prior to chemical etching and it is not effective for application to small coins that have been heavily exposed. Some powerful non-destructive methods are required in our case. XRD is a method that reveals the phase composition of alloy samples giving a proper qualitative characterization and allowing for a semi-quantitative analysis of the composition of binary systems [21,22]. It is often coupled with the SEM-EDS elemental investigation method due to the high precision of the elemental quantification related to the high-resolution imaging of the investigated site [22,23]. Modern SEM-EDS allows for the high-resolution complex elemental mapping of the microstructure, which in turn enables the visualization of the micro-structural constituents within the Ag–Cu binary system without polishing and chemical etching. This would be a great achievement for the present research. Therefore, we aim to use a combination of XRD and high-precision SEM-EDS analysis to elucidate the mysteries regarding the silver depreciation of the 3-Polkercoins produced during the reign of Sigismund III of Poland during 1619–1627.

2. Materials and Methods

2.1. Samples' Description and Preparation

All samples investigated in the present research are 3-Polker coins issued by Sigismund III of Poland during 1619–1627 from the collection of Zalau Museum of History and Arts, Zalau City, Transylvania, Romania. Each coin is presented in Figure 1 with obverse and reverse sides grouped on the same indicative.

Figure 1. The 3-Polker coins issued by Sigismund III of Poland investigated in the current research. Minting years: (**a**) 1619, (**b**) 1620, (**c**) 1621, (**d**) 1622, (**e**) 1623, (**f**) 1624, (**g**) 1625, (**h**) 1626, and (**i**) 1627. Coins' dimensional characteristics are given in Table 1.

Table 1. Coins physical characteristics.

Minting Year	Grand Treasurer of the Crown	Mass, g	Diameter, mm	Tickness, mm
1619	Nicolas Danilowicz	1.3679	19.84	0.65
1620	Nicolas Danilowicz	1.3054	19.57	0.65
1621	Nicolas Danilowicz	1.0341	19.38	0.50
1622	Nicolas Danilowicz	1.0279	19.32	0.54
1623	Nicolas Danilowicz	1.0842	19.28	0.53
1624	Nicolas Danilowicz	0.9773	19.67	0.55
1625	Nicolas Danilowicz	1.0618	19.23	0.59
1626	Hermann Ligenza	1.2629	19.17	0.59
1627	Hermann Ligenza	1.1632	19.33	0.58

Note: All coins were minted at Bydgoszcz Mint.

The coins were selected from several hoards and numismatic accumulations discovered in Transylvania (Salaj and Cluj counties) and hosted by the Zalau museum. The coins issued in 1619 and 1620 are from Salajeni Hoard; coins issued in 1621–1622 are from Verveghiu Hoard; coins issued in 1623 are from Hoard No. 1 of Zalau (the earliest 3-Polker hoard found in Transylvania); coins issued in 1624 and 1625 belong to the Mineu Hoard; the coins issued in 1626 and 1627 were selected from the Aghires Hoard (Cluj County). Each mentioned hoard contains only few issue years from the considered range. Therefore, we selected coins from five different hoards. It is noteworthy to mention that the earliest issue years for this currency (e.g., 1614–1618) are very rare in the hoards found in Transylvania and were not available for this study.

Coins' physical characteristics such as mass, diameter, and thickness were measured and centralised in Table 1. Each coin was weighed on an electronic analytical balance with the precision of four digits; diameter and thickness were measured with a standard micrometer device.

Each coin was individually inspected in terms of their good preservation and absence of defects caused by long-term exposure to soil weathering; additionally, the coins were properly cleansed. Data in literature describe an alternative cleaning method using high-frequency cold plasma cleaning [24] to remove all impurities. However, the coins were previously restored and conserved according to the museum's procedures. Current cleaning was effectuated using standard calcium bicarbonate detergent (15 min soaking and 5 min of gentle washing with cotton cloth)followed by intensive rinsing with bi-distilled water in order to cleansethe coin to properly expose the metallic surface to the non-destructive investigation methods.

2.2. Optical Microscopy

The general aspect of the coins' surface was investigated by optical transmitted light microscopy using lateral light source dark-field observation on a Laboval 2 microscope, Carl Zeiss, Oberkochen, Germany, equipped with a digital capture camera Kodak 10 Mpx, Rochester, NY, USA.

Metallographic investigation was on an Olympus BX3M metallographic microscope, Shinjuku, Tokyo, Japan. The metallographic investigation was performed on 3-Polker coin fragments from years 1619 and 1625. Półtoraks' broken fragments were discovered during archaeological digging. The year numerals"9" and"5" were observed on the right side of the cross, allowing for the identification of the issue years. The metallographic investigation was effectuated on the reverse side (the side with the numeral). The fragments were embedded into epoxy resin and successively grinded with abrasive discs with granulation of 800, 1200, 2000, and 4000 and followed by alumina polishing. They were chemically etched with Potassium Dichromate and Hydrochloric Acid for 45s and rinsed with bi-distilled water and soaked on the filter paper.

2.3. X-ray Diffraction (XRD)

X-ray diffraction (XRD) was effectuated on the coins' reverse side (face with the years' numerals) using a Bruker D8 Advance diffractometer with CuKα1 monochromatic radiation (λ = 1.54056 Å) at room temperature. The XRD patterns were recorded in the range 30–100 of 2-theta degree with a speed of 1 degree/minute. The phase identification was effectuated with Match software, Crystal Impact Company, Bonn, Germany, using the PDF database.

2.4. Scanning Electron Microscopy and Elemental Analysis (SEM-EDX)

SEM images were obtained on the coins' reverse side (face with the year numerals), with a Hitachi SU8230 microscope, equipped with an Energy-Dispersive Spectroscopy (EDS) detector X-Max 1160 EDX (Oxford Instruments). The microscopic and elemental analyses wereconducted on at least three different spots on the coins' surface.

3. Results

3.1. Optical Microscopy

All the coins involved in present research were well-conserved in the museum collections preventing their surface depreciation. The cleaning procedure removed any traces of patina on their surface, there by allowing for the metal core to be observed. The success of the cleaning operation and the general aspect of the coins' surfaces were monitored by dark field optical microscopy, as shown in Figure 2. Thus, the cleaning procedure removed all the impurities from all the coins from 1619 to 1627.

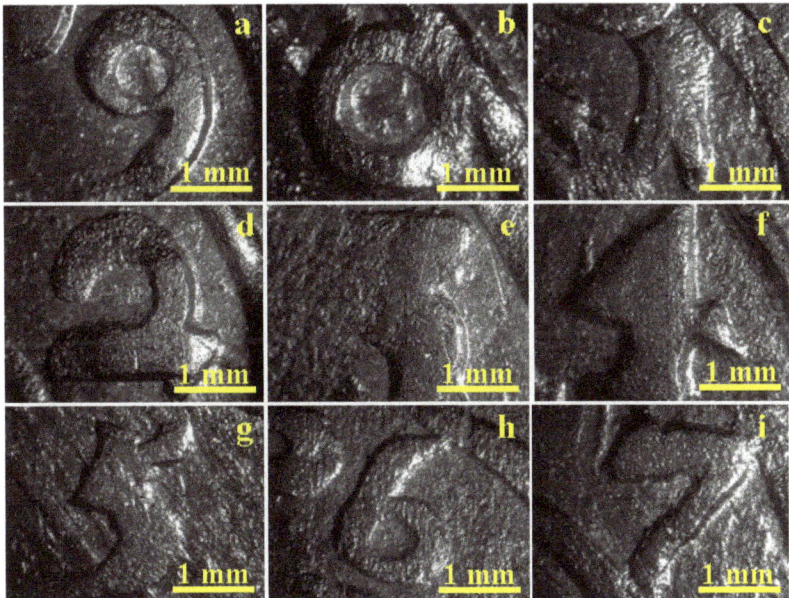

Figure 2. Dark field opticalmicroscopy images of the investigated 3-Polker coins from the minting years: (**a**) 1619, (**b**) 1620, (**c**) 1621, (**d**) 1622, (**e**) 1623, (**f**) 1624, (**g**) 1625, (**h**) 1626, and (**i**) 1627.

The lateral lighting used in the dark field observation enabled the better observation of the coins' reliefs. There are two topographic planes on the coins: the upper one formed by the top of the inscriptions and the lower one formed by the base generated by the die surface. The wear from circulation affected the top of the inscription while the base irregularities are more related to the die quality.

The coin minted in 1619 presents a very good quality of the inscription without scratches and blunted details as well as a smooth base (Figure 2a). No traces of impurities such as dirt or oxide conglomerate are observed. It proves that the coin was engraved with very good quality dies and is less circulated, thereby preserving the engraved details.

The coins issued in 1620 and 1621 present some scratches on the top of their inscriptions, shown in Figure 2b,c, evidencing its more intense circulation and the good quality of the base, thus indicating that they were engraved with good quality dies. The coins issued in 1622 feature less wear from circulation but there are some irregularities on the base indicating an advanced usage of the dies (Figure 2d).

The numeral three is blunted on the left side of the coin from 1623 and presents significant scratches on the right, as shown in Figure 2e, due to intense wear. The base also presents blunted areas along with micro-structural irregularities. This indicates an intense usage of the dies when the coin was engraved combined with significant wear from circulation.

The coins issued between 1624 and 1627, show in Figure 2f–i, reveal a moderate degree of wear from circulation on the top of the inscription; the numerals observed are well-preserved and provide good details. Their base is affected by some pitches, which indicate an advanced usage of the dies when the coins were engraved. Some small darkened spots are observed at the inscription's junction with the base for the coins from 1624, 1625, and 1627 (Figure 2f,g,i). These spots might be oxide traces; their nature will be revealed by XRD analysis.

A metallographic investigation was conducted on some of the fragments of the three-Polker coins from 1619 and 1625 discovered in an archaeological dig. The obtained microstructures are presented in Figure 3. The coin issued in 1619 presents a typical microstructure of a good quality silver alloy, shown in Figure 3a,b. α phase grains representing pure silver and eutectic grains representing a mixture of about 71.9% Ag and 28.1% Cu were identified. The grains' shape is elongated in the lamination direction of the silver plate used for coin blanks. This observed aspect is in good agreement with the data in the literature [25,26].

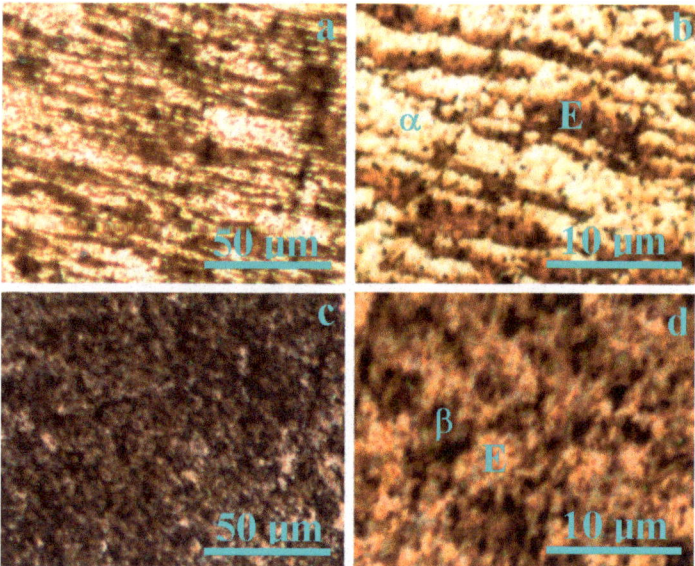

Figure 3. Metallographic microstructures of 3-Polker coin issued in 1619: (**a**) general view and (**b**) microstructure detail and of coin issued in 1625: (**c**) general view and (**d**) microstructure detail, where: E—eutectic, α—silver phase, and β—copper phase.

The coin issued in 1625 features a hypereutectic microstructure corresponding to a poor silver quality. Grains of a β phase (pure copper), mixed with eutectic Ag–Cu, were identified(Figure 3c,d). Fact is in good agreement with the data in literature [27,28]. The grain structure is very fragmented due to the intensive deformation during the plate lamination but does not show elongation. This is explained by the large presence of the β phase. It will be interesting to correlate these facts with the XRD investigation.

3.2. X-ray Diffraction (XRD)

The XRD patterns of the investigated coins are presented in Figure 4. They feature very intense and narrow peaks corresponding to a crystalline state of the samples. The obtained XRD results prove that the coins correspond to the Ag–Cu binary system.

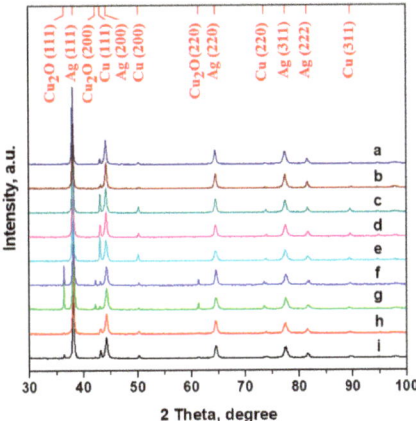

Figure 4. XRD spectra for the investigated coinsissued in: (a) 1619, (b) 1620, (c) 1621, (d) 1622, (e) 1623, (f) 1624, (g) 1625, (h)1626, and (i) 1627.

The samples were analysed by XRD, and it was found that they contain Ag (PDF:2871), Cu (PDF:89-2838), and some trace amounts of Cu_2O (PDF:78-2076).The Reference Intensity Ratio (RIR) method was used to determine the percentage content of each phase. If we have two phases, a and b, then:

$$\frac{I_a}{I_b} = \left(\frac{xa}{xb}\right)\left(\frac{cfa}{cfb}\right). \tag{1}$$

where I_a and I_b are the most intense diffraction peaks for phases a and b, respectively; xa and xb are the mass percentage concentrations for phases a and b, respectively; and cfa, cfb are the corundum factors for the two phases. The corundum factors are 18 for Ag, 8.86 for Cu, and 8.28 for Cu_2O (cuprite). Using equation 1, the mass concentration ratio for the two phases can be determined. A phase analysis was performed using the RIR method and the results are presented in Table 2. The dominant peak corresponds to Ag in all cases but an increasein the Cu peaks was observed for certain issuing years such as the range from 1621–1625.

Table 2. Phase composition of the investigated 3-Polker coins.

Minting Year	Ag (%)	Cu (%)	Cu_2O (%)
1619	86.97	13.03	-
1620	92.49	7.51	-
1621	69.15	30.65	-
1622	71.36	28.64	-
1623	61.61	38.39	-
1624	59.14	6.86	34.06
1625	53.76	5.00	41.24
1626	88.03	11.97	-
1627	80.40	15.35	4.25

Several Cu_2O peaks were identified for the coins issued in 1624, 1625, and 1627, which are correlated with the very small dark dots observed by optical microscopy in Figure 1. The presence of cuprite affects the copper content determined by the RIR method from the XRD patterns but does not affect the silver content. The results sustain a significant depreciation of the silver title between 1621 and 1625, a fact that is in good agreement with the historical information about the monetary inflation but that does not agree with the very low title mentioned in the historical sources.

The data in the literature mention the cuprite formation on the restored coins after a long-time storage [29]. Therefore, a fresh cleaning is required every time a coin is subjected to a material analysis. However, a small amount of copper oxides such as cuprite and tenorite may occur in the XRD patterns, for example, as in one of our previous studies on a freshly cleaned lot of Dyrrachium drachmas from the Cehei hoard [30].

3.3. Scanning Electron Microscopy and Elemental Analysis (SEM-EDX)

The SEM investigation associated with the elemental analysis was very complex and investigated several morphologic features of the coins' surfaces. We found some very rich silver areas on the bases of the coins' surfaces that might have occurred due to the initial restoration effectuated at the moment of the hoard's preservation in the museum treasury or due to some actions from sometime in the past. This aspect will be expanded upon and discussed in a further paper. The tops of the inscriptions are considerably worn compared to the bases and further expose the material inside the coin, which is ideal for the morphological observation as well as for the EDS spectroscopy.

Figure 5a reveals the morphology of the 1619 coin with a uniform surface and a compact microstructure. Small pores were observed with diameters in the range of 1.5 to 5 μm. These are related to the poor-quality workmanship of the hammered coins' manufactured in medieval times, which is often associated with micro-mineral contaminants on the dies' surfaces. Several corrugations with dendrite shapes and lamellar inner structures were observed. These are related to the lamellar Ag–Cu eutectic grains, which are more predisposed to the mechanical erosion.. A similar microstructure is observed for the coin issued in 1620, shown in Figure 5b, which displays smaller pores with diameters between 1–2.5 μm and some superficial scratches related to significantly intense circulation.

The morphological aspect is significantly changed starting from 1621 (Figure 5c). The surface is not very uniform, which indicates that it was more affected by soil weathering. There appear to be a few pores with diameters between 2 to 8 μm and enlarged corrugated dendrite areas related to the eutectic presence. The microstructural change is more likely caused by the increase in the amount of copper.

The microstructures of the coins issued from 1621 to 1625 present similar aspects as those observed in Figure 5d–e. There are relatively few pores with an equivalent diameter in the range of 1 to 3 μm. The surfaces are relatively irregular due to the alternation of the compact and lamellar grains corresponding to a typical hypereutectic microstructure [27].

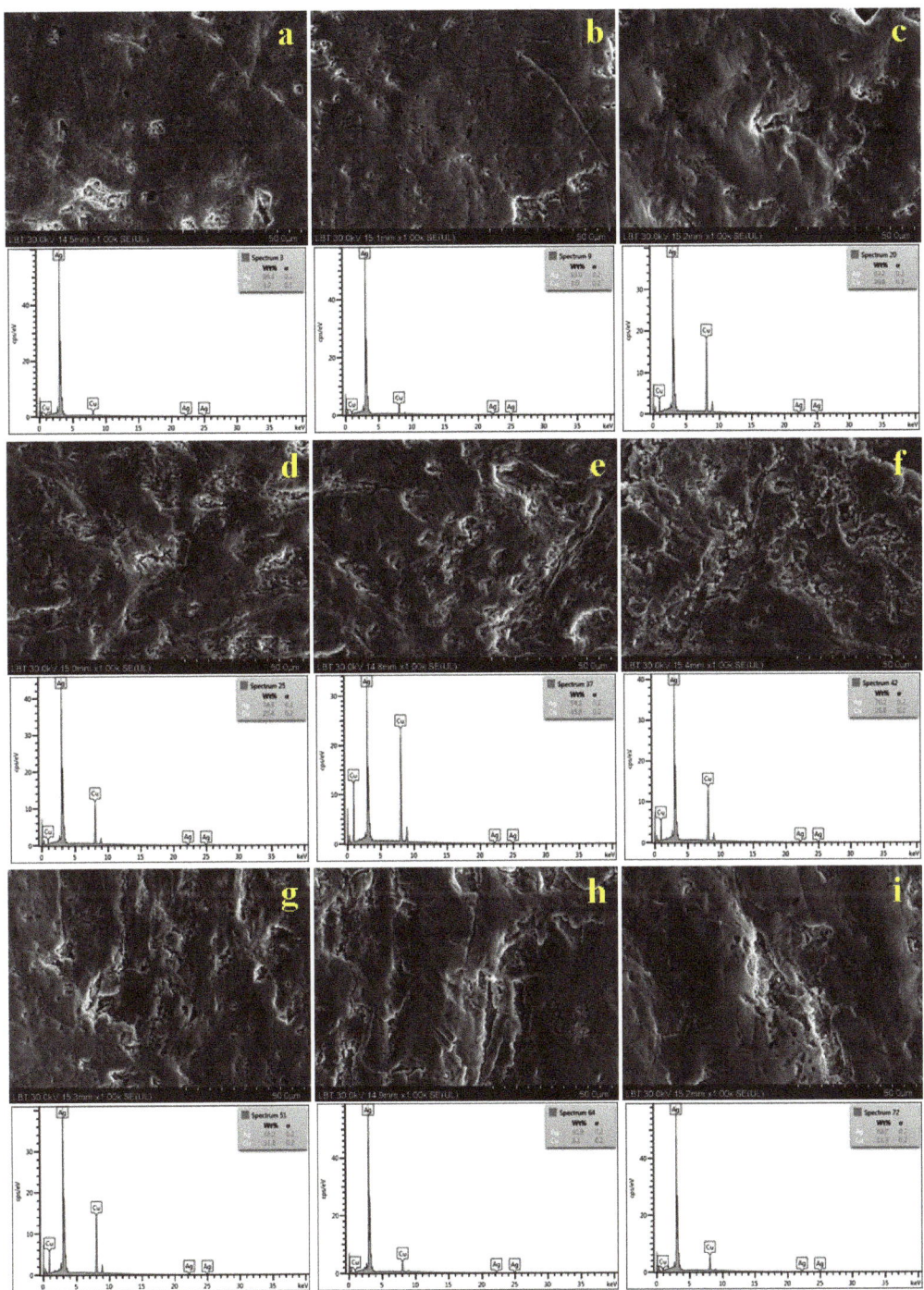

Figure 5. SEM images and EDS spectra resulting from the investigated 3-Polker coins issued in: (**a**) 1619, (**b**) 1620, (**c**) 1621, (**d**) 1622, (**e**) 1623, (**f**) 1624, (**g**) 1625, (**h**) 1626, and (**i**) 1627.

Another microstructural change was observed for the coins issued in 1626 and 1627, shown in Figure 5h,i. Their morphologies closely resemble those observed for the coins issued in 1619 and 1620. Large compact grains with wear marks and eutectic dendrite—partially corrugated due to the coins' circulation—were observed. This might be a microstructural clue regarding the silver title's restoration.

The microstructural observations are coupled with the XDS spectra obtained for the observed surface. The resulting elemental composition is summarised in Table 3. It is in good agreement with the XRD results in Table 2.

Table 3. Elemental compositions of the investigated 3-Polker coins.

Minting Year	Ag		Cu		Traces
	Wt.%	S.D.	Wt.%	S.D.	Wt.%
1619	96.8	4.84	3.2	0.22	undetectable
1620	93.0	4.65	7.0	0.49	undetectable
1621	63.2	3.16	36.8	2.57	undetectable
1622	74.6	3.73	25.4	1.77	undetectable
1623	54.2	2.71	45.8	3.20	undetectable
1624	70.2	3.51	29.8	2.08	undetectable
1625	68.2	3.41	31.8	2.22	undetectable
1626	91.9	4.59	8.1	0.56	undetectable
1627	88.7	4.43	11.3	0.79	undetectable

The coins issued in 1619 and 1620 have a good silver title, even better than the reference for the German Thaler, which is 83.5%. These are situated in the hypoeutectic domain of the Ag–Cu binary system from the metallographic point of view. This means that the microstructure contains α phase and eutectic grains. This fact is sustained by the elemental map in Figure 6a,b: the α-phase grains present a green aspect dotted with yellow spots due to the rich silver amount and the eutectic grains present brown spots due to the presence of copper. This result corresponds to the metallographic observation in Figure 3b.

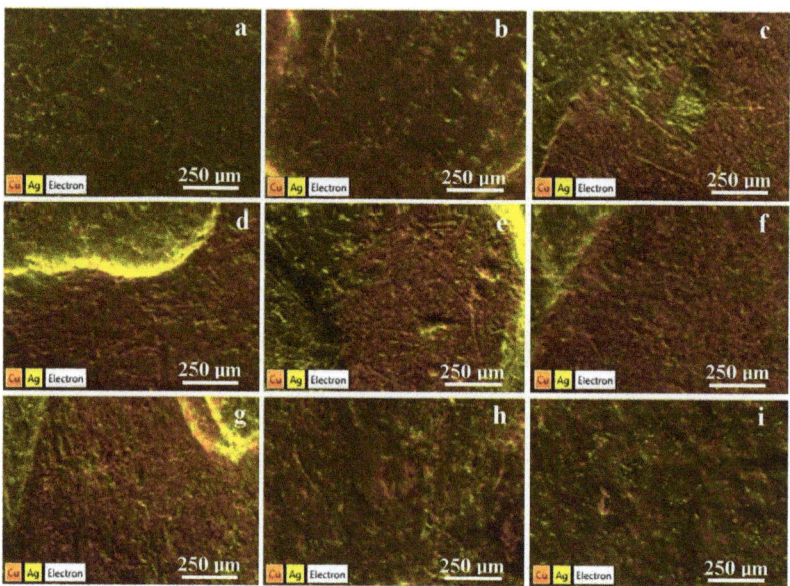

Figure 6. Elemental maps of the 3-Polker coins issued in: (a) 1619, (b) 1620, (c) 1621, (d) 1622, (e) 1623, (f) 1624, (g) 1625, (h) 1626, and (i) 1627.

A significant depreciation of the silver title was found for the coins issued from 1621 to 1625. The coin issued in 1622 is very close to a eutectic composition, which means that the microstructure contains a large number of eutectic grains and only a few grains of the α phase. This fact is sustained by the EDS map in Figure 6c, wherein the copper amount corresponding to the eutectic region is intense and tends to form compact dendrite areas.

The other coins issued in 1621, 1623, 1624, and 1625 are situated in the hypereutectic domain of the Ag–Cu binary system. A significant amount of β copper-rich phase grains mixed with eutectic grains occurs in these coins. This fact is sustained by the EDS maps presented in Figure 6d–g. We observed compact areas (brown-coloured) with a dendrite shape, which correspond to the β phase grains. The green areas dotted with small brown spots correspond to the eutectic grains. The elemental map of the coin from 1625, shown in Figure 6b, corresponds to the metallographic aspect evidenced in Figure 3d.

Some yellow bands can be observed in the EDS maps in Figure 6d,e,g. These bands are not situated on the top of the inscription but rather appear in some areas situated at the base of the inscription and are silver-rich areas which correspond to the silver's segregation. These aspects are not related to the bulk material of the coins. Rather, they are related to a possible physicochemical reaction between the preservation compounds and the coin surface. The compound was certainly removed by the cleaning procedures, but it seems that some silver segregation still remains in the mentioned places. These facts require more detailed investigation, which will be the subject of a further article.

The coins issued in 1626 and 1627 present a high amount of silver situated in the hypoeutectic region of the Ag–Cu binary system and present in the elemental maps—shown in Figure 6h,i—with a strong resemblance to the observation made for the coins issued in 1619 and 1620. These facts indicate that the silver title was properly restored in 1626 when the mint master was changed.

The depreciated coins may have traces of elements such as Sn, Pb, and Zn, as impurities resulted from the improper metallurgical process or were deliberately placed in the composition to falsify the silver content. The elemental analysis was effectuated with high accuracy, and trace elements such as Sn, Pb, and Zn were searched for but not detected. This result is in good agreement with the XRD observation, where no odd phases were detected.

The EDS spectra along with the distribution maps allow for the identification of the microstructures' constituent grains without polishing and chemical etching but by obtaining the proper details from the secondary electron images. The eutectic grains within the good silver coins have an elongated dendritic shape with a fine lamellar structure, but the lamellas are broken due to the coins' surface wear (Figure 7a,b,h,i). They are surrounded by very compact α-phase grains that also have an elongated shape.

The eutectic grains are more affected by the wear in the depreciated coins (Figure 7c,e–g). The lamellas are also affected, implying a significant material loss corresponding to the formation of local depressions. These are surrounded by compact β-phase grains, which also present an elongated shape and wear marks on their surface.

A special microstructure was revealed for the coin issued in 1622 because it is situated in the hypoeutectic domain but very close to a eutectic composition. Therefore, the eutectic grains are very numerous, their shape is dendritic and elongated, and many of them are affected by the coins' surface wear, as shown in Figure 7d. Some compact α-grains with diameters below 5 μm can be observed. They are primarily crystallized from the melted alloy during cooling, while the eutectic grains are formed when the temperature decreases to about 779 °C.

The coins' silver amount values measured by XRD and SEM–EDX are displayed in Figure 8. The border line between good and depreciated coins is on the title of 80% Ag. It is clear from the results that the coins issued in 1619, 1620, 1626, and 1627 are made of good silver.

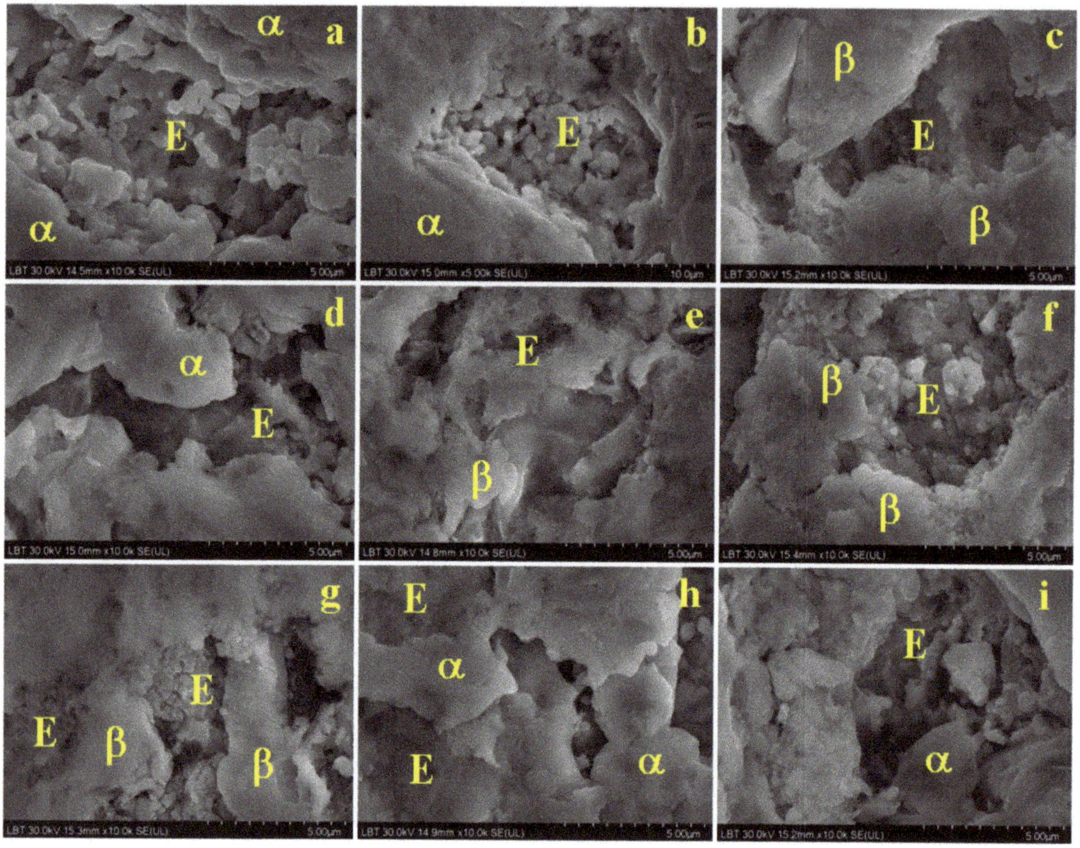

Figure 7. SEM morphologic details revealed on the surface of investigated 3-Polker coins issued in: (**a**) 1619, (**b**) 1620, (**c**) 1621, (**d**) 1622, (**e**) 1623, (**f**) 1624, (**g**) 1625, (**h**) 1626, and (**i**) 1627, where E—eutectic, α—silver phase, and β—copper phase.

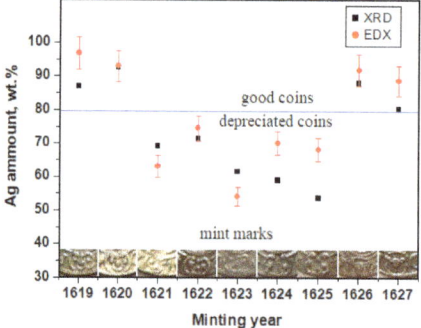

Figure 8. Silver content variation during the minting time evidenced by XRD and SEM-EDS compared with the aspect of the mint marks.

The coins issued in 1621, 1622, 1623, 1624, and 1625 have depreciated silver titles. Despite the significant silver portion decreasing the observed depreciation, this effect is not as significant as expected based on the historical information. The mint mark of each coin was also placed in Figure 8 because some of the graphical aspects tend to be correlated with the depreciation. The connection between the reduced silver title and mintmark variation will be discussed below.

4. Discussion

The coins investigated in the present research are very interesting because they have corresponding currencies in two monetary systems based on the silver title. Their face value and nominal equivalence are given by expression (2):

$$1 Płtorak(e.g.\ 3\ Polker) = 3\ denars = 1.5\ Gross = 1.5\ Kreutzer = \frac{1}{24} Thaler \qquad (2)$$

where *Denars* are a Hungarian small silver denomination, *Gross* are a typical Polish small silver denomination, and *Kreutzer* are the imperial small silver denomination (of the Holy Roman Empire). The silver standard at that time was the *Thaler* of 28.83 g, containing 24.31 g of fine silver according Gumovski [5], which is in good agreement with Toma [4]. It corresponds to a title of 84.3%; consequentially, a 3-Polker coin must have a weight of 1.20 g of silver alloy with a title of 84.3% corresponding to 1.01 g of fine silver as a standard reference.

The correlation of the data in Tables 1–3 demonstrate that coins issued in 1619 and 1620 have a title significantly higher than the standard reference and a weight in excess of about 0.1 g and 0.16 g. In consequence, these coins fit the standard mentioned by Gumovski [5]. The coins issued in 1626 and 1627 have a weight and title around the standard value, proving their quality. The coins issued from 1621 to 1625 are lighter with about 0.12 to 0.2 g and their silver title is situated in the range of 63.2–74.6%, a fact which indicates a certain depreciation compared to the standard.

The mintmark of Nicolas Danilowicz, Grand Treasurer of the Crown, features a horizontal moon crescent with a vertical arrow that separates two stars and is placed in round brackets below the cross orb on the coin's obverse side. The stars are horizontally aligned for the good title coins and the right star is placed slightly lower for all depreciated coins. As a supposition, this might be a sign for an expert to easily identify the depreciated coins.

Therefore, the coin collector's supposition of the continuous depreciation of the silver title with the issuing year has been rejected by the obtained results. The material evidence indicates that the depreciation is situated in the period of 1621–1625, which is in good agreement with the historical data but also disagrees with the mentioned title of 40.6% [4,5]. Our findings confirm the theory of historian Buza János [8] that the 3-Polker coins issued by Sigismund III of Poland cannot be depreciated coins with a title of 40.6%.

This historical contradiction resides in the medieval expression that "copper-rich Polturak" came from Silesia, which describes coins with a silver title about 40.6%. Silesia was certainly not a Polish possession and the 3-Polker coins afflux from this region is unlikely. The explanation resides in the medieval crisis known as, "Kipper- und Wipperzeit", which starts in 1619 along with the 30-years war [31,32]. The small coin title was debased from the nominal value of 84.3% to about 40.6% in several states such as Silesia. However, they did not debase their own coins, preferring to forge other countries' coins, such as the Polish 3-Polker, and passing them on to other territories. Massive inflation arose in those conditions and the face value of the coins was not respected. Each individual coin was inspected and weighed; the good ones were kept for the treasury and the depreciated ones were discarded.

Many of these coins arrived in Transylvania and were subjected to the Cluj and Bistrita diet regulations in 1622. In addition, the Transylvanian issues of Gabriel Bethlen were debased during 1619–1623, causing monetary confusion [4]. In such conditions, people managed to identify good 3-Polker coins issued by Poland and treasure them and thus try to circumvent the use of bad coins. It seems that the Kipper- und Wipperzeit crisis had less of an effect on the original Polish coins issued during 1621–1625, being slowly and weakly depreciated compared to the Transylvanian and Silesian coins, which were strongly depreciated. This fact caused the preference for the hoarding of 3-Polker coins along with other denominations such as the triple and sextuple gross issued in Poland at that time.

The forged 3-Polker coins with a title of 40.6% were never found in a hoard and they are very rare in Transylvania because they were collected and melted down during the monetary reform effectuated by Gabriel Bethlen in 1626. Our findings prove that the silver title restoration for the Polish 3-Polker starts in 1626 and continues until 1627 when the issue stops because the triple and sextuple gross become more popular.

5. Conclusions

The materials science investigation (XRD and SEM –EDX) effectuated on the 3-Polker coins issued by Sigismund III Vasa, King of Poland during 1619 and 1627 evidenced a certain depreciation of the silver title from about 84.3% to a range of 63.2–74.6% with respect to the coins issued between 1621–1625. Therefore, the material information confirms the historical data about the inflation during 1621–1625 but the depreciated title is not as low as that recorded by historical sources. This fact is sustained by the archaeological records that prove that the investigated coins were hoarded, which means that they were considered good at that time. An important conclusion derived from the observation within the current research is that the "copper rich Półtorak" is definitely not equivalent to the 3-Polker coins issued by Sigismund III of Poland. The coins issued in 1619, 1620, 1626, and 1627 have a very good title, with most of them exceeding 83.5%, proving the title's restoration after 1626.

Author Contributions: Conceptualization, I.P. and E.P.; methodology, I.P.; software, G.B.; validation, E.P. and I.P.; formal analysis, G.A.P.; investigation, G.B., L.B.T. and I.P.; resources, E.P.; data curation, I.P. and G.A.P.; writing—original draft preparation, I.P. and G.A.P.; writing—review and editing, I.P. and G.A.P.; visualization, L.B.T. and I.P.; supervision, I.P. All authors have read and agreed to the published version of the manuscript.

Funding: This research received no external funding.

Institutional Review Board Statement: Not applicable.

Informed Consent Statement: Not applicable.

Data Availability Statement: Data sharing is not applicable.

Conflicts of Interest: The authors declare no conflict of interest.

References

1. Kokabian, P. Black Currency of Middle Ages and Case for Complementary Currency. *J. Risk Financ. Manag.* **2020**, *13*, 114. [CrossRef]
2. Borges, R.; Silva, R.J.C.; Alves, L.C.; Araújo, M.F.; Candeias, A.; Corregidor, V.; Vieira, J. European Silver Sources from the 15th to the 17th Century: The Influx of "New World" Silver in Portuguese Currency. *Heritage* **2018**, *1*, 453–467. [CrossRef]
3. Young, A.T. Consent or Coordination? Assemblies in Early Medieval Europe. *Int. Rev. Law Econ.* **2022**, *72*, 106096. [CrossRef]
4. Toma, C. *Monetary Treasures and Transylvanian Treasuring during XV/XVI–XVII Centuries*, 1st ed.; Oradea Museum Publishing House: Oradea, Romania, 2016; p. 471.
5. Gumowski, M. *Handbuchder Polnischen Nuimismatik (Polish Numismatic Handbook)*; Akademische Druck und Verlagsanstalt: Graz, Austria, 1960; p. 227.
6. Sandor, S. *Erdélyi Országgűlési Emlékek (Monumenta Comitialia Regni Transylvaniae; Monuments of the Transylvania Shire)*; Volume VIII (1621–1629); Akademia Konyvkiado-Hivatala: Budapest, Hungary, 1882; pp. 94–95.
7. Mihailescu, S.G. *Nagy SzabóFerencz of Târgu Mureș Memorial (1580–1658). Translation, Notes and Introductive Study*; Kriterion Publishing House: Bucharest, Romania, 1993; p. 245.
8. Buza, J. Inflációspénzforgalomtörvényesegédlettel; az 1622. évi 77. törvénycikk (Inflationary money circulation is legal with assistance; Article 77 of 1622). In *A rendtartótörténészTanulmányoklmrehIstvánszületésénekszázadikévfordulójára*; Tamas, F., Emoke, G., Eds.; ErdélyiMúzeum Publishing House: Budapest, Hungary, 2021; pp. 147–160.
9. Chirila, E.; Gudea, N.; Cabuz, I. Salajeni Hoard XVII—Century. In *Monetary Hoards in the Northen Transylvania between XVI-XVIII Centuries*; Zalau Museum Publishing House: Zalau, Romania, 1970; pp. 33–43.
10. Chirila, E.; Bajusz, I. Monetary hoard of Verveghiu XVI-XVII centuries. *Acta Musei Porolisensis* **1978**, *9*, 119–127.
11. Pripon, E. *Medieval and Pre-Modern Monetary Hoards Discovered on the Area of Zalau City*; Mega Publishing House: Cluj-Napoca, Romania, 2018; p. 99.

12. Chirila, E. *Mineu Monetary Hoard XVII Centrury, Monetary Hoards in the Northen Transylvania between XVI-XVIII Centuries*; Zalau Museum Publishing House: Zalau, Romania, 1970; pp. 55–63.
13. Chirila, E.; Lucacel, V. Feudal hoard of Aghires XVI-XVII centuries. *Stud. Univ. Babeş-Bolyai series Historia.* **1965**, *2*, 33–55.
14. imić, K.; Soljačić, I.; Mudronja, D.; PetrovićLeš, T. Metal Content and Structure of Textiles in Textile Metal Threads in Croatia from 17th to 20th Century. *Materials* **2022**, *15*, 251.
15. liwka-Kaszyńska, M.; Ślebioda, M.; Brillowska-Dąbrowska, A.; Mroczyńska, M.; Karczewski, J.; Marzec, A.; Rybiński, P.; Drążkowska, A. Multi-Technique Investigation of Grave Robes from 17th and 18th Century Crypts Using Combined Spectroscopic, Spectrometric Techniques, and New-Generation Sequencing. *Materials* **2021**, *14*, 3535. [CrossRef]
16. Kouřil, M.; Boháčková, T.; Strachotová, K.C.; Švadlena, J.; Prošek, T.; Kreislová, K.; Fialová, P. Lead Corrosion and Corrosivity Classification in Archives, Museums, and Churches. *Materials* **2022**, *15*, 639. [CrossRef]
17. Taylor, S.L. An Investigation of the Mechanical and Physical Properties of Copper-Silver Alloys and the Use of these Alloys in Pre-Columbian America. Bachelor Dissertation, Massachusetts Institute of Technology, Cambridge, MT, USA, 2013.
18. Mecking, O. The colours of archaeological copper alloys in binary and ternary copper alloys with varying amounts of Pb, Sn and Zn. *J. Archaeol. Sci.* **2020**, *121*, 105199. [CrossRef]
19. Krupiński, M.; Smolarczyk, P.E.; Bonek, M. Microstructure and Properties of the Copper Alloyed with Ag and Ti Powders Using Fiber Laser. *Materials* **2020**, *13*, 2430. [CrossRef]
20. Korneva, A.; Straumal, B.; Kilmametov, A.; Chulist, R.; Cios, G.; Baretzky, B.; Zięba, P. Dissolution of Ag Precipitates in the Cu-8wt.%Ag Alloy Deformed by High Pressure Torsion. *Materials* **2019**, *12*, 447. [CrossRef] [PubMed]
21. Tayyari, J.; Emami, M.; Agha-Aligol, D. Identification of microstructure and chemical composition of a silver object from Shahrak-e Firouzeh, Nishapur, Iran (~2nd millennium BC). *Surf. Interfaces* **2021**, *25*, 101168. [CrossRef]
22. Northover, S.; Northover, J. Microstructures of ancient and modern cast silver–copper alloys. *Mater. Charact.* **2014**, *90*, 173–184. [CrossRef]
23. Salem, Y.; Mohamed, E.H. The role of archaeometallurgical characterization of ancient coins in forgery detection. *Nucl. Instrum. Methods Phys. Res. Sect. B Beam Interact. Mater. At.* **2019**, *15*, 247–255. [CrossRef]
24. Ioanid, E.G.; Ioanid, A.; Rusu, D.E.; Doroftei, F. Surface investigation of some medieval silver coins cleaned in high-frequency cold plasma. *J. Cult. Heritage* **2010**, *12*, 220–226. [CrossRef]
25. Wu, B.; Kong, L.; Li, J. Abnormal dynamic behavior and structural origin of Cu-Ag eutectic melt. *Acta Mater.* **2021**, *207*, 116705. [CrossRef]
26. Dong, H.; Chen, Y.; Zhang, Z.; Shan, G.; Zhang, W.; Liu, F. Mechanisms of eutectic lamellar destabilization upon rapid solidification of an undercooled Ag-39.9 at.% Cu eutectic alloy. *J. Mater. Sci. Technol.* **2020**, *59*, 173–179. [CrossRef]
27. Zuo, X.; Han, K.; Zhao, C.; Niu, R.; Wang, E. Precipitation and dissolution of Ag in ageing hypoeutectic alloys. *J. Alloy. Compd.* **2015**, *622*, 69–72. [CrossRef]
28. Xie, M.; Huang, W.; Chen, H.; Gong, L.; Xie, W.; Wang, H.; Yang, B. Microstructural evolution and strengthening mechanisms in cold-rolled Cu–Ag alloys. *J. Alloy. Compd.* **2020**, *851*, 156893. [CrossRef]
29. Pripon, E. Removing the copper deposits from the surface of coins made of low content silver alloys. *Acta Musei Porolisensis* **2013**, *35*, 343–348.
30. Arghir, G.; Petean, I.; Pripon, E.; Pascuta, P. On the mathematical algorithm of some ancient drachmas minting time. *Acta Tech. Napoc. Ser. Appl. Math. Mech. Eng.* **2014**, *57*, 141–148.
31. Boubaker, H.; Cunado, J.; Gil-Alana, L.A.; Gupta, R. Global crises and gold as a safe haven: Evidence from over seven and a half centuries of data. *Phys. A Stat. Mech. its Appl.* **2019**, *540*, 123093. [CrossRef]
32. Karaman, K.K.; Pamuk, Ş.; Yıldırım-Karaman, S. Money and monetary stability in Europe, 1300–1914. *J. Monetary Econ.* **2020**, *115*, 279–300. [CrossRef]

Article

Mechanical Properties of Orthodontic Cements and Their Behavior in Acidic Environments

Cristina Iosif [1], Stanca Cuc [2,*], Doina Prodan [2], Marioara Moldovan [2], Ioan Petean [3,*], Anca Labunet [1], Lucian Barbu Tudoran [4,5], Iulia Clara Badea [6], Sorin Claudiu Man [7], Mîndra Eugenia Badea [6] and Radu Chifor [6]

1. Department of Prosthetic Dentistry and Dental Materials, "Iuliu Hatieganu" University of Medicine and Pharmacy, 32 Clinicilor Street, 400006 Cluj-Napoca, Romania
2. Department of Polymer Composites, Institute of Chemistry "Raluca Ripan", University Babes-Bolyai, 30 Fantanele Street, 400294 Cluj-Napoca, Romania
3. Faculty of Chemistry and Chemical Engineering, University Babes-Bolyai, 11 Arany János Street, 400028 Cluj-Napoca, Romania
4. Department of Molecular Biology and Biotechnology, Electron Microscopy Laboratory, Biology and Geology Faculty, Babes-Bolyai University, 5–7 Clinicilor Str., 400006 Cluj-Napoca, Romania
5. Electron Microscopy Integrated Laboratory, National Institute for Research and Development of Isotopic and Molecular Technologies, 65-103 Donath Street, 400293 Cluj-Napoca, Romania
6. Dental Prevention Department, Faculty of Dental Medicine, "Iuliu Hatieganu" University of Medicine and Pharmacy, Avram Iancu 31, 400083 Cluj-Napoca, Romania
7. Mother and Child Department, 3Rd Department of Paediatrics, "Iuliu Hatieganu" University of Medicine and Pharmacy, 2-4 Campeni Street, 400217 Cluj-Napoca, Romania
* Correspondence: stanca.boboia@ubbcluj.ro (S.C.); ioan.petean@ubbcluj.ro (I.P.)

Abstract: The present research is focused on three different classes of orthodontic cements: resin composites (e.g., BracePaste); resin-modified glass ionomer RMGIC (e.g., Fuji Ortho) and resin cement (e.g., Transbond). Their mechanical properties such as compressive strength, diametral tensile strength and flexural strength were correlated with the samples' microstructures, liquid absorption, and solubility in liquid. The results show that the best compressive (100 MPa) and flexural strength (75 Mpa) was obtained by BracePaste and the best diametral tensile strength was obtained by Transbond (230 MPa). The lowestvalues were obtained by Fuji Ortho RMGIC. The elastic modulus is relatively high around 14 GPa for BracePaste, and Fuji Ortho and Transbond have only 7 GPa. The samples were also subjected to artificial saliva and tested in different acidic environments such as Coca-Cola and Red Bull. Their absorption and solubility were investigated at different times ranging from 1 day to 21 days. Fuji Ortho presents the highest liquid absorption followed by Transbond, the artificial saliva has the best absorption and Red Bull has the lowest absorption. The best resistance to the liquids was obtained by BracePaste in all environments. Coca-Cola presents values four times greater than the ones observed for artificial saliva. Solubility tests show that BracePaste is more soluble in artificial saliva, and Fuji Ortho and Transbond are more soluble in Red Bull and Coca-Cola. Scanning electron microscopy (SEM) images evidenced a compact structure for BracePaste in all environments sustaining the lower liquid absorption values. Fuji Ortho and Transbond present a fissure network allowing the liquid to carry out in-depth penetration of materials. SEM observations are in good agreement with the atomic force microscopy (AFM) results. The surface roughness decreases with the acidity increasing for BracePaste meanwhile it increases with the acidity for Fuji Ortho and Transbond. In conclusion: BracePaste is recommended for long-term orthodontic treatment for patients who regularly consume acidic beverages, Fuji Ortho is recommended for short-term orthodontic treatment for patients who regularly consume acidic beverages and Transbond is recommended for orthodontic treatment over an average time period for patients who do not regularly consume acidic beverages.

Keywords: orthodontic cements; acidic environment; mechanical properties; surface roughness

Citation: Iosif, C.; Cuc, S.; Prodan, D.; Moldovan, M.; Petean, I.; Labunet, A.; Barbu Tudoran, L.; Badea, I.C.; Man, S.C.; Badea, M.E.; et al. Mechanical Properties of Orthodontic Cements and Their Behavior in Acidic Environments. *Materials* **2022**, *15*, 7904. https://doi.org/10.3390/ma15227904

Academic Editors: Andrei V. Petukhov and Rosalia Maria Leonardi

Received: 4 October 2022
Accepted: 6 November 2022
Published: 9 November 2022

Publisher's Note: MDPI stays neutral with regard to jurisdictional claims in published maps and institutional affiliations.

Copyright: © 2022 by the authors. Licensee MDPI, Basel, Switzerland. This article is an open access article distributed under the terms and conditions of the Creative Commons Attribution (CC BY) license (https://creativecommons.org/licenses/by/4.0/).

1. Introduction

Orthodontic problems often require bracket usage as a long-term treatment [1,2]. They are subjected to mastication forces which induce shearing stress into the bonding interface but they must also transmit the orthodontic forces to the teeth [3,4]. The factors involved in the successful transmission of forces during orthodontic treatment are the preparation of the surface of enamel, the characteristics of the bracket and the type of cement used for bonding [5]. Therefore, the properties and characteristics of the involved dental materials need to be considered and g an appropriate product for successful performance needs to be chosen.

Materials used for orthodontic bonding usually fail as a result of micro fissures infiltrated by oral bacteria, or complete bracket debonding [6]. The ideal cement should perform enough retention to provide resistance during normal masticator forces, to transmit orthodontic forces and to be easily removable without causing damages to the surface of the enamel [7]. The two materials frequently used for bonding are composite resins [8,9] and glass ionomers [10,11].Composite resins were introduced by Buonocore and are most frequently used for bonding brackets due to their bond strength that increases and doubles within 24 h [12]. The composites' resins need micro-mechanical retentions of an acid-etched surface of the tooth with a primer-bonding agent to aid coupling of the two surfaces. The disadvantage of this method is the risk of enamel demineralization and the probability of enamel defects after bracket removal. That is the reason why composite resins were replaced with novel orthodontic cements based on fluoride-releasing materials, such as glass-ionomers cements [10,11].

Glass ionomers can release fluoride and prevent enamel demineralization, but they have a lower adhesion to the surface of enamel and they determine frequent debonding of the brackets, with a negative effect on orthodontic treatment. Glass ionomers modified with resins (RMGIC) were developed as a hybrid of glass ionomers and composite resins to ensure a better bond strength [13,14]. RMGIC have the advantages of composite resins and of glass ionomers: a better bond strength, fluoride-releasing properties and they can be used when it is difficult to dry the teeth (impacted canine, lower second molar, palatal/lingual surface, etc.). RMGIC preparation is based on an acid chemical reaction assisted by a light-activated additional polymerization of hydroxyethyl-methacrylate. Photo-polymerizable cements cure in approximately 30 s if exposed to a light source of the correct wavelength [15,16]. The acid-base reaction does not end when the photo polymerization process is complete, which means that the mechanical properties of the material are further improved. The difficulties of hand-mixing powder-liquid can provide us with errors, with a negative effect on the physical properties of the glass ionomers. This problem is solved by the use of encapsulation with fixed proportions by the manufacturer and mechanical mixing [15].

These material types are not new but the continuous challenge to improve their microstructure and mechanical properties is always of the great interest. The filler particles refinement and their embedding into the polymer matrix are constantly improved to release novel materials for dental use. Examples of orthodontic cement with newly improved formulation are: resin composites (e.g., BracePaste); resin-modified glass ionomer RMGIC (e.g., Fuji Ortho) and resin cement (e.g., Transbond).

A very important requirement of dental materials is to resist to the loads during mastication such as: shear, compression and flexing. The bonding layer between the enamel and metallic brackets is affected in a complex manner by compression, shear and flexing solicitations [12,14]. This complex behaviour can be quantified using standardized determination of the mechanical properties that include: diametral tensile strength, compressive strength, shear bond strength, and flexural strength. The results of these standard tests allow specialists to choose the proper orthodontic cement type to the specificity of the intended treatment. The orthodontic treatment time is one the most important parameters because the cement bonding must ensure the desired mechanical properties during this whole time period. A novel approach of the present research is to compare the behaviour of the three types of orthodontic adhesives in acidic conditions similar to the ones in the patient's mouth.

The real conditions from the patient's mouth represent an acidic environment caused by the low pH of food and drinks which may affect all three elements of the orthodontic treatment: enamel surface, bracket material and the bonding layer [17,18]. The bonding layer is very sensitive to the acid erosion only if it presents increased liquid absorbtion or a significant solubility in certain liquids such as acidic soft drinks [18,19]. The most popular acidic beverages among youngsters are Coca-Cola acidified with phosphoric acid and Red Bull containing a mixture of acids, with the most representative being citric acid [20]. If the microfissures occur in the adhesive layer they will generate microleackages [21] which may be decay accelerators when discussingthe acidic environment. Therefore, the orthodontic cements require the investigation of liquid absorbtion and solubility, microstructural aspects and surface topography in order to provide a completecharacterization. It is necessary for the dental practitioners to choose the proper cement for each orthodontic case.

The aim of present research is to realize a comparative investigation of mechanical properties such as: compressive strength, diametral tensile strength and flexural strength related to the liquid absorbtion, solubility and microscopic characterization of the following orthodontic cements: BracePaste, Fuji Ortho and Transbond. Shear bond strength comparative investigation for these materials in conditions similar to the patient'smouth is the challenge for the article.

The null hypothesis states that the mechanical properties are the same for all three of the investigated materials and that there are no microstructure, absorption and solubility differences between the exposure environments: artificial saliva, Coca-Cola and Red Bull.

2. Materials and Methods

2.1. Materials Description

The materials used in the present research are described in Table 1as follows: commercial product name, producer and the complete composition displayed on the product label. There is no other additional information regarding the used materials.

Table 1. Materials characteristics.

Product Name	Producer	Composition
BracePaste	American Orthodontics, Sheboygan, WI, USA	Methacrylic acid ester, activator, Ethoxylated Bisphenol A, Dimethacrylate, Tetramethylene Dimethacrylate, Diphenyl (2,4,6-trimethylbenzoyl) phosphine oxide.
Fuji Ortho LC	GC Company, Tokio, Japan	20% Polyacrilic acid Fluoro-aluminium-silicate glass Polyacrilic acid, HEMA, UDMA, silicon dioxide, distilled water, initiators, pigment.
Transbond Colour Change	3M Unitek, St.Paul, MN, USA	35%Phosphoric acid Primer- bis-GMA, TEGMA Adhesive paste- bis-GMA, TEGMA, Silane, treated quartz, amorphous silica, camphor quinone.

The complexity of the current paper requires various samples prepared from the materials described in Table 1 following the standard protocol of each investigation method. Therefore, the sample preparation is described for each experimental method in its subchapter.

2.2. Mechanical Properties

2.2.1. Compressive Strength

The samples for compressive test were prepared in a Teflon/PTFE matrix with the shape of the disk being 0.6 mm thick, composed of two pieces with a cylindrical hole in the middle—

with a diameter of 0.4 cm and a height of 0.6 cm. After polymerization, the resulting specimens were kept submerged in water for 24 h at a constant temperature of 37 °C.

Compressive strength was tested in a load range of 400–1100 N and the loading rate of 1 mm/min according to the American Dental Association (ADA) Standard No. 27. The value of compressive strength is the mean of 6 tests out of a total of 10. The specimens that exceeded the mean value by 15% were excluded. If more than 4 specimens fell outside the mean range, the testing for that series was repeated.

2.2.2. Diametral Tensile Strength

The general technique for preparing and testing the specimens is similar to that used for compressive strength testing, with the exception of the specimen dimensions: 0.3 cm thickness and 0.6 cm diameter. The specimens are compressed along their diameter according to the American Dental Association (ADA) Standard No. 27. The application of loading force on the cylinder determines the stress on the vertical axial plane, as it is held between the two apparatus plates. Diametral tensile strength was tested at between 400–1100 N and the loading rate of 0.5 mm/min. The value of diametral tensile strength is the mean value of at least 6 determinations.

2.2.3. Flexural Strength

The specimens used for flexural strength testing were polymerized in a Teflon/PTFE matrix in a parallelepiped shape—dimensions: 25 mm length, 2 mm width, 2 mm height. After polymerization, the specimens were submerged in distilled water for 24 h at the temperature of 37 °C.

To determine the flexural strength, the specimens were placed symmetrically on two supports with a 2 mm diameter, and the distance between the axes was l = 20 mm according to the ISO 4049/2000 Standard. The load F (N) that bends the specimen is applied centrally through a 2mm diameter cylinder. A3-point bending test was run with a loading rate of 0.5 mm/min. The value of flexural strength is the mean value of at least 6 determinations.

2.2.4. Statistical Analysis

The results of the mechanical testing were analyzed using descriptive statistics (mean, median, standard deviation) and inferential statistics. One-way ANOVA of the 3 test groups was run for each group, and the significance level was set at $\alpha = 0.05$. The statistical difference was analyzed using the Tukey test, with the use of Origin2019b Graphing and Analysis software.

2.3. Liquid Absorbtion and Samples Solubility

The composites are polymerized in a Teflon/PTFE matrix that produces specimens of the following dimensions: 15 ± 1 mm diameter and 1 mm thickness, to determine liquid absorption. The specimens were kept in a desiccator at 23 °C, before the initial weighing, until a constant weight was obtained (m1). Fifteen disk-shaped specimens were produced for each of the 3 sample groups, as per the ISO 4049/2000 standard. Five specimens of each material were placed in Group A and were immersed in a Cola-type beverage, another 5 were placed in Group B and were immersed in Red Bull, while the remaining 5 were the control sample and were immersed in artificial saliva. The specimens were desiccated and weighed immediately afterwards (m2), then after 2 h of recovery time in the desiccator (m3) at certain intervals (1 day, 2 days, 5 days, 6 days, 7 days, 14 days, 21 days). The liquid absorption (A) is calculated with Formula (1) and the samples solubility (S) is determined with Equation (2).

$$A = 100 \times \frac{m2 - m1}{m1} \tag{1}$$

$$S = 100 \times \frac{m1 - m3}{m1} \tag{2}$$

The same method of weighing the specimen disks immersed in the three liquids solutions used to determine absorption was used to determine the solubility of polymerizable dental composites, as per the ISO 4049/2000 standard.

2.4. Scanning Electron Microscopy SEM

The perfectly dried specimens used for the liquid absorption test were used for the scanning electron microscopy investigation. SEM images were obtained on the dried sample discs using a Hitachi SU8230 microscope, Hitachi Hi-Tec Corporation, Tokyo, Japan. The secondary electron images were obtained on the uncoated samples using an acceleration voltage of 30 kV and high vacuum chamber.

2.5. Atomic Force Microscopy AFM

Samples used for the atomic force microscopy are the dried specimens that were used for the liquid absorption. They were effectuated using a JEOL microscope (JSPM 4210, Tokyo, Japan). All the samples were investigated in tapping mode using NSC 15 cantilevers produced by MikroMasch, Bulgaria Headquarters, Sofia. The cantilever resonant frequency was 330 kHz, and the spring constant was 48 N/m. Three separate macroscopic areas were scanned for each sample at a scan size of 20 µm × 20 µm. The images were processed using the WinSPM 2.0 JEOL software (Tokyo, Japan) in accordance with standard procedures, presenting 2D topographic images, 3D images, and the Ra and Rq surface roughness parameters were measured. The 3D images are also called 3D profiles; they are a graphic representation of the depth profile of the enamel surface, closely related to the measured values of surface roughness.

3. Results

3.1. Mechanical Properties

Compressive strength results are displayed in Figure 1a with the statistical analysis results. The best resistance was obtained by BracePaste around 100 MPa followed by Transbond around 80 MPa. The least resistant was the RGMIC sample, Fuji Ortho, having only 73 MPa. The differences in the compressive strength values are due to the particle-size distribution, where very fine powder particles insert themselves into larger ones, which lead to a decrease in the interstitial spaces between them. These inserted particles support the compressive stress, thus reducing the frequency of breakage.

Diametral tensile strength results are presented in Figure 1b and the statistical analysis results are displayed on the graph. The best compression strength was obtained for Transbond which was about 230 MPa. BracePaste and Fuji Ortho prove to be weaker under tensile load. The macroscopic examination of the samples after testing indicates a brittle fracture generated by the granular matter failure under axial effort. Therefore, the elastic modulus was calculated, Figure 1c. The results show that BracePaste and Fuji Ortho are more elastic than Transbond. This might be an important aspect concerning the material behavior under masticator forces.

Shear stress generated by the mastication induces flexural solicitation of the bonding layer. The flexural strength results are displayed in Figure 1d. The best result was obtained for BracePaste which was about 75 MPa and the lowest value was obtained by Fuji Ortho at about 27 MPa. This means that the higher concentration of filler determines the increased values of flexural strength.

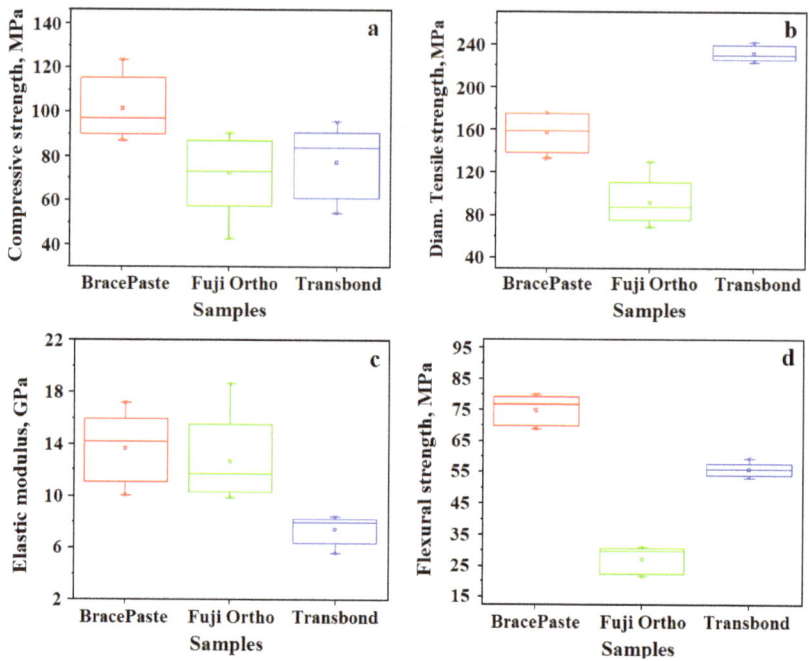

Figure 1. Mechanical properties variation for the initial cement samples with statistical analysis: (**a**) compressive strength, (**b**) diametraltensile strength, (**c**) elastic modulus and (**d**) flexural strength.

3.2. Liquid Absorbtion

Five polymerized specimens of each tested material type (BracePaste, Fuji Ortho and Transbond) were used for each immersion environment (Coca-Cola, Red Bull and artificial saliva). Absorption and solubility were tested at the following intervals: 24 h, 48 h, 5 days, 6 days, 7 days, 14 days, and 21 days. The study found that liquid absorption in the tested materials is dependent upon their chemical composition, the time factor, and the immersion medium, as shown in Figure 2. In the first 2 days, liquid absorption continues to significantly increase for all the tested materials, followed by a tendency to become stable until day 7, followed by another great increase in the majority of cases until the 21st day.

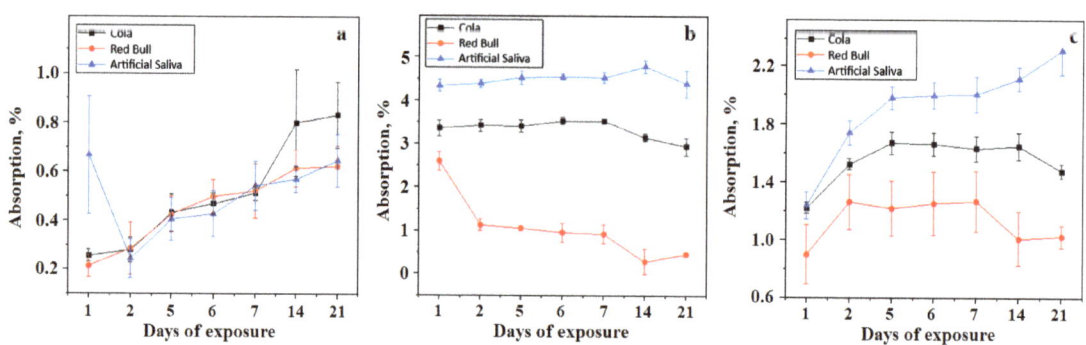

Figure 2. Absorption curveresults for the tested orthodontic cements: (**a**) BracePaste, (**b**) Fuji Ortho, and (**c**) Transbond.

BracePaste absorbs more artificial saliva after 1 day of exposure and less Red Bull and the absorption is almost the same after 2 days of exposure. The situation evolves in a similar manner for all the exposure environments until 14 days, as shown in Figure 2a. After this point, we observed that the higher absorption occurs for Coca-Cola and the lower absorption occurs for the artificial saliva and Red Bull. This clearly indicates BracePaste sensibility to the phosphoric acid.

The results for Fuji Ortho, Figure 2b, and Transbond, Figure 2c, reveal that they absorbed more artificial saliva and less Red Bull. This behaviour may instead be explained by liquid viscosity, with Red Bull being more viscous than saliva than to its acidity.

Analyzing the absorption of the three types of materials in the same immersion medium, we can posit that all three materials have distinctly different behaviours, with the values of p much lower than 0.05. However, when immersed in Red Bull ($p = 0.02$), the glass ionomer material did not present any statistical differences when compared to the other two materials tested. Finally, the results show that BracePaste absorbs less liquid and Fuji Ortho presents the highest liquid absorption among the tested materials.

3.3. Solubility in Liquid Environment

The bonding cement solubility is one of the most important physicochemical aspects regarding the integrity of the bracket adhesion on the teeth. If the adhesive is dissolved by the acid in food and drinks, bonding weakening takes place and this even affects the adhesion dissolution over time [22,23]. The percentage solubility is determined by samples weight and using Equation (2). The obtained values were centralized in Figure 3. BracePaste is more susceptible to being soluble in artificial saliva and less soluble in Coca-Cola over longer terms of exposure up to 14 and 24 days, as shown in Figure 3a. This is a very unusual aspect because the behavior in saliva must feature a strong stability not a predisposition to solubility [24–26], a fact which will be debated in the discussion section. However, the obtained values are below zero, and as a consequence, there is no dissolution danger to the patients during normal eating.

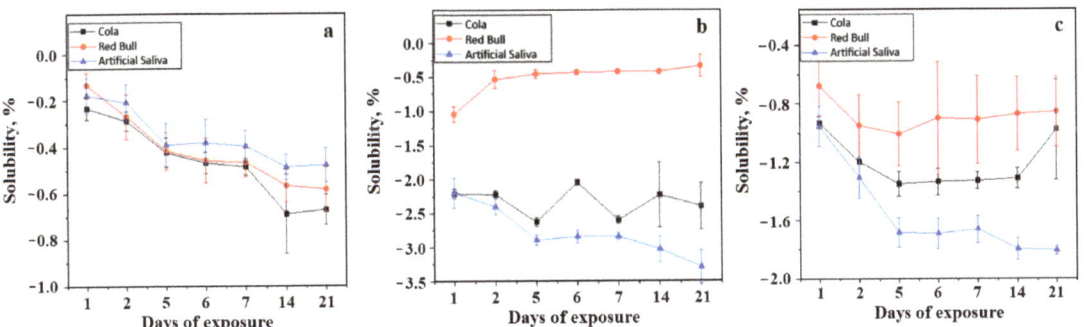

Figure 3. Solubility curve results for the tested orthodontic cements: (a) BracePaste, (b) Fuji Ortho, and (c) Transbond.

Fuji Ortho and Transbond present a similar behavior of being more susceptible to being dissolved in Red Bull and being less soluble in artificial saliva, as shown in Figure 3b,c. There is strong evidence that exposure to an acidic environment may facilitate local dissolution of the specific micro-structural components of these orthodontic cements.

By analyzing the solubility of the three types of materials in the same immersion medium, we can posit that all three materials have distinctly different behaviours when immersed in Coca-Cola ($p = 1.99377 \times 10^{-12}$); there are significant statistical differences ($p = 2.52524 \times 10^{-4}$) when they are immersed in Red Bull, without any differences between BracePaste and glass ionomer; there are significant statistical differences when they are immersed in artificial saliva ($p = 3.30099 \times 10^{-11}$) for all three of the tested materials.

3.4. Scanning Electron Microscopy SEM

SEM images were taken at average magnification to observe all the morphological details regarding the samples' morphology. The interaction with the exposure environment is followed.

The initial BracePaste sample presents a relatively uniform and compact morphology due to the optimal cohesion between the polymer and mineral filler particles. Unfortunately, several pores occur: a few of them are bigger with rounded shapes and diameters of about 30 to 75 μm and several smaller pores with dendritic shapes are present, which thus means a more detailed topographic investigation is required, Figure 4A(a). The exposure to artificial saliva ensures a proper wetting of the sample surface (similar to the mouth conditions), a fact which improves the microstructure quality, Figure 4A(b). Exposure to an acidic environment reveals a good preservation of the samples' microstructure, with only the outermost unevenness being eroded, Figure 4A(c,d).

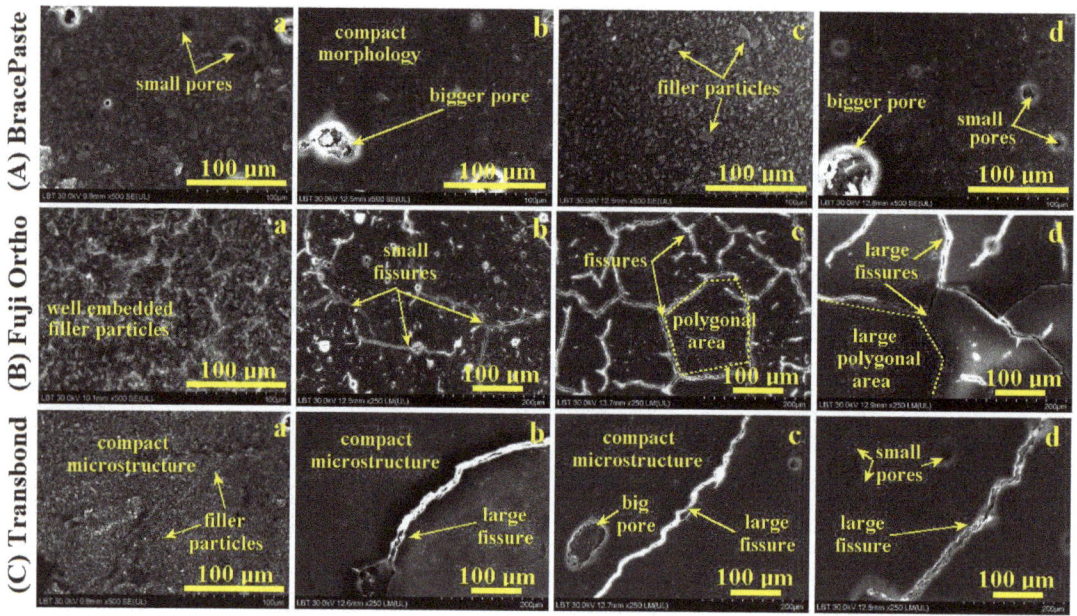

Figure 4. SEM images of the investigated samples: BracePaste (**A**); Fuji Ortho (**B**) and Transbond (**C**) exposed at different liquid environment: (**a**) initial—unexposed, (**b**) artificial saliva, (**c**) Coca-Cola and (**d**) Red Bull.

Fuji Ortho RMGIC cement presents a typical microstructure with fluoro-aluminum-silicate glass particles very well dispersed in the polymer matrix. The silica nanoparticles are not visible, but their presence ensures the good consistence of the sample, Figure 4B(a). Proper cohesion between the filler particles and the polymer matrix is observed. The exposure to artificial saliva, Figure 4B(b), reveals a relative enhancement of the filler particle exposure to wet conditions in the outer most layers due to the liquid absorption. It generates a network of small fissures on the sample surface. Exposure to Coca-Cola generates more definite fissures propagated into a network which isolates polygonal areas with diagonal areas ranging from about 75 to 300 μm, Figure 4B(c). Exposure to Red Bull also generates thicker fissures propagated in a network, Figure 4B(d), that isolates polygonal areas with a greater diagonal area (e.g., 300–500 μm) than the one observed for Coca-Cola.

The initial Transbond sample, Figure 4C(a), revealed a compact and uniform microstructure which deals with refined filler particles such as amorphous silica. The exposure to the wet environment generates fewer strong fissures propagated over the samples surface, Figure 4C(b–d). A progressive corrugation of the outermost layer of the microstructure is observed especially for the exposure to Coca-Cola, Figure 4C(c), and to Red Bull, Figure 4C(d).

The surface topography of the initial BracePaste sample, Figure 5A(a), reveals a relatively uniform surface randomly punctured by the presence of some pores with a dendritic shape and dimensions ranging from 5 to 7 µm. They are more visible in the tridimensional profile given below the topographic image in Figure 5A(a) and they determine a relatively high value of the roughness. The exposure to artificial saliva tends to preserve the microstructural aspects, the dendritic pores' presence is significantly reduced as observed in Figure 5A(b). This leads to the roughness decreasing. Exposure to an acidic environment contributes to an enhanced regularization of the topography with an attenuation of the pore depth as observed for Coca-Cola in the left lower corner in Figure 5A(c) and for Red Bull in the right upper corner of Figure 5A(d). This fact induces a strong decrease in the surface roughness, Figure 6a.

Initial Fuji Ortho presents a complex topography due to its various granular constituents. There are remarkable fluoro-aluminum-silicate glass particles with boulder shapes and diameters ranging from 3 to 5 µm and small submicron silica particles with diameters of 200–500 nm depending on their local arrangements, Figure 5B(a). The compactness between the polymer and mineral filler associated to a uniform surface ensures the relatively low value of the roughness. The exposure to the artificial saliva generally preserves the topographic features of the Fuji Ortho sample, with only some small alterations occurring due to the relative liquid sorption, Figure 5B(b). Therefore, the roughness increases slowly. The exposure to Coca-Cola presents an altered topography due to the acid erosion. Some of the fluoro-aluminum-silicate glass particles were delaminated from the polymer bonding and dislocated from the microstructure, Figure 5B(c), leading to the roughness increasing. The erosive aspect is more prone after Red Bull exposure, Figure 5B(d), which contributes to a severe increase in the roughness as observed in Figure 6b.

Transbond initial topography is presented in Figure 5C(a) and corresponds to a very uniform and compact surface. The filler particles are nanostructured and some of them form submicron clusters very well embedded into the polymer matrix. This fact determines a very low roughness value. The exposure to the artificial saliva facilitates liquid absorption on the superficial layers which causes a mild corrugation of the surface that unveils the filler particles, Figure 5C(b), which slowly increases the roughness value. Exposure to the phosphoric acid within Coca-Cola causes significant acid erosion which leads to the development of topographical depressions generated by filler particle loss, Figure 5C(c). Red Bull proves to be more erosive, Figure 5C(d), by generation of enlarged and deeper depressions on the sample surface. The acid effect causes a greater increase in the surface roughness as observed in Figure 6c.

Statistical analysis performed on the roughness measured for all the samples involved in the present research was not relevant due to the opposite behavior of BracePaste samples compared to Fuji Ortho and Transbond. This particular behavior is in strong connection with the observations on liquid absorption and microstructure local solubility.

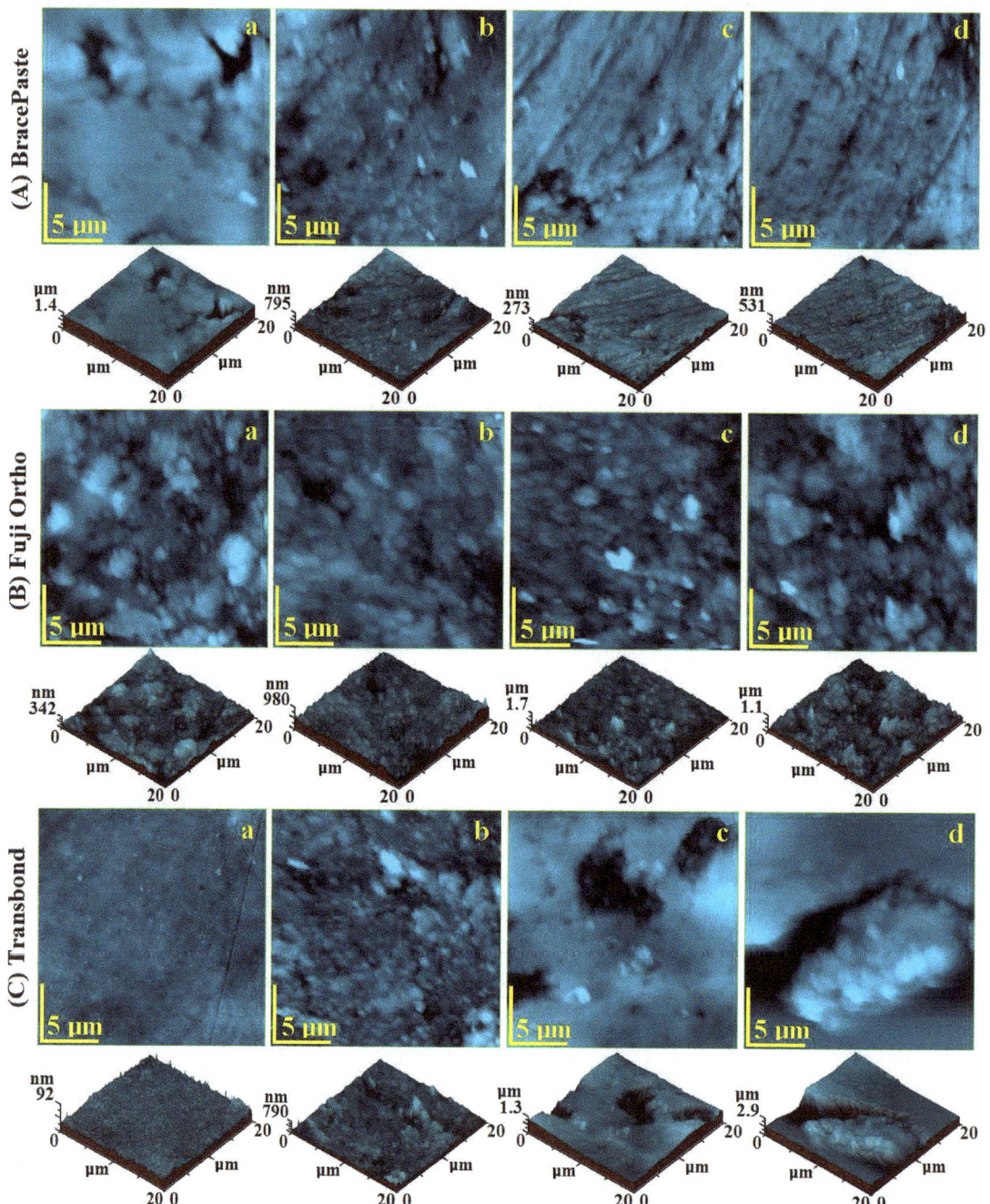

Figure 5. AFM images of the investigated samples: BracePaste (**A**); Fuji Ortho (**B**) and Transbond (**C**) exposed to different liquid environments: (**a**) initial—unexposed, (**b**) artificial saliva, (**c**) Coca-Cola and (**d**) Red Bull.

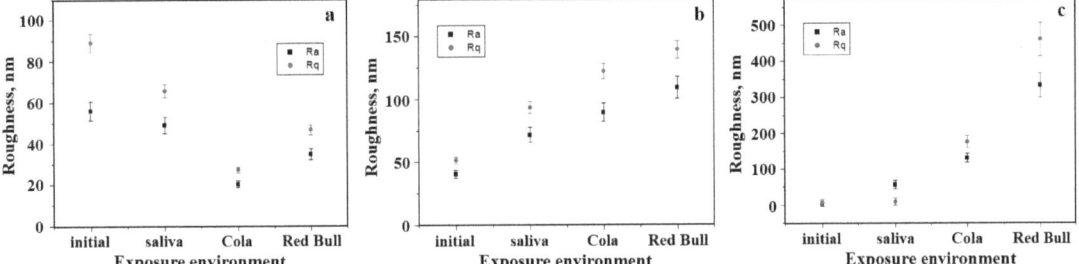

Figure 6. Roughness evolution after acidic environment exposure for the tested orthodontic cements: (a) BracePaste, (b) Fuji Ortho and (c) Transbond.

4. Discussion

The materials used in dentistry must be durable and provide long-term adhesion, despite the hostile nature of the environment—there is permanent contact with saliva, a bacteria-rich fluid that contains multiple other organic and inorganic substances [24–26]. The interaction between saliva and the structural elements of the cements can cause the degradation or dissolution of superficial strata, the release of unbound or insufficiently bound components in the cement structure, or water absorption in the cement itself [27].

The mechanical properties of the materials included in this study depend on the following characteristics of the cements' components (organic and inorganic): mechanical strength of the filling material, dispersed phase consistency, dispersion nanoparticle geometry, dispersed phase orientation, dispersed and continuous phase compositions and ratios, as well as the relationships between the two phases. Our previous study indicates that the refined filler particles are better dispersed onto the polymer matrix thus assuring a good cohesion of the microstructure [28,29], a fact which is in good agreement with the data in the literature [10,23]. There is a correlation between the physical–mechanical properties of the cements used for bonding and the mass and ratio of inorganic phase components as observed by AFM microscopy, which is in good agreement with the SEM observations. This indicates that the filler amount is an important factor but also the particles' refinement and their proper dispersion into the fluid polymer before solidification plays a key role in the mechanical properties of the investigated materials. Therefore, BracePaste has the best compressive and flexural strength. High flexural strength is correlated with an increased elasticity modulus that ensures proper behavior during mastication. On the other hand, good compressive strength ensures good cohesion into the bonding layer, which is in good agreement with the literature [30]. Therefore, BracePaste is the best configuration for a strong adhesion between the enamel surface and brackets. BracePaste's average diametral tensile strength does not affect long-term orthodontic treatment unless the brackets are accidentally pulled out. Transbond has the best diametral tensile strength, a good flexural strength, weaker compressive strength and low elasticity modulus. Its rigidity may affect the bonding layer behavior under long-term orthodontic treatment. However, the flexural strength of about 58 MPa ensures successful orthodontic treatment over an average time period, a fact which is in good agreement with the data in the literature [31].

On the other hand, Fuji Ortho presents low values of mechanical properties especially compressive and flexural strength because of the bigger filler particles (e.g., fluoro-aluminum-silicate glass) which are relatively brittle and tend to crack during the force load and the resulting fragments tend to further section the polymer matrix that acts as a failure promoter. However, the heterogeneous filler mixture observed in the Fuji Ortho RMGIC proves to be beneficial to achieving a good value of the elastic modulus, which is in good agreement with the requirements for a bracket adhesive [32,33]. Therefore, the first part of the null hypothesis regarding mechanical properties was rejected.

In the case of resin-based cements, liquid absorption can have beneficial effects and the possibility of lessening the residual polymerization contraction stress inside the organic matrix [16]. Results show that BracePaste has a greater liquid absorption in saliva [24–26] than in Red Bull, a fact that is related to the liquid viscosity and saliva being less viscous than energy drinks. Oberhofer et al. noticed this aspect in his study regarding the impact of energy drinks on young people and children [34]. The lower penetration of energy drinks among the filler particles proves to be protective against acid erosion propagation on the adhesion layer bulk. In the case of photo-polymerizable glass ionomer cements, some studies have shown that liquid absorption occurs for several months after the initial polymerization, which lessens their rigidity and strength [17]. Resin-modified ionomer cements undergo a rapid and marked expansion if immersed in an aqueous medium after hardening. This expansion compensates for the initial contraction, but may also induce fissures on the sample surface as observed in the SEM images. It has been demonstrated that RMGIC adsorbed more liquid than the composite resin [26] because of the hydroxyethyl methacrylate (HEMA) presence in the structure which is hydrophilic. Even if liquid absorption is characteristic of resin-based materials, the obtained results demonstrated that it has a negative effect on the mechanical properties, especially by decreasing flexural strength. The null hypothesis regarding liquid absorption and samples' solubility was rejected.

The data in the literature mention the negative effect of acid erosion on the mechanical properties of the orthodontic cements' adhesion between enamel and bracket. There are a few evidenced ways of erosion such as: enamel–cement interface decay, bracket–cement interface failure or cement layer internal failure [35–37]. The AFM observation proves that BracePaste presents a good behavior under acidic environments such as Coca-Cola and Red Bull because of the strong cohesion between the filler particles and the polymer matrix which ensures proper insulation of the mineral particles from contact with acids in the mentioned soft drinks. Fuji Ortho is significantly affected by acid erosion and Transbond is the most affected by acid erosion, a fact sustained by the surface roughness variation. This is due to the combination between the erosive action of the acid on the mineral filler particles and the fissures observed on their microstructures. We noticed the aggressive effect of phosphoric acid from Coca-Cola and citric acid in Red Bull in a previous study regarding CR and RMGIC materials used for inlays and crown adhesion [38]. This corresponds to the data mentioned in the literature [39,40]. It is a matter of the filler particles' distribution in the polymer matrix and their interaction with the exposure environment. The roughness decrease under acid exposure sustains the BracePaste resistance against in-depth penetration of the erosive agent and therefore ensures long-term protection of the bonding layer within the orthodontic treatment. The fissure and pores' development under acid exposure evidenced for Fuji Ortho and Transbond causes the surface roughness to increase as a consequence of the erosive effect. The different behavior observed on the investigated materials under acidic conditions rejects the null hypothesis. Therefore, it is necessary that the dentist asks the patient what their diet habits are (e.g., if they regularly consume acidic beverages or not) prior to establishing the cement to be used for the orthodontic treatment.

5. Conclusions

The investigated orthodontic cements present a complex relationship between their mechanical properties, microstructures and their behavior in acidic condition. They allow dentists to choose the optimal bonding of the brackets considering both the length of time of the orthodontic treatment and the patient's feeding preferences:

1. BracePaste presents the best combination of mechanical properties and erosion resistance in various acidic environments, with it being recommended for long-term orthodontic treatment especially when the patient regularly consumes acidic drinks such as Coca-Cola and Red Bull. Prolonged exposure to acid overtime will make the debonding procedure at the end of the orthodontic treatment easy.

2. Fuji Ortho RMGIC features the best handling associated with easy bracket procedures and presents low erosive resistance overtime. Therefore, it is easy to be applied at the beginning of treatment and it is easy to use in debonding at the end of the orthodontic treatment with minimal discomfort for the patient. Beside these important aspects, it is recommended only for short-term orthodontic treatment and is suitable for patients who regularly consume acidic beverages such as Coca-Cola and Red Bull.
3. Transbond presents the highest diametral tensile strength but its compressive and flexural strength are inferior to BracePaste, and it also presents poor erosive resistance in acidic environments. This adhesive may ensure good adhesion of the brackets for orthodontic treatment over an average period of time for patients who do not regularly consume acidic beverages.

Author Contributions: Conceptualization, M.M. and M.E.B.; methodology, C.I.; software, D.P.; validation, S.C.M. and R.C.; formal analysis, A.L.; investigation, I.P., S.C. and L.B.T.; resources, I.C.B.; data curation, L.B.T.; writing—original draft preparation, C.I.; writing—review and editing, I.P., S.C.; visualization, S.C. and I.P.; supervision, M.M.; project administration, M.E.B.; funding acquisition, C.I. All authors have read and agreed to the published version of the manuscript.

Funding: This research received no external funding.

Institutional Review Board Statement: Not applicable.

Informed Consent Statement: Not applicable.

Data Availability Statement: The data presented in this study are available on request from the corresponding author.

Acknowledgments: The authors acknowledge the Research Centre in Physical Chemistry "CECHIF" of Babes Bolyai University for AFM assistance.

Conflicts of Interest: The authors declare no conflict of interest.

References

1. Chee, S.; Mangum, J.; Teeramongkolgul, T.; Tan, S.; Schneider, P. Clinician preferences for orthodontic bracket bonding materials: A quantitative analysis. *Australas. Orthod. J.* **2022**, *38*, 173–182. [CrossRef]
2. Vartolomei, A.-C.; Serbanoiu, D.-C.; Ghiga, D.-V.; Moldovan, M.; Cuc, S.; Pollmann, M.C.F.; Pacurar, M. Comparative Evaluation of Two Bracket Systems' Kinetic Friction: Conventional and Self-Ligating. *Materials* **2022**, *15*, 4304. [CrossRef]
3. Urichianu, M.; Makowka, S.; Covell, D., Jr.; Warunek, S.; Al-Jewair, T. Shear Bond Strength and Bracket Base Morphology of New and Rebonded Orthodontic Ceramic Brackets. *Materials* **2022**, *15*, 1865. [CrossRef]
4. Sfondrini, M.F.; Pascadopoli, M.; Gallo, S.; Ricaldone, F.; Kramp, D.D.; Valla, M.; Gandini, P.; Scribante, A. Effect of Enamel Pretreatment with Pastes Presenting Different Relative Dentin Abrasivity (RDA) Values on Orthodontic Bracket Bonding Efficacy of Microfilled Composite Resin: In Vitro Investigation and Randomized Clinical Trial. *Materials* **2022**, *15*, 531. [CrossRef]
5. Fricker, J.P. Therapeutic properties of glass-ionomer cements: Their application to orthodontic treatment. *Aust. Dent. J.* **2021**, *67*, 12–20. [CrossRef]
6. Naranjo, A.A.; Triviño, M.L.; Jaramillo, A.; Betancourth, M.; Botero, J.E. Changes in the subgingival microbiota and periodontal parameters before and 3 months after bracket placement. *Am. J. Orthod. Dentofac. Orthop.* **2006**, *130*, 275.e17–275.e22. [CrossRef]
7. Mickenautsch, S.; Yengopal, V.; Banerjee, A. Retention of orthodontic brackets bonded with resin-modified GIC versus composite resin adhesives—A quantitative systematic review of clinical trials. *Clin. Oral. Investig.* **2012**, *16*, 1–14. [CrossRef]
8. Bilgrami, A.; Maqsood, A.; Alam, M.K.; Ahmed, N.; Mustafa, M.; Alqahtani, A.R.; Alshehri, A.; Alqahtani, A.A.; Alghannam, S. Evaluation of Shear Bond Strength between Resin Composites and Conventional Glass Ionomer Cement in Class II Restorative Technique—An In Vitro Study. *Materials* **2022**, *15*, 4293. [CrossRef]
9. Chin, A.; Ikeda, M.; Takagaki, T.; Nikaido, T.; Sadr, A.; Shimada, Y.; Tagami, J. Effects of Immediate and Delayed Cementations for CAD/CAM Resin Block after Alumina Air Abrasion on Adhesion to Newly Developed Resin Cement. *Materials* **2021**, *14*, 7058. [CrossRef]
10. Nica, I.; Stoleriu, S.; Iovan, A.; Tărăboanță, I.; Pancu, G.; Tofan, N.; Brânzan, R.; Andrian, S. Conventional and Resin-Modified Glass Ionomer Cement Surface Characteristics after Acidic Challenges. *Biomedicines* **2022**, *10*, 1755. [CrossRef]
11. Kaga, N.; Nagano-Takebe, F.; Nezu, T.; Matsuura, T.; Endo, K.; Kaga, M. Protective Effects of GIC and S-PRG Filler Restoratives on Demineralization of Bovine Enamel in Lactic Acid Solution. *Materials* **2020**, *13*, 2140. [CrossRef]

12. Bilgrami, A.; Alam, M.K.; Qazi, F.u.R.; Maqsood, A.; Basha, S.; Ahmed, N.; Syed, K.A.; Mustafa, M.; Shrivastava, D.; Nagarajappa, A.K.; et al. An In-Vitro Evaluation of Microleakage in Resin-Based Restorative Materials at Different Time Intervals. *Polymers* **2022**, *14*, 466. [CrossRef]
13. Vicente, A.; Rodríguez-Lozano, F.J.; Martínez-Beneyto, Y.; Jaimez, M.; Guerrero-Gironés, J.; Ortiz-Ruiz, A.J. Biophysical and Fluoride Release Properties of a Resin Modified Glass Ionomer Cement Enriched with Bioactive Glasses. *Symmetry* **2021**, *13*, 494. [CrossRef]
14. Thepveera, W.; Potiprapanpong, W.; Toneluck, A.; Channasanon, S.; Khamsuk, C.; Monmaturapoj, N.; Tanodekaew, S.; Panpisut, P. Rheological Properties, Surface Microhardness, and Dentin Shear Bond Strength of Resin-Modified Glass Ionomer Cements Containing Methacrylate-Functionalized Polyacids and Spherical Pre-Reacted Glass Fillers. *J. Funct. Biomater.* **2021**, *12*, 42. [CrossRef]
15. Sokolowski, K.; Szczesio-Wlodarczyk, A.; Bociong, K.; Krasowski, M.; Fronczek-Wojciechowska, M.; Domarecka, M.; Sokolowski, J.; Lukomska-Szymanska, M. Contraction and Hydroscopic Expansion Stress of Dental Ion-Releasing Polymeric Materials. *Polymers* **2018**, *10*, 1093. [CrossRef]
16. Gorseta, K.; Borzabadi-Farahani, A.; Vrazic, T.; Glavina, D. An In-Vitro Analysis of Microleakage of Self-Adhesive Fissure Sealant vs. Conventional and GIC Fissure Sealants. *Dent. J.* **2019**, *7*, 32. [CrossRef]
17. Sajadi, S.S.; EslamiAmirabadi, G.; Sajadi, S. Effects of two soft drinks on shear bond strength and adhesive remnant index of orthodontic metal brackets. *J. Dent.* **2014**, *11*, 389–397.
18. Santos, C.N.; Souza Matos, F.; Mello Rode, S.; Cesar, P.F.; Nahsan, F.P.S.; Paranhos, L.R. Effect of two erosive protocols using acidic beverages on the shear bond strength of orthodontic brackets to bovine enamel. *Dental Press J. Orthod.* **2018**, *23*, 64–72. [CrossRef]
19. Panpan, L.; Chungik, O.; Hongjun, K.; Chen-Glasser, M.; Park, G.; Jetybayeva, A.; Yeom, J.; Kim, H.; Ryu, J.; Hong, S. Nanoscale effects of beverages on enamel surface of human teeth: An atomic force microscopy study. *J. Mech. Behav. Biomed. Mater.* **2020**, *110*, 103930. [CrossRef]
20. Torres-Gallegos, I.; Zavala-Alonso, V.; Patino-Marin, N.; Martinez-Castanon, G.A.; Anusavice, K.; Loyola-Rodriguez, J.P. Enamel roughness and depth profile after phosphoric acid etching of healthy and fluorotic enamel. *Aust. Dent. J.* **2012**, *57*, 151–156. [CrossRef]
21. Pulgaonkar, R.; Chitra, P. Stereomicroscopic analysis of microleakage, evaluation of shear bond strengths and adhesive remnants beneath orthodontic brackets under cyclic exposure to commonly consumed commercial "soft" drinks". *Indian J. Dent. Res.* **2021**, *32*, 98–103. [CrossRef]
22. Chanachai, S.; Chaichana, W.; Insee, K.; Benjakul, S.; Aupaphong, V.; Panpisut, P. Physical/Mechanical and Antibacterial Properties of Orthodontic Adhesives Containing Calcium Phosphate and Nisin. *J. Funct. Biomater.* **2021**, *12*, 73. [CrossRef]
23. Hasan, L.A. Evaluation the properties of orthodontic adhesive incorporated with nano-hydroxyapatite particles. *Saudi Dent. J.* **2021**, *33*, 1190–1196. [CrossRef]
24. Voina, C.; Delean, A.; Muresan, A.; Valeanu, M.; Mazilu Moldovan, A.; Popescu, V.; Petean, I.; Ene, R.; Moldovan, M.; Pandrea, S. Antimicrobial activity and the effect of green tea experimental gels on teeth surfaces. *Coatings* **2020**, *10*, 537. [CrossRef]
25. Esberg, A.; Johansson, L.; Berglin, E.; Mohammad, A.J.; Jonsson, A.P.; Dahlqvist, J.; Stegmayr, B.; Johansson, I.; Rantapää-Dahlqvist, S. Oral Microbiota Profile in Patients with Anti-Neutrophil Cytoplasmic Antibody–Associated Vasculitis. *Microorganisms* **2022**, *10*, 1572. [CrossRef]
26. Kunrath, M.F.; Dahlin, C. The Impact of Early Saliva Interaction on Dental Implants and Biomaterials for Oral Regeneration: An Overview. *Int. J. Mol. Sci.* **2022**, *23*, 2024. [CrossRef]
27. Daniele, V.; Macera, L.; Taglieri, G.; Di Giambattista, A.; Spagnoli, G.; Massaria, A.; Messori, M.; Quagliarini, E.; Chiappini, G.; Campanella, V.; et al. Thermoplastic Disks Used for Commercial Orthodontic Aligners: Complete Physicochemical and Mechanical Characterization. *Materials* **2020**, *13*, 2386. [CrossRef]
28. Sarosi, C.; Moldovan, M.; Soanca, A.; Roman, A.; Gherman, T.; Trifoi, A.; Chisnoiu, A.M.; Cuc, S.; Filip, M.; Gheorghe, G.F.; et al. Effects of Monomer Composition of Urethane Methacrylate Based Resins on the C=C Degree of Conversion, Residual Monomer Content and Mechanical Properties. *Polymers* **2021**, *13*, 4415. [CrossRef]
29. Pastrav, M.; Chisnoiu, A.M.; Pastrav, O.; Sarosi, C.; Pordan, D.; Petean, I.; Muntean, A.; Moldovan, M.; Chisnoiu, R.M. Surface Characteristics, Fluoride Release and Bond Strength Evaluation of Four Orthodontic Adhesives. *Materials* **2021**, *14*, 3578. [CrossRef]
30. Mishnaevvsky, J.L. Nanostructured interfaces for enhancing mechanical properties of composites: Computational micromechanical studies. *Compos. Part B Eng.* **2015**, *68*, 75–84. [CrossRef]
31. Ghoubril, V.; Ghoubril, J.; Khoury, E. A comparison between RMGIC and composite with acid-etch preparation or hypochlorite on the adhesion of a premolar metal bracket by testing SBS and ARI: In vitro study. *Int. Orthod.* **2020**, *18*, 127–136. [CrossRef]
32. Najeeb, S.; Khurshid, Z.; Zafar, M.S.; Khan, A.S.; Zohaib, S.; Martí, J.M.N.; Sauro, S.; Matinlinna, J.P.; Rehman, I.U. Modifications in Glass Ionomer Cements: Nano-Sized Fillers and Bioactive Nanoceramics. *Int. J. Mol. Sci.* **2016**, *17*, 1134. [CrossRef]
33. Zheng, B.-W.; Cao, S.; Ali-Somairi, M.A.; He, J.; Liu, Y. Effect of enamel-surface modifications on shear bond strength using different adhesive materials. *BMC Oral Health* **2022**, *22*, 224. [CrossRef]
34. Oberhoffer, F.S.; Li, P.; Jakob, A.; Dalla-Pozza, R.; Haas, N.A.; Mandilaras, G. Energy Drinks Decrease Left Ventricular Efficiency in Healthy Children and Teenagers: A Randomized Trial. *Sensors* **2022**, *22*, 7209. [CrossRef]

35. Caixeta, R.V.; Berger, S.B.; Lopes, M.B.; Paloco, E.A.C.; Faria-Junior, E.M.; Contreras, E.F.R.; Gonini-Junior, A.; Guiraldo, R.D. Evaluation of enamel roughness after the removal of brackets bonded with different materials: In vivo study. *Braz. Dent. J.* **2021**, *32*, 30–40. [CrossRef]
36. Kessler, P.; Turp, J.P. Influence of Coca-Cola on orthodontic material. *Swiss Dent. J. SSO* **2020**, *130*, 777–780.
37. Oncag, G.; Tuncer, A.V.; Tosun, Y.S. Acidic Soft Drinks Effects on the Shear Bond Strength of Orthodontic Brackets and a Scanning Electron Microscopy Evaluation of the Enamel. *Angle Orthod.* **2005**, *75*, 247–253. [CrossRef]
38. Tisler, C.E.; Moldovan, M.; Petean, I.; Buduru, S.D.; Prodan, D.; Sarosi, C.; Leucuța, D.-C.; Chifor, R.; Badea, M.E.; Ene, R. Human Enamel Fluorination Enhancement by Photodynamic Laser Treatment. *Polymers* **2022**, *14*, 2969. [CrossRef]
39. Šimunović, L.; Blagec, T.; Vrankić, A.; Meštrović, S. Color Stability of Orthodontic Ceramic Brackets and Adhesives in Potentially Staining Beverages—In Vitro Study. *Dent. J.* **2022**, *10*, 115. [CrossRef]
40. Goracci, C.; Di Bello, G.; Franchi, L.; Louca, C.; Juloski, J.; Juloski, J.; Vichi, A. Bracket Bonding to All-Ceramic Materials with Universal Adhesives. *Materials* **2022**, *15*, 1245. [CrossRef]

Article

Structural Aspects and Intermolecular Energy for Some Short Testosterone Esters

Alexandru Turza [1], Violeta Popescu [2], Liviu Mare [2] and Gheorghe Borodi [3,*]

[1] Mass Spectrometry, Chromatography and Applied Physics Department, National Institute for R&D of Isotopic and Molecular Technologies, 67-103 Donat, 400293 Cluj-Napoca, Romania
[2] Physics & Chemistry Department, Technical University of Cluj-Napoca, 28 Memorandumului Str., 400114 Cluj-Napoca, Romania
[3] Molecular and Biomolecular Physics Department, National Institute for R&D of Isotopic and Molecular Technologies, 67-103 Donat, 400293 Cluj-Napoca, Romania
* Correspondence: borodi@itim-cj.ro

Abstract: Testosterone (17β-hydroxyandrost-4-en-3-one) is the primary naturally occurring anabolic–androgenic steroid. The crystal structures of three short esterified forms of testosterone, including propionate, phenylpropionate, and isocaproate ester, were determined via single-crystal X-ray diffraction. Furthermore, all the samples were investigated using powder X-ray diffraction, and their structural features were described and evaluated in terms of crystal energies and Hirshfeld surfaces. They were also compared with the base form of testosterone (without ester) and the acetate ester. Moreover, from a pharmaceutical perspective, their solubility was evaluated and correlated with the length of the ester.

Keywords: 17β-hydroxyandrost-4-en-3-one; testosterone; ester; crystal structure; lattice energy; solubility

1. Introduction

Testosterone (17β-hydroxyandrost-4-en-3-one) is a cholesterol derivative and a naturally occurring anabolic steroid. It can be viewed as a derivative of the androstane group and the primary male sex hormone. It plays a major role in the development of male reproductive tissues and the maintenance of secondary male characteristics [1]. Testosterone has been shown to impact overall health and well-being [2] and prevent osteoporosis [3]. By binding to the androgen receptor, it exerts anabolic and androgenic properties that are the specific common characteristic of all derivatives belonging to this class [4]. In a medical context, testosterone is used to relieve symptoms of low testosterone in men (male hypogonadism) and breast cancer in women, as well as for hormone therapy in transgender men [5]. Testosterone targets androgen receptors [6], and previous studies have shown that normal testosterone levels in older men have an overall positive impact on health, decreasing body and visceral fat, increasing lean body mass, and improving cholesterol panel and carbohydrate metabolism [7]. Since it is an anabolic–androgenic steroid, testosterone is often used by athletes to increase performance [8]. Furthermore, medically, it can be used to relieve or treat protein degradation in certain catabolic states [9].

It is known that the testosterone base (Figure 1) has a short half-life of roughly a few hours; thus, it is often subjected to esterification in order to increase the half-life by intramuscular injections and avoid daily administration [10]. The esterified forms of testosterone possess a half-life ranging from around less than 1 day for testosterone acetate, 1 day for propionate, 2.5 days for phenylpropionate, and up to 3.1 days for testosterone isocaproate [11]. In this regard, the length of the ester can be correlated with the length of the carbon chain; thus, the longer the ester, the longer the half-life.

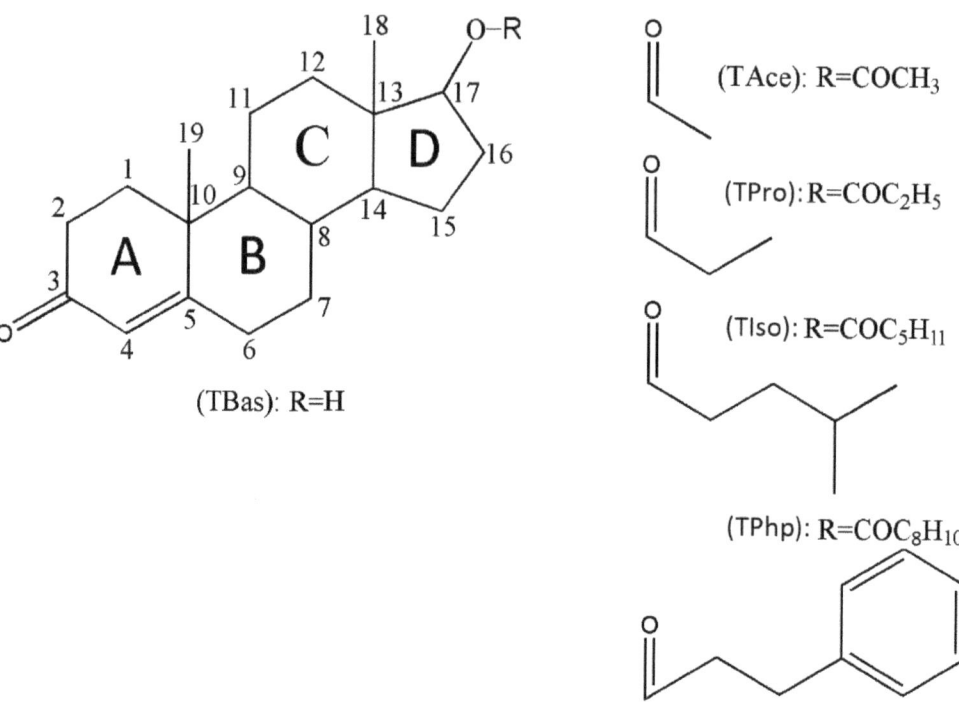

Figure 1. Chemical structures of 17β-hydroxyandrost-4-en-3-one (testosterone) displaying the steroid backbone labelling system and other studied testosterone-based steroids.

The scheme of atoms and the labelling of steroid skeleton rings was made according to the established notations for the compounds belonging to this group [12] (Figure 1).

The current study aimed to investigate the characterisation of four testosterone pro-drugs as follows:

(i) Testosterone acetate (Androst-4-en-17β-ol-3-one 17β-acetate; TAce);
(ii) Testosterone propionate (Androst-4-en-17β-ol-3-one 17β-propionate; TPro);
(iii) Testosterone isocaproate (Androst-4-en-17β-ol-3-one 17β-4-methylpentanoate; TIso);
(iv) Testosterone phenylpropionate (Androst-4-en-17β-ol-3-one 17β-phenylpropionate; TPhp).

For the last three, the crystal structures were determined and reported.

The literature reports the crystal structure of a form of testosterone propionate that has a slightly different unit cell and does not have the positions of the hydrogen atoms reported [13], and another paper presents only the approximate parameters of the unit cell [14]. Other testosterone-based crystal structures that have been reported are testosterone buciclate [15], testosterone acetate [16], and testosterone base (the basic form without ester attached) [17]. This manuscript is aimed to investigate the structural features of these testosterone esters.

A complete structural characterisation was carried out by means of X-ray single-crystal diffraction (for TPro, TPhp, TIso), X-ray powder diffraction, and the conformational analysis of steroid rings, and a quantitative measure of intermolecular interactions was accomplished by computation of lattice energies using the Coulomb–London–Pauli model corroborated with Hirshfeld surface analysis. Furthermore, computational methods were also applied to the testosterone base in order to be compared with the four esters.

As many pharmaceutical compounds, including various esterified forms of steroids, are labelled as poorly water-soluble but lipophilic agents, they might be dissolved in lipid-based preparations [18]. Based on this, the solubility in solutions of various organic oils was measured. The formulations of various drugs are currently available on the market, and all use oils as vehicles for certain compounds, including deoxycorticosterone, progesterone/oestradiol esters, testosterone esters with their analogues/derivatives, and vitamins such as K and E as well [19,20].

2. Materials and Methods

2.1. Materials and Crystallisation Experiments

Crystalline, white powders of esters for scientific research purposes, were received from Wuhan Shu Mai Technology Co., Wuhan, China and solvents from Merck, Taufkirchen, Germany.

All the investigated steroids were obtained at room temperature as white crystalline powders, which possess the possibility to be subjected to various recrystallisation methods. Suitable single crystals for X-ray data collection were successfully obtained in alcohols: methanol (TPro), ethanol (Tiso), and isopropyl alcohol (TPhp).

Oils meeting the requirements of United States Pharmacopeia were received from Sigma-Aldrich (Taufkirchen, Germany), Tex Lab supply (USA), and Med Lab supply (USA).

2.2. X-ray Powder Diffraction (XRPD)

The samples were scanned on a Bruker D8 Advance diffractometer (Karlsruhe, Germany) having the X-ray tube set at 40 kV and 40 mA. The diffractometer is equipped with a germanium (1 1 1) monochromator used to obtain only the desired CuKα1 radiation and an LYNXEYE position-sensitive detector. X-ray diffraction patterns were recorded in the 3–40° (2θ) range using the DIFFRAC plus XRD Commander program, employing a scanning speed of 0.02°/s.

2.3. Single-Crystal X-ray Diffraction and Structural Refinement

The experimental single-crystal X-ray diffraction intensities were collected using a SuperNova diffractometer (Rigaku, Tokyo, Japan) equipped with dual microsources (Mo and Cu) and the X-ray tube operating at 50 kV and 0.8 mA. Data collection and reduction, Lorentz, polarisation, and absorption effect corrections were all performed with CrysAlis PRO software (Yarnton, Oxfordshire, UK) [21]. A multi-scan method using spherical harmonics in the SCALE3 abspack algorithm was applied for the empirical absorption correction. The crystal structures were solved as follows: Tiso and TPhp were solved with SHELXT [22] program using intrinsic phasing, and TPro was solved with direct methods by SHELXS [23]. Steroid structures were further refined via least squares minimisation with SHELXL [24] refinement package, and all programs were incorporated into Olex2 software (Durham, UK) [25].

Hydrogen atoms were geometrically located, treated, and refined as riding atoms, with the isotropic displacement parameter Uiso(H) = 1.2 Ueq(C) for ternary CH groups (C-H = 0.93 Å), secondary CH_2 groups (C-H = 0.97 Å), and 1.5 Ueq(C) considered for all methyl CH_3 groups (C-H = 0.96 Å).

2.4. Crystal Lattice Energy Computation and Hirshfeld and Fingerprint Plot Analyses

Based on the positions of atoms in the unit cell (determined by single-crystal X-ray diffraction technique), the classical atom–atom potential was calculated using the Coulomb–London–Pauli (CLP) model (Milan, Italy) [26]. The method evaluates crystal energies that can be divided into three distinct attraction terms, namely Coulombic, polarisation, and dispersion energies, and a fourth term that represents the repulsive component.

Energy computation via the Coulomb–London–Pauli (CLP) approach involves pairs of individual atoms (i, j) that belong to different molecules and is the sum of four interaction terms according to Relation (1):

$$E_{ij} = 1/(4\pi\varepsilon_0)(q_i q_j) R_{ij}^{-1} - F_P P_{ij} R_{ij}^{-4} - F_D D_{ij} R_{ij}^{-6} + F_R T_{ij} R_{ij}^{-12} \tag{1}$$

$$q_i = F_Q q_i^0 \tag{2}$$

The Coulombic energy is the first term, polarisation energy is the second term, the dispersion term is the third one, and the last term is repulsion.

The F_Q, F_P, F_D, and F_R coefficients involved in Relations (1) and (2) are empirically disposable scaling parameters and the P_{ij}, D_{ij}, and T_{ij} coefficients depend on the local vicinity of the atom in the molecule.

The Coulombic component is treated according to Coulomb's law, the polarisation term is estimated in the approximation of the linear dipole, the dispersion energy is approximated as the inverse of the distance at the sixth power, and the repulsive term is due to the modulation of the overlapping wave function.

Molecular 3D Hirshfeld surfaces and their related 2D fingerprint plots were generated by CrystalExplorer software (Perth, Australia) [27] based on the d_{norm} function, which can be expressed in Relation (3).

$$d_{norm} = \frac{d_i - r_i^{vdW}}{r_i^{vdW}} + \frac{d_e - r_e^{vdW}}{r_e^{vdW}} \tag{3}$$

where d_e is the distance from the surface to the nearest external nucleus, while d_i represents the distance from the surface to the nearest nucleus inside the surface. The fingerprint plots are a 2D diagram, where d_i and d_e are represented in order to identify the nature and types of different intermolecular contacts [28].

2.5. Solubility Check

The solubility for the four esters (mg/mL) was measured in solutions of various organic oils: medium-chain triglyceride (MCT), grape seed oil (GSO), castor oil, cottonseed oil, apricot oil, and sesame oil.

Each solution was composed of a mixture of benzyl benzoate, benzyl alcohol, and oil having a volumetric ratio of 78% oil, 20% benzyl benzoate, and 2% benzyl alcohol. In various pharmaceutical preparations of lipophilic compounds, including various steroids, benzyl benzoate is used as a solubiliser (co-solvent), benzyl alcohol acts as a solvent and at the same time prevents microbial growth and increases the lipid solubility of various esterified compounds, while the oils are used as carriers.

The solubility evaluation was performed in multiple steps at room temperature (25 °C) by successively adding small amounts of raw materials (2–5 mg each step), and the solution was stirred for up to several hours until dissolved. When it was found that excess raw material remained (in suspension), small amounts of solution (mixture of benzyl benzoate, benzyl alcohol, and oil) were added until the resulting solution became perfectly transparent and clear. In order to obtain good accuracy, three such procedures were carried out, and their average was used.

3. Results and Discussion

3.1. Crystal Structures and Supramolecular Descriptions

The good agreement generated between the experimental powder X-ray diffraction patterns and the simulated patterns based on the positions of the atoms in the unit cell shows a good structural homogeneity and that the studied single crystals are representative of the entire bulk of samples (see Figure S1, Supplementary Materials).

The details with regard to single-crystal data and refinement for the studied esters are given in Table 1.

Table 1. Crystal structures and refinement data of investigated esters.

Identification Code	TPro (Testosterone Propionate)	TIso (Testosterone Isocaproate)	TPhp (Testosterone Phenylpropionate)
Empirical formula	$C_{22}H_{32}O_3$	$C_{25}H_{38}O_3$	$C_{28}H_{36}O_3$
Formula weight	344.47	386.57	420.57
Temperature/K	293(2)	293(2)	293(2)
Crystal system	orthorhombic	monoclinic	monoclinic
Space group	$P2_12_12_1$	$P2_1$	$P2_1$
a/Å	7.57470(16)	7.2877(3)	13.4097(7)
b/Å	12.6768(2)	12.3741(5)	5.9105(3)
c/Å	20.4038(4)	13.1272(6)	15.4054(8)
$\alpha/°$	90	90	90
$\beta/°$	90	103.305(4)	95.073(5)
$\gamma/°$	90	90	90
Volume/Å^3	1959.23(6)	1152.02(9)	1216.22(11)
Z	4	2	2
$\rho calc\, g/cm^3$	1.168	1.106	1.148
μ/mm^{-1}	0.594	0.552	0.568
F(000)	752.0	418.0	456.0
Crystal size/mm	0.09 × 0.08 × 0.07	0.1 × 0.03 × 0.01	0.09 × 0.09 × 0.01
Radiation	CuKα (λ = 1.54184)	CuKα (λ = 1.54184)	CuKα (λ = 1.54184)
2Θ range/°	8.212 to 141.254	9.952 to 141.104	5.76 to 141.522
Index ranges	$-8 \leq h \leq 9, -15 \leq k \leq 15, -24 \leq l \leq 24$	$-8 \leq h \leq 8, -15 \leq k \leq 15, -15 \leq l \leq 16$	$-16 \leq h \leq 16, -7 \leq k \leq 7, -18 \leq l \leq 18$
Reflections collected	28,102	15,810	14,176
Independent reflections	3723 [R_{int} = 0.0243, R_{sigma} = 0.0119]	4322 [R_{int} = 0.0246, R_{sigma} = 0.0185]	4529 [R_{int} = 0.0796, R_{sigma} = 0.0508]
Data/restraints/parameters	3723/0/229	4322/1/257	4529/1/282
Goodness-of-fit on F2	1.044	1.076	1.078
Final R indexes [I ≥ 2σ (I)]	R_1 = 0.0421, wR_2 = 0.1187	R_1 = 0.0621, wR_2 = 0.1739	R_1 = 0.0762, wR_2 = 0.1784
Final R indexes [all data]	R_1 = 0.0440, wR_2 = 0.1217	R_1 = 0.0740, wR_2 = 0.1900	R_1 = 0.1102, wR_2 = 0.1959
Largest diff. peak/hole/e Å^{-3}	0.17/−0.17	0.33/−0.21	0.25/−0.21
Flack parameter	0.04(6)	0.06(8)	−0.4(3)

The aim was to determine the absolute configurations for each of the testosterone esters investigated. The values of Flack parameters of 004(6) for TPro and 0.06(8) for TIso confirm the correctness of the absolute configurations; on the other hand, for TPhp, the negative value of the Flack parameter shows that this parameter has no meaning.

3.1.1. TAce (Testosterone Acetate)

The acetate ester is the shortest esterified steroid ester available and is also the shortest testosterone ester available. The CSD database contains one entry reporting only the cell parameters for this particular testosterone ester [16] and one entry reporting the unit cell parameters and atomic coordinates [14].

The acetate ester was found to crystallise in the noncentrosymmetric orthorhombic $P2_12_12_1$ space group with one molecule in the asymmetric unit (Figure S2a, Supplementary Materials) and four in the unit cell. The carbonyl O1 oxygen of the ketone group is involved in C-H•••O bifurcated hydrogen bonds and contributes to crystal stability. One is made with a neighbouring five-membered ring (C15-H15A•••O3) and one towards the CH_3 of terminal methyl in the acetate group (Figure S2b, Supplementary Materials). Hydrogen bonding distances are presented in Table S1 (Supplementary Materials).

3.1.2. TPro (Testosterone Propionate)

Compared with testosterone acetate, the propionate ester is characterised by an ester chain with an extra carbon atom. The crystal structure of propionate ester was previously reported [13] but has slightly smaller unit cell parameters and lacks hydrogen atoms. Similar to TAce, it crystallises in the orthorhombic $P2_12_12_1$ space group, with one molecule in the asymmetric unit (Figure 2a), the unit cell hosting four such molecules. Considering the H atoms located in idealised positions via X-ray crystallography, apparently, it seems that only the C6-H6B•••O1 interaction between the O1 ketone group and the neighbouring D ring in the structure has a separating distance shorter than the sum of van der Waals radii and plays a role in stability. In reality, having the H atoms normalised (the C-H distance is 1.089 Å), the ketone O1 oxygen is involved in bifurcated C-H•••O interactions, with the second being C16-H16A•••O1 with d(H•••O) = 2.6185 Å. This distance is close to the sum of the van der Waals radii with 1.20 Å for hydrogen and 1.52 Å for oxygen [29]. An overall packing perspective is shown in Figure 2b.

3.1.3. TIso (Testosterone Isocaproate)

The asymmetric unit (Figure 3a) consists of only one steroid molecule and was found to crystallise in the noncentrosymmetric $P2_1$ monoclinic space group. Ketone O1 participates in the formation of supramolecular self-assemblies, being involved in the trifurcated C-H•••O hydrogen bonding. One interaction is formed between a neighbouring six-membered B ring (C6-H6B•••O1), one binds the O1 carbonyl oxygen with a five-membered ring (C16-H16B•••O1), while the last bridges the methyl CH_3 group (C19-H19A•••O1); all these interactions are detailed in Table S1 (Supplementary Materials). An overall packing diagram shows the self-arrangements of steroid molecules in layers (Figure 3b).

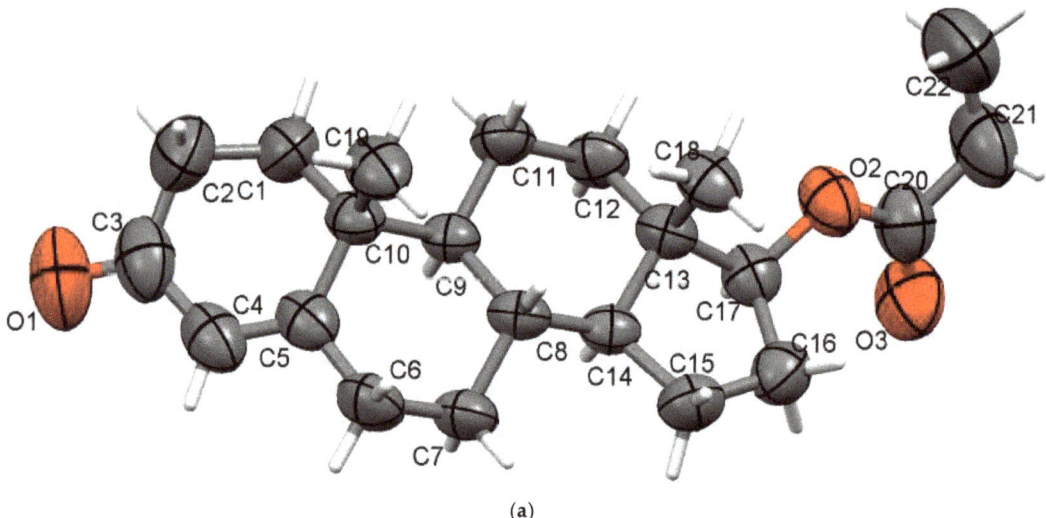

(a)

Figure 2. Cont.

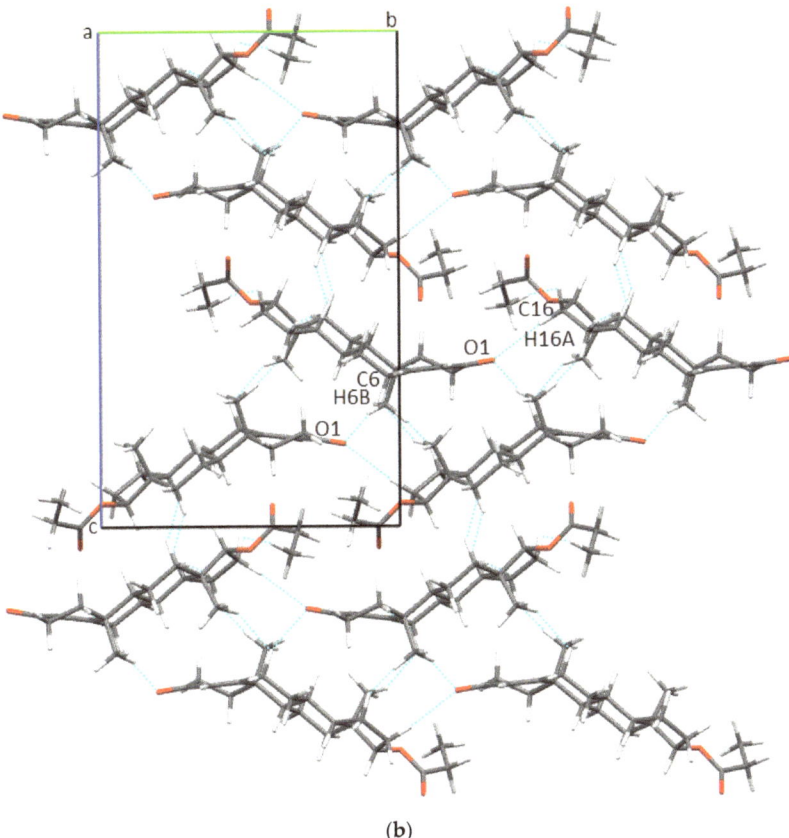

(b)

Figure 2. Asymmetric unit of TPro presenting nonhydrogen atoms at 50% probability level (**a**) and overall packing diagram along a-axis (**b**).

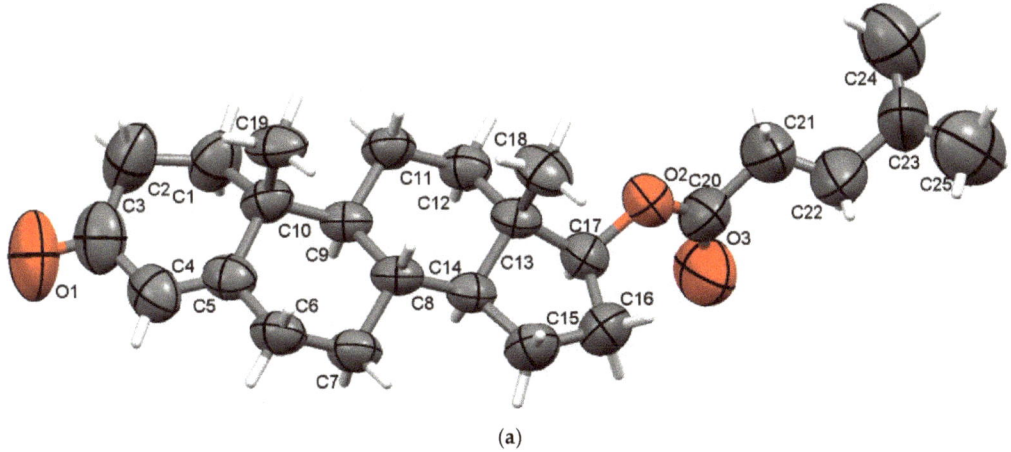

(a)

Figure 3. *Cont.*

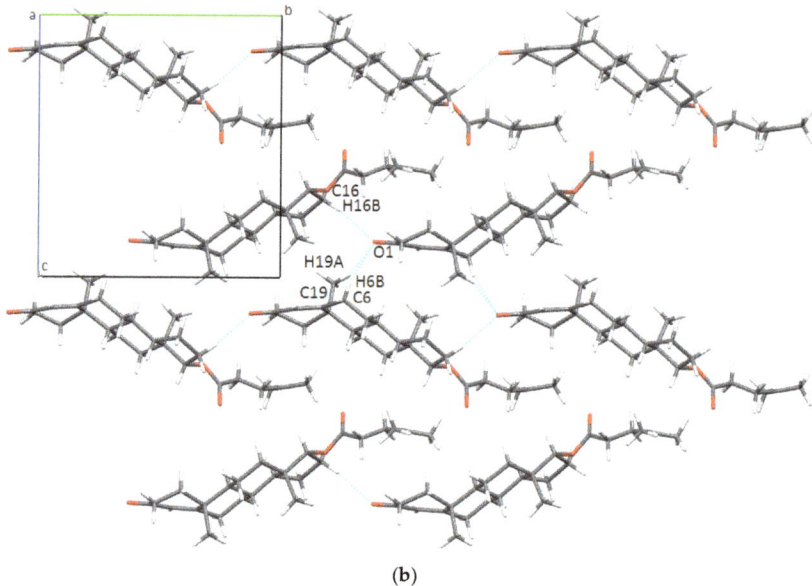

(b)

Figure 3. Asymmetric unit of TIso presenting nonhydrogen atoms at 50% probability level (**a**) and overall packing diagram along a-axis (**b**).

3.1.4. TPhp (Testosterone Phenylpropionate)

It was found that the steroid crystallises in the monoclinic P2$_1$ space group with one molecule in the asymmetric unit (Figure 4a) and two in the unit cell. Similar to testosterone propionate, considering the H positions determined via X-ray diffraction, it seems only one C-H•••O interaction exists in the crystal lattice (C24-H24•••O1 between the terminal phenyl ring and the ketone O1 oxygen, Table S1) that is shorter than the sum of van der Waals radii. After the normalisation of the C-H distances, there is C2-H2A•••O3 interaction between ring A and carbonyl O3 oxygen, which has d(H•••O) = 2.714 Å, situated just at the limit distance of 2.72 Å [29]. The overall packing perspective of testosterone phenylpropionate is presented along the ob-axis (Figure 4b).

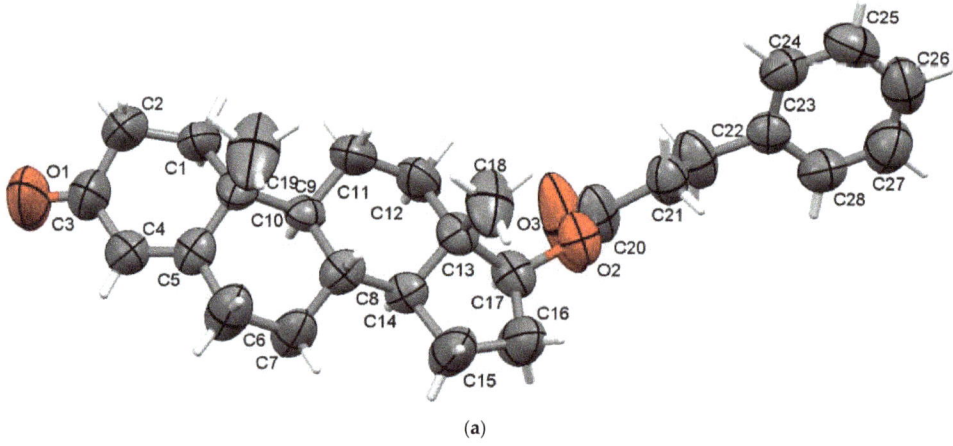

(a)

Figure 4. *Cont.*

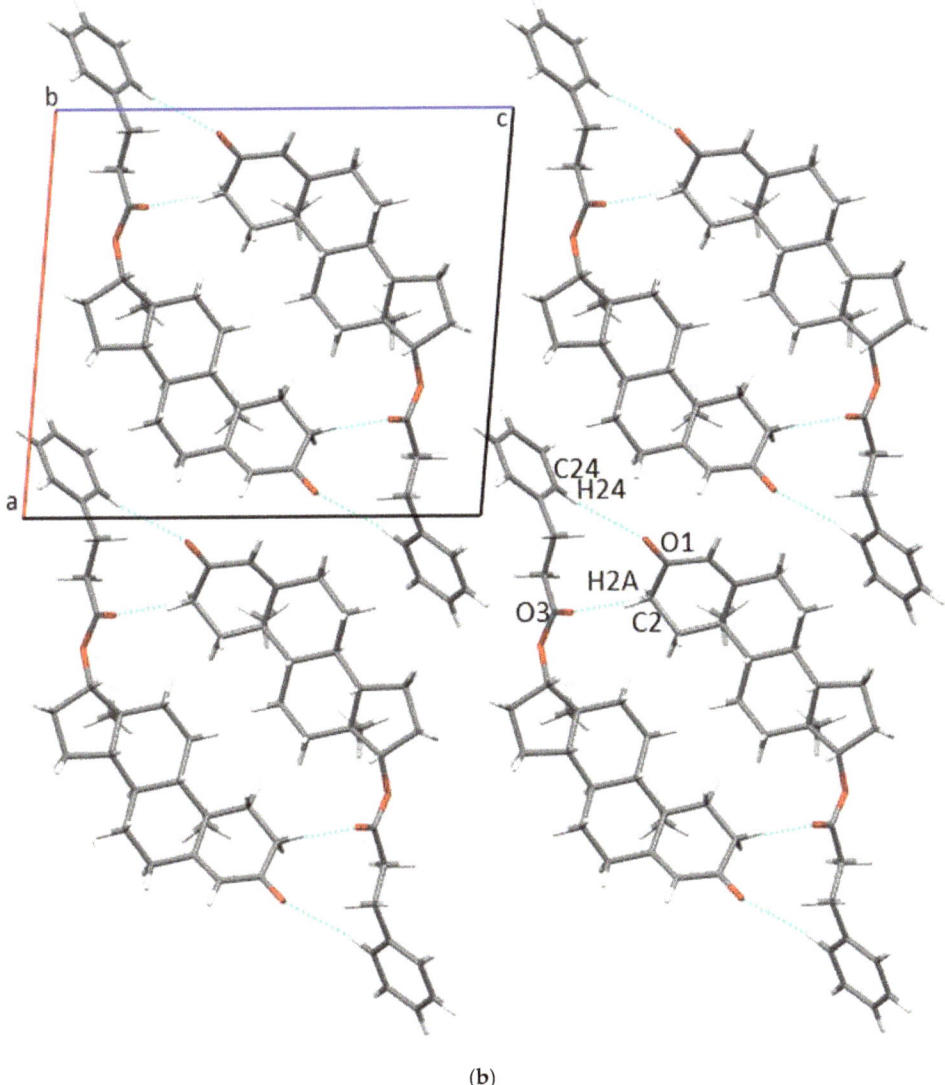

(b)

Figure 4. Asymmetric unit of TPhp presenting nonhydrogen atoms at 50% probability level (**a**) and overall packing diagram along b-axis (**b**).

From the analysis of the crystal structures, the following conclusions can be summarised:
(i) Asymmetric units are characterised by a single molecule for each ester;
(ii) The formations of supramolecular 3D assemblies are to some extent driven by the C-H•••O interactions, although the dispersion energy has the greatest weight, as will be shown in the crystal energies analysis section; donor–acceptor separation distances show similar values to those of other crystals driven by C-H•••O interactions and belong to the steroid family [30–34];
(iii) The six-membered A rings are found in the intermediate sofa-half-chair geometry, and the B and C rings show chair conformations, while the five-membered D rings adopt

intermediate envelope-half-chair geometry. Similar geometries of skeleton rings have been reported in the crystal structure of its C-17 methylated form [35].

In Figure 5, the overlap of the molecular structures is exemplified. For example, (TPro, Tphp), which differ only by the extra phenyl ring of TPhp, show a totally different orientation of the tails, as in the case of the (TAce, TIso) pair as well. It can be seen that the part of the molecules representing testosterone, the base of the ester structures, overlaps very well in all pairs. Instead, there are differences in the orientation of carbon tails. Thus, for the pair (TPro, TPhp), the C17-O2-C20-O3 torsion angle is 2.89° for TPro and −3.78° for TPhp. The angle between the planes defined by the O3-C20-O2 atoms for the pair (TPro, TPhp) is 64.93°. For the pair (TAce, TIso), the C17-O2-C20-O3 torsion angle is 3.73° for TAce and 4.93° for TIso, and the angle between the planes defined by the O3-C20-O2 atoms for the two structures is 75.03°.

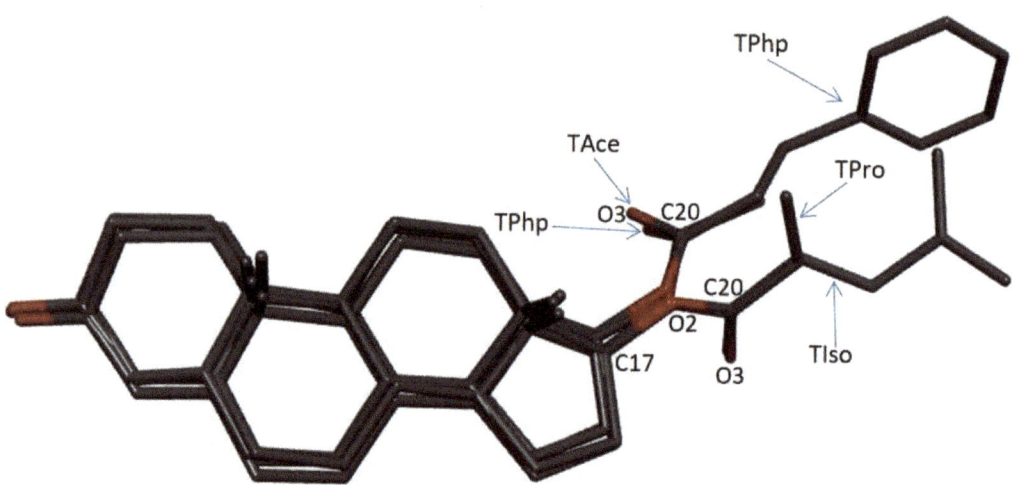

Figure 5. Molecular overlaps of the investigated esters.

The overlap of (TAce, TPhp) and (TPro, TIso) pairs show a better match; thus, the angle between the O3-C20-O2 planes is 25.35° for (TAce, TPhp) and 11.22° for (TPro, TIso).

3.2. Crystal Energy Analysis

The total crystal lattice energies, as well as the nature and magnitudes of the individual four energy components, were computed using the atom–atom Coulomb–London Pauli model, and the results are shown in Table 2. Moreover, the energies of the four testosterone esters were compared with those of the testosterone base [17] deposited in CSD (denoted Tbas), which is not esterified.

All four steroid structures are characterised by large values of dispersion energies, and this component is dominant. As a general trend, the dispersion energies were found to be more significant, as the ester chain is longer; thus, TAce, which is the shorter ester, has a value of −126.1 kJ/mol, whereas TPhp, which represents the longest ester, is −149.0 kJ/mol.

Due to the fact that the structures lack strong hydrogen bonds, the Coulombic energy contributes the least in the crystal packings, with similar values through all the crystals, which are in the range of −15.5 kJ/mol for TAce and −21.3 kJ/mol for TIso. Carbonyl•••hydrogen interactions were found in the Coulombic term and for the derivatives under study, the Coulombic energy has a weight between 9.7% (for TAce) and 12.6% (for TIso).

Table 2. Crystal energies of base form and its four esters.

Structure	Molar Mass g/mol	E_{coul} (kJ/mol)	E_{pol} (kJ/mol)	E_{disp} (kJ/mol)	E_{rep} (kJ/mol)	E_{latt} (kJ/mol)
TBas	288.43	−33.3	−47.5	−130.9	60.9	−150.8
TAce	330.46	−15.5	−55.0	−126.1	37.1	−159.5
TPro	344.49	−18.3	−55.5	−126.4	33.9	−166.3
TIso	386.57	−21.3	−57.7	−141.9	52.7	−168.2
TPhp	420.59	−19.3	−52.3	−149.0	36.1	−184.5

E_{coul}: the Coulombic term; E_{pol}: the polarisation term; E_{disp}: the dispersion term; E_{rep}: the repulsive term; E_{latt}: total crystal lattice energy.

By contrast, the Coulombic component (−33.3 kJ/mol) in the testosterone base (without ester), which presents strong classical O-H•••O hydrogen bonds, contributes more to the lattice energy.

Polarisation and repulsive components do not display a particular trend, but on the other hand, the total lattice energy becomes lower as the steroid molecular mass increases. The shortest ester (TAce) has a total lattice energy of −159.5 kJ/mol and TPhp, which is the longest ester, has an energy of −184.5kJ/mol and the most bound structure of all five steroids.

Similar crystal lattice behaviour in the sense that the dispersion term dominates the crystals, and the total energy becomes greater with the increase in ester length has been previously reported in other anabolic–androgenic agents from the steroid group [36–39].

3.3. Hirshfeld and Fingerprint Plot Analysis

The molecular 3D Hirshfeld surfaces of the studied esters (Figure S3b, Supplementary Materials) were generated based on d_{norm} and were compared with those of the base form (TBas) (Figure S3a, Supplementary Materials). As the asymmetric unit of TBas is characterised by two individual molecules, they were analysed separately.

The surfaces are interactively illustrated with arrows for the intermolecular C-H•••O and O-H•••O contacts with distances shorter than the sum of the van der Waals radii, which are listed in Table S1 (Supplementary Materials).

The surfaces can be understood by colour code; specifically, the red areas illustrate the intermolecular contacts with distances smaller than the sum of the van der Waals radii, the white indicates separation distances approximately equal to vdW radii, and the blue areas show the contacts with longer distances.

For each crystal, a 2D fingerprint plot (Figure 6) is generated, which is a transposition of the 3D Hirshfeld surface. The fingerprint plot of the base form (TBas) shows two fingerprint diagrams.

The analyses of 3D Hirshfeld surfaces, their related 2D fingerprint plots shapes, and (d_e and d_i) distances summarise the following structural features:

(i) Fingerprint plots of esterified forms (Figure 6b) (TAce, TPro, TPhp, and TIso) show symmetry in the spikes, which is a particular feature for the crystals with one molecule in asymmetric units, while the plots of TBas (Figure 6a) are asymmetric due to the different molecular environment in the crystal;

(ii) The diagrams of Tace, TPro, and TIso illustrate protruding H•••O/O•••H spikes, denoting the presence of C-H•••O hydrogen bonds, while for TPhp, the lack of H•••O/O•••H spikes shows that the separation distances of the C-H•••O interactions fall in a range closer to the sum of vdW radii;

(iii) The fingerprint plots of Tbas show more protruding H•••O/O•••H spikes compared with its esterified forms and suggest that strong O-H•••O interactions play more important roles in packing; this feature is seen in the evaluation of crystal energies where the Coulombic energy becomes more significant in TBas due to the presence of strong O-H•••O contacts;

(iv) The quantitative breakdown of fingerprint diagrams (Table 3) in all five crystals reveals a high percentage of H•••H contacts, medium contribution by O•••H/H•••O intercontacts and considerably smaller for C•••H/H•••C, respectively;
(v) The large percentages in H•••H contacts for all five structures (fingerprint plots breakdown in Table 3) corroborated with the crystal energies (Table 2) are suggesting that dispersion effects play the major role.

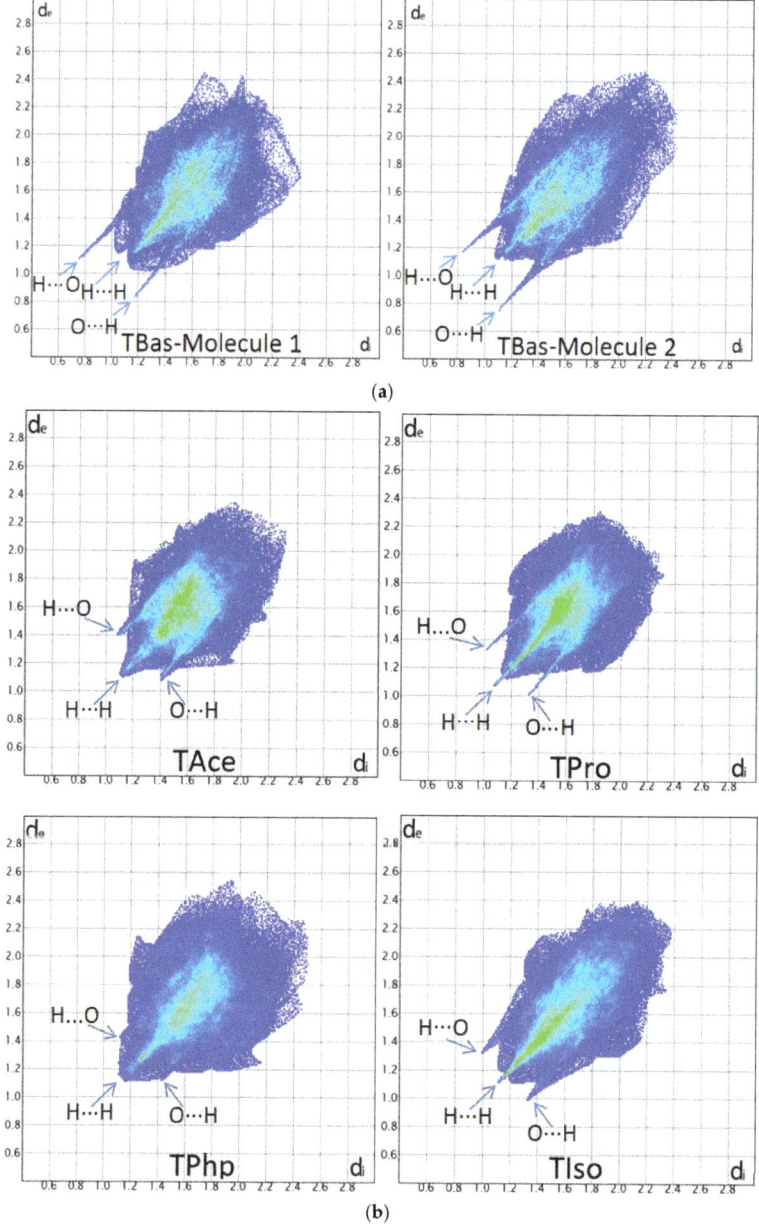

Figure 6. Fingerprint plots displaying close contacts in studied crystals: TBas (**a**) and studied esters (**b**).

Table 3. Contributions to the Hirshfeld surfaces for various intercontacts.

Structure	H•••H	O•••H/H•••O	C•••H/H•••C	C•••O/O•••C	O•••O	C•••C
TBas Mol A	78.8%	17.2%	4.0%	-	-	-
TBas Mol B	76.4%	19.7%	3.9%	-	-	-
TAce	75.4%	20.2%	2.8%	1.2%	-	0.4%
TPro	76.5%	19.4%	4.0%	-	0.1%	-
TIso	80.4%	16.3%	3.3%	-	-	-
TPhp	72.7%	15.2%	12.1%	-	-	-

3.4. Solubility Check

The values obtained by solubility evaluation are summarised in Table S2 (Supplementary Materials) and are graphically represented in Figure 7.

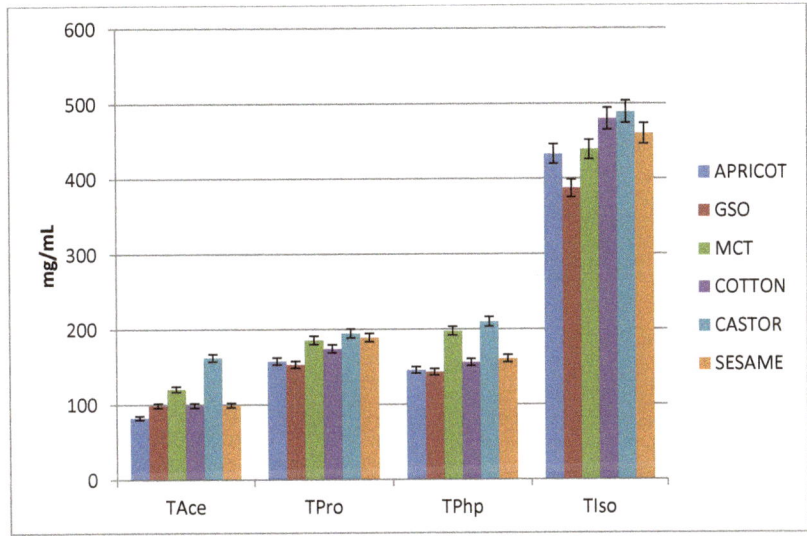

Figure 7. Graphical representations of ester solubility.

Depending on the length of the ester, it is observed that the shortest ester (the acetate) has the lowest solubility, while the longest ester (isocaproate) has a roughly four-fold greater solubility. Propionate and phenylpropionate esters have similar and slightly higher values than acetate. Although phenylpropionate has six more carbon atoms than propionate, the solubilities are similar, so it can be noted that what matters in increasing solubility is the length of the chain (the phenyl ring does not lead to an increase in solubility).

It is worth mentioning that solubility correlates with the half-life of the prodrug, so acetate, which has the shortest half-life, has the lowest solubility, and isocaproate, which has the longest half-life, has the highest solubility.

Out of the six mixtures analysed, it is observed that castor oil can support the highest solubility without crashing (crystallisation of compound) and to a lesser extent MCT (for TAce, TPro, and TPhp), at the same time behaving as solvents as well. This is interesting because MCT is characterised by the lowest value of the viscosity coefficient, while castor oil is the most viscous.

The literature reports the solubility of hydroxyprogesterone caproate polymorphs in castor oil at a temperature of 20 °C, which are 278 mg/mL and 301 mg/mL [40]. Another method used for the development of the preparations of drugs with poor water solubility (including various esterified forms of steroids) is by oil-in-water (o/w) microemulsions. For example, a previous study shows that a microemulsion based on soybean oil and dimethoxytetraethylene glycol supports concentrations of 3.42 mg/mL (for testosterone propionate), 31.5 mg/mL (for testosterone enanthate), and 2.16 mg/mL (for medroxyprogesterone acetate) in soybean oil. In dimethoxytetraethylene glycol, higher concentrations of 12 mg/mL (for testosterone propionate) 91.2 mg/mL (for testosterone enanthate) and 1.32 mg/mL (for medroxyprogesterone acetate) were obtained [41].

4. Conclusions

The crystal structures of three testosterone-based esters (propionate, phenylpropionate, and isocaproate) were determined; they belong to the noncentrosymmetric monoclinic $P2_1$ and orthorhombic $P2_12_12_1$ space groups.

These three esters were further analysed and compared with the acetate ester and with the nonesterified base form. Backbone steroid rings possess similar geometry in all compounds, the A steroid rings adopt intermediate sofa-half-chair conformations, and the B and C rings have chair conformations, while ring D depicts an intermediate envelope-half-chair conformation. By overlapping the esters, a very good match of the steroid skeleton rings (the base of the ester structures) emerges, and the major structural differences are manifested in the orientation of the tails.

Computational methods showed that in all crystal structures, the supramolecular arrangements and crystal stability are characterised and assured by dominant dispersion effects, and the total lattice energies are greater in absolute terms, as the ester chain is longer, while C-H•••O hydrogen bonds in all esters play a less important role.

The solubility of the four derivatives was tested to evaluate the changes based on the added ester functionalities, and it was found that the shortest ester (acetate) has the lowest solubility, while the longest ester (isocaproate) is roughly four times greater; meanwhile, propionate and phenylpropionate are between the two and show similar values.

Supplementary Materials: The following supporting information can be downloaded at: https://www.mdpi.com/article/10.3390/ma15207245/s1, Figure S1: Experimental and simulated XRPD patterns comparison: TAce (a), TPro (b), TIso (c), TPhp (d); Figure S2: Asymmetric unit of TAce presenting non-hydrogen atoms at 50% probability level (a) Overall packing diagram along a-axis (b); Figure S3: Views of the Hirshfeld surfaces mapped with d_{norm} ilustrating the intermolecular contacts referred in Table S1: TBas (a) and studied esters (b); Table S1: Hydrogen bond geometry for analyzed crystals (Å, °); Table S2: Solubility of analyzed testosterone esters in various oil solutions. CIF files of studied esters have been deposited with the Cambridge Crystallographic Data Centre with the deposition numbers as follows: 2192706 (TPro); 2192707 (TIso); 2192708 (TPhp). Copies of them can be obtained free of charge on written application to CCDC, 12 Union Road, Cambridge CB2 1EZ, UK (fax: +44-12-2333-6033); on request via e-mail to deposit@ccdc.cam.uk or by access to http://www.ccdc.cam.ac.uk (accessed on 9 September 2022).

Author Contributions: Conceptualisation, A.T. and G.B.; methodology, G.B.; software, A.T.; validation, G.B., L.M. and V.P.; formal analysis, L.M.; investigation, A.T., G.B., L.M. and V.P.; resources, A.T. and G.B.; data curation, G.B.; writing—original draft preparation, A.T., V.P. and L.M.; writing—review and editing, A.T.; visualisation, G.B.; supervision, V.P.; project administration, V.P.; funding acquisition, A.T. and G.B. All authors have read and agreed to the published version of the manuscript.

Funding: This research was funded by the Romanian Ministry of Research and Innovation, grant number 19 35 02 02.

Institutional Review Board Statement: Not applicable.

Informed Consent Statement: Not applicable.

Data Availability Statement: Data is contained within the article or supplementary material.

Conflicts of Interest: The authors declare no conflict of interest.

References

1. Mooradian, A.D.; Morley, J.E.; Korenman, S.G. Biological actions of androgens. *Endocr. Rev.* **1987**, *8*, 1–28. [CrossRef] [PubMed]
2. Bassil, N.; Alkaade, S.; Morley, J.E. The benefits and risks of testosterone replacement therapy: A review. *Ther. Clin. Risk Manag.* **2009**, *5*, 427–448. [CrossRef] [PubMed]
3. Tuck, S.P.; Francis, R.M. Testosterone, bone and osteoporosis. *Front. Horm Res.* **2009**, *37*, 123–132. [CrossRef] [PubMed]
4. Luetjens, C.M.; Weinbauer, G.F. *Chapter 2: Testosterone: Biosynthesis, Transport, Metabolism and (Non-Genomic) Actions*, 4th ed.; Cambridge University Press: Cambridge, UK, 2012; pp. 15–32.
5. Pappas, I.I.; Craig, W.Y.; Spratt, L.V.; Spratt, D.I. Efficacy of Sex Steroid Therapy Without Progestin or GnRH Agonist for Gonadal Suppression in Adult Transgender Patients. *J. Clin. Endocrinol. Metab.* **2021**, *106*, E1290–E1300. [CrossRef]
6. Quigley, C.A.; Bellis, A.D.; Marschke, K.B.; Awady, M.K.; Wilson, E.M.; French, F.S. Androgen receptor defects: Historical, clinical, and molecular perspectives. *Endocr. Rev.* **1995**, *16*, 271–321. [CrossRef]
7. Stanworth, R.D.; Jones, T.H. Testosterone for the aging male; current evidence and recommended practice. *Clin. Interv. Aging.* **2008**, *3*, 25–44. [CrossRef]
8. Institute of Medicine (US). Committee on Assessing the Need for Clinical Trials of Testosterone Replacement Therapy. In *Testosterone and Aging: Clinical Research Directions*; National Academies Press: Washington, DC, USA, 2004.
9. Braunstein, G.D. The influence of anabolic steroids on muscular strength. *Princ. Med. Biol.* **1997**, *8*, 465–474. [CrossRef]
10. Vermeulen, A. Longacting steroid preparations. *Acta Clin. Belg.* **1975**, *30*, 48–55. [CrossRef]
11. Forsdahl, G.; Erced, D.; Geisendorfer, T.; Turkalj, M.; Plavec, D.; Thevis, M.; Tretzele, L.; Gmeinera, G. Detection of testosterone esters in blood. *Drug Test. Anal.* **2015**, *7*, 983–989. [CrossRef]
12. Elks, J.; Ganellin, C.R. *The Dictionary of Drugs: Chemical Data: Chemical Data, Structures and Bibliographies*, 1st ed.; Springer: Easton, PA, USA, 1990; p. 652.
13. Reisch, J.; Eki-Gucer, N.; Takacs, M.; Henkel, G. Photochemische Studien, 54. Photodimerisierung von Testosteronpropionat in kristallinem Zustand und Kristallstruktur von Testosteronpropionat. *Liebigs Ann. Der Chem.* **1989**, *1989*, 595–597. [CrossRef]
14. Griffiths, P.J.F.; James, K.C.; Rees, M. Crystallographic data for some testosterone esters. *Acta Cryst.* **1965**, *19*, 149. [CrossRef]
15. Alcock, N.W.; Sanders, K.J.; Rodger, A. Potential injectable contraceptive steroids: Testosterone buciclate. *Acta Cryst.* **2004**, *E60*, 348–349. [CrossRef]
16. Böcskei, Z.; Gérczei, T.; Bodor, A.; Schwartz, R.; Náray-Szabó, G. Three Testosterone Derivatives. *Acta Cryst.* **1996**, *C52*, 2899–2903. [CrossRef]
17. Roberts, P.J.; Pettersen, R.C.; Sheldrick, G.M.; Isaacs, N.W.; Kennard, O. Crystal and molecular structure of 17β-hydroxyandrost-4-en-3-one (testosterone). *J. Chem. Soc. Perkin Trans.* **1973**, *2*, 1978–1984. [CrossRef]
18. Land, L.M.; Li, P.; Baummer, P.M. The Influence of Water Content of Triglyceride Oils on the Solubility. *Pharm. Res.* **2005**, *5*, 784–788. [CrossRef]
19. Remington. *The Science and Practice of Pharmacy*, 23rd ed.; Adejare, A., Ed.; Elsevier: Philadelphia, PA, USA, 2020.
20. Rowe, R.C.; Sheskey, P.J.; Quinn, M.E. (Eds.) *Handbook of Pharmaceutical Excipients*, 5th ed.; Pharmaceutical Press: London, UK; American Pharmacists Association: Washington, DC, USA, 2006.
21. CrysAlis PRO. *Agilent Technologies UK Ltd., Oxford Diffraction*; Agilent Technologies UK Ltd.: Yarnton, Oxfordshire, UK, 2015.
22. Sheldrick, G.M. SHELXT—Integrated space-group and crystal-structure determination. *Acta Cryst.* **2015**, *A71*, 3–8. [CrossRef]
23. Sheldrick, G.M. A short history of SHELX. *Acta Cryst.* **2008**, *A64*, 112–122. [CrossRef]
24. Sheldrick, G.M. Crystal structure refinement with SHELXL. *Acta Cryst.* **2015**, *C71*, 3–8. [CrossRef]
25. Dolomanov, O.V.; Bourhis, L.J.; Gildea, R.J.; Howard, J.A.K.; Puschmann, H. OLEX2: A complete structure solution, refinement and analysis program. *J. Appl. Cryst.* **2009**, *42*, 339–341. [CrossRef]
26. Gavezzotti, A. Efficient computer modeling of organic materials. The atom–atom, Coulomb–London–Pauli (AA-CLP) model for intermolecular electrostatic-polarization, dispersion and repulsion energies. *New J. Chem.* **2011**, *35*, 1360–1368. [CrossRef]
27. Turner, M.J.; McKinnon, J.J.; Wolff, S.K.; Grimwood, D.J.; Spackman, P.R.; Jayatilaka, D.; Spackman, M.A. *CrystalExplorer17*; University of Western Australia: The Nedlands, Australia, 2017.
28. Spackman, M.A.; McKinnon, J.J. Fingerprinting intermolecular interactions in molecular crystals. *CrystEngComm* **2002**, *4*, 378–392. [CrossRef]
29. Alvarez, S. A cartography of the van der Waals territories. *Dalton Trans.* **2013**, *42*, 8617–8636. [CrossRef] [PubMed]
30. Ohrt, J.; Haner, B.A.; Norton, D.A. Crystal data (II) for some androstanes. *Acta Cryst.* **1965**, *19*, 479. [CrossRef]
31. Turza, A.; Borodi, G.; Pop, M.M.; Ulici, A. Polymorphism and β-cyclodextrin complexation of methyldrostanolone. *J. Mol. Struct.* **2022**, *1250*, 131852. [CrossRef]
32. Turza, A.; Ulici, A.; Pop, M.M.; Borodi, G. Solid forms and β-cyclodextrin complexation of turinabol. *Acta Cryst.* **2022**, *C78*, 305–313. [CrossRef]
33. Rajnikant, V.; Dinesh, D.; Aziz, N.; Gupta, B.D. Analysis of C–H•••O Intermolecular Interactions in 4-Androstene-3,17-dione. *J. Chem. Crystallogr.* **2009**, *39*, 24–27. [CrossRef]
34. Rajnikant, V.; Gupta, V.K.; Khan, E.H.; Shafi, S.; Hashmi, S.; Shafiullah, B. Varghese, Dinesh, Crystal structure of cholest-4-ene-3,6-dione: A steroid. *Crystallogr. Rep.* **2001**, *46*, 963–966. [CrossRef]
35. Gaedecki, Z. Structure of 17-α-methyl-testosterone semihydrate $C_{20}H_{30}O_2 \cdot \frac{1}{2} H_2O$. *J. Crystallogr. Spectrosc. Res.* **1989**, *19*, 577–587. [CrossRef]

36. Turza, A.; Miclaus, M.O.; Pop, A.; Borodi, G. Crystal and molecular structures of boldenone and four boldenone steroid esters. *Z. Kristallogr. Cryst. Mater.* **2019**, *234*, 671–683. [CrossRef]
37. Borodi, G.; Turza, A.; Camarasan, P.A.; Ulici, A. Structural studies of Trenbolone, Trenbolone Acetate, Hexahydrobenzylcarbonate and Enanthate esters. *J. Mol. Struct.* **2020**, *1212*, 128127. [CrossRef]
38. Borodi, G.; Turza, A.; Bende, A. Exploring the Polymorphism of Drostanolone Propionate. *Molecules* **2020**, *25*, 1436. [CrossRef] [PubMed]
39. Turza, A.; Borodi, G.; Pop, A.; David, M. Structural studies of some androstane based prodrugs. *J. Mol. Struct.* **2022**, *1248*, 131440. [CrossRef]
40. Caplette, J.; Frigo, T.; Jozwiakowski, M.; Shea, H.; Mirmehrabi, M.; Müller, P. Characterization of new crystalline forms of hydroxyprogesterone caproate. *Int. J. Pharm.* **2017**, *527*, 42–51. [CrossRef] [PubMed]
41. Molcomson, C.; Lawrence, M.J. A comparison of incorporation of model steroids into non-ionic micellar and microemulsion systems. *J. Pharm. Pharmacol.* **1993**, *45*, 141–143. [CrossRef]

Article

Influence of Anodizing Conditions on Biotribological and Micromechanical Properties of Ti–13Zr–13Nb Alloy

Agnieszka Stróż [1], Joanna Maszybrocka [1], Tomasz Goryczka [1], Karolina Dudek [2], Patrycja Osak [1] and Bożena Łosiewicz [1,*]

1. Institute of Materials Engineering, Faculty of Science and Technology, University of Silesia in Katowice, 75 Pułku Piechoty 1A, 41-500 Chorzów, Poland
2. Refractory Materials Center, Institute of Ceramics and Building Materials, Łukasiewicz Research Network, Toszecka 99, 44-100 Gliwice, Poland
* Correspondence: bozena.losiewicz@us.edu.pl; Tel.: +48-32-3497-527

Abstract: The biomedical Ti–13Zr–13Nb bi-phase (α + β) alloy for long-term applications in implantology has recently been developed. The porous oxide nanotubes' (ONTs) layers of various geometries and lengths on the Ti–13Zr–13Nb alloy surface can be produced by anodizing to improve osseointegration. This work was aimed at how anodizing conditions determinatine the micromechanical and biotribological properties of the Ti–13Zr–13Nb alloy. First-generation (1G), second-generation (2G), and third-generation (3G) ONT layers were produced on the Ti–13Zr–13Nb alloy surface by anodizing. The microstructure was characterized using SEM. Micromechanical properties were investigated by the Vickers microhardness test under variable loads. Biotribological properties were examined in Ringer's solution in a reciprocating motion in the ball-on-flat system. The 2D roughness profiles method was used to assess the wear tracks of the tested materials. Wear scars' analysis of the ZrO_2 ball was performed using optical microscopy. It was found that the composition of the electrolyte with the presence of fluoride ions was an essential factor influencing the micromechanical and biotribological properties of the obtained ONT layers. The three-body abrasion wear mechanism was proposed to explain the biotribological wear in Ringer's solution for the Ti–13Zr–13Nb alloy before and after anodizing.

Keywords: anodizing; biomaterials; biotribology; oxide nanotubes; Ti–13Zr–13Nb alloy

1. Introduction

Medical implants, which are a substitute for bone parts in the human body, are accompanied by the action of friction, which includes a set of phenomena occurring in the contact area of two bodies moving relative to each other, as a result of which resistance to movement arises [1–5]. As a consequence, the continuity of the passive layer or protective coating is broken, and then the native structure of the alloy changes. Thus, the protection against the corrosive environment is lost [6–14]. Biotribology deals with the search for minimizing the effects of friction, which would involve reducing the wear of the surfaces involved in the friction of the elements and reducing the energy accompanying the process [15–23].

Currently, four basic groups of metallic biomaterials can be distinguished, among which the most widely used are austenitic steels [4]. The second group consists of shape memory alloys belonging to smart materials [4,14]. For long-term implants, the service life of which should not exceed 15 years, cobalt alloys are applied [4]. Titanium and titanium alloys are most often used due to their best biocompatibility, and their service life may exceed 20 years [1–13,15–17,19,21]. The market of long-term implants is dominated by a commercially pure titanium alloy (Cp Ti) and a bi-phase (α + β) Ti–6Al–4V alloy known also as Grade 5 Titanium or Ti 6-4, which has excellent strength, a low modulus of elasticity, high corrosion resistance, good weldability, and is heat treatable [19,23]. Al and V alloy additions increase the hardness of titanium and improve its physical and mechanical properties. However, the Ti–6Al–4V alloy shows unfavorable tribological properties, and its Young's modulus (~110 GPa) is higher than that of bone (~10–64 GPa). Aluminum has well-documented toxicity in the serum or urine of patients; moreover, it has a causal relationship with neurotoxicity and senile dementia of the Alzheimer type. In addition, vanadium is thermodynamically unstable in conditions corresponding to the tissue environment and is considered a toxic alloying additive, similarly to vanadium oxide (V_2O_5), which has relatively good solubility and high toxicity in living organisms. Therefore, growing doubts about the cytotoxicity of the Ti–6Al–4V alloy have prompted investigations into the development of modern vanadium-free Ti alloys containing biocompatible elements such as Mo, Nb, Zr, and Ta, which are able to stabilize the β structure in titanium [10,24–26].

One of the newest classes of biomedical alloys which excludes harmful elements in the form of Al and V from the composition is the Ti–13Zr–13Nb alloy developed by Davidson and Kovacs [24]. The Ti–13Zr–13Nb alloy combines a low modulus with high strength and excellent hot and cold serviceability. Research on this alloy has shown that its mechanical properties can be controlled to a large extent by hot working, heat treating, and cold working. The Ti–13Nb–13Zr alloy also shows high corrosion resistance in a physiological environment, which is one of the most important factors that has a decisive impact on the use of a given biomaterial for implants [7,11,12,24,27]. Like titanium, niobium is an element with high corrosion resistance, which is due to its susceptibility to being covered with a self-passive Nb_2O_5 oxide layer [28]. Zirconium has similar physicochemical properties to those of titanium and can be processed by similar methods. In some strength parameters, this element even surpasses titanium [23]. Zr is not ferromagnetic, which allows the use of implants made of the Ti–13Zr–13Nb alloy in patients undergoing nuclear magnetic resonance imaging. The Ti–13Nb–13Zr alloy is characterized by favorable mechanical properties for implant applications and a low Young's modulus. The alloys used for the production of biomaterials should have mechanical properties close to those they replace, therefore, the modulus of elasticity of Ti–13Zr–13Nb may vary between 41 and 83 GPa.

The natural and uneven oxide layer present on titanium and its alloys does not sufficiently protect the implant in the environment of body fluids, therefore the surfaces of these materials are often subjected to modification [6–8,11–14,29,30]. One of the currently most popular electrochemical methods of modifying the oxide film to form self-assembled nanotubular oxide structures is anodizing, which allows the production of oxide layers in the form of a matrix of ordered, vertically arranged oxide nanotubes (ONTs) [11–13,17–19,28,29,31–40]. The ONTs exhibit numerous unique properties compared to ultrathin oxide films formed spontaneously. Increasing the surface roughness at the nanoscale by producing ONTs contributes to a better adhesion tendency of bone-forming cells. The porous surface of the ONT layers has a beneficial effect on the osseointegration process and ensures faster tissue growth and stronger bonding of the bone with the implant due to the chemical and morphological similarity of the nanotube oxide layer to the structure of bone tissue. The type of electrolyte in which the anodizing process is carried out has the greatest impact on the microstructure and properties of the ONTs obtained on titanium oxide and its alloys [11–13,17–19,28,29,31–40]. The prospect of using ONTs dominated the efforts of researchers at the expense of understanding the mechanical properties of the nanotubular oxide layer [21,23,36,41–43]. Most biocompatibility studies of ONTs focus on their use in dentistry, orthopedics, and cardiovascular surgery due to their high affinity for bone cell adhesion and differentiation, hydroxyapatite formation, and outstanding biochemical inertia [1–5]. ONT layers on the surface of titanium and its alloys are considered promising bionanomaterials for controlled drug delivery systems to suppress local inflammation after the implantation process [29]. The possibility of selecting anodizing parameters when obtaining ONTs with desired, predetermined morphological features can be used in practical applications in personalized medicine. ONTs used as intelligent drug carriers enable the delivery of poorly water-soluble drugs and prevent the first-pass effect through the liver. ONTs can be produced on a variety of substrates, including three-dimensional, nonplanar, and curved surfaces such as, for example, thin, long surgical wires and bone fixation needles, allowing ONTs to be clinically used on the surface of implants or surgical supports in orthopedics [31–40]. However, despite numerous studies on ONTs, relatively little is known about their impact on biotribological wear of modern titanium alloys in a biological environment due to the dominance of research in dry sliding conditions [17,19].

This work is a continuation of our interest in the surface functionalization of the biomedical Ti–13Zr–13Nb alloy through the production of ONTs of the first, second, and third generation. The main purpose of the undertaken research is to determine for the first time the effect of anodizing conditions on biotribological wear in Ringer's solution and micromechanical properties of the Ti–13Zr–13Nb alloy for the development of long-term implants. This work brings new insights into the relationship between the anodizing conditions and biotribological wear of the Ti–13Zr–13Nb alloy under wet sliding conditions.

2. Materials and Methods

2.1. Substrate Surface Treatment

The substrate material was a rod with a diameter of 9 mm and a length of 1 m made of a bi-phase Ti–13Zr–13Nb alloy composed of a mixture of α and β phases (BIMO TECH, Wrocław, Poland) with a chemical composition in wt% according to the standard ASTM F1713-08(2021)e1 [44]. One side of the cut samples in the shape of disks with a thickness of 5 mm was subjected to wet grinding and polishing using a metallographic grinding and polishing machine Metkon Forcipol 102 (Metkon Instruments Inc., Bursa, Turkey), equipped with an automatic header. The samples were embedded into conductive PolyFast resin (Struers, Cleveland, OH, USA) using an ATM Opal 400 hot mounting press (Spectrographic Ltd., Guiseley, Leeds, UK) at 180 °C for 10 min. A mirror-like surface of the substrate was obtained on a wheel used for grinding at 250 rpm with water-based silicon carbide abrasive papers of P600 to P2500 gradations (Buehler Ltd., Lake Bluff, IL, USA). Diamond suspensions with 6 to 1 μm grain size (Buehler, Waukegan, IL, USA) were used for further polishing. Polishing was finished using a polishing cloth (Buehler, Waukegan, IL, USA) and a colloidal SiO_2 suspension with a grain size of 0.04 μm (Struers, Cleveland, OH, USA).

The polished samples were cleaned for 20 min in an ultrasonic cleaner USC-TH (VWR International, Radnor, PA, USA) with acetone (Avantor Performance Materials Poland S.A., Gliwice, Poland) and then in ultrapure water with resistivity of 18.2 MΩ cm (Milli-Q Advantage A10 Water Purification System, Millipore SAS, Molsheim, France) with two Milli-Q water changes. As a result of high-frequency ultrasonic waves propagating in the liquid, vacuum bubbles were formed, under the influence of which rapid evaporation of the liquid and the formation of water vapor bubbles took place. During the implosion of the vacuum bubbles, a local increase in temperature and pressure was generated. The imploding bubbles, located at the contaminated surface, detached the pollutants.

2.2. Anodizing Conditions of Ti–13Zr–13Nb Alloy

Anodes made of the Ti–13Zr–13Nb alloy with a one-sided geometric surface area of 3.14 cm^2 were prepared in accordance with the detailed information provided in our earlier work [29]. Immediately before anodizing, the prepared electrodes were immersed in 25% v/v HNO_3 (Avantor Performance Materials Poland S.A., Gliwice, Poland) for 10 min at room temperature to remove oxides from the surface of the Ti–13Zr–13Nb alloy, and then sonicated in Milli-Q water for 20 min.

The ONTs on the Ti–13Zr–13Nb electrode surface were produced in a two-electrode system by one-step anodizing at room temperature under the conditions shown in Table 1.

Table 1. Conditions for anodic production of 1G [32], 2G [34], and 3G [31] ONTs on the Ti–13Zr–13Nb alloy.

Conditions	ONTs on Ti–13Zr–13Nb		
	I Generation (1G)	II Generation (2G)	III Generation (3G)
Voltage (V)	20	20	50
Time (min)	120	120	80
Electrolyte	0.5% HF	1M $(NH_4)_2SO_4$ + 2% NH_4F	1M $C_2H_6O_2$ + 4% NH_4F
Inner diameter of ONT (nm)	71(7)	61(11)	218(39)
Outer diameter of ONT (nm)	87(10)	103(16)	362(44)
Length of ONT (µm)	0.94(9)	3.9(2)	9.7(6)

Hydrofluoric acid (48 wt.% in H_2O, ≥99.99% trace metals basis), ammonium sulfate (for molecular biology, ≥99.0%), ammonium fluoride (≥99.99% trace metals basis), and ethylene glycol (anhydrous, 99.8%) were supplied by Sigma-Aldrich (Saint Louis, MO, USA). A PWR800H high-current power supply (Kikusui Electronics Corporation, Yokohama, Japan) was used. The anodes with freshly prepared ONTs were placed in Milli-Q water, which was vigorously stirred for 5 min.

2.3. Physicochemical Characteristics of ONTs on Ti–13Zr–13Nb Alloy

The microstructure of the Ti–13Zr–13Nb alloy before and after anodizing was studied using a TESCAN Mira 3 LMU scanning electron microscope (SEM, TESCAN ORSAY HOLDING, Brno-Kohoutovice, Czech Republic). The secondary electrons (SE) were used for imaging that were generated by the incoming electron beam as they entered the surface. The use of SE allowed to obtain a high-resolution signal with a resolution that was only limited by the electron beam diameter. The SEM examinations were performed on the samples covered by an ultrathin layer of chromium deposited using a Quorum Q150T ES Sputter Coater (Quorum Technologies, East Sussex, UK).

2.4. Microhardness of ONTs on Ti–13Zr–13Nb Alloy

The microhardness of the Ti–13Zr–13Nb alloy before and after anodizing was determined using the Vickers method by means of a Wilson®–WolpertTM Microindentation Tester 401MVD (Wilson Instruments, LLC, Carthage, TX, USA). A hardness scale of HV = 0.1 was used. A Vickers indenter was applied, which was a square-based pyramidal-shaped diamond indenter with face angles of 136° according to the ISO 6507-1 standard [45]. The Vickers microhardness measurements were performed with variable loads of 490, 980, 1560, and 4900 mN, respectively. Indentation diagonal lengths were between 0.020 and 1.400 mm. The Vickers hardness number was calculated based on Equation (1):

$$\mu HV = \frac{1854.4 \times P}{d^2} \quad (1)$$

where P—force (gf) and d—mean diagonal length of the indentation (µm).

A direct method of checking and calibrating the microhardness tester, indenter, and diagonal length of the measuring system was used according to the ISO 6507-2:2018 standard [46].

2.5. Biotribology of ONTs on Ti–13Zr–13Nb Alloy in Ringer's Solution

Measurements of biotribological wear of the Ti–13Zr–13Nb alloy in the initial state and after anodizing were performed in a reciprocating motion in the ball-on-flat system using a tribometer (Anton Paar Polska, Warsaw, Poland) shown in Figure 1a. The countersample in the test was a ZrO_2 ball with a diameter of 6 mm. The tests were carried out in Ringer's solution with the chemical composition (g cm^{-3}): 8.60—NaCl, 0.30—KCl, and 0.33—$CaCl_2 \cdot 2H_2O$. To adjust the pH of the solution to 7.4(1), 4% NaOH and 1% $C_3H_6O_3$

were used. Chemical reagents of recognized analytical grade (Avantor Performance Materials Poland S.A., Gliwice, Poland) and Milli-Q water were used to prepare the solution. The normal force (F_n) in the friction node was 1 N. A sliding rate (r) of 2.5 cm s^{-1} and stroke length (l) of 4 mm were applied. The biotribological test consisted of 3749 cycles (back and forth = 1 cycle), which corresponded to a total friction distance (s) of 30 m.

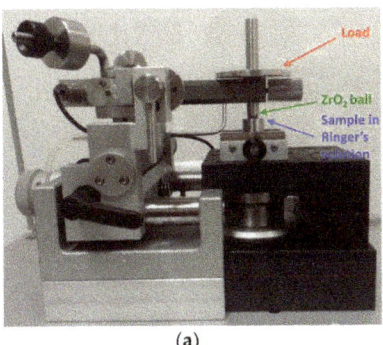 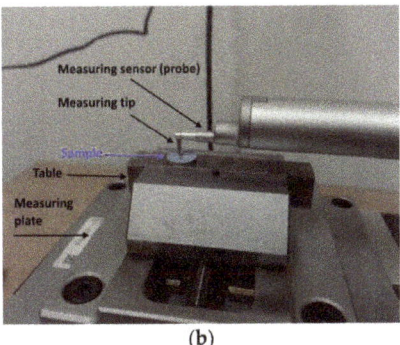

Figure 1. Setup for testing biotribological wear: (**a**) tribometer for measurements in the ball-on-flat system with a load of 1 N; (**b**) method of arranging the sample before testing on the contact profilograph.

The assessment of biotribological wear of the tested samples was carried out based on profilometric analysis of wear scars using the SURFTEST SJ-500 profilometer (Mitutoyo Corporation, Kanagawa, Japan) shown in Figure 1b.

The specific wear rate (Vv) in mm^3 N^{-1} m^{-1} was determined according to Equation (2):

$$V_V = \frac{A \cdot l}{F_n \cdot s} \tag{2}$$

where A—the wear scar area (mm^2), l—the stroke length (mm), F_n—the normal force (N), and s—the friction distance (m).

The biotribological wear analysis of the ZrO$_2$ ball was carried out based on measurements of the diameter of the wear scar on the counter-sample using a BX51 optical microscope (Olympus, Shinjuku, Tokyo, Japan). The microstructure was analyzed, with particular emphasis on the surface of the wear scar of the tested samples, resulting from prior biotribological testing.

Five samples of each type were tested, and the values of the determined parameters are the average values with their standard deviations (SD).

3. Results and Discussion

3.1. Microstructure of ONTs on Ti–13Zr–13Nb Alloy

Figure 2 shows SEM images of the microstructure of the self-assembled ONTs obtained on the surface of the Ti–13Zr–13Nb alloy by anodizing under different conditions (see Table 1). The detailed mechanism of the electrochemical formation of ONT layers on the surface of the Ti–13Zr–13Nb alloy in electrolytes containing fluoride ions that form water-soluble complexes with titanium, zirconium, and niobium was described in our earlier paper [33,35]. A strong influence of the type of electrolyte on the surface morphology of the obtained ONT layers could be observed. The on-top general view of the Ti–13Zr–13Nb alloy with the ONTs of 1G (Figure 2a), 2G (Figure 2b), and 3G (Figure 2c) revealed the presence of well-developed and evenly distributed oxide nanotubes with single and very smooth walls.

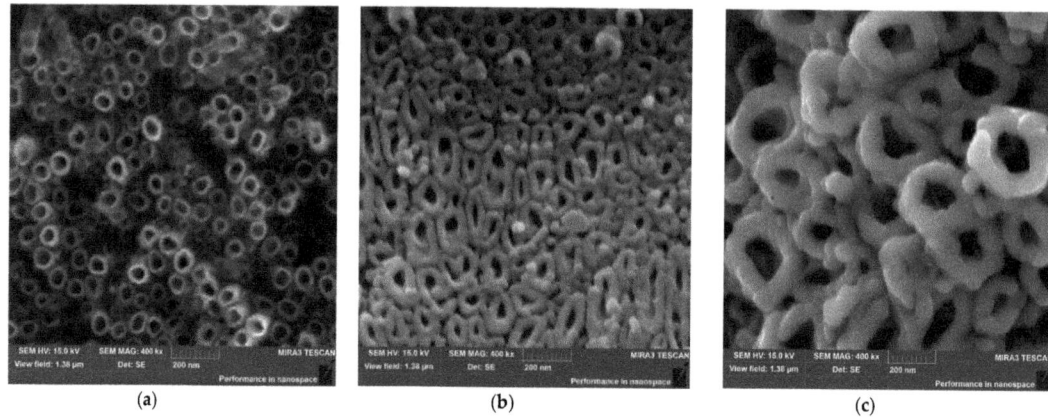

Figure 2. SEM image of the microstructure of ONT layer on the Ti–13Zr–13Nb alloy obtained by anodizing in: (**a**) 0.5% HF electrolyte (1G); (**b**) 1M $(NH_4)_2SO_4$ + 2% NH_4F electrolyte (2G); (**c**) 1M $C_2H_6O_2$ + 4% NH_4F electrolyte (3G).

The 1G ONTs had the smallest inner diameter of 71(7) nm, an outer diameter of 87(10) nm, and nanotube length of 0.94(9) μm [32]. The wall thickness of the 1G ONTs was also the smallest among all the obtained generations of ONTs on the surface of the Ti–13Zr–13Nb alloy. The ONTs formed in the 0.5% HF electrolyte at 20 V for 120 min had a circular or elliptical cross-section (Figure 2a). The 2G ONTs were characterized by higher values of geometrical parameters of the produced nanotubes on the Ti–13Zr–13Nb alloy, which were determined based on SEM images from selected areas of the oxide surfaces (Figure 2b). The ONTs with an inner diameter of 61(11) nm and an external diameter of 103(16) nm were obtained in the electrolyte of 1M $(NH_4)_2SO_4$ + 2% NH_4F at 20 V for 120 min [34]. The length of vertically positioned 2G ONTs with a circular cross-section was 3.9(2) μm. The 2G ONTs were the most densely packed and tended to form clusters of oxide nanotubes of various lengths. The obtained results showed that using the organic electrolyte 1M $C_2H_6O_2$ + 4% NH_4F at 50 V for 80 min allowed obtaining an over three-fold increase in inner diameter and an about four-fold increase in outer diameter of 3G ONTs compared to that of the ONTs of 1G and 2G [31]. The ONTs of 3G with a circular cross-sectional shape had the thickest nanotube walls. The length of 3G ONTs increased more than 10-fold and two-fold compared to the length of 1G ONTs and 2G ONTs, respectively.

The obtained results indicate that the modification of the surface of the Ti–13Zr–13Nb alloy made it possible to obtain porous layers of ONTs with different morphological parameters and lengths. The surface morphology of all obtained generations of ONTs was very similar to the structure of trabecular bone, which will affect the ability to transfer load by implants made of the Ti–13Zr–13Nb alloy with an ONT layer applied. It was signaled in the literature that ONT layers on titanium and its alloys show the ability to accelerate osseointegration, leveling the inflammatory states and preventing the penetration of harmful corrosion products into the biological environment of the body [10,41,43]. The ONT layers can also be used as intelligent carriers of medicinal substances in drug delivery systems, especially for personalized medicine [29].

Long-term in vitro corrosion resistance studies of the Ti–13Zr–13Nb alloy in saline solution showed the influence of the electrode immersion time on the change in the thickness of the self-passive oxide layer [7]. In studies using the electrochemical impedance spectroscopy method, it was shown that the thickness of the self-passive oxide layer increased from 0.6 to 2.3 nm for 20 days of immersion. It should be emphasized that after 20 days of immersion tests, no pitting was observed on the surface of the Ti–13Zr–13Nb alloy. Comparative assessment of the determined corrosion resistance parameters showed that surface modification of the Ti–13Zr–13Nb alloy by anodizing increased its corrosion resistance in a saline solution for the 2G ONT layers [12] and slightly decreased for the 3G ONT layer [31] as compared to the non-anodized Ti–13Zr–13Nb alloy surface. It was shown that the in vitro corrosion resistance of ONT layers depended on their structure and morphological parameters. In the case of both the Ti–13Zr–13Nb alloy before and after the production of 2G and 3G ONT layers, no breakdown potential was revealed in potentiodynamic tests conducted up to 9.5 V, which indicates that the tested biomaterials had excellent resistance to pitting corrosion and can be promising biomaterials for long-term use in implantology.

3.2. Micromechanical Properties of ONTs on Ti–13Zr–13Nb Alloy

The assessment of the effect of anodizing conditions on the micromechanical properties of the Ti–13Zr–13Nb alloy was carried out based on the Vickers microhardness tests. Microhardness, which is the hardness of the material exposed to low applied loads, is particularly useful for assessing the structural integrity of the bone tissue on the surface of the porous layer of ONTs and the bone tissue that surrounds the implant in the process of osseointegration. For this reason, it is important to compare the micromechanical properties of the Ti–13Zr–13Nb alloy before and after anodizing. The quantitative assessment of the anodizing effect in the microscale consisted in determining the micromechanical properties of the Ti–13Zr–13Nb alloy before and after the formation of ONT layers in various conditions of electrochemical oxidation (Figure 3).

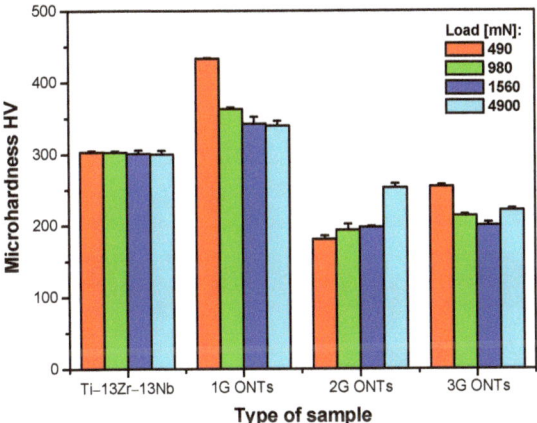

Figure 3. Vickers microhardness of the Ti–13Zr–13Nb alloy surface before and after anodizing with variable loads.

The results presented in Figure 3 indicate that Vickers microhardness of the Ti–13Zr–13Nb alloy changed as a result of anodizing. The value of Vickers microhardness for the non-anodized alloy determined under variable loads in the range from 490 to 4900 mN was constant within the error limits and amounted to 302(1). In the case of three generations of ONT layers, Vickers microhardness depended on the applied load, which resulted from the different lengths of the tested oxide nanotubes and their morphological parameters (Figure 2). Compared to the Ti–13Zr–13Nb alloy in the initial state, an increase in the microhardness was observed in the case of the 1G ONT layer, for which Vickers microhardness value decreased in the range from 433(1) to 340(7) with increasing load. Such micromechanical properties of the 1G ONT layer resulted from the smallest outer diameter of oxide nanotubes and their shortest length among all obtained generations of ONTs (Table 1). The 1G ONT layer had the largest number of individual nanotubes on the surface, and thus had the greatest load carrying capacity [18,19]. Vickers microhardness value dropped for the 2G and 3G ONT layers compared to that of the substrate surface and 1G ONT layer (Figure 3). The 2G ONT layers showed the smallest Vickers microhardness ranging from 181(5) to 252(6) with increasing load in the range of 490 to 4900 mN. The 3G ONT layers revealed a slightly higher Vickers microhardness than the 2G ONT layers did, which varied from 254(3) to 221(3) with increasing load. Larger outer diameters of the 2G and 3G ONTs and their greater length cause easier deformation and cracking of the obtained oxide nanotubes [17]. With the increase in the ONTs diameter on the Ti–13Zr–13Nb alloy, the number of oxide nanotubes carrying loads in the contact area of the tested surfaces with the diamond indenter decreased. On the other hand, ONTs are able to compensate for the large hardness defect of the biomedical Ti–13Zr–13Nb alloy used for the production of implants, eliminate implant–bone stress mismatch, and minimize "stress shielding" [35].

3.3. Biotribological Properties of ONTs on Ti–13Zr–13Nb Alloy

Biotribological wear resistance tests and friction coefficient measurements were carried out under sliding friction conditions in the presence of Ringer's solution, which was a biological lubricating fluid. The Ti–13Zr–13Nb alloy in the initial state and with the 1G, 2G, and 3G ONT layers was subjected to biotribological tests in reciprocating motion in the ball-on-flat system, after which microscopic analysis of wear scars of the ZrO_2 ball was performed (Figure 4).

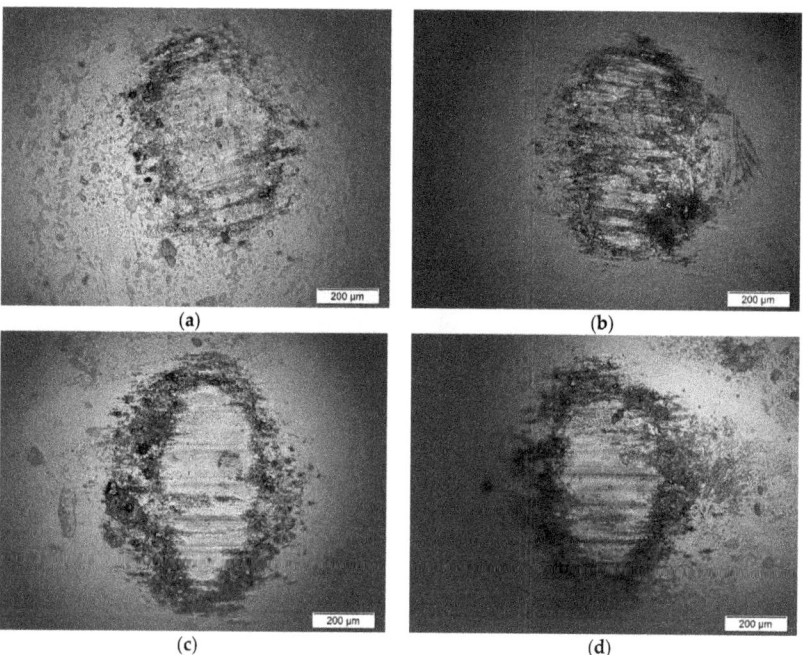

Figure 4. ZrO_2 ball wear scar after the ball-on-flat biotribological test against (**a**) Ti–13Zr–13Nb substrate; (**b**) 1G ONT layer; (**c**) 2G ONT layer; (**d**) 3G ONT layer.

On the microscopic images of counter-sample wear scars after the biotribological wear test, the direction of damage to the ball from top to bottom was observed for all tested materials. Residual abrasion of materials on the surface of the ZrO_2 ball was also visible. The wear scar of the ZrO_2 ball in combination with the surface of the Ti–13Zr–13Nb alloy was characterized by the smoothest surface with a small number of impurities transferred to the surface of the counter-sample compared to the tested ONT layers, which indicated the lowest biotribological wear of the alloy substrate. During the friction process of the 1G, 2G, and 3G ONT layers, numerous scratches appeared on the surface of the ZrO_2 ball. Such an effect was caused by the presence of residual abrasion products (debris) in the form of particles of the ONTs, which were subject to abrasion during the friction process and constituted an additional factor damaging the surface, causing an increase in friction.

Based on microscopic observations, the average value of the ZrO_2 ball wear scar diameter (d_{av}) was determined, the values of which are shown in Figure 5a. The d_{av} value for the Ti–13Zr–13Nb alloy in the initial state was 650(11) μm. The d_{av} parameter assumed higher values in the presence of ONT layers produced on the alloy substrate, and thus indicated an increase in the specific wear of the counter-sample in the form of a ZrO_2 ball (V_b). Figure 5b shows a comparison of the V_b values determined for all tested materials. The ZrO_2 ball wear scar for the 2G ONT layer had the largest width of d_{av} = 784(11) μm (Figure 5a) and showed the largest value of V_b equal to $2.07(10) \cdot 10^{-4}$ mm^3 N^{-1} m^{-1} (Figure 5b), showing a more than two-fold increase in the biotribological wear of the counter-sample compared to V_b obtained for the ZrO_2 ball-alloy Ti–13Zr–13Nb combination.

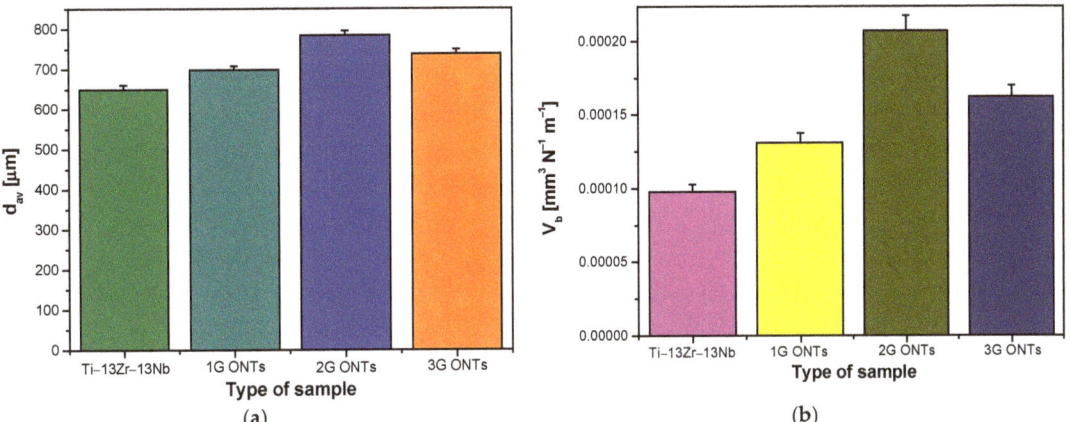

Figure 5. Parameters obtained after the biotribological wear test in the ball-on-flat system for the Ti–13Zr–13Nb alloy before and after anodizing: (**a**) ZrO_2 ball wear scar (d_{av}); (**b**) ZrO_2 ball-specific wear (V_b).

The results of profilometric tests of the wear track on the surface of the tested materials after the biotribological wear test in the ball-on-flat system are shown in Figure 6. Based on the obtained cross-sectional profiles of wear tracks for the Ti–13Zr–13Nb alloy before and after anodizing, it can be concluded that biotribological wear of the material surface depended on the anodizing conditions, while the surface of the Ti–13Zr–13Nb alloy in the initial state showed less material wear compared to anodized surfaces.

The width of the wear track for the surface of the Ti–13Zr–13Nb alloy determined based on the data in Figure 6 was 665(7) μm, and for electrochemically oxidized surfaces it increased, reaching the largest value of 827(8) μm for the 2G ONT layer. Larger values of the wear track width were observed for all the tested materials compared to the width of the ZrO_2 ball wear scar (Figure 5a), which suggests a higher biotribological wear of the Ti–13Zr–13Nb alloy before and after anodizing compared to the ZrO_2 ball used in a ball-on-flat combination.

Figure 7b presents wear track depth obtained after the biotribological wear test for the Ti–13Zr–13Nb alloy before and after electrochemical oxidation. The smallest wear track depth was shown by the alloy substrate, the value of which was 14(1) μm. All generations of ONT layers showed greater wear track depth compared to that of the non-anodized substrate, with the greatest wear track depth value being 25(1) μm for the 2G ONT layer.

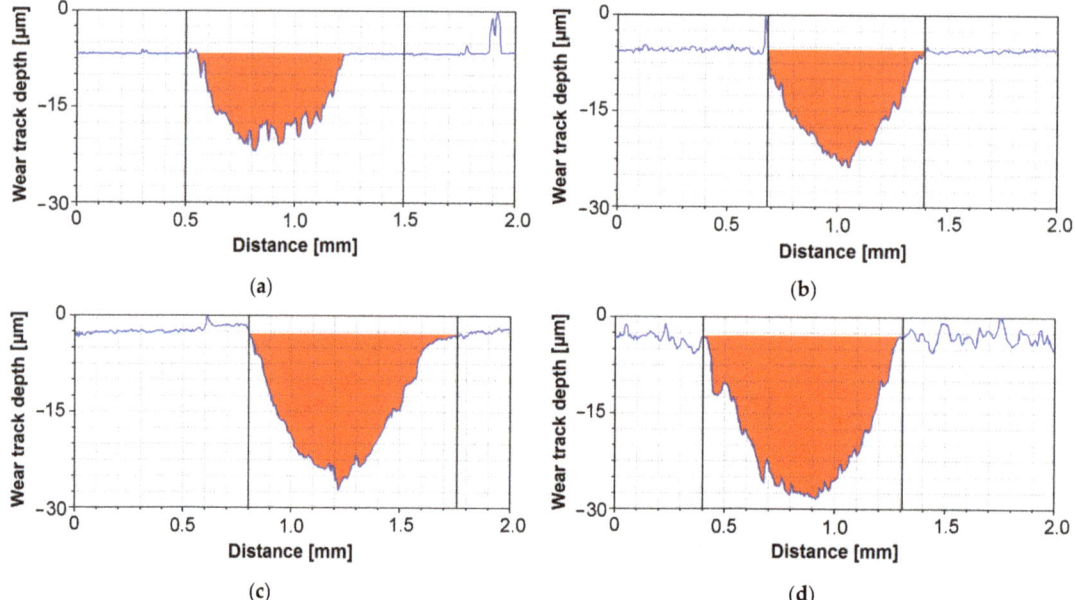

Figure 6. Cross-sectional profiles of wear tracks after the biotribological wear test in the ball-on-flat system for the Ti–13Zr–13Nb alloy before and after anodizing: (**a**) Ti–13Zr–13Nb alloy; (**b**) 1G ONT layer; (**c**) 2G ONT layer; (**d**) 3G ONT layer.

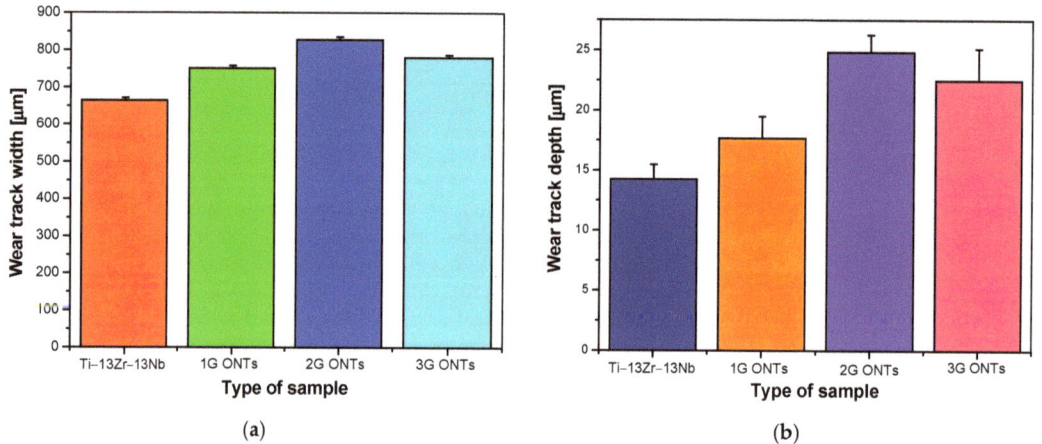

Figure 7. Parameters obtained after the biotribological wear test in the ball-on-flat system for the Ti–13Zr–13Nb alloy before and after anodizing: (**a**) wear track width; (**b**) wear track depth.

Based on the d_{av} parameter (Figure 5a), which strongly depends on the type of the tested surface used in the friction node, the average wear surface area (A_{av}) was determined after the biotribological test, which took the smallest value equal to 6814(64) μm^2 for the Ti–13Zr–13Nb alloy in the initial state (Figure 8a). Along with the increase in the average outer diameter of ONTs and their length, an upward trend of A_{av} was observed. The A_v value determined for the non-anodized substrate was more than two times lower than the average wear surface area value for the 2G and 3G ONT layers.

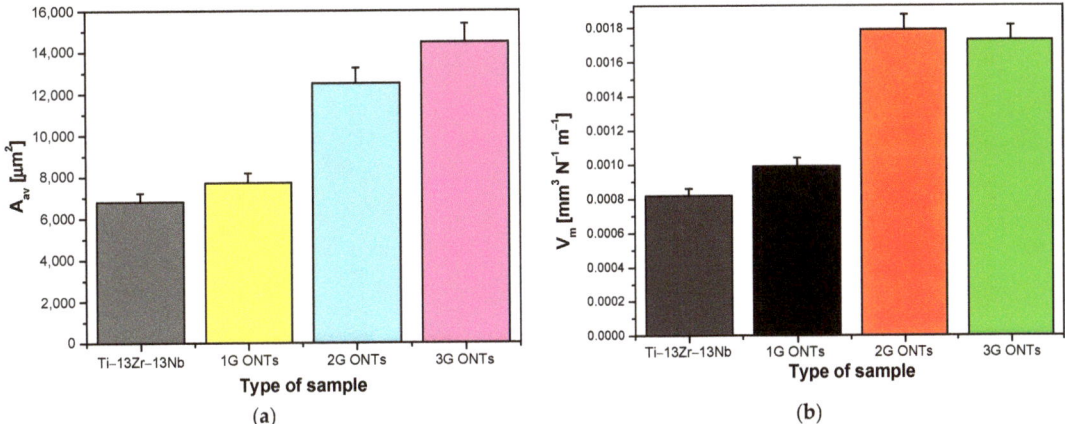

Figure 8. Parameters obtained after the tribological wear test in the ball-on-flat system for the Ti–13Zr–13Nb alloy before and after anodizing: (**a**) average wear surface area (A_{av}); (**b**) average material volume consumption (V_m).

Average material volume consumption (V_m) took the lowest value of $8.20(4) \cdot 10^{-4}$ mm^3 N^{-1} m^{-1} for the electrochemically unoxidized surface of the Ti–13Zr–13Nb alloy (Figure 8b). The obtained V_m value was twice as high as compared to the average material volume consumption for mechanically polished grade 4 titanium [15], 16 times higher than V_m for sandblasted grade 4 titanium [15], and 11 times higher than for sandblasted and steam-sterilized grade 4 titanium [16], subjected to biotribological wear test under comparable conditions in protein-free artificial saliva. The obtained results indicate that the value of V_m increased after the anodizing process of the Ti–13Zr–13Nb alloy in a solution of 0.5% HF, 1M (NH$_4$)$_2$SO$_4$ + 2% NH$_4$F, and 1M C$_2$H$_6$O$_2$ + 4% NH$_4$F. The highest value of Vm equal to $1.79(9) \cdot 10^{-3}$ mm^3 N^{-1} m^{-1} was observed for the 2G ONT layer, which was very close to the V_m obtained for the 3G ONT layer within the limit of error (Figure 8b). Such a significant consumption of ONTs resulted from the porous structure of the tested layers. Oxide nanotubes are a kind of hollow tubes that carry load. During the friction process, the thin walls of 2G and 3G ONTs with a lower microhardness compared to that of the alloy substrate and 1G ONTs (Figure 3) break more easily, resulting in higher material consumption.

Figure 9 shows the course of the friction coefficient as a function of the sliding distance for the Ti–13Zr–13Nb alloy and its surface after anodizing. In the conducted biotribological tests, the coefficient of friction was a measure of the resistance of the tested materials in the process of friction against the counter-sample penetrating the material under study.

In the graph of the coefficient of friction shown in Figure 9, the initial course of friction was attributed to the initial oxide layer, which was then systematically removed [20]. The initial lower friction coefficient values were mainly due to two factors. Firstly, the wear debris produced in the first stage filled the pores of the outer layer, which increased the contact area between the ZrO$_2$ ball and the tested surface. The outermost layer of ONTs was easy to remove and was a carrier of particles that affected friction processes. Secondly, as the porous outer layer was gradually worn away, the counter-sample had increased contact with the denser inner layer. At the end of the friction process, particles formed in the friction process (wear debris) accumulated and the friction coefficient increased. Long nanotubes contributed to the accumulation of a large amount of worn material and thus a higher value of the friction coefficient [21,22]. On the graph of the friction coefficient for 2G ONTs, the smoothest course along the entire length of the sliding distance was visible, showing the easiest wear of the material (Figure 9e,f). Based on the friction coefficient, the kinetic coefficient of friction (μ_k) was determined, which took the smallest value of

0.86(8) for the surface of the Ti–13Zr–13Nb alloy with a 2G ONT layer (Figure 10). A similar value of μ_k of 0.86(6) and 0.87(3) was determined for sandblasted [15] and sandblasted and sterilized [16] grade 4 titanium, respectively.

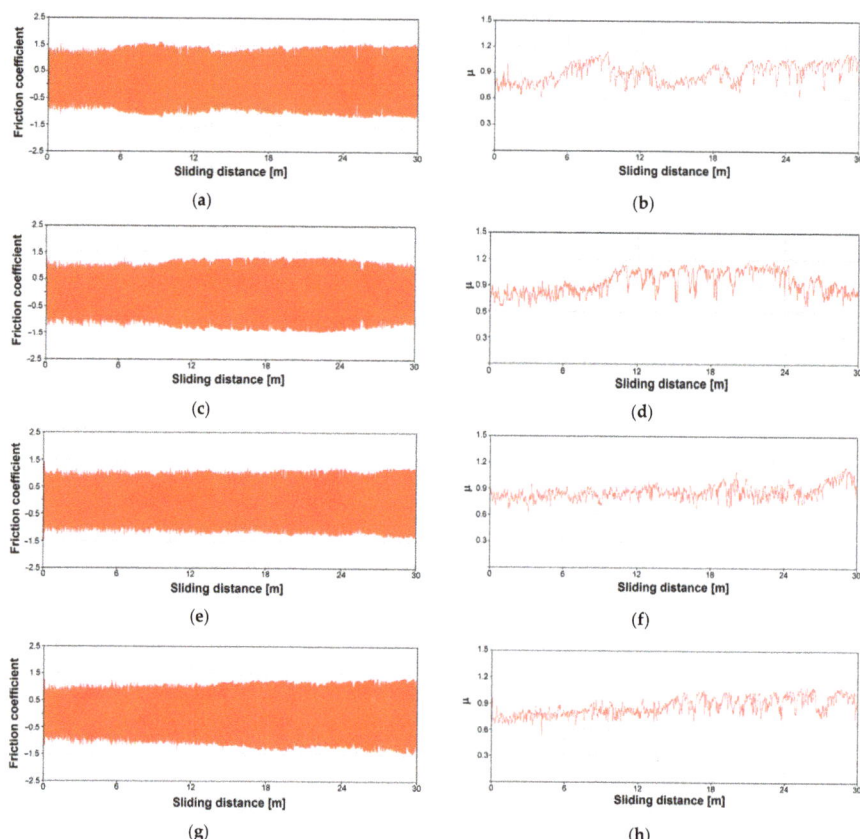

Figure 9. Friction coefficient (μ) as a function of the sliding distance in the ball-on-flat system for: (**a**,**b**) Ti–13Zr–13Nb alloy; (**c**,**d**) 1G ONT layer; (**e**,**f**) 2G ONT layer; (**g**,**h**) 3G ONT layer.

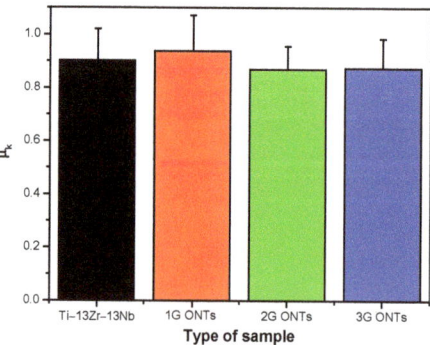

Figure 10. Kinetic coefficient of friction (μ_k) obtained in the tribological wear test using the ball-on-flat system for the Ti–13Zr–13Nb alloy before and after anodizing.

Figure 11 shows exemplary microscopic images of wear tracks of the tested materials after the biotribological test in Ringer's solution performed in the center region of wear tracks (Figure 11a,c,e,g) and in the border of wear tracks and untested surfaces (Figure 11b,d,f,h).

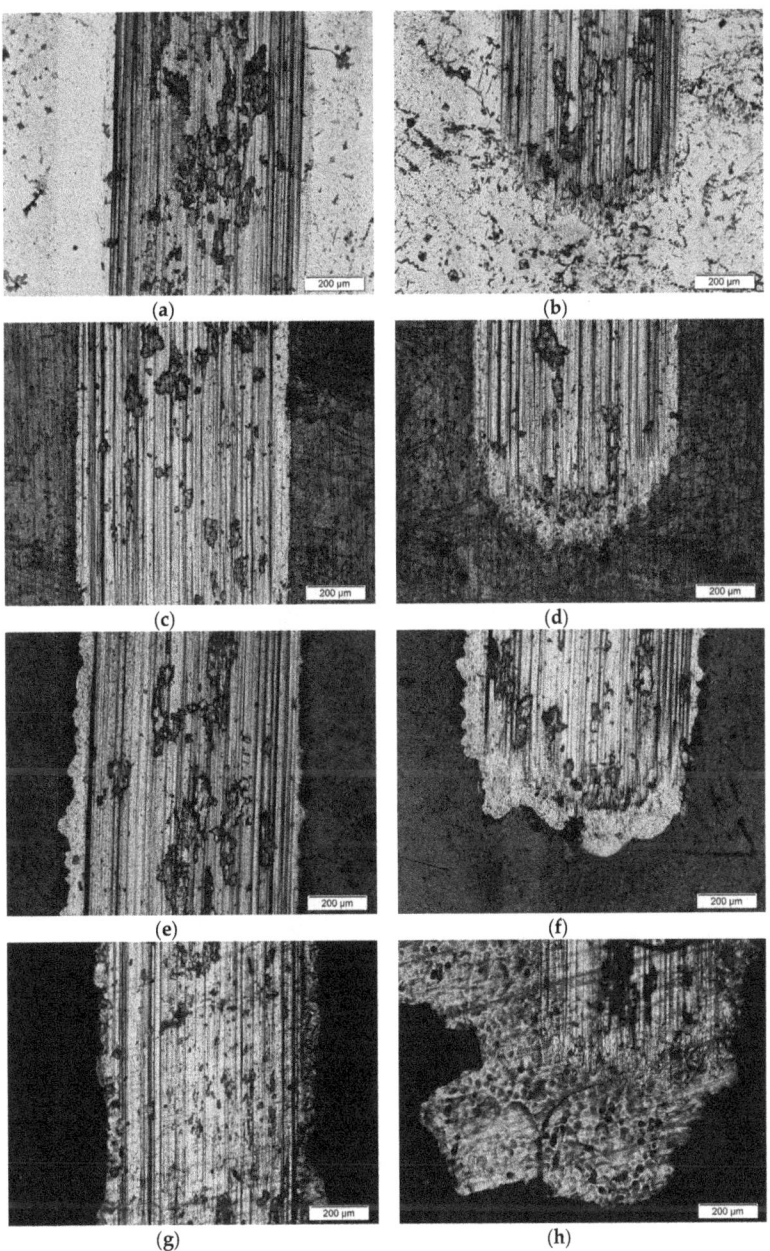

Figure 11. Microscopic image of wear track after the biotribological test in the ball-on-flat combination system for: (**a**,**b**) Ti–13Zr–13Nb alloy; (**c**,**d**) 1G ONT layer; (**e**,**f**) 2G ONT layer; (**g**,**h**) 3G ONT layer.

Visible scratches and furrows in Figure 11 were the result of the movement of wear products along the working path of the counter-sample. The surface of the Ti–13Zr–13Nb alloy in the initial state had numerous material losses and pits filled with remnants of the abraded material and lubricant in the form of Ringer's solution (Figure 11a,b). The surface of the 1G ONT layer (Figure 11c,d) and 2G ONT layer (Figure 11e,f) also showed numerous material losses. Accumulations of abraded material were visible in the contact area of the counter-sample with the track of abrasion. The surface with 3G ONTs showed delamination of the oxide layer (Figure 11g,h). In addition, cracks in the 3G ONT layer were caused by cyclic loading during the biotribological test. Transverse cracks visible in the wear track image for the 3G ONT layer indicated additional fatigue wear (Figure 11g,h). The analysis of microscopic wear tracks showed that abrasive wear was the dominant mechanism.

3.4. Wear Mechanism of Ti–13Zr–13Nb Alloy before and after Anodizing in Ringer's Solution

The mechanism of biotribological wear of ONT layers on the Ti–13Zr–13Nb alloy substrate in Ringer's solution is based on the breaking of oxide nanotubes and their densification in the outer part of the oxide layers according to the wear mechanism of three-body abrasive wear [15,16]. In the proposed mechanism, between the surface of the Ti–13Zr–13Nb alloy with a layer of ONTs (body 1) and the surface of the ZrO_2 ball (body 2), there are particles of worn material (body 3), that act as a carrier abrasive (Figure 12). The wear results from the gradual loss of material in the contact area of the interacting surfaces as body 1 and body 2 move relative to each other.

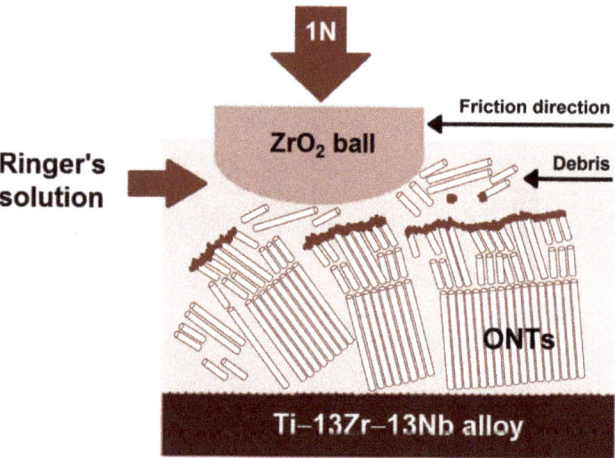

Figure 12. Three-body abrasion wear mechanism of the Ti–13Zr–13Nb alloy before and after anodizing in Ringer's solution.

Wear debris is mainly formed in the form of single or aggregated ONTs as a result of cracking of the oxide layers in various places, which results in the appearance of ONT fragments of various sizes. It was previously observed in the literature that densification of ONTs is accompanied by wear and cracking [41]. The increase in the indentation depth of ONTs causes cracking of oxide nanotubes and bending and cracking of adjacent ONTs resulting in gradual densification of small fragments of ONT layers [47]. As a consequence of the detachment of the ONT layer from the substrate, the remains of the oxide layer are released in the contact area, which can be pushed out of the contact or trapped in it. As soon as sliding starts, it can be expected that wear debris in the contact area will be exposed to mechanical and electrochemical influences, which may occur sequentially or at the same time, contributing to increased wear of the material. It was reported that as a result of continuous smashing and densification of the wear debris in the central area of the wear track, a compact oxide layer is formed [41]. The tribolayer obtained in this way may reveal protective properties against corrosion and biotribological wear of the substrate. Simultaneously with the formation of the tribolayer along with the movement of the counter-sample, part of the wear debris is pushed to the ONT layer surrounding the sliding contact area. This is a probable reason for inducing cyclical compressive stresses, which cause damage to the structure of the surface and subsurface parts of the ONT layer by initiation and propagation of cracks leading to delamination. This revealed that the ONT layers were brittle and had poor adhesion to the Ti–13Zr–13Nb alloy substrate. Xu and co-authors [47] suggested that brittle ONT layers bend elastically to a very small strain and consequently collapse. Through the emerging cracks in the ONT layer, the electrolyte can penetrate into the substrate, which induces electrochemical corrosion and additionally affects the detachment of the oxide layer.

The micromechanical and biotribological properties of the porous ONT layers on the Ti–13Zr–13Nb alloy under wet sliding in Ringer's solution strongly depend on the anodizing conditions, among which the composition of the electrolyte containing fluoride ions plays a key role.

4. Conclusions

The assessment of the effect of anodizing conditions on the micromechanical properties of the Ti–13Zr–13Nb alloy shows that Vickers microhardness determined under variable loads changed depending on the type of electrolyte and applied voltage–time parameters of electrochemical oxidation. Vickers microhardness for the non-anodized alloy was independent of the load used and amounts to 302(1). For 1G, 2G, and 3G ONT layers, the dependence of Vickers microhardness on applied load was revealed due to the differences in the morphological parameters and lengths of the ONTs. For the 1G ONT layer, an increase in Vickers microhardness in the range from 433(1) to 340(7) with increasing load was observed, which was related to the smallest outer diameter of ONTs with the shortest nanotube length. Vickers microhardness decreased from 181(5) to 252(6) and from 254(3) to 221(3) with increasing load for 2G and 3G ONT layers, respectively, compared to the alloy substrate.

Based on the biotribological tests carried out in Ringer's solution in a reciprocating motion in the ball-on-flat system for the Ti–13Nb–13Zr alloy before and after anodizing, it was found that the non-anodized alloy was characterized by the highest wear resistance for which the average material volume consumption was $8.20(4) \cdot 10^{-4}$ mm^3 N^{-1} m^{-1}. The resistance to abrasive wear decreased for 1G, 2G, and 3G ONT layers, taking the highest value of the average material volume consumption of $1.79(9) \cdot 10^{-3}$ mm^3 N^{-1} m^{-1} for the 2G ONT layer. It was ascertained that the lower the coefficient of friction, the greater the volumetric wear, i.e., the lower the resistance to abrasive wear. The kinetic coefficient of friction determined based on the friction coefficient, took the smallest value of 0.86(8) for the 2G ONT layer. The highest coefficient of kinetic friction of 0.94(1) was characterized by the surface of the 1G ONT layer. Based on the results obtained, a three-body abrasion wear mechanism was proposed for biotribological wear of the Ti–13Zr–13Nb alloy before and after anodizing in Ringer's solution.

In this study, the in vitro biotribological properties of the tested biomaterials were studied in protein-free simulated body fluid. In order to create studies under wet sliding more similar to in vivo conditions, future research will focus on the determination of the wettability of ONT layers and biotribology wear assessment in a simulated body fluid with the addition of proteins.

Author Contributions: Conceptualization, B.Ł.; methodology, B.Ł., J.M., A.S. and T.G.; investigation, A.S., J.M., T.G., P.O. and K.D.; formal analysis, B.Ł., A.S., J.M., T.G., K.D. and P.O.; writing—original draft preparation, B.Ł., A.S., P.O. and J.M.; writing—review and editing, K.D. and T.G; visualization, A.S., J.M., P.O., T.G. and K.D.; funding acquisition, B.Ł. All authors have read and agreed to the published version of the manuscript.

Funding: This research received no external funding.

Institutional Review Board Statement: Not applicable.

Informed Consent Statement: Not applicable.

Data Availability Statement: Not applicable.

Conflicts of Interest: The authors declare no conflict of interest.

References

1. Ti-Based Biomaterials. *Ti-Based Biomaterials*; MDPI: Basel, Switzerland, 2020.
2. Anene, F.; Jaafar, C.A.; Zainol, I.; Hanim, M.A.; Suraya, M. Biomedical materials: A review of titanium based alloys. *Proc. Inst. Mech. Eng. Part C J. Mech. Eng. Sci.* **2020**, *235*, 095440622096769. [CrossRef]
3. *Titanium in Medical and Dental Applications*; Elsevier: Amsterdam, The Netherlands, 2018.
4. Chen, Q.; Thouas, G.A. Metallic implant biomaterials. *Mater. Sci. Eng.* **2015**, *87*, 1–57. [CrossRef]
5. Abdel-Hady Gepreel, M.; Niinomi, M. Biocompatibility of Ti-alloys for long-term implantation. *J. Mech. Behav. Biomed. Mater.* **2013**, *20*, 407–415. [CrossRef]
6. Łosiewicz, B.; Osak, P.; Maszybrocka, J.; Kubisztal, J.; Stach, S. Effect of autoclaving time on corrosion resistance of sandblasted Ti G4 in artificial saliva. *Materials* **2020**, *13*, 4154. [CrossRef] [PubMed]
7. Szklarska, M.; Dercz, G.; Simka, W.; Łosiewicz, B.A.C. impedance study on the interfacial properties of passivated Ti13Zr13Nb alloy in physiological saline solution. *Surf. Interface Anal.* **2014**, *46*, 698–701. [CrossRef]
8. Aniołek, K.; Łosiewicz, B.; Kubisztal, J.; Osak, P.; Stróż, A.; Barylski, A.; Kaptacz, S. Mechanical properties, corrosion resistance and bioactivity of oxide layers formed by isothermal oxidation of Ti—6Al—7Nb alloy. *Coatings* **2021**, *11*, 505. [CrossRef]
9. Osak, P.; Maszybrocka, J.; Kubisztal, J.; Ratajczak, P.; Łosiewicz, B. Long-Term Assessment of the In Vitro Corrosion Resistance of Biomimetic ACP Coatings Electrodeposited from an Acetate Bath. *J. Funct. Biomater.* **2021**, *12*, 12. [CrossRef]
10. Costa, B.C.; Tokuhara, C.; Rocha, L.A.; Oliveira, R.C.; Lisboa-Filho, P.N.; Pessoa, J. Vanadium ionic species from degradation of Ti-6Al-4V metallic implants: In vitro cytotoxicity and speciation evaluation. *Mater. Sci. Eng. C* **2019**, *96*, 730–739. [CrossRef]
11. Ossowska, A.; Zieliński, A.; Supernak, M. Formation of High Corrosion Resistant Nanotubular Layers on Titanium Alloy Ti13Nb13Zr. *Solid State Phenom.* **2011**, *183*, 137–142. [CrossRef]
12. Smołka, A.; Dercz, G.; Rodak, K.; Łosiewicz, B. Evaluation of corrosion resistance of nanotubular oxide layers on the Ti13Zr13Nb alloy in physiological saline solution. *Arch. Metall. Mater.* **2015**, *60*, 2681–2686. [CrossRef]
13. Aïnouche, L.; Hamadou, L.; Kadri, A.; Benbrahim, N.; Bradai, D. Interfacial barrier layer properties of three generations of TiO$_2$ nanotube arrays. *Electrochim. Acta* **2014**, *133*, 597–609. [CrossRef]
14. Dudek, K.; Dulski, M.; Łosiewicz, B. Functionalization of the NiTi Shape Memory Alloy Surface by HAp/SiO$_2$/Ag Hybrid Coatings Formed on SiO$_2$-TiO$_2$ Glass Interlayer. *Materials* **2020**, *13*, 1648. [CrossRef] [PubMed]
15. Osak, P.; Maszybrocka, J.; Zubko, M.; Rak, J.; Bogunia, S.; Łosiewicz, B. Influence of Sandblasting Process on Tribological Properties of Titanium Grade 4 in Artificial Saliva for Dentistry Applications. *Materials* **2021**, *14*, 7536. [CrossRef] [PubMed]
16. Osak, P.; Maszybrocka, J.; Kubisztal, J.; Łosiewicz, B. Effect of amorphous calcium phosphate coatings on tribological properties of titanium grade 4 in protein-free artificial saliva. *Biotribology* **2022**, *32*, 100219. [CrossRef]
17. Luz, A.R.; de Souza, G.B.; Lepienski, C.M.; Siqueira, C.J.M.; Kuromoto, N.K. Tribological properties of nanotubes grown on Ti-35Nb alloy by anodization. *Thin Solid Films* **2018**, *660*, 529–537. [CrossRef]
18. Shen, X.-J.; Pei, X.-Q.; Liu, Y.; Fu, S.-Y. Tribological performance of carbon nanotube–graphene oxide hybrid/epoxy composites. *Compos. Part B Eng.* **2014**, *57*, 120–125. [CrossRef]
19. Sarraf, M.; Zalnezhad, E.; Bushroa, A.R.; Hamouda, A.M.S.; Rafieerad, A.R.; Nasiri-Tabrizi, B. Effect of microstructural evolution on wettability and tribological behavior of TiO$_2$ nanotubular arrays coated on Ti–6Al–4V. *Ceram. Int.* **2015**, *41*, 7952–7962. [CrossRef]
20. Dervishi, E.; McBride, M.; Edwards, R.; Gutierrez, M.; Li, N.; Buntyn, R.; Hooks, D.E. Mechanical and tribological properties of anodic Al coatings as a function of anodizing Conditions. *Surf. Coat. Technol.* **2022**, *444*, 128652. [CrossRef]
21. Li, Z.; Bao, Y.; Wu, L.; Cao, F. Oxidation and tribological properties of anodized Ti45Al8.5Nb alloy. *Trans. Nonferrous Met. Soc. China* **2021**, *31*, 3439–3451. [CrossRef]
22. Chen, Z.; Ren, X.; Ren, L.; Wang, T.; Qi, X.; Yang, Y. Improving the Tribological Properties of Spark-Anodized Titanium by Magnetron Sputtered Diamond-Like Carbon. *Coatings* **2018**, *8*, 83. [CrossRef]
23. Zhang, S.; Qin, J.; Yang, C.; Zhang, X.; Liu, R. Effect of Zr addition on the microstructure and tribological property of the anodization of Ti-6Al-4V alloy. *Surf. Coat. Technol.* **2018**, *356*, 38–48. [CrossRef]
24. Davidson, J.A.; Kovacs, P. New Biocompatible, Low Modulus Titanium Alloy for Medical Implants. U.S. Patent No. 5,169,597, 8 December 1992.
25. Bălțatu, M.S.; Vizureanu, P.; Bălan, T.; Lohan, M.; Țugui, C.A. Preliminary Tests for Ti-Mo-Zr-Ta Alloys as Potential Biomaterials. *IOP Conf. Ser. Mater. Sci. Eng.* **2018**, *374*, 012023. [CrossRef]
26. Baltatu, I.; Sandu, A.V.; Vlad, M.D.; Spataru, M.C.; Vizureanu, P.; Baltatu, M.S. Mechanical Characterization and In Vitro Assay of Biocompatible Titanium Alloys. *Micromachines* **2022**, *13*, 430. [CrossRef]

27. Lee, M.; Kim, I.-S.; Moon, Y.H.; Yoon, H.S.; Park, C.H.; Lee, T. Kinetics of Capability Aging in Ti-13Nb-13Zr Alloy. *Crystals* **2020**, *10*, 693. [CrossRef]
28. Stróż, A.; Goryczka, T.; Łosiewicz, B. Electrochemical formation of self-organized nanotubular oxide layers on niobium (Review). *Curr. Nanosci.* **2019**, *15*, 42–48. [CrossRef]
29. Łosiewicz, B.; Stróż, A.; Osak, P.; Maszybrocka, J.; Gerle, A.; Dudek, K.; Balin, K.; Łukowiec, D.; Gawlikowski, M.; Bogunia, S. Production, Characterization and Application of Oxide Nanotubes on Ti–6Al–7Nb Alloy as a Potential Drug Carrier. *Materials* **2021**, *14*, 6142. [CrossRef] [PubMed]
30. Agour, M.; Abdal-hay, A.; Hassan, M.K.; Bartnikowski, M.; Ivanovski, S. Alkali-Treated Titanium Coated with a Polyurethane, Magnesium and Hydroxyapatite Composite for Bone Tissue Engineering. *Nanomaterials* **2021**, *11*, 1129. [CrossRef]
31. Stróż, A.; Łosiewicz, B.; Zubko, M.; Chmiela, B.; Balin, K.; Dercz, G.; Gawlikowski, M.; Goryczka, T. Production, structure and biocompatible properties of oxide nanotubes on Ti13Nb13Zr alloy for medical applications. *Mater. Charact.* **2017**, *132*, 363–372. [CrossRef]
32. Stróż, A.; Dercz, G.; Chmiela, B.; Łosiewicz, B. Electrochemical synthesis of oxide nanotubes on biomedical Ti13Nb13Zr alloy with potential use as bone implant. *AIP Conf. Proc.* **2019**, *2083*, 030004. [CrossRef]
33. Smołka, A.; Rodak, K.; Dercz, G.; Dudek, K.; Łosiewicz, B. Electrochemical Formation of Self-Organized Nanotubular Oxide Layers on Ti13Zr13Nb Alloy for Biomedical Applications. *Acta Phys. Pol.* **2014**, *125*, 932–935. [CrossRef]
34. Stróż, A.; Dercz, G.; Chmiela, B.; Stróż, D.; Łosiewicz, B. Electrochemical formation of second generation TiO$_2$ nanotubes on Ti13Nb13Zr alloy for biomedical applications. *Acta Phys. Pol.* **2016**, *130*, 1079–1080. [CrossRef]
35. Łosiewicz, B.; Skwarek, S.; Stróż, A.; Osak, P.; Dudek, K.; Kubisztal, J.; Maszybrocka, J. Production and Characterization of the Third-Generation Oxide Nanotubes on Ti-13Zr-13Nb Alloy. *Materials* **2022**, *15*, 2321. [CrossRef] [PubMed]
36. Durdu, S.; Cihan, G.; Yalcin, E.; Altinkok, A. Characterization and mechanical properties of TiO$_2$ nanotubes formed on titanium by anodic oxidation. *Ceram. Int.* **2021**, *47*, 10972–10979. [CrossRef]
37. Ossowska, A.; Olive, J.-M.; Zielinski, A.; Wojtowicz, A. Effect of double thermal and electrochemical oxidation on titanium alloys for medical applications. *Appl. Surf. Sci.* **2021**, *563*, 150340. [CrossRef]
38. Ossowska, A.; Zieliński, A.; Olive, J.-M.; Wojtowicz, A.; Szweda, P. Influence of Two-Stage Anodization on Properties of the Oxide Coatings on the Ti–13Nb–13Zr Alloy. *Coatings* **2020**, *10*, 707. [CrossRef]
39. Handzlik, P.; Gutkowski, K. Synthesis of oxide nanotubes on Ti13Nb13Zr alloy by the electrochemical method. *J. Porous Mater.* **2019**, *26*, 1631–1637. [CrossRef]
40. Stępień, M.; Handzlik, P.; Fitzner, K. Electrochemical synthesis of oxide nanotubes on Ti6Al7Nb alloy and their interaction with the simulated body fluid. *J. Solid State Electrochem.* **2016**, *20*, 2651–2661. [CrossRef]
41. Schneider, S.G.; Nunes, C.A.; Rogero, S.P.; Higa, O.Z.; Bressiani, J.C. Mechanical properties and cytotoxic evaluation of the Ti—13Nb—13Zr alloy. *Biomecánica* **2000**, *8*, 84–87. [CrossRef]
42. Lee, T. Variation in Mechanical Properties of Ti—13Nb—13Zr Depending on Annealing Temperature. *Appl. Sci.* **2020**, *10*, 7896. [CrossRef]
43. Wu, S.; Wang, S.; Liu, W.; Yu, X.; Wang, G.; Chang, Z.; Wen, D. Microstructure and properties of TiO$_2$ nanotube coatings on bone plate surface fabrication by anodic oxidation. *Surf. Coat. Technol.* **2019**, *374*, 362–373. [CrossRef]
44. ASTM F1713-08(2021)e1; Standard Specification for Wrought Titanium-13Niobium-13Zirconium Alloy for Surgical Implant Applications (UNS R58130). ASTM: West Conshohocken, PA, USA, 2021.
45. ISO 6507-1:2018; Metallic Materials—Vickers Hardness Test—Part 1: Test Method. ISO: Geneva, Switzerland, 2018.
46. ISO 6507-2:2018; Metallic Materials—Vickers Hardness Test—Part 2: Verification and Calibration of Testing Machines. ISO: Geneva, Switzerland, 2018.
47. Xu, Y.N.; Liu, M.N.; Wang, M.C.; Oloyede, A.; Bell, J.M.; Yan, C. Nanoindentation study of the mechanical behavior of TiO$_2$ nanotube arrays. *J. Appl. Phys.* **2015**, *118*, 145301. [CrossRef]

Disclaimer/Publisher's Note: The statements, opinions and data contained in all publications are solely those of the individual author(s) and contributor(s) and not of MDPI and/or the editor(s). MDPI and/or the editor(s) disclaim responsibility for any injury to people or property resulting from any ideas, methods, instructions or products referred to in the content.

Article

In Vitro Bioelectrochemical Properties of Second-Generation Oxide Nanotubes on Ti–13Zr–13Nb Biomedical Alloy

Agnieszka Stróż [1], Thomas Luxbacher [2], Karolina Dudek [3], Bartosz Chmiela [4], Patrycja Osak [1] and Bożena Łosiewicz [1,*]

[1] Faculty of Science and Technology, Institute of Materials Engineering, University of Silesia in Katowice, 75 Pułku Piechoty 1A, 41-500 Chorzów, Poland
[2] Anton Paar GmbH, Street 20, 8054 Graz, Austria
[3] Refractory Materials Center, Institute of Ceramics and Building Materials, Łukasiewicz Research Network, Toszecka 99, 44-100 Gliwice, Poland
[4] Insitute of Materials Science, Silesian University of Technology, Z. Krasińskiego 8, 40-019 Katowice, Poland
* Correspondence: bozena.losiewicz@us.edu.pl; Tel.: +48-32-3497-527

Abstract: Surface charge and in vitro corrosion resistance are some of the key parameters characterizing biomaterials in the interaction of the implant with the biological environment. Hence, this work investigates the in vitro bioelectrochemical behavior of newly developed oxide nanotubes (ONTs) layers of second-generation (2G) on a Ti–13Zr–13Nb alloy. The 2G ONTs were produced by anodization in 1 M $(NH_4)_2SO_4$ solution with 2 wt.% of NH_4F. The physical and chemical properties of the obtained bamboo-inspired 2G ONTs were characterized using scanning electron microscopy with field emission and energy dispersive spectroscopy. Zeta potential measurements for the examined materials were carried out using an electrokinetic analyzer in aqueous electrolytes of potassium chloride, phosphate-buffered saline and artificial blood. It was found that the electrolyte type and the ionic strength affect the bioelectrochemical properties of 2G ONTs layers. Open circuit potential and anodic polarization curve results proved the influence of anodizing on the improvement of in vitro corrosion resistance of the Ti–13Zr–13Nb alloy in PBS solution. The anodizing conditions used can be proposed for the production of long-term implants, which are not susceptible to pitting corrosion up to 9.4 V.

Keywords: anodizing; corrosion resistance; oxide nanotubes; Ti–13Zr–13Nb alloy; zeta potential

Citation: Stróż, A.; Luxbacher, T.; Dudek, K.; Chmiela, B.; Osak, P.; Łosiewicz, B. In Vitro Bioelectrochemical Properties of Second-Generation Oxide Nanotubes on Ti–13Zr–13Nb Biomedical Alloy. *Materials* **2023**, *16*, 1408. https://doi.org/10.3390/ma16041408

Academic Editors: Thomas Dippong, Chunguang Yang and Hendra Hermawan

Received: 31 December 2022
Revised: 4 February 2023
Accepted: 6 February 2023
Published: 8 February 2023

Copyright: © 2023 by the authors. Licensee MDPI, Basel, Switzerland. This article is an open access article distributed under the terms and conditions of the Creative Commons Attribution (CC BY) license (https://creativecommons.org/licenses/by/4.0/).

1. Introduction

Innovative biomaterials inspired by nature are the answer to the key challenges of modern medicine. Current scientific trends in medicine concern the use of intelligent bionanomaterials for the needs of dynamically developing regenerative medicine, tissue engineering and targeted therapy [1–3]. The latest generation of bionanomaterials that can give hope to hundreds of thousands of patients waiting in queues for their health and life include oxide nanotubes (ONTs) inspired by the structure of bamboo obtained on titanium and titanium alloys intended for long-term implants [4–21]. Currently, titanium and its alloys are commonly used metallic biomaterials in medicine due to their unique properties, which include high corrosion resistance [1,4–6,11,15,20,22–27], biological inertness [1,11,28], low specific gravity and excellent mechanical properties [1,7,16,24,29–34]. In vitro corrosion resistance of metallic biomaterials affects their functionality and durability. It is also the main factor that determines biocompatibility. According to the fundamental paradigm of metallic biomaterials, which does not apply only to biodegradable metals, the more corrosion-resistant a biomaterial is, the greater its biocompatibility [27]. The Ti–13Zr–13Nb alloy belongs to the newest group of titanium alloys, which do not contain allergenic nickel or aluminum and carcinogenic vanadium [4,5,11–15,18,19,23,32–35]. This bi-phase (α + β) alloy shows high biotolerance and very good corrosion resistance and is classified as a long-lasting biomaterial [4,5,11–15,18,19,23,32,34]. The biocompatibility of titanium and its alloys results from the presence of a native oxide layer on its surface, which is characterized by thermodynamic stability and low electronic conductivity. Its thickness ranges from 2 to 10 nm and provides high corrosion resistance [23]. In the biological environment of the human body, in the oxide layer on the surface of titanium and its alloys, processes involving the incorporation of elements from the fluids of cells and tissues surrounding the implant takes place. Moreover, an in vitro hemocompatibility study of the Ti–13Zr–13Nb alloy before and after anodizing in 1 M ethylene glycol solution with 4 wt.% of NH_4F revealed the hemolytic index of 0.30 (8) and 0.00 before and after surface modification, respectively [11]. The obtained results confirmed that anodizing of the Ti–13Zr–13Nb biomedical alloy allowed the complete elimination of hemolysis.

To increase the biocompatibility of the Ti–13Zr–13Nb alloy and improve its biological activity, we propose modifying the surface of this alloy using the anodizing method, making it possible to produce self-organized ONTs of various geometry and length. Bamboo-like ONTs layers were obtained on the Ti–13Zr–3Nb alloy using inorganic and organic electrolytes [4,5,12–15,18,19]. The porous layers of ONTs can be additionally modified by enriching them with electrolyte components, e.g., phosphates, giving them bioactivity features. The ONTs can also be saturated with therapeutic agents, bactericides, active substances or tissue-forming hydroxyapatite, thanks to which they can be used in orthopedics, dentistry and intelligent drug delivery systems [10]. The bamboo-like ONTs enable the strengthening of bone functions at the boundary of the implant and bones. In the case of the rough surface of titanium implants, an increase in osteogenic properties such as cell proliferation, protein adsorption and deposition of calcium have been reported, which favor osseointegration [1–3].

As a result of surface modification, the physical properties change, which also affects the chemical properties of the surface. The implant affects the surrounding tissues through its surface. This is due to the interaction of implant surfaces and body fluids, which is often mediated by adsorbed proteins [5,11,20,22–26]. The features of the implant surface, regarding its roughness, topography and surface chemistry, are then "translated" by the protein layer into information that is understandable by the cells.

A sensitive indicator for the actual surface charge of a biomaterial in contact with a biological environment is the zeta potential (ζ) [36–38]. Wettability and surface roughness of ONTs formed on the Ti–13Zr–13Nb alloy implant surfaces are characterized in the literature, while the surface charge is still unavailable [39,40]. Studying the zeta potential of the biocompatible Ti–13Zr–13Nb alloy before electrochemical modification and with vertically oriented ONTs will allow learning the mechanism by which ions will adhere better to smooth surfaces and will prefer a porous structure. The technique of streaming potential is based on the phenomenon of creating an electric field when the electrolyte flows, which remains tangential to the stationary, charged surface of the tested material. The ζ value is calculated from the generated streaming potential. The local surface charge is related to the surface roughness of ONTs in the nanoscale. On the edges of ONTs with different internal and external diameters, there is a high surface charge density, which is the binding site for monovalent and divalent ions, as well as proteins mediating osteoblast adhesion [39,40]. Despite this, there is a lack of data in the literature regarding this important bioelectrochemical property of charged solid–liquid interfaces for the Ti–13Zr–13Nb alloy in body fluids. Therefore, the main purpose of this work was an evaluation of the anodizing effect on ζ and in vitro corrosion resistance of the Ti–13Zr–13Nb alloy in artificial body fluid. This work brings a new contribution to the description of the relationship between the new anodizing conditions and in vitro bioelectrochemical properties of the latest generation Ti–13Zr–13Nb alloy, which has been intensively researched in recent years due to its unique properties.

2. Materials and Methods

2.1. Substrate Treatment

The material under study was Ti–13Zr–13Nb (wt.%) alloy (BIMO TECH, Wrocław, Poland). Disc-shaped samples with a thickness of 5 mm were cut from a wire with a diameter of 20 mm and a length of 1 m. A specification covering chemical, mechanical and metallurgical requirements for wrought Ti–13Nb–13Zr alloy for surgical implant applications is provided in standard ASTM F1713-08(2021)e1 [41]. The samples were subjected to one-sided wet grinding on a metallographic grinding and polishing machine Forcipol 202 (Metkon Instruments Inc., Bursa, Turkey) at 250 rpm of the grinding wheel with soft start and soft stop. SiC abrasive papers of P600, P1200 and P3000 gradations (Buehler Ltd., Lake Bluff, IL, USA) were used. The ground samples with a mirror-like surface were rinsed thoroughly under tap water and sonicated for 20 min in acetone (Avantor Performance Materials Poland S.A., Gliwice, Poland) and then in ultrapure water with resistivity of 18.2 MΩ cm (Milli-Q Advantage A10 Water Purification System, Millipore SAS, Molsheim, France). The cleaning procedure in ultrapure water was repeated twice with a change of water.

2.2. Production of ONTs on Ti–13Zr–13Nb Alloy

A detailed method of preparing anodes was described in our previous work [10]. The self-passive oxide layer on the anode surface was removed immediately before electrochemical oxidation by dissolving in 25% v/v HNO$_3$ (Avantor Performance Materials Poland S.A., Gliwice, Poland) for 10 min at room temperature. The depassivated anodes were cleaned with Milli-Q water in an ultrasonic bath for 20 min.

To produce the ONTs layers, the prepared anodes were subjected to one-step anodizing. Electrochemical oxidation was carried out in 1 M (NH$_4$)$_2$SO$_4$ solution with 2 wt.% of NH$_4$F at 20 V for 120 min at room temperature. Ammonium sulfate ($\geq$99.0% purity) and ammonium fluoride ($\geq$99.99% trace metals basis) were supplied by Sigma-Aldrich (Saint Louis, MI, USA). The mechanism of obtaining 2G ONTs layers on the surface of the Ti–13Zr–13Nb alloy under the applied anodizing conditions was described in detail in our earlier work [13]. Anodizing was conducted in a two-electrode system in which the anode was a sample tested in a Teflon holder, while the 4 cm^2 platinum mesh served as a counter electrode. The geometric surface of the anode subjected to electrochemical oxidation was 0.64 cm^2. The distance between the cathode and anode was 25 mm. After anodizing, each anode was immersed for 5 min in Milli-Q water subjected to vigorous agitation.

2.3. Physicochemical Characteristics of ONTs on Ti–13Zr–13Nb Alloy

Surface morphology and thickness of 2G ONTs layers obtained on the Ti–13Zr–13Nb alloy were examined using a scanning electron microscope with field emission (FE-SEM) Hitachi HD-2300A (Hitachi Ltd., Tokyo, Japan) under low-vacuum conditions of 50 Pa at an accelerating voltage of 15 kV. FE-SEM images were collected by secondary electrons (SE). Before microscopic examinations, a 5 nm chromium layer was deposited on the surface of the tested samples using an ion sputtering machine Quorum Q150T ES equipment (Quorum Technologies, East Sussex, UK) with argon as the ion extracting source. Local chemical composition with the surface distribution of elements was performed using an Energy Dispersive Spectrometer (EDS, Oxford Instruments, Abingdon, UK).

2.4. In Vitro Surface Characteristics of ONTs on Ti–13Zr–13Nb Alloy in Body Fluids

The surface zeta potential for the non-anodized and anodized Ti–13Zr–13Nb alloy was measured in aqueous electrolytes with different ionic strengths. The following solutions were applied: KCl (0.001 mol L^{-1}) as the background electrolyte, PBS (0.001 mol L^{-1} and 0.01 mol L^{-1}) [42] and artificial blood (0.01 mol L^{-1}) [43] in a wide pH range from over 3 to 9 at 37(2) °C. The pH of these aqueous electrolytes was adjusted with 0.05 mol L^{-1} HCl and 0.05 mol L^{-1} NaOH, respectively. Streaming current measurements were performed with an electrokinetic analyzer SurPASS 3 (Anton Paar GmbH, Graz, Austria) for surface testing of materials with automatic ζ analysis. The Adjustable Gap Cell shown in Figure 1a was used. During the measurement of ζ, a pair of the same samples were used, which were fixed in the holders with a cross-section of 10 × 10 mm using double-sided adhesive tape (Figure 1b). The sample holders were inserted into the Adjustable Gap Cell in such a way that the tested surfaces of the discs faced each other. Both tested surfaces were located at a distance of about 100 μm. The SurPASS 3 instrument enabled the use of both the classic streaming potential method and the streaming current method for direct analysis of the surface zeta potential. The surface conductance of the investigated sample influenced the surface zeta potential of tested surfaces evaluated from streaming potential measurements. In turn, the streaming current method required that the geometry of the flow channel be known (Figure 1c). The ζ of planar samples was preferably determined from the measurement of the streaming current due to their electrical conductivity.

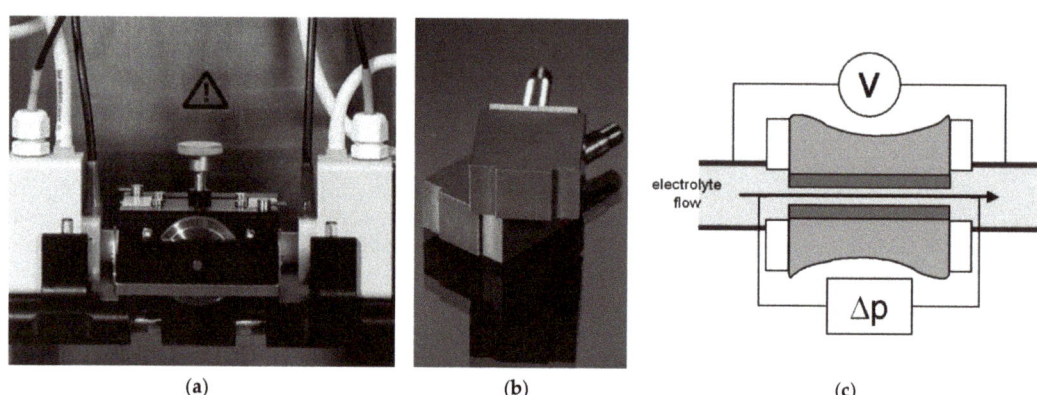

Figure 1. Adjustable Gap Cell mounted between electrodes (**a**); Sample holder (10 × 10 mm, center) (**b**); Measuring principle (**c**).

Scheme of the rectangular slit channel between adjacent solid samples with a planar surface indicating its dimensions is presented in Figure 2. L, W and H stand for the length, width and height of the flow channel, respectively. The blue arrows indicate the direction of electrolyte flow during the ζ measurement.

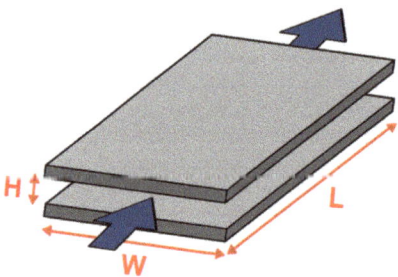

Figure 2. Scheme of the rectangular slit channel between adjacent solid samples with a planar surface. L, W and H are the length, width and height of the flow channel, respectively.

The fundamental Helmholtz–Smoluchowski dependence for evaluating the ζ from streaming current measurements is given by Equation (1), which relates electrokinetic phenomena like the streaming potential and the streaming current to the zeta potential [39,40]:

$$\zeta = \frac{dI_{str}}{d\Delta p} \cdot \frac{\eta}{\varepsilon_{rel} \cdot \varepsilon_0} \cdot \frac{L}{A} \qquad (1)$$

In Equation (1), the measured streaming current coupling coefficient $dI_{str}/d\Delta p$ is related to the cell constant L/A of the flow channel, which is the gap between adjacent solid samples. As shown schematically in Figure 2, L denotes the length of the rectangular slit channel formed between two planar surfaces and A is its cross-section, $A = W \times H$, with W denoting channel width and H gap height. In Equation (1), η is the viscosity of the electrolyte, ε_{rel} is related to the dielectric coefficient of the electrolyte and ε_0 denotes the vacuum permittivity. For dilute aqueous solutions, η and ε_{rel} of water were used. The application of Equation (1) requires a thorough knowledge of the flow channel geometry, defined by the L/A cell constant. In case of the rectangular slot channel, the L and W parameters are determined by the size of the solid sample. The value of the parameter H, which is the gap height, can be calculated based on the measured volume flow rate of electrolyte passing through the flow channel driven by the applied differential pressure.

2.5. In Vitro Corrosion Resistance of ONTs on Ti–13Zr–13Nb Alloy in PBS

In vitro corrosion resistance measurements of the non-anodized Ti–13Zr–13Nb alloy with 2G ONTs layers were conducted in PBS at 37(2) °C using the method of open circuit potential (OCP) and anodic polarization curves. For the preparation of PBS containing 8.0 g L^{-1} NaCl, 0.2 g L^{-1} KCl, 1.42 g L^{-1} Na$_2$HPO$_4$ and 0.24 g L^{-1} KH$_2$PO$_4$ [42] analytically pure reagents (Avantor Performance Materials Poland S.A., Gliwice, Poland) and Milli-Q water were used. The pH of the PBS solution was adjusted to 7.4(1) using 4% NaOH and 1% C$_3$H$_6$O$_3$. Prior to each measurement, a fresh portion of PBS was deaerated using an Ar flow (UHP Ar, 99.999%) for 20 min. Corrosion behavior of the tested electrodes was characterized in a single-chamber electrochemical cell using a conventional three-electrode system. The cathode was the Ti–13Zr–13Nb alloy without and with the ONTs layer. The anode was platinum foil with dimensions of $40 \times 20 \times 2$ mm. The reference electrode was the saturated calomel electrode (SCE) immersed in PBS electrolyte using Luggin capillary. Open circuit potential (E_{OC}) was stabilized for 2 h. Then, the anodic polarization curves in the range of potentials from E_{OC} minus 150 mV to 9.4 V were registered at the polarization rate of $v = 1$ mV s^{-1}. All electrochemical tests were conducted using the Autolab/PGSTAT20 computer-controlled electrochemical system (Metrohm Autolab B.V., Utrecht, The Netherlands) equipped with the General Purpose Electrochemical System software.

3. Results and Discussion

3.1. FE-SEM/EDS Studies of ONTs on Ti–13Zr–13Nb Alloy

The surface morphology of the Ti–13Zr–13Nb alloy after anodizing in 1 M (NH$_4$)$_2$SO$_4$ solution with 2 wt.% of NH$_4$F at 20 V for 120 min can be observed in Figure 3. The SE FE-SEM image in Figure 3a presents the on-top general view of the 2G ONTs layer. Figure 3b shows the surface morphology of the obtained bamboo-like ONTs in more detail. Oxide nanotubes with a circular cross-section and single walls are evenly distributed over the observed surface and arranged vertically. Bundles of ONTs that grew in local areas corresponding to the presence of α and β phases are also visible [15]. The multi-step process of 2G ONTs layer formation on the surface of the Ti–13Zr–13Nb alloy in aqueous solutions containing fluoride ions was discussed in detail in our earlier work [13].

Figure 3c presents an exemplary SE FE-SEM image of the mechanically scratched region of the 2G ONTs layer formed on the Ti–13Zr–13Nb substrate. The top view of the mechanically fractured oxide layer reveals micro-areas with invisible and visible bamboo-like nanotubes, which form an ordered matrix of vertically oriented ONTs. The high degree of ordering of smooth-walled ONTs may increase the corrosion resistance of the Ti–13Zr–13Nb alloy in body fluids [4,5,11]. The inner diameter, outer diameter and length of the obtained 2G ONTs layer were determined based on the FE-SEM images recorded from the selected areas of the Ti–13Zr–13Nb alloy surface in our preliminary results [14]. Empirical distribution histograms of the ONTs' diameters allowed determining the average values of the morphological parameters. It was found that under the proposed conditions, bamboo-like ONTs with an inner diameter of 61(11) nm and an outer diameter of 103(16) nm were obtained. The average value of 2G ONTs length was 3.9(2) µm, and the specific surface area per cm^2 was equal to 15.6 cm^2 cm^{-2}.

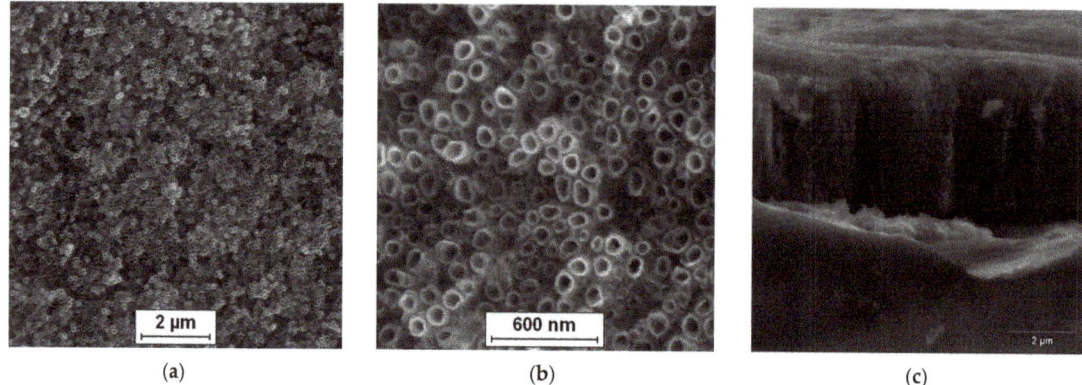

Figure 3. SE FE-SEM image of the Ti–13Zr–13Nb alloy after anodizing in 1 M $(NH_4)_2SO_4$ solution with 1 wt.% of NH_4F at 20 V for 120 min: (**a**) On-top general view of 2G ONTs layer; (**b**) View of 2G ONTs layer in a selected micro-region; (**c**) Fracture of 2G ONTs layer [14].

The X-ray structural investigations of the 2G ONTs layer obtained in 1 M $NH_4(SO_4)_2$ solution with 2 wt.% NH_4F on the Ti–13Zr–13Nb alloy surface were carried out in the previous work [14]. The grazing incidence X-ray diffraction (GIXD) results shown in Figure 4 revealed the presence of α-Ti and β-Ti phases for the bi-phase Ti–13Zr–13Nb alloy substrate and an amorphous halo related to the 2G ONTs layer.

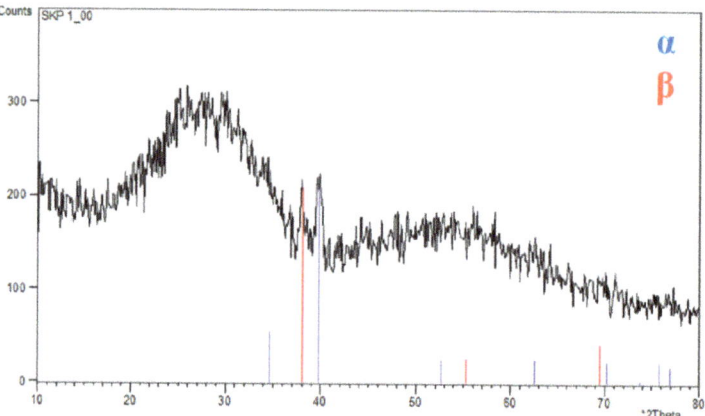

Figure 4. GIXD pattern of Ti–13Zr–13Nb after anodization in 1 M $(NH_4)_2SO_4$ with 2 wt.% NH_4F [14].

A new generation of biomedical titanium alloy was selected for research in this work. By eliminating toxic elements such as vanadium, aluminum and nickel from the composition of the titanium alloy, the appropriate requirements for medical implant applications were ensured [32]. The chemical composition of the Ti–13Zr–13Nb alloy was investigated in our previous work [15]. The results of the analysis of the local chemical composition of the Ti–13Zr–13Nb alloy with the obtained 2G ONTs layer are shown in Figure 5.

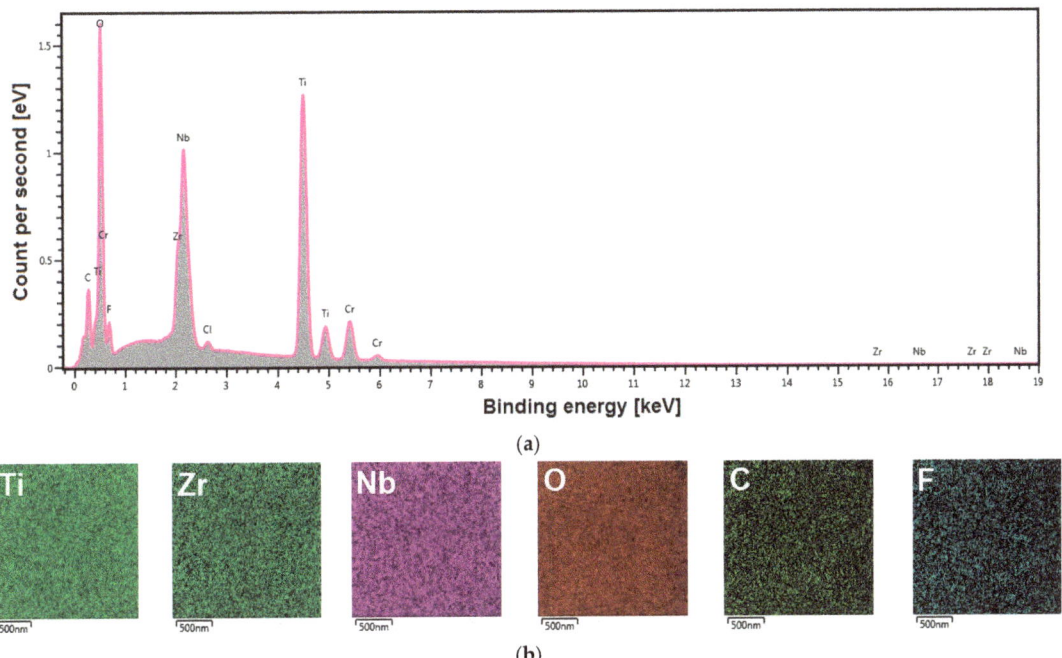

Figure 5. Local chemical composition of the Ti–13Zr–13Nb alloy with the 2G ONTs layer: (**a**) Energy dispersive spectrum in the micro-region; (**b**) EDS maps of elements distribution (Ti, Zr, Nb, O, C, F) in the micro-region.

A representative energy dispersive spectrum in the micro-region on the surface of the anodized Ti–13Zr–13Nb alloy shows the relationship between the count per second and the binding energy (Figure 5a). Peaks from Ti, Zr and Nb alloying elements are visible. The presence of an oxygen-derived peak of high intensity in the EDS spectrum in Figure 5a testifies that, on the surface of the tested biomaterial, an oxide layer is present. The obtained EDS spectrum also reveals the trace amounts of the elements included in the electrolyte used for anodizing (F, Cl), an ultra-thin layer applied to improve the conductivity of the tested sample (Cr) or impurities (C).

Figure 5b show the corresponding distribution maps of chemical elements in the micro-region. EDS distribution maps for individual elements have been recorded in different colors, which allows distinguishing the location of elements such as Ti, Zr, Nb, O, C and F. The obtained EDS distribution maps show that all identified elements are evenly distributed on the surface in the studied micro-region, and C and F occur in small amounts.

3.2. In Vitro Bioelectrochemical Characteristics in Body Fluids

The zeta potential is related to the surface charge at the biomaterial | electrolyte interface, and its knowledge allows for characterizing surface properties and designing new biomaterials. To determine the ζ parameter for the Ti–13Zr–13Nb alloy before and after anodizing, the measurement of streaming current was performed alternatively in both flow directions. Figure 6 shows exemplary pressure ramps as streaming current vs. differential pressure for the Ti–13Zr–13Nb alloy after anodizing at different pH of the electrolyte. One can see that the dependence representing the streaming current on the applied differential pressure is strictly linear, and the linear regression coefficients take values higher than $R^2 = 0.99$.

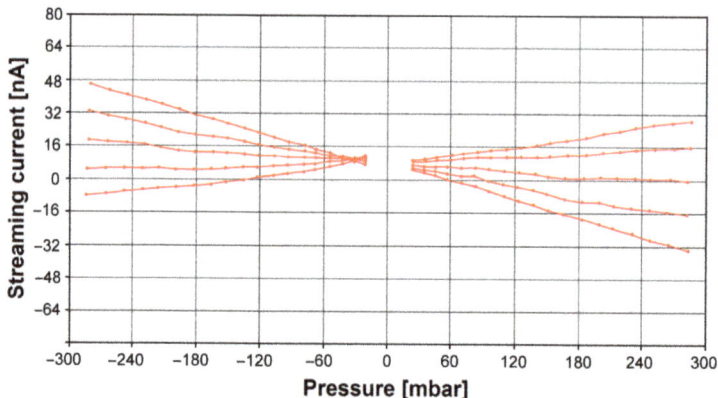

Figure 6. Streaming current vs. differential pressure for the Ti–13Zr–13Nb alloy after anodizing at various pH of the electrolyte.

Figure 7 shows the flow behavior of electrolytes passing through the gap between exemplary sample surfaces for the Ti–13Zr–13Nb alloy after anodizing expressed as volume flow rate vs. differential pressure. For all series of measurements, a linear dependence that indicated laminar flow behavior was found.

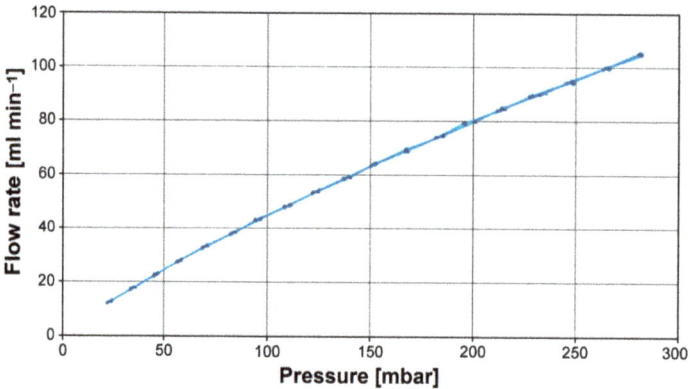

Figure 7. Flow rate vs. differential pressure for the Ti–13Zr–13Nb alloy after anodizing.

The determined zeta potential for the Ti–13Zr–13Nb alloy with and without the ONTs layer in the presence of the inert aqueous solution of 0.001 mol L^{-1} KCl was compared. Figure 8 shows the dependence of electrolyte pH in the range of pH 3–9 on zeta potential, where the corresponding isoelectric points (IEPs) were found. Note that the pH was first changed to low pH by starting close to pH 6 (the native pH of a freshly prepared KCl solution) and adding acid.

The measuring cell was then rinsed with Milli-Q water, thereby keeping the Ti–13Zr–13Nb alloy disks mounted. Afterward, the electrolyte was exchanged, and the titration proceeded to high pH. Note the coincidence of repetitive measurements close to pH 6, which indicates the stability of the Ti–13Zr–13Nb alloy surfaces in the presence of the aqueous solution in the pH range investigated. For the polished Ti–13Zr–13Nb alloy disks, a pH dependence on zeta potential with an IEP 4.2 typical for surfaces with little or no functional groups was found [44]. We assumed a native oxide layer present on the Ti–13Zr–13Nb substrate whose thickness was in the nanometer range [23]. This behavior was also found, e.g., for gold, stainless steel or polymer surfaces [44]. The growth of the ONTs on the Ti–13Zr–13Nb substrate is approved by the significant shift of the IEP to pH 5.4 and renders the surface of the anodized sample amphoteric. It was revealed that the 2G ONTs on the Ti–13Zr–13Nb substrate represent the amorphous oxide phase [14].

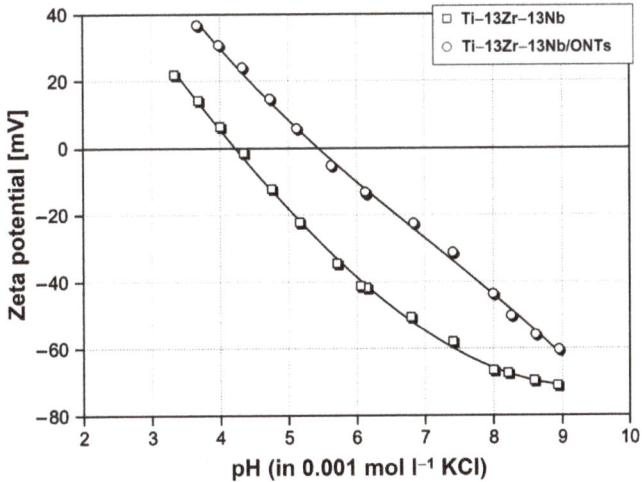

Figure 8. Dependence of electrolyte pH on zeta potential for the Ti–13Zr–13Nb alloy without and with the ONTs layer in 0.001 mol l^{-1} KCl.

In the next step, the effect of different buffer solutions on the zeta potential of the Ti–13Zr–13Nb alloy before and after anodizing was compared. Figure 9 shows the zeta potential at pH 7.4 for the tested materials in the presence of 0.001 mol L^{-1} KCl, 0.001 mol L^{-1} PBS, 0.01 mol L^{-1} PBS and 0.01 mol L^{-1} artificial blood, respectively. When exchanging the inert 1:1 electrolyte by PBS with comparable ionic strength, a significant increase in the negative ζ for the Ti–13Zr–13Nb/ONTs sample while the zeta potential remains almost unaffected for the Ti–13Zr–13Nb sample within the experimental error was found. We thus assumed a strong interaction of phosphate ions with the amorphous oxide surface.

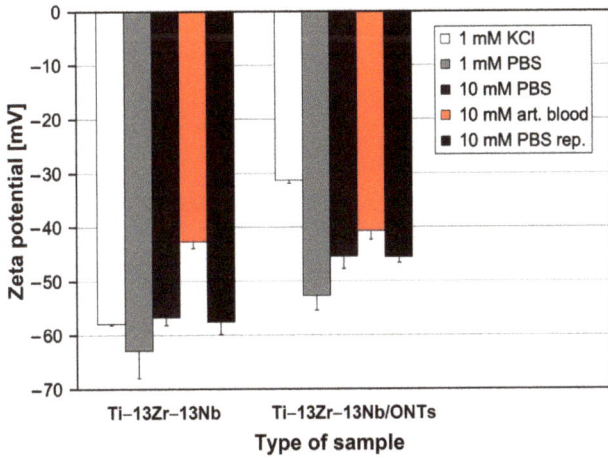

Figure 9. Zeta potential at pH 7.4 for the Ti–13Zr–13Nb alloy before and after anodizing in the presence of different buffer solutions, where the number of experiments for the error bars was n = 3.

By further increasing the ionic strength of PBS to 0.01 mol L^{-1}, a decrease in the zeta potential for samples Ti–13Zr–13Nb and Ti–13Zr–13Nb/ONTs was noted, which follows the prediction of the model of the double layer. Increasing ionic strength compresses the diffuse layer of surface-charge compensation counter-ions and thus reduces the magnitude of the zeta potential. This prediction assumes the absence of selective interaction between the electrolyte ions with the solid surface. An explanation for this observation is simply a smaller slope of the dependence of (negative) zeta potential on the ionic strength [44]. However, when exchanging the PBS buffer with artificial blood, thereby maintaining the ionic strength, a decrease in the negative zeta potential for all samples was found. This decrease is higher for the Ti–13Zr–13Nb sample (25%) as compared to the sample Ti–13Zr–13Nb/ONTs (10%). It was concluded that the complex ions contained in artificial blood have a stronger affinity to hydrophobic surfaces than to more hydrophilic ones.

After completing the analysis in artificial blood, the disk samples were kept mounted in the Adjustable Gap Cell and rinsed with Milli-Q water. The repetitive measurement in the presence of 0.01 mol L^{-1} PBS for samples Ti–13Zr–13Nb and Ti–13Zr–13Nb/ONTs confirmed that the adsorption of complex ions contained in artificial blood was fully reversible.

Since the zeta potential at pH 7.4 was negative for all samples and to determine the IEP, as a final measurement step, another titration starting at physiological pH and proceeding towards low pH was performed. Figure 10 shows the results, which confirm that the complex ions contained in the simulated body fluid (SBF) adsorb on the Ti–13Zr–13Nb alloy before and after anodizing and shift the IEP to low pH.

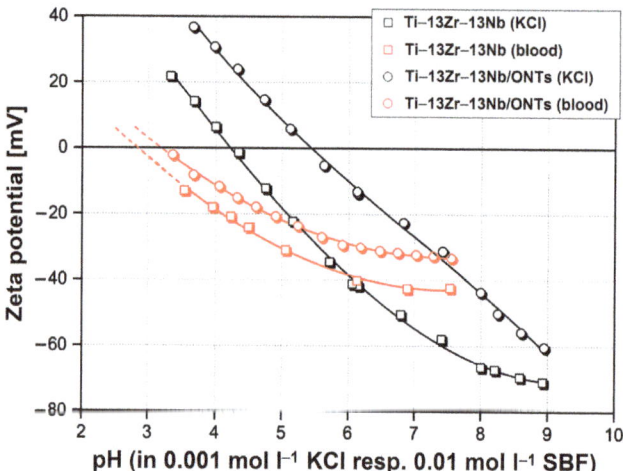

Figure 10. Dependence of electrolyte pH on zeta potential for the Ti–13Zr–13Nb alloy without and with the ONTs layer in 0.001 mol L^{-1} KCl and 0.01 mol L^{-1} simulated body fluid (SBF).

It is interesting to note that the effect of SBF is smaller for unoxidized Ti–13Zr–13Nb alloy. After finishing the pH titration, the sample disks were rinsed with Milli-Q water, and measurements were repeated for both types of samples in the presence of 0.01 mol L^{-1} PBS in order to investigate whether the adsorption of complex ions was reversible or permanent. The corresponding zeta potential at pH 7.4 is shown in Figure 9.

3.3. In Vitro Open Circuit Potential Characteristics in Body Fluids

The open circuit potential (E_{OC}) parameter was used to determine the initial in vitro corrosion resistance of the Ti–13Zr–13Nb electrode before and after anodizing under conditions similar to those in the human body. Due to the fact that inflammation occurs immediately after implantation, which is associated with a decrease in pH in the acidic direction in the tissues surrounding the implant, electrochemical E_{OC} measurements were carried out in the PBS electrolyte at physiological and acidic pH. Changes in pH affect the risk of reduced corrosion resistance of the implant. Therefore, the OCP method was used to assess the protective properties of the obtained 2G ONTs layers on the Ti–13Zr–13Nb alloy. Figure 11 shows the course of E_{OC} for the Ti–13Zr–13Nb electrode without and with 2G ONTs layer for 2 h of immersion in PBC with pH 7.4 and 5.5 at 37 °C.

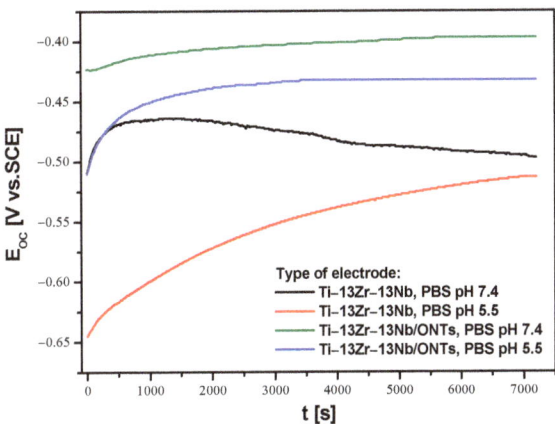

Figure 11. Open circuit potential (E_{OC}) in the function of immersion time (t) for the Ti–13Zr–13Nb electrode without and with 2G ONTs layer in PBS at 37 °C.

The ionic–electron equilibrium at the interfacial boundary between the electrode surface and electrolyte was reached after 7200 s of immersion. It should be noted that E_{OC} stabilized more slowly on the non-oxidized Ti–13Zr–13Nb electrode in both neutral and acidic PBS. This phenomenon is probably related to the self-passivation of the Ti–13Zr–13Nb electrode surface after immersion in the electrolyte and the thickening of the ultrathin oxide film [23]. In the presence of 2G ONTs layers with protective properties, E_{OC} stabilization occurred faster. The stabilized E_{OC} value, treated in further studies as an approximate value of corrosion potential (E_{cor}), was negative for all tested electrodes. The non-anodized Ti–13Zr–13Nb electrode showed lower corrosion resistance in PBS in comparison to the Ti–13Zr–13Nb/ONTs electrode in both neutral and acidic electrolytes. Such changes in the E_{OC} indicate a decrement in the thermodynamic tendency to the corrosion of anodized Ti–13Zr–13Nb electrode. The lowest corrosion resistance was demonstrated by the non-anodized Ti–13Zr–13Nb electrode in PBS of pH 5.5, for which the average value of E_{OC} was −0.513(30) V. It means that the increase in the content of aggressive chloride ions in the electrolyte accelerated the corrosion processes. The highest average E_{OC} of −0.397(19) V was determined for the Ti–13Zr–13Nb/ONTs electrode in PBS of pH 7.4, which indicates that the obtained 2G ONTs layer has stronger barrier properties as compared to the native oxide layer. These results confirm that the application of the anodizing process under the conditions used can significantly improve the corrosion resistance of the Ti–13Zr–13Nb electrode in body fluids.

3.4. In Vitro Susceptibility to Pitting Corrosion in Body Fluids

Analysis of the anodic polarization curves shown in the semi-log form in Figure 12 revealed a similar course for all investigated electrodes with apparent passive anodic behavior. A shift towards cathode potentials for both Ti–13Zr–13Nb and Ti–13Zr–13Nb/ONTs electrodes in acidic PBS is observed as compared to physiological PBS. The reason for this is a decrease in the corrosion resistance caused by an increase in the aggressiveness of the corrosive environment. On the other hand, it can be seen that the production of the 2G ONTs layers by anodizing caused the desired shift of the $\log |j| = f(E)$ curves towards anodic potentials in both physiological and acid PBS in comparison with the non-anodized Ti–13Zr–13Nb electrode. In the range of potentials corresponding to the cathode branch with values lower than E_{cor}, the tested electrodes are resistant to corrosion. At potentials equal to E_{cor}, oxidation processes begin at the anode branch. The higher passive current densities of the order of 10^{-4} A cm^{-2} were observed for the Ti–13Zr–13Nb electrode in PBS of pH 5.5, which means the highest corrosion rate among the tested electrodes. The decrease in anodic current densities for the anodized Ti–13Zr–13Nb electrode in both PBS of pH 7.4 and 5.5 is associated with the presence of 2G ONTs layers which are characterized by greater stability as compared to the self-passive oxide layer on the Ti–13Zr–13Nb electrode surface.

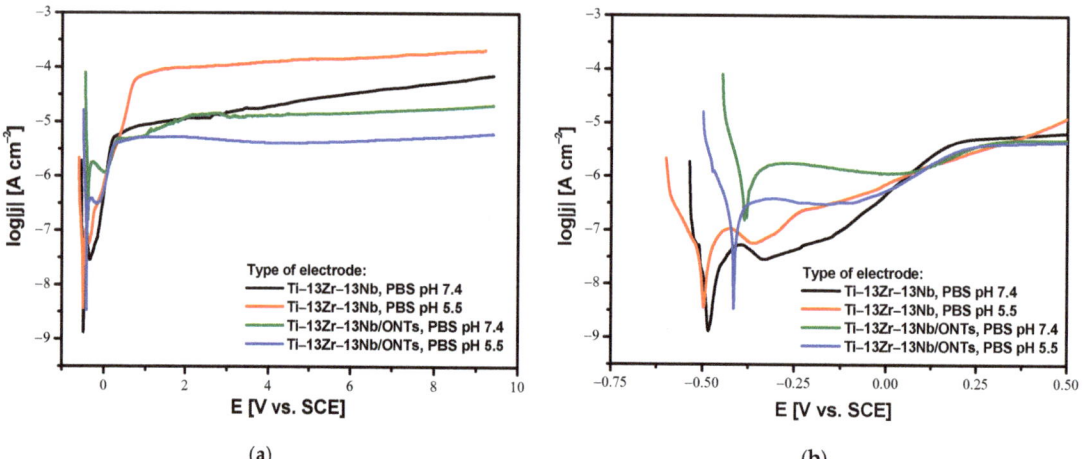

Figure 12. Anodic polarization curves for the Ti–13Zr–13Nb electrode without and with 2G ONTs layer in physiological and acidic PBS at 37 °C: (**a**) In the whole range of tested potentials; (**b**) The inset in the range of potentials corresponding to the cathode-anode transition.

It is worth noting that the Ti–13Zr–13Nb alloy, both before and after anodizing, did not show susceptibility to pitting corrosion because, in the range of the tested potentials up to 9.4 V, there was no breakdown of the oxide layers to the substrate. Registration of anode polarization curves for potentials above 9.4 V was impossible due to the limitations of the apparatus. The obtained results prove the excellent corrosion resistance of the tested biomaterials in environments containing chlorides, which can be proposed for the production of long-term implants. Compared to conventional steel biomaterials, the Ti–13Zr–13Nb alloy, both before and after anodizing, shows exceptional electrochemical properties in PBS solution [42]. An anodic behavior study of different types of stainless steel in PBS solution revealed that drastically lower breakdown potential (E_{bd}) values were observed. At the E_{bd} of 0.208 V for AISI 304L and 0.559 V for AISI 316L austenitic stainless steels, as well as 1.120 V for 2205 duplex stainless steel, a sharp increase in current density on anodic polarization curves was observed, indicating much easier destruction of the oxide layer.

4. Conclusions

The obtained results confirm that the proposed conditions of anodizing for Ti–13Zr–13Nb alloy allow for obtaining bamboo-like 2G ONTs layers. It was found that the effect of electrolyte pH on the zeta potential of the examined surfaces occurred. In a neutral aqueous KCl solution, the 2G ONTs layer moves the isoelectric point from 4.2 for the non-anodized Ti–13Zr–13Nb alloy, which is typical for the surface without a functional group to pH of 5.4, which is characteristic for amorphous oxide phase. Comparison of the influence of different electrolytes such as KCl, PBS and artificial blood on the zeta potential at pH of 7.4 for the Ti–13Zr–13Nb alloy before and after anodizing revealed a strong reaction of calcium anions with amorphous surfaces. The complex ions contained in artificial blood have demonstrated a stronger affinity to the hydrophobic surface before anodizing than the hydrophilic one after electrochemical oxidation. The increase in corrosion resistance of the anodized Ti–13Zr–13Nb electrode in PBS as compared with the non-anodized Ti–13Zr–13Nb electrode was due to the presence of a stable 2G ONTs layer. For both types of electrodes, no susceptibility to pitting corrosion up to 9.4 V was found in potentiodynamic studies.

Knowledge of the surface charge of the biomaterial is of fundamental importance for predicting the biological response of the organism to the implant, especially immediately after the implantation procedure. However, the zeta potential method used in these in vitro studies could not be used in vivo due to technical limitations. Moreover, determination of the breakdown potential of the 2G ONTs layer on the Ti–13Zr–13Nb alloy in PBS was not possible due to the technical limitations of the potentiostat to the tested potential range of 10 V.

The conducted research encourages further research on surface modification of the biomedical Ti–13Zr–13Nb alloy by anodizing in new electrochemical conditions. In the near future, in vitro and in vivo biological tests are planned, which will make it possible to obtain the CE certificate required to implement the developed surface modification technology. Knowledge about the kinetics of drug release from the obtained ONTs will facilitate the future development of personalized implants that are carriers of tissue-forming and therapeutic substances, supporting the process of osseointegration of the implant in the human body.

Author Contributions: Conceptualization, B.Ł.; methodology, B.Ł., A.S. and T.L.; investigation, A.S., T.L., K.D., B.C. and P.O.; formal analysis, B.Ł., A.S., T.L., K.D., B.C. and P.O.; writing—original draft preparation, B.Ł., A.S., T.L. and P.O.; writing—review and editing, B.C. and K.D.; visualization, A.S., T.L., K.D., B.C. and P.O.; funding acquisition, B.Ł. All authors have read and agreed to the published version of the manuscript.

Funding: This research received no external funding.

Institutional Review Board Statement: Not applicable.

Informed Consent Statement: Not applicable.

Data Availability Statement: Not applicable.

Conflicts of Interest: The authors declare no conflict of interest.

References

1. Anene, F.; Jaafar, C.A.; Zainol, I.; Hanim, M.A.; Suraya, M. Biomedical materials: A review of titanium based alloys. *Proc. Inst. Mech. Eng. Part C J. Mech. Eng. Sci.* **2020**, *235*, 3792–3805. [CrossRef]
2. *Ti-Based Biomaterials*; MDPI AG: Basel, Switzerland, 2020.
3. *Titanium in Medical and Dental Applications*; Elsevier BV: Amsterdam, The Netherlands, 2018.
4. Ossowska, A.; Zieliński, A.; Supernak, M. Formation of High Corrosion Resistant Nanotubular Layers on Titanium Alloy Ti13Nb13Zr. *Solid State Phenom.* **2011**, *183*, 137–142. [CrossRef]
5. Smołka, A.; Dercz, G.; Rodak, K.; Łosiewicz, B. Evaluation of corrosion resistance of nanotubular oxide layers on the Ti13Zr13Nb alloy in physiological saline solution. *Arch. Metall. Mater.* **2015**, *60*, 2681–2686. [CrossRef]
6. Aïnouche, L.; Hamadou, L.; Kadri, A.; Benbrahim, N.; Bradai, D. Interfacial barrier layer properties of three generations of TiO_2 nanotube arrays. *Electrochim. Acta* **2014**, *133*, 597–609. [CrossRef]
7. Luz, A.R.; de Souza, G.B.; Lepienski, C.M.; Siqueira, C.J.M.; Kuromoto, N.K. Tribological properties of nanotubes grown on Ti-35Nb alloy by anodization. *Thin Solid Film.* **2018**, *660*, 529–537. [CrossRef]
8. Sarraf, M.; Zalnezhad, E.; Bushroa, A.R.; Hamouda, A.M.S.; Rafieerad, A.R.; Nasiri-Tabrizi, B. Effect of microstructural evolution on wettability and tribological behavior of TiO_2 nanotubular arrays coated on Ti-6Al-4V. *Ceram. Int.* **2015**, *41*, 7952–7962. [CrossRef]
9. Stróż, A.; Goryczka, T.; Łosiewicz, B. Electrochemical formation of self-organized nanotubular oxide layers on niobium (Review). *Curr. Nanosci.* **2019**, *15*, 42–48. [CrossRef]

10. Łosiewicz, B.; Stróż, A.; Osak, P.; Maszybrocka, J.; Gerle, A.; Dudek, K.; Balin, K.; Łukowiec, D.; Gawlikowski, M.; Bogunia, S. Production, Characterization and Application of Oxide Nanotubes on Ti–6Al–7Nb Alloy as a Potential Drug Carrier. *Materials* **2021**, *14*, 6142. [CrossRef]
11. Stróż, A.; Łosiewicz, B.; Zubko, M.; Chmiela, B.; Balin, K.; Dercz, G.; Gawlikowski, M.; Goryczka, T. Production, structure and biocompatible properties of oxide nanotubes on Ti13Nb13Zr alloy for medical applications. *Mater. Charact.* **2017**, *132*, 363–372. [CrossRef]
12. Stróż, A.; Dercz, G.; Chmiela, B.; Łosiewicz, B. Electrochemical synthesis of oxide nanotubes on biomedical Ti13Nb13Zr alloy with potential use as bone implant. *AIP Conf. Proc.* **2019**, *2083*, 030004. [CrossRef]
13. Smołka, A.; Rodak, K.; Dercz, G.; Dudek, K.; Łosiewicz, B. Electrochemical Formation of Self-Organized Nanotubular Oxide Layers on Ti13Zr13Nb Alloy for Biomedical Applications. *Acta Phys. Pol.* **2014**, *125*, 932–935. [CrossRef]
14. Stróż, A.; Dercz, G.; Chmiela, B.; Stróż, D.; Łosiewicz, B. Electrochemical formation of second generation TiO_2 nanotubes on Ti13Nb13Zr alloy for biomedical applications. *Acta Phys. Pol.* **2016**, *130*, 1079–1080. [CrossRef]
15. Łosiewicz, B.; Skwarek, S.; Stróż, A.; Osak, P.; Dudek, K.; Kubisztal, J.; Maszybrocka, J. Production and Characterization of the Third-Generation Oxide Nanotubes on Ti-13Zr-13Nb Alloy. *Materials* **2022**, *15*, 2321. [CrossRef] [PubMed]
16. Durdu, S.; Cihan, G.; Yalcin, E.; Altinkok, A. Characterization and mechanical properties of TiO_2 nanotubes formed on titanium by anodic oxidation. *Ceram. Int.* **2021**, *47*, 10972–10979. [CrossRef]
17. Ossowska, A.; Olive, J.-M.; Zielinski, A.; Wojtowicz, A. Effect of double thermal and electrochemical oxidation on titanium alloys for medical applications. *Appl. Surf. Sci.* **2021**, *563*, 150340. [CrossRef]
18. Ossowska, A.; Zieliński, A.; Olive, J.-M.; Wojtowicz, A.; Szweda, P. Influence of Two-Stage Anodization on Properties of the Oxide Coatings on the Ti–13Nb–13Zr Alloy. *Coatings* **2020**, *10*, 707. [CrossRef]
19. Handzlik, P.; Gutkowski, K. Synthesis of oxide nanotubes on Ti13Nb13Zr alloy by the electrochemical method. *J. Porous Mater.* **2019**, *26*, 1631–1637. [CrossRef]
20. Stępień, M.; Handzlik, P.; Fitzner, K. Electrochemical synthesis of oxide nanotubes on Ti6Al7Nb alloy and their interaction with the simulated body fluid. *J. Solid State Electrochem.* **2016**, *20*, 2651–2661. [CrossRef]
21. Wu, S.; Wang, S.; Liu, W.; Yu, X.; Wang, G.; Chang, Z.; Wen, D. Microstructure and properties of TiO_2 nanotube coatings on bone plate surface fabrication by anodic oxidation. *Surf. Coat. Technol.* **2019**, *374*, 362–373. [CrossRef]
22. Łosiewicz, B.; Osak, P.; Maszybrocka, J.; Kubisztal, J.; Stach, S. Effect of autoclaving time on corrosion resistance of sandblasted Ti G4 in artificial saliva. *Materials* **2020**, *13*, 4154. [CrossRef]
23. Szklarska, M.; Dercz, G.; Simka, W.; Łosiewicz, B. A.c. impedance study on the interfacial properties of passivated Ti13Zr13Nb alloy in physiological saline solution. *Surf. Interface Anal.* **2014**, *46*, 698–701. [CrossRef]
24. Aniołek, K.; Łosiewicz, B.; Kubisztal, J.; Osak, P.; Stróż, A.; Barylski, A.; Kaptacz, S. Mechanical properties, corrosion resistance and bioactivity of oxide layers formed by isothermal oxidation of Ti–6Al–7Nb alloy. *Coatings* **2021**, *11*, 505. [CrossRef]
25. Osak, P.; Maszybrocka, J.; Kubisztal, J.; Ratajczak, P.; Łosiewicz, B. Long-Term Assessment of the In Vitro Corrosion Resistance of Biomimetic ACP Coatings Electrodeposited from an Acetate Bath. *J. Funct. Biomater.* **2021**, *12*, 12. [CrossRef] [PubMed]
26. Dudek, K.; Dulski, M.; Łosiewicz, B. Functionalization of the NiTi Shape Memory Alloy Surface by HAp/SiO_2/Ag Hybrid Coatings Formed on SiO_2-TiO_2 Glass Interlayer. *Materials* **2020**, *13*, 1648. [CrossRef] [PubMed]
27. Tanji, A.; Feng, R.; Lyu, Z.; Sakidja, R.; Liaw, P.K.; Hermawan, H. Passivity of AlCrFeMnTi and AlCrFeCoNi high-entropy alloys in Hanks' solution. *Corros. Sci.* **2023**, *210*, 110828. [CrossRef]
28. Costa, B.C.; Tokuhara, C.; Rocha, L.A.; Oliveira, R.C.; Lisboa-Filho, P.N.; Pessoa, J. Vanadium ionic species from degradation of Ti-6Al-4V metallic implants: In vitro cytotoxicity and speciation evaluation. *Mater. Sci. Eng. C* **2019**, *96*, 730–739. [CrossRef]
29. Osak, P.; Maszybrocka, J.; Zubko, M.; Rak, J.; Bogunia, S.; Łosiewicz, B. Influence of Sandblasting Process on Tribological Properties of Titanium Grade 4 in Artificial Saliva for Dentistry Applications. *Materials* **2021**, *14*, 7536. [CrossRef]
30. Osak, P.; Maszybrocka, J.; Kubisztal, J.; Łosiewicz, B. Effect of amorphous calcium phosphate coatings on tribological properties of titanium grade 4 in protein-free artificial saliva. *Biotribology* **2022**, *32*, 100219. [CrossRef]
31. Zhang, S.; Qin, J.; Yang, C.; Zhang, X.; Liu, R. Effect of Zr addition on the microstructure and tribological property of the anodization of Ti-6Al-4V alloy. *Surf. Coat. Technol.* **2018**, *356*, 38–48. [CrossRef]
32. Schneider, S.G.; Nunes, C.A.; Rogero, S.P.; Higa, O.Z.; Bressiani, J.C. Mechanical properties and cytotoxic evaluation of the Ti–13Nb–13Zr alloy. *Biomecánica* **2000**, *8*, 84–87. [CrossRef]
33. Lee, T. Variation in Mechanical Properties of Ti–13Nb–13Zr Depending on Annealing Temperature. *Appl. Sci.* **2020**, *10*, 7896. [CrossRef]
34. Davidson, J.A.; Kovacs, P. New Biocompatible, Low Modulus Titanium Alloy for Medical Implants. U.S. Patent No. 5,169,597, 8 December 1992.
35. Lee, M.; Kim, I.-S.; Moon, Y.H.; Yoon, H.S.; Park, C.H.; Lee, T. Kinetics of Capability Aging in Ti-13Nb-13Zr Alloy. *Crystals* **2020**, *10*, 693. [CrossRef]
36. ISO 13099-1:2012; Colloidal Systems—Methods for Zeta Potential Determination—Part 1: Electroacoustic and Electrokinetic Phenomena. International Organization for Standardization (ISO): Geneva, Switzerland, 2012.
37. ISO 13099-2:2012; Colloidal Systems—Methods for Zeta Potential Determination—Part 2: Optical Methods. International Organization for Standardization (ISO): Geneva, Switzerland, 2012.

38. *ISO 13099-3:2012*; Colloidal Systems—Methods for Zeta Potential Determination—Part 3: Acoustic Methods. International Organization for Standardization (ISO): Geneva, Switzerland, 2012.
39. Lorenzetti, M.; Gongadze, E.; Kulkarni, M.; Junkar, I.; Iglič, A. Electrokinetic Properties of TiO_2 Nanotubular Surfaces. *Nanoscale Res. Lett.* **2016**, *11*, 378. [CrossRef] [PubMed]
40. Lorenzetti, M.; Luxbacher, T.; Kobe, S.; Novak, S. Electrokinetic behaviour of porous TiO_2-coated implants. *J. Mater. Sci. Mater. Med.* **2015**, *26*, 191. [CrossRef]
41. *ASTM F1713-08(2021)e1*; Standard Specification for Wrought Titanium-13Niobium-13Zirconium Alloy for Surgical Implant Applications (UNS R58130). ASTM: West Conshohocken, PA, USA, 2021.
42. Gudić, S.; Nagode, A.; Šimić, K.; Vrsalović, L.; Jozić, S. Corrosion Behavior of Different Types of Stainless Steel in PBS Solution. *Sustainability* **2022**, *14*, 8935. [CrossRef]
43. Liu, L.; Qiu, C.L.; Chen, Q.; Zhang, S.M. Corrosion behavior of Zr-based bulk metallic glasses in different artificial body fluids. *J. Alloys Compd.* **2006**, *425*, 268–273. [CrossRef]
44. Ferraris, S.; Cazzola, M.; Peretti, V.; Stella, B.; Spriano, S. Zeta Potential Measurements on Solid Surfaces for in Vitro Biomaterials Testing: Surface Charge, Reactivity upon Contact with Fluids and Protein Absorption. *Front. Bioeng. Biotechnol.* **2018**, *6*, 60. [CrossRef] [PubMed]

Disclaimer/Publisher's Note: The statements, opinions and data contained in all publications are solely those of the individual author(s) and contributor(s) and not of MDPI and/or the editor(s). MDPI and/or the editor(s) disclaim responsibility for any injury to people or property resulting from any ideas, methods, instructions or products referred to in the content.

Article

Extending the Protection Ability and Life Cycle of Medical Masks through the Washing Process

Julija Volmajer Valh [1], Tanja Pušić [2,*], Mirjana Čurlin [3] and Ana Knežević [2]

1. Faculty of Mechanical Engineering, University of Maribor, Smetanova 17, 2000 Maribor, Slovenia
2. Faculty of Textile Technology, University of Zagreb, Prilaz Baruna Filipovića 28a, 10000 Zagreb, Croatia
3. Faculty of Food Technology and Biotechnology, University of Zagreb, Pierottijeva 6, 10000 Zagreb, Croatia
* Correspondence: tanja.pusic@ttf.unizg.hr

Abstract: The reuse of decontaminated disposable medical face masks can contribute to reducing the environmental burden of discarded masks. This research is focused on the effect of household and laboratory washing at 50 °C on the quality and functionality of the nonwoven structure of polypropylene medical masks by varying the washing procedure, bath composition, disinfectant agent, and number of washing cycles as a basis for reusability. The barrier properties of the medical mask were analyzed before and after the first and fifth washing cycle indirectly by measuring the contact angle of the liquid droplets with the front and back surface of the mask, further by measuring air permeability and determining antimicrobial resistance. Additional analysis included FTIR, pH of the material surface and aqueous extract, as well as the determination of residual substances—surfactants—in the aqueous extract of washed versus unwashed medical masks, while their aesthetic aspect was examined by measuring their spectral characteristics. The results showed that household washing had a stronger impact on the change of some functional properties, primarily air permeability, than laboratory washing. The addition of the disinfectant agent, didecyldimethylammonium chloride, contributes to the protective ability and supports the idea that washing of medical masks under controlled conditions can preserve barrier properties and enable reusability.

Keywords: medical masks; washing; detergent; didecyldimethylammonium chloride; air permeability; antimicrobial activity; residuals

1. Introduction

The challenges of the COVID-19 epidemic/pandemic and the efforts against the spread of the virus require both short- and long-term recommendations and measures. One of the recommendations and types of personal protection is wearing a mask. In practice, three types of masks with different characteristics and degrees of protection have become widely used, namely face masks, medical masks, and filtering masks. Their characteristics and protective effects are usually highlighted in the attached product specifications. The face mask can prevent the spread of infectious diseases in public areas by preventing both the inhalation of infectious droplets and their subsequent exhalation and transmission [1].

Home textiles have been shown to play an important role in spreading infections caused by pathogenic microorganisms: viruses, bacteria, and fungi [2–4].

The results of the research show that cleaning and washing textiles is one of the key factors in ensuring hygiene in the home [5–7]. There are other types of disinfection, e.g., ultraviolet (UV) light with destructive activity against pathogenic bacteria, including Clostridioides difficile spores [8], hydrogen peroxide, UV-rays, moist heat, dry heat, and ozone [7].

The main aim of household laundering is to remove visible and invisible dirt and unpleasant odors to achieve freshness, a satisfactory aesthetic appearance, and a high level of cleanliness. This can be achieved by optimizing the factors of the Sinner's circle, compensating for an increase in the proportion of one of the factors by reducing the others [9,10].

The key factor in decontamination of microorganisms on surfaces is increased temperature. Considering the environmental guidelines to lower the washing temperature and the increased proportion of synthetic textiles, the chemical effect based on the composition of the detergent is of utmost importance. This makes bleaching agents and their activators important factors for disinfection during washing [11,12]. In applying disinfectants, the principle of efficiency is the most important requirement, while at the same time the principle of sustainability and environmental performance must be observed.

Numerous challenges and the pandemic period of SARS-CoV-2 affecting people's respiratory system [13] have raised certain questions and doubts about the extent to which consumer laundry detergents can reasonably ensure the level of disinfection during washing. There are also questions about the extent to which washing and rinsing aids can improve disinfection efficacy [14,15].

During the global COVID-19 pandemic, cationic surfactants were extensively used as antiseptics and disinfectants [16]. Recent studies have shown that the growth in the production and use of surfactants was +196%, for biocides it was +152%, and for cationic quaternary ammonium surfactants (used as surfactants and biocides) it was +331% [17].

Research has confirmed that disposable face masks belong not only to personal protective equipment (PPE) that greatly contributes to the protection of people [18–20] but also to the generation of textile waste [21,22]. The problem of the accumulation of textile waste gave rise to the idea of reusing medical masks after decontaminating them in the process of washing them with detergents at 60 °C and treating them with disinfectants, such as hydrogen peroxide, UV rays, moist heat, dry heat, ozone, ethanol, and sodium hypochlorite [7,19,20,23,24]. Decontamination procedures, along with effective pathogen reduction, should preserve the protective features of medical masks (filtration efficiency and breathability) without the harmful chemical effect on the user [18,19].

In this research, the influence of household and laboratory washing procedures on the properties of disposable medical masks made of three-layer nonwoven polypropylene was investigated. The novelty of this investigation can be seen in several aspects. The first is the reduction in the washing temperature from 60 °C to 50 °C. Washing the masks with commercial color detergent at home and in the laboratory with standard detergents to verify reusability and extend the life cycle distinguishes this investigation from others. In addition, the influence of the factors of the washing process at home and in the laboratory on the degree of protection against respiratory droplets and the human ecological properties of disposable protective masks was investigated. Washing in the laboratory was performed with a standard detergent, and the rinsing process was varied. After washing with detergent, part of the medical mask samples was rinsed with water in four cycles, while the remaining samples were rinsed with water in three cycles, and disinfection with didecyldimethylammonium chloride (DDAC) was performed in the fourth cycle.

The barrier properties of the medical mask—the effectiveness of "protection against respiratory problems" before and after the first and fifth washing cycle in the specified conditions were analyzed indirectly by measuring the contact angle of the liquid droplets with the front and back surface of the mask, measuring air permeability, and determining antimicrobial resistance.

The reusability and washability of medical masks were analyzed by FTIR, by measuring the pH of the material surface and the aqueous extract. The new concept for characterization of masks includes the analysis of surfactants in an aqueous extract, which may have irritating potential. The aspect of the aesthetic component of the masks was studied by measuring their spectral properties.

2. Materials and Methods

Disposable medical face masks, model KZ020, manufactured by Chuzhou Qiao Dong Industrial Co., Ltd, (Quzhou, China) were purchased through the distribution channel of Drogerie Markt (dm), Coburg, Germany. The nonwoven structure of the masks is made of polypropylene fibers (PP), with the proportion of PP microfibers of 30%, produced by meltblown technology and integrated between two spunbond layers [25]. The material is declared as hypoallergenic, and the mask is classified as type IIR, whose filtration efficiency is 98%, and the properties meet the requirements of HRN EN 14683:2020. The front surface of the medical mask is blue, and the back is white, with ear loops and a reinforcement in the area of the nose (nose wire).

The masks were washed in a Hanseatic household washing machine, model HWT 8614A, at 50 °C, with a liquid color detergent (100 g) containing anionic surfactants (5–15%), nonionic surfactants ($\leq$5%), phosphonates, polycarboxylates, zeolites, enzymes, fragrance hexyl cinnamal, and water. After the washing cycle, 4 rinsing cycles with inter-centrifugations according to the washing machine program were performed. The described procedure was repeated 5 times.

The washing in the laboratory was performed in a laboratory device PolyColor, Werner Mathis AG, Oberhasli, Switzerland, at 50 °C with a standard ECE A detergent containing linear sodium alkylbenzene sulfonate (9.7%), sodium soap (5.2%), defoamer (4.5%), sodium aluminosilicate (32.5%), sodium carbonate (11.8%), sodium salt copolymer of acrylic and maleic acid (5.2%), sodium silicate (3.4%), carboxymethyl cellulose (1.3%), diethylenetri-amine penta (methylene phosphonic acid 5.2%), sodium sulfate (9.8%), and water (12.2%).

The detergent was dosed in a concentration of 1.25 g/L with a bath ratio of 1:50.

After the washing process in the laboratory device, the medical mask samples from the container with plastic tweezers were transferred to a laboratory beaker, where they were rinsed in two ways: (i) 4 individual rinsing cycles in water; (ii) 3 individual rinsing cycles in water, and subsequent disinfection in water with 1 mL/L of disinfectant for 30 min as recommended by the supplier. The main ingredient of this disinfection agent is the cationic active substance, didecyldimethylammonium chloride (DDAC, according to IUPAC, N-decyl-NN-dimethyldecane-1-ammonium chloride). It is specified that 100 g of this agent contains 2.49 g of DDAC.

After the individual washing and rinsing, the samples were air dried in a protected area and placed in plastic zip bags to avoid possible contamination. The sample labels before and after the 1st (1×) and 5th (5×) washing according to the described procedures are shown in Table 1.

Table 1. Labeling of medical masks.

Label	Medical mask
N	Pristine
U	Cut
L 1×	Laboratory-washed—1 cycle
L 5×	Laboratory-washed—5 cycles
L_{DS} 1×	Laboratory-washed with addition of DDAC—1 cycle
L_{DS} 5×	Laboratory-washed with addition of DDAC—5 cycles
H 1×	Home-washed—1 cycle
H 5×	Home-washed—5 cycles

2.1. Methods

The medical masks listed in Table 1 were analyzed by the methods for the purpose of monitoring their properties. Fourier transform infrared spectroscopy was used to characterize the composition of the medical masks. Spectrum 100S FT-IR UATR + TG /IR Interface TL8000 (RedShift), Perkin Elmer, Waltham, MA, USA was used for the analysis.

The spectral characteristics of the front of the medical masks were measured at four different locations using a Spectraflash SF300 computer-controlled dual-channel remission spectrophotometer, Datacolor AG, Rotkreuz, Switzerland, with an aperture of 20 mm, standard light D65, and d/8° geometry. Results are presented as differences in lightness (dL^*), hue (dH^*), and chroma (dC^*) of the washed samples compared to the pristine samples.

The total color difference (dE) of the washed samples compared to the original medical mask sample is calculated as follows in Equation (1):

$$dE = \sqrt{(dL^*)^2 + (dC^*)^2 + (dH^*)^2} \tag{1}$$

dL*—difference in lightness (dL* = L*washed sample − L*pristine)
dC*—difference in chroma (dC* = C*washed sample − C*pristine)
dH*—difference in hue

Instrumental evaluation of color change to determine the gray scale rating was additionally performed according to the rating system ISO A05 and AATCC. The method also specifies an instrumental method for evaluating the color change in a washed mask compared to an untreated mask as a reference. Calculations were performed to convert the instrumental measurements into a gray scale rating, with a score of 5 indicating excellent color fastness and a score of 1 indicating poor color fastness [26].

The whiteness of the back surface of medical masks before and after washing was determined by spectral measurement (Spectraflash SF300, Datacolor AG, Rotkreuz, Switzerland), with an aperture size of 20 mm, under standard illumination D65) according to ISO 105-J02:1997 Textiles—Colour fastness testing—Part J02: Instrumental assessment of relative whiteness [27]. The measurement results were expressed as the average of four individual measurements.

The analysis of the presence of aerobic bacteria on medical mask samples was performed according to the standard procedure for water analysis [28].

This standard method was adapted to a medical mask sample inoculated onto microbiological agar in a Petri dish. Samples were incubated at a temperature of 37 °C for 48 h or at a temperature of 22 °C for 72 h. After incubation, all colonies present in each inoculated dish were determined visually. Since this method was modified and adapted to mask samples, the presence of colonies on the plates was taken as a visual result only and is presented by images for the unwashed sample the home-washed sample and the sample washed under laboratory conditions and rinsed with a disinfectant.

After determining the contact angles of the liquid droplets with the surface of the medical masks using a goniometer (DataPhysics, Filderstadt, Germany), the hydrophilicity/hydrophobicity of the samples was evaluated before and after the washing cycle. The static contact angle (SCA) was measured with a drop of liquid lying on the surface. At least 3 measurements were taken for each sample, and even more individual measurements were taken for some samples. A goniometer with OCA 20 software was used to determine the SCA at room temperature with ultrapure water. A drop with a volume of 3 µL was carefully placed on the surface of the sample.

The barrier properties of medical masks were established indirectly by determining the air permeability according to EN ISO 9273 (Textiles—Test methods for nonwovens—Part 15: Determination of air permeability) to evaluate the influence of washing conditions. To measure air permeability, a device with four working heads produced by the Karl Schroeder company was used (power supply U = 220 V; frequency f = 50 Hz; current I = 5 A; power P = 1100 W; with a three-phase measurement range: 5...90 L/m²xs, 50...550 L/m²xs, 500...5600 L/m²s). The air permeability of the medical masks before and after the process was measured at three different points (at the center of the mask and two measurements at a radius of 1 cm around the center of the mask).

The nonwoven structure of the front and back surface of the medical mask was characterized by measuring the pH of the sample at four different locations using the contact electrode of the Multimeter SevenCompact™ Duo S213 (Mettler-Toledo GmbH, Greifensee, Switzerland).

The pH of the aqueous extract of the nonwoven structure was determined in accordance with EN ISO 3071:2020 (Textiles—Determination of pH of aqueous extract) such that the elastics (ear loops) and flexible nose piece were removed from the samples. For this analysis, 6 samples were excluded, half of which were cut into strips of 5 mm (C), while the remaining 3 were left intact (N). The procedure was performed by placing 2 g of the sample in an Erlenmeyer flask and adding 100 mL of KCl solution. The flasks were exposed to mechanical shaking for 2 h. After mixing, the pH of the aqueous extract was measured using the previously described multimeter.

Considering the composition of detergents in washing and of DDAC in rinsing, the content of residual surfactants in the aqueous extract of the medical masks was analyzed: anionic and nonionic ones from detergents and cationic ones from DDAC. Potentiometric titrations with the Metrohm Autotitrator 736 GP Titrino and electrodes, all from Metrohm, Herisau, Switzerland, were used to determine the residual surfactants. For the determination of anionic and cationic surfactants, the High Sense Surfactant Electrode (6.0504.150) was used as an ion-selective surfactant electrode in combination with the reference, the Ag/AgCl electrode (6.0733.100). The NIO surfactant electrode (6.0507.010) in combination with the same reference electrode was used for the determination of the nonionic surfactant. The determination of a surfactant residue in the aqueous extract by potentiometric titrations was preceded by the preparation of calibration curves. Selected residues: (i) solutions of the anionic surfactant sodium dodecyl sulphate (NLS) were titrated with Hyamine 1622 at pH 3; (ii) solutions of the cationic surfactant (Hyamine 1622) were titrated with NLS at pH 10; (iii) solutions of the nonionic surfactant (Triton X-100) were titrated with sodium tetraphenyl borate and the addition of BaCl2 [29,30].

3. Results and Discussion
3.1. Medical Masks Characterization
3.1.1. Spectral Characteristics

Medical masks have different colors; the front surface is blue and the back is white. Accordingly, the change in spectral characteristics of the front surface was monitored by the differences in lightness (dL^*), chromaticity (dC^*), hue (dH^*), total color difference (dE), and color fastness grades of laboratory and home-washed versus unwashed masks according to ISO and AATCC, Table 2.

Table 2. Average values of spectral properties of the front surface of medical masks after the first and fifth washing cycle compared to the unwashed sample.

Sample	dL^*	dC^*	dH^*	dE	ISO-A05	AATCC
H 1×	−0.447	1.383	0.269	1.573	4	4
H 5×	1.567	−4.083	−0.141	4.526	3	3
L 1×	−2.087	1.888	−0.219	3.291	3	3
L 5×	−0.907	−0.015	−0.782	2.076	4	4
L_{DS} 1×	0.028	−2.190	−0.722	2.341	4	3–4
L_{DS} 5×	−0.515	−0.264	−0.777	1.655	4	4

The changes in spectral properties of medical masks after the first and fifth washing depend on the washing conditions. The household washing was carried out with a liquid color detergent and the laboratory washing with a standard detergent. These detergents have different compositions, which consequently affects the pH of their solutions. The results of the total difference in color (dE) indicate that the effect of the first household wash (H) is slightly weaker compared to the laboratory washing for both variants (L and L_{DS}). However, the cumulative effect of the household washing over five cycles is more noticeable (rating 3) than that of the laboratory washing (rating 4), which can be attributed to mechanical agitation, including interval centrifugation in the washing machine [31]. The laboratory wash with the addition of DDAC in the final rinse cycle (L_{DS}) has a mild protective effect on the hue, as evidenced by the lower dE values and color fastness ratings compared to the laboratory wash alone (L).

The back of the medical mask is a white nonwoven structure. The influence of washing was analyzed in terms of whiteness (W_{CIE} and Y) and hue (TV, TD), with the average values shown in Table 3.

Table 3. Average whiteness of the back of medical masks before and after washing.

Sample	W_{CIE}	TV	TD	Y
N	98.5	5.7	-	80.1
H 1×	98.8	5.4	-	81.0
H 5×	99.8	7.6	GG	77.1
L 1×	86.8	7.7	GG	72.3
L 5×	86.5	7.3	GG	73.4
L_{DS} 1×	86.1	6.5	GG	74.2
L_{DS} 5×	87.1	7.7	GG	73.0

The baseline whiteness of the back side of an unwashed medical mask is 98.5, indicating that no optically active components are present. Household washing was performed with a liquid color detergent containing no optical bleaching agents, to which only slightly different values for the whiteness of the medical mask (H) were indicated.

Laboratory washing was performed with a standard powder detergent that also did not contain optical brightening agents. There is a decrease in whiteness by more than 10 units for the samples washed 1× and 5× compared to the unwashed sample. These results may indicate the presence of residual substances from the laboratory washing process for both variants.

3.1.2. Antimicrobial Properties

To achieve the antimicrobial activity of the synthetic fibers, of which the tested medical masks are made, an antimicrobial agent can be incorporated into the polymer before spinning or mixed into the fibers during manufacture. This ensures maximum durability, as the agent is physically embedded in the fiber structure and is released at a certain rate during use. Since no antimicrobial agents embedded in the analyzed medical masks were either known or detected, the antimicrobial efficacy using the usual and standardized method for textile materials was not determined.

The disinfectant agent contained a cationic active component, didecyldimethylammonium chloride (DDAC). This agent, from the supplier dm GmbH, was dosed into the final rinse water, where it was left to sit for at least 30 min. According to the manufacturer the described procedure removes 99.9% of bacteria, fungi, and viruses. This agent has been dermatologically tested, does not affect the feel of textiles, and has no irritating effect on the skin. However, the supplier points out that people with sensitive skin must strictly follow the dosage instructions.

The aim of this study was to verify the effect of the disinfectant DDAC and to compare the microbiological characteristics of unwashed masks, masks washed in the household with detergent, and masks washed in the laboratory with detergent only and detergent with the addition of DDAC. The presence of colonies of aerobic microorganisms on all samples was tested with the method for the microbiological analysis of water, adapted to the rigid sample of masks. The analysis results after incubation for 72 h at 22 °C and for 72 and 48 h at 37 °C are shown in Figures 1–4.

Figure 1 shows that no aerobic bacteria were present after incubation at 22 °C and 37 °C, indicating a certain purity (sterility) of the purchased masks, which were packaged in a plastic bag.

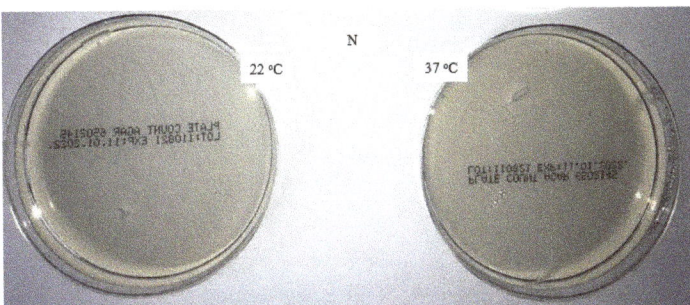

Figure 1. Colonies of aerobic microorganisms on a pristine medical mask (N).

On a medical mask washed in household conditions, a certain number of colonies of aerobic microorganisms was detected (Figure 2). The reason for the appearance of microorganisms was the fact that the masks were washed in the washing machine together with the other laundry, so that it is assumed that a cross-contamination occurred either from other textiles or from contamination in the washing machine.

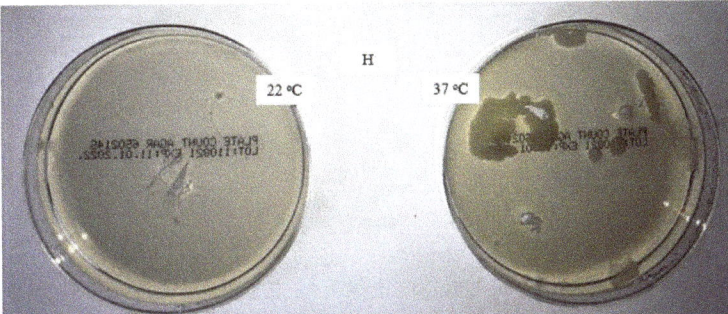

Figure 2. Colonies of aerobic microorganisms on a medical mask after household washing (H 5×).

Figure 3 shows the presence of microorganisms on medical masks washed in the laboratory with a standard detergent. The amount of microorganisms was smaller compared to medical masks washed at home. These samples point to a contamination of the washing system, or the ineffectiveness of the standard detergent for this activity.

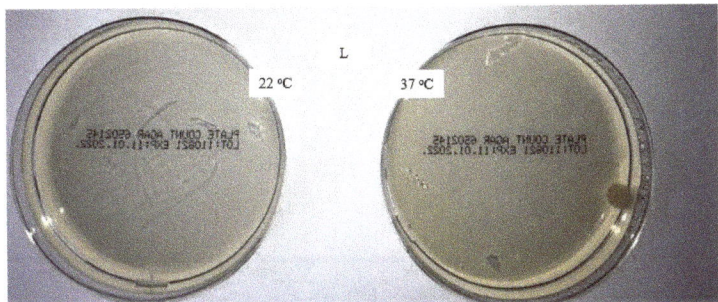

Figure 3. Colonies of aerobic microorganisms on a medical mask after laboratory washing (L 5×).

The addition of the disinfectant DDAC in the fourth rinse cycle has a significant effect on preventing the appearance of microorganisms on the mask samples washed in laboratory tests, as can be seen in Figure 4, where no colonies of microorganisms are visible.

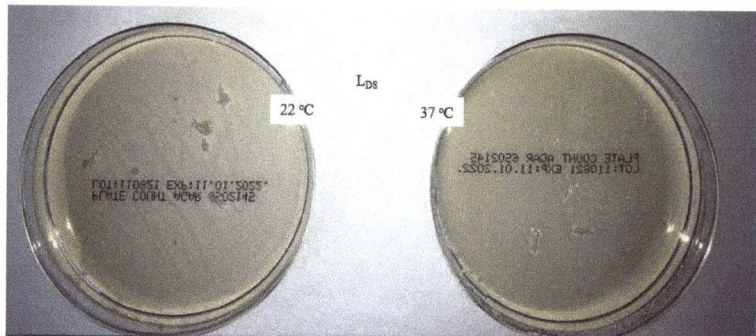

Figure 4. Colonies of aerobic microorganisms on a medical mask after laboratory washing with the addition of DDAC (L$_{DS}$ 5×).

The final results regarding the presence of microorganisms on samples of unwashed and washed masks after different procedures point to the disinfecting effect in the washing process with DDAC. These results justify the washability of medical masks at 50 °C through five cycles and the possibility of reuse. These results are comparable to the functional properties of medical masks washed at 60 °C [20] with the addition of a disinfectant [7,32], whose ecological performance is questionable.

Further research will be focused on testing the effectiveness of DDAC in household washing where, based on the results, it is necessary to prevent cross-contamination of the system.

3.1.3. Hydrophobic Properties

The behavior of droplets on the textile surface (shape, spilling, and sorption) depends on the surface, the type of liquid, the interaction of cohesive forces (between liquid molecules), and adhesive forces (between fibers and liquid) [33,34]. Medical masks are waterproof to a certain extent [35]. The influence of washing conditions on the changes in the adhesive properties of the front and back surface of the medical mask was analyzed by determining the static contact angle (SCA). The average values of the SCA on the front and back surfaces are highlighted in Tables 4 and 5. In determining the SCA on all the samples of the medical masks, there was the problem of wrinkles on the unwashed samples and breaks on the washed samples due to the mechanical agitation in the washing process.

Table 4. The average SCA of the liquid droplet with the front and back surface of the medical mask before and after washing.

Sample	SCA (°)						
	N	H 1×	H 5×	L 1×	L 5×	L$_{DS}$ 1×	L$_{DS}$ 5×
Front surface	131.24	117.38	122.11	118.04	123.61	115.64	129.28
Back surface	122.30	109.48	113.78	123.02	117.77	123.84	112.46

Table 5. Air permeability of medical masks before and after washing.

Sample	N	H 1×	H 5×	L 1×	L 5×	L$_{DS}$ 1×	L$_{DS}$ 5×
v (L/m^{2}s)	31.0	181.0	90.0	87.3	104.8	109.5	98.4

The average SCA of the unwashed medical mask (N) was 131.24°, which demonstrates the expected hydrophobicity of the medical mask, considering that it is made of polypropylene (PP). The values of the individual measurements of the unwashed sample were considerably scattered, which could be expected given the fact that the analyzed structure is nonwoven.

The influence of washing on the SCA depended mainly on the number of cycles performed, again with a large scattering of the individual measurements. In all washing processes the contact angle of the samples was smaller after the first wash cycle than after the fifth one. The reason for this could be the deposition of inorganic substances, as the washing process was carried out in hard water.

The result analysis of the SCA of the liquid droplets with the back of the medical mask proves the previously highlighted differences in the individual measurements. The changes in the SCA of the droplets with the back do not completely comply with the number of cycles on the contact angle of the droplets with the front of the medical mask.

The photographs of the liquid droplets with the surface of the medical mask sample are shown in Figures 5 and 6.

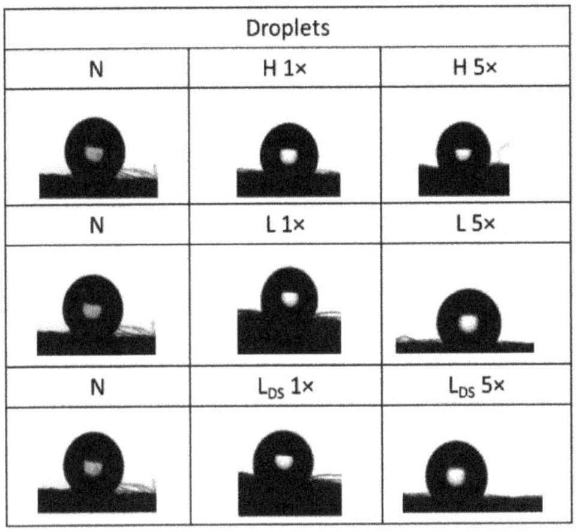

Figure 5. Droplets of the liquid with the front surface of the medical mask before and after washing.

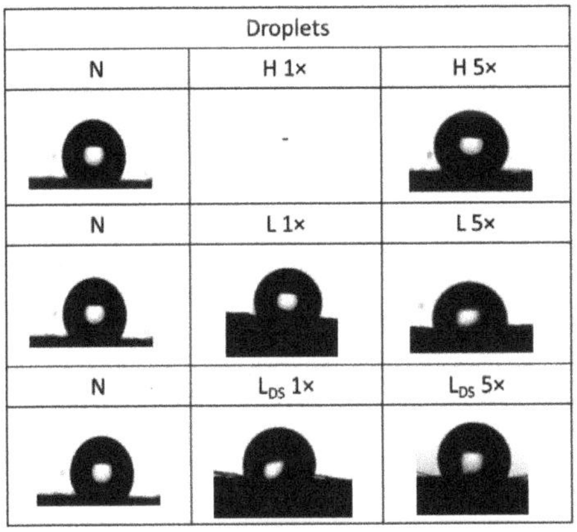

Figure 6. Droplets of the liquid with the surface of the back surface of the medical mask before and after laboratory washing with a standard detergent and rinsing with a disinfectant.

The pictures of droplets confirm the hydrophobicity of the front/back surface of the medical mask before and after washing. There are differences in the appearance of the surface of the examined samples, in the form of protruding fibers. Their appearance from the nonwoven structure is to be expected, while their distribution depends on numerous parameters of washing, drying, and testing.

3.1.4. Air Permeability

Information about air permeability properties of certain textiles makes it possible to evaluate and compare the performance characteristics of specific products, such as raincoats, tents, shirting fabrics, sails, industrial filters, and pillowcases. Testing this property is important when a softener is included in the finish or care [36,37]. The use of different types of softeners causes significant differences in air permeability [38].

Air permeability was tested according to EN ISO 9237 (Textiles—Determination of air permeability of fabrics). According to this source, the air permeability should be at least 96 L/m²s at a vacuum pressure of 100 Pa [39].

The air permeability of medical masks measured on samples before and after the washing cycles is shown in Table 5.

The air permeability of an unwashed medical mask at a pressure of 1013 mbar ($\approx$105 Pa) was 31.0 L/m²s. The air permeability of the washed samples in the wash processes (H and L) increased compared to the unwashed sample. The increase in air permeability of the samples washed at home was greater than in the laboratory. The values of the samples washed at home (H 1x) increased six and three times (H 5x) compared to the unwashed sample (N). The differences in air permeability of the medical mask samples after the first and fifth cycle of washing in the laboratory (L and L_{DS}) were insignificant; they were about three times higher than for the unwashed sample (N).

3.1.5. pH Analysis

The pH analysis of textiles after various processing and care procedures is useful because it provides information about the condition of the surface. In this work, the pH of the front and back surfaces of unwashed and washed medical mask samples was tested using a contact electrode (Tables 6 and 7).

Table 6. pH value of the front surface of whole medical masks before and after washing.

Sample	pH	T (°C)
N	5.79	27.68
H 1×	5.72	27.68
H 5×	5.69	27.7
L 1×	5.07	25.3
L 5×	6.97	22.8
L_{DS} 1×	5.05	25.5
L_{DS} 5×	6.97	23.2

Table 7. pH value of the back surface of medical masks before and after washing.

Sample	pH	T (°C)
N	5.75	27.68
H 1×	5.71	27.73
H 5×	5.71	28.85
L 1×	5.06	25.6
L 5×	7.00	23.23
L_{DS} 1×	5.05	25.8
L_{DS} 5×	6.87	22.7

The front surface of the unwashed mask had a pH of 5.79. This value was almost the same for the home-washed samples, which was expected given the lower pH of the color liquid detergent solution. Washing under laboratory conditions for five cycles resulted in an increase in the pH of the surface (pH ~ 7.00), owing to the standard detergent, the solution of which is strongly alkaline (pH ~ 10.50).

The changes on the back surface were similar to the changes on the front of the medical mask, which was expected given the washing conditions.

3.1.6. FTIR Analysis

The pristine (N) medical mask was characterized by FTIR spectroscopy to confirm the composition of the raw material (Figure 7). The spectra of the inner (back) and outer (face) nonwoven layers of the medical mask, the ear loops, and nose pads were recorded.

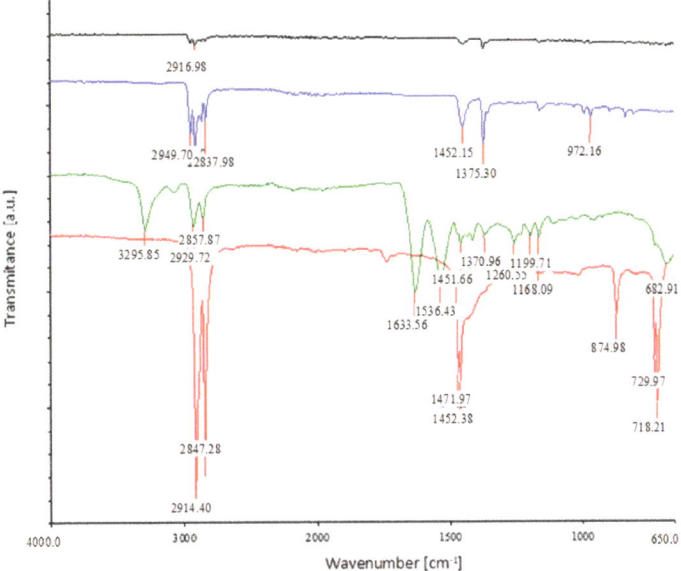

Figure 7. FTIR spectra of the back (black spectrum) and front (blue spectrum) nonwoven layer, the ear loops (green spectrum), and nose pads (red spectrum) of the medical mask.

The spectra of the inner (back) and outer (face) layers showed the following bands: multiple signals in the wavenumber range from 3000 to 2800 cm^{-1} and two large signals in the range from 1452 to 1375 cm^{-1}. The signals in the range from 3000 to 2800 cm^{-1} were attributable to asymmetric and symmetric stretching vibrations of CH_2 groups, while the signals at 2950 and 2850 cm^{-1} were due to the asymmetric and symmetric stretching vibrations of CH_3. The peak at 1456 cm^{-1} indicates the asymmetric CH_3 vibrations or CH_2 scissor vibrations, while the signal at 1375 cm^{-1} was the result of the symmetric CH_3 deformation [22]. All mentioned signals represent typical signals for polypropylene materials.

When comparing the FTIR spectrum of the ear loops with the spectrum of the synthetically produced polyurethane, it was found that the signals overlap at the following wavenumbers: 3290, 1700, 1630, and 1540 cm^{-1} [40]. There are minor differences, most likely due to the use of different additives.

The FTIR spectrum of the nose pads gave only two signals in the range from 3000 to 2800 cm^{-1}, namely, at the wavenumber 2914 and 2847 cm^{-1}, signals at 1471 and 1462 cm^{-1}, small peak at 874 cm^{-1}, and two signals at 729 and 718 cm^{-1}. These signals are attributed to polyethylene [22].

Figure 8 shows the spectra of the front nonwoven layers of the medical mask before and after the first and fifth washing cycle with detergent in two washing variants, whereas Figure 9 shows the spectra of the back nonwoven layers of the medical mask before and after the first and fifth washing cycle with detergent in two washing variants.

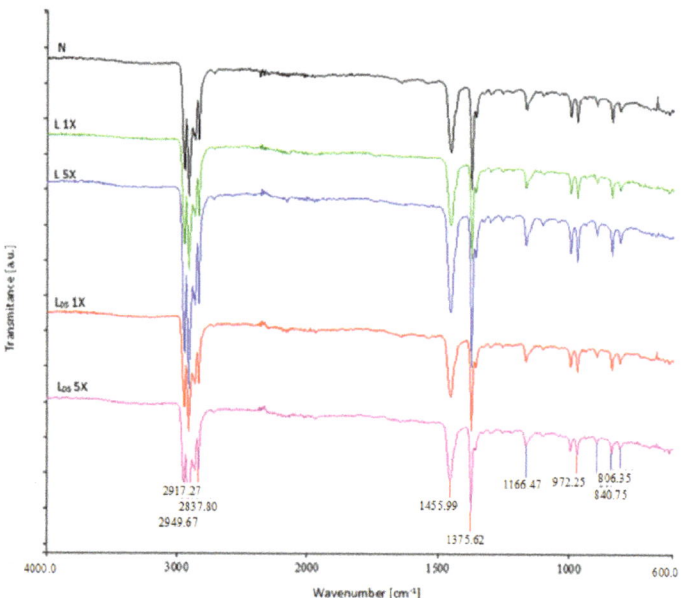

Figure 8. FTIR spectra of the front nonwoven layers of the medical mask before and after the first and fifth washing cycle.

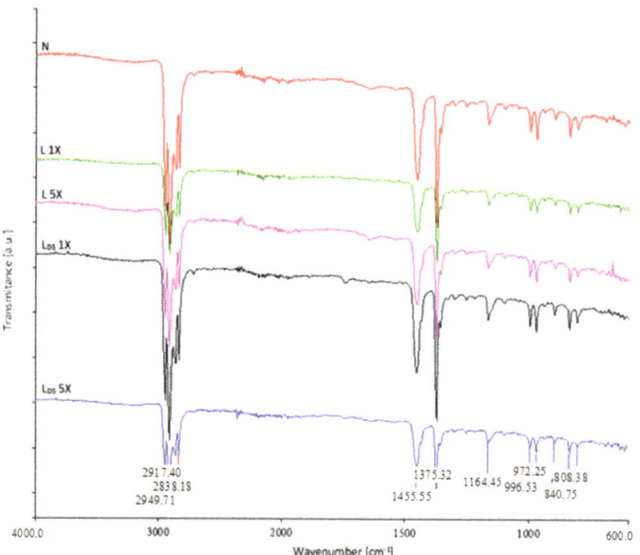

Figure 9. FTIR spectra of the back nonwoven layers of the medical mask before and after the first and fifth washing cycle.

According to the FTIR analysis performed, it was found that the washed (back and face) nonwoven layers of the medical mask do not indicate the influence of detergent or disinfectant on the chemical structure of the nonwoven layers in washing.

3.2. Characterization of the Aqueous Extract

In addition, the pH of the aqueous extract was tested. Variations in the preparation of the sample were made, leaving a portion of the medical masks whole and the other cut into strips (U), Tables 8 and 9.

Table 8. pH of the aqueous extract of the whole mask samples before and after washing.

Sample	pH	T (°C)
N	7.01	23.6
H 1×	6.90	23.9
H 5×	8.74	25.2
L 1×	8.87	23.7
L 5×	8.95	23.1
L_{DS} 1×	8.93	23.9
L_{DS} 5×	8.81	22.9
KCl	5.73	23.8

Table 9. pH of the aqueous extract of medical mask samples cut into strips before and after washing.

Sample	pH	T (°C)
N U	6.49	23.9
H U 1×	8.01	24.4
H U 5×	8.63	24.8
L U 1×	9.12	23.9
L U 5×	8.72	22.9
L_{DS} U 1×	8.97	23.9
L_{DS} U 5×	9.21	22.7
KCl	5.91	22.7

The aqueous extract of the whole mask samples before and after washing depends on the washing conditions. The aqueous extract of medical masks washed at home through one to five cycles has a different pH. A sample of the medical mask washed through five cycles has an almost 1.8 pH units higher value than after a single wash. The aqueous extract of medical masks washed through one to five cycles under laboratory conditions is alkaline, with all the samples having almost the same value (pH 8.9).

Washing with a standard detergent increased the pH of the water extract, which was slightly higher than the limits of the control system for laundries RAL-GZ 992 (RAL Gütezeichen RAL-GZ 992 Sicherheit durch professionellen Wäscheservice) [41].

Table 9 shows the pH results of the aqueous extract for the medical mask samples cut into strips (U).

The pH of the aqueous extract of the cut mask samples before washing at the same temperature was 0.5 pH units lower than that of the whole mask sample. The obtained results show that the washed cut mask samples have a more alkaline aqueous extract compared to the whole mask samples, which could be due to residual alkalis within the nonwoven structure, which has a higher migration potential away from the structure when the samples are cut.

3.2.1. Surfactants Residuals Characterization

In accordance with the hypotheses highlighted in [19], decontamination methods should not only demonstrate effective pathogen reduction but also preserve the properties of medical masks without harmful chemical effects for the user. In accordance with that, the whole and cut mask samples were analyzed for the presence and content of surfactants before and after the first and fifth wash cycle, depending on the composition of the applied detergents and DDAC.

Medical masks not subjected to washing (N) were analyzed by determining all surfactants, ionic (anionic and cationic), and nonionic.

The home-washed samples were analyzed by determining the anionic and nonionic surfactants, while the laboratory-washed samples with the addition of DDAC in the rinse were analyzed by determining the anionic, cationic, and nonionic surfactants.

The results of potentiometric determination of residues of anionic surfactants in the aqueous extract of whole medical masks before and after washing are shown in Table 10.

Table 10. Amounts of anionic, nonionic, and cationic surfactants in the aqueous extract of medical masks before and after washing—whole samples.

Sample	Surfactant Residuals in the Sample (µg/g)		
	Anionic	Nonionic	Cationic
N	1175.98	68.42	46.32
H 1×	361.01	-	-
H 5×	-	-	-
L 1×	-	348.9	-
L 5×	50.10	-	-
L_{DS} 1×	112.5	-	44.24
L_{DS} 5×	284.43	-	130.14

The results of the amount of surfactants in the whole mask samples, Table 10, show the presence of ionic and nonionic surfactants only in some samples. The whole mask samples contained all surfactants before washing, with an extremely high amount of anionic surfactants and a much lower amount of nonionic and cationic surfactants compared to the anionic ones. The reason for this can indeed be attributed to contamination during production/packaging and/or an analysis error. The obtained amounts of cationic and nonionic surfactant per mass of the sample do not differ significantly, and in terms of residuals their values are lower than the permissible values according to the quality system, RAL-GZ 992.

In the 1× home-washed sample (H 1×), the anionic surfactant was isolated. Cumulative washing cycles had no effect on the increase in the residual amount of surfactants but eliminated them completely (H 5×).

There was a change after the laboratory washing of the whole mask samples compared to the unwashed whole mask samples, which can be concluded from the presence of the anionic surfactant in the 1× washed sample (L 1×) and a nonionic surfactant in the 5× washed sample (L 5×). Washing in the laboratory with a standard detergent and rinsing with the addition of DDAC (L_{DS}) affected the state of the surface so that anionic and cationic surfactants were present after the first and fifth cycles, and their amount increased with the number of cycles, indicating the cumulative aggregation of surfactants.

Considering the potential and actual interactions of ionic surfactants (anionic and cationic) in the same bath, their combination on textiles is not common. However, in the performed procedure, where DDAC was added as a cationic surfactant in the fourth rinse cycle (after washing with standard detergent and the third rinse cycle in water), it partially reduced the amount of anionic surfactant, without eliminating it completely.

A conceptually identical analysis was performed on the cut mask samples (U), with the combined results of the surfactant analysis on the medical mask samples presented in Table 11.

Table 11. Amounts of anionic, nonionic, and cationic surfactants in the aqueous extract of cut medical masks before and after washing.

Sample	Surfactant Residuals in the Sample (µg/g)		
	Anionic	Nonionic	Cationic
N U	-	-	28.24
H U 1×	201.42	-	-
H U 5×	339.47	-	-
L U 1×	34.00	110.75	-
L U 5×	281.95	-	-
L_{DS} U 1×	91.73	93.31	-
L_{DS} U 5×	285.17	196.83	-

Surfactant residuals were found in the unwashed and washed samples, and their values were within the limits set by the control system RAL-GZ 992. It is better to perform the analysis of the parameters on whole samples than on cut samples. The surfactant content results of the cut medical mask samples listed in Table 11 are significantly different from those listed in Table 10. Cationic surfactants were found only in the unwashed medical mask sample, and their amount was twice smaller than for the whole sample. According to this analysis, the cationic surfactant identified in the unwashed sample of the whole and cut medical masks indicates the possible addition of this surfactant during production as an additive to provide antimicrobial activity. The anionic surfactant was found on all the cut samples of the medical masks, and its amount was below the permissible levels (400 µg/g) according to RAL-GZ 992.

The nonionic surfactant was found in three samples (L U 1×, L_{DS} S U 1×, L_{DS} S U 5×), and its amount was not higher than the permissible amount (200 µg/g) according to RAL-GZ.

In summary, the FTIR analysis results confirmed that the back and front nonwoven layers of medical masks were made of polypropylene, and that the washing procedures had an unfavorable influence on the spectral values of the medical masks (fastness evaluation 3–4), through which the aesthetic properties of the masks were slightly damaged.

The new packaged masks are microbiologically safe; significant contamination occurs during home washing, whereas it is considerably lower during laboratory washing. The addition of a disinfectant in the fourth rinse has a favorable effect on microbiological efficiency, and no microorganisms are present. The increase in the pH of the surface of the medical masks washed through five cycles under laboratory conditions is due to the alkalinity of the standard detergent solution. The aqueous extract of the medical masks washed at home and under laboratory conditions through five cycles is alkaline, with almost the same value (pH 8.9) found for all samples.

It is better to perform the analysis on the whole samples than on the cut ones.

The nonwoven structure of the medical mask and the wrinkles that form during washing make it difficult to perform the analysis.

By applying the hierarchical cluster analysis (HCA) [42] to the results of air permeability, contact angle, and whiteness of the samples, a group distribution was obtained, Figure 10.

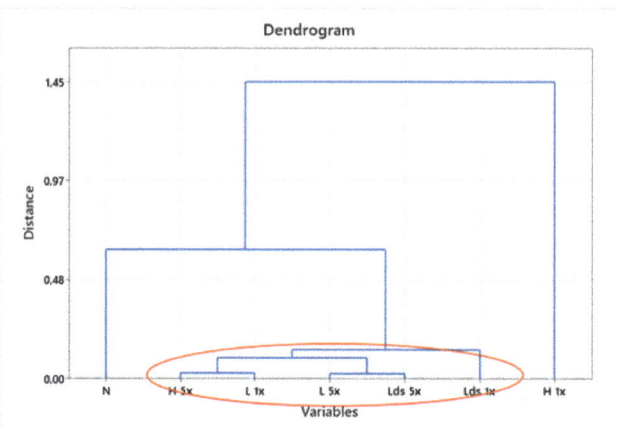

Figure 10. Dendrogram of Euclidean distance for contact angle, air permeability, and whiteness as variables of unwashed and 1× and 5× washed samples. Red circle indicates samples with very small distances.

The group distribution highlights the unwashed sample, and there is a greater distance between the samples washed 1× in the household washing machine. The highlighting of these samples confirms the earlier claim about the significantly different influence of the mechanics of the household washing machine compared to the laboratory washing device, Figure 10. The influence of the number of washing cycles on the observed parameters was additionally confirmed.

The similarities and differences between the groups show and confirm the claims about the impact of washing on the contact angle, which primarily depends on the number of wash cycles performed.

The static contact angle, air permeability, and FTIR results demonstrated the reusability and washability of medical masks at 50 °C in home and laboratory washing procedures. It was also confirmed that washed masks contain minimal amounts of residue, so contact with skin is not hazardous. The spectral properties of the washed masks changed slightly due to bleeding of the color during the washing process. The added benefit of the proposed approach was the use of the disinfectant didecyldimethylammonium chloride in the washing process.

4. Conclusions

This work is focused on the evaluation of different procedures of washing medical masks with detergents and the disinfectant DDAC. One of the hypotheses of this work was that medical masks can maintain filtration properties and breathability after washing at 50 °C.

The hydrophobicity of medical masks determined through the static contact angle depends on the number of cycles carried out. The static contact angle of the samples after the first cycle is lower than after the fifth cycle in all procedures. Images of ultrapure water drops on the surface confirm the hydrophobicity of the front/back of the medical mask before and after washing. The air permeability of washed samples in household and laboratory washing procedures through five cycles increases (~100 L/m^2s) compared to the unwashed sample (~30 L/m^2s). The spectral values of the washed medical masks confirmed slightly impaired aesthetic features.

The obtained results indicate that the performed washing procedures adversely affect some properties of the analyzed medical masks. Despite this, the washability of medical masks at 50 °C gives a basis for their reuse and the partial reduction in mask waste, whose disposal has not been adequately resolved.

The application of the disinfectant didecyldimethylammonium chloride in rinsing proved to be a useful method for maintaining the initial protective properties of medical masks.

Author Contributions: Conceptualization, T.P., M.Č. and J.V.V.; methodology, A.K., M.Č., J.V.V. and T.P.; formal analysis, A.K., J.V.V. and M.Č. investigation, A.K., M.Č., J.V.V. and T.P.; writing—original draft preparation, T.P., M.Č. and J.V.V. All authors have read and agreed to the published version of the manuscript.

Funding: This research was partly supported by the Slovenian Research Agency (research programme P2-0118).

Institutional Review Board Statement: Not applicable.

Informed Consent Statement: Not applicable.

Data Availability Statement: Not applicable.

Acknowledgments: Part of the research was performed on equipment purchased by K.K.01.1.1.02.0024 project "Modernization of Textile Science Research Centre Infrastructure" (MI-TSRC).

Conflicts of Interest: The authors declare no conflict of interest.

References

1. Davies, A.; Thompso, K.A.; Giri, K.; Kafatos, G.; Walker, J.; Bennet, A. Testing the Efficacy of Homemade Masks: Would They Protect in an Influenza Pandemic? *Disaster Med. Public Health Prep.* **2013**, *7*, 413–418. [CrossRef] [PubMed]
2. Chua, M.; Weiren, C.; Simin Goh, S.; Kong, J.; Bing, L.; Jason, Y.; Lim, C.; Mao, L.; Wang, S.; Xue, K.; et al. Face Masks in the New COVID-19 Normal: Materials, Testing and Perspectives. *Research* **2020**, *3*, 1–40. [CrossRef] [PubMed]
3. Scott, E. Prevention of the Spread of Infection—The Need for a Family-Centred Approach to Hygiene Promotion. *Am. J. Infect. Control* **2010**, *38*, 1–3. [CrossRef] [PubMed]
4. Giedraitienė, A.; Ruzauskas, M.; Šiugždinienė, R.; Tučkutė, S.; Milcius, D. Antimicrobial Properties of CuO Particles Deposited on a Medical Mask. *Materials* **2022**, *15*, 7896. [CrossRef]
5. Neral, B. Kvaliteta pranja u kućanskom stroju za pranje s ozonizatorom. *Tekstil* **2016**, *65*, 241–251.
6. Gonçalves, A. Ozone—An Emerging Technology for the Seafood Industry. *Braz. Arch. Biol. Technol.* **2009**, *52*, 1527–1539. [CrossRef]
7. Rubio-Romero, J.-C.; Pardo-Ferreira, M.-C.; Torrecilla-García, J.-A.; Calero-Castro, S. Disposable masks: Disinfection and sterilization for reuse, and non-certified manufacturing, in the face of shortages during the COVID-19 pandemic. *Saf. Sci.* **2020**, *129*, 104830. [CrossRef]
8. Attia, F.; Whitener, C.; Mincemoyer, S.; Houck, J.; Julian, K. The effect of pulsed xenon ultraviolet light disinfection on healthcare-associated Clostridioides difficile rates in a tertiary care hospital. *Am. J. Infect. Control* **2020**, *48*, 1116–1118. [CrossRef]
9. Sinner, H. Über das Waschen mit Haushaltwaschmaschinen: In Welchem Umfang Erleichtern Haushaltwaschmaschinen und -Geräte das Wäschehaben im Haushalt? Haus und Heim-Verlag: Hamburg, Germany, 1960.
10. Soljačić, I.; Pušić, T. *Njega Tekstila–I dio.: Čišćenje u Vodenim Medijima*; Sveučilište u Zagrebu Tekstilno-tehnološki fakultet: Zagreb, Hrvatska, 2005; ISBN 953-7105-08-3. (in Croatian)
11. Gerhardts, A. Testing of the Adhesion of Herpes Simplex Virus on Textile Substrates and Its Inactivation by Household Laundry Processes. *J. Biosci. Med.* **2016**, *4*, 111–125. [CrossRef]
12. Fijan, S.; Šostar-Turk, S. Antimicrobial Activity of Selected Disinfectants Used in a Low Temperature Laundering Procedure for Textiles. *Fibres Text. East. Eur.* **2010**, *18*, 89–92.
13. Dietz, L.; Horve, P.F.; Coil, D.A.; Fretz, M.; Eisen, J.A.; Wymelenberg, K.V.D. Novel Coronavirus (COVID-19) Pandemic: Built Environment Considerations to Reduce Transmission. *mSystems* **2022**, *7*, e00375-20.
14. McQuery, M.; Easter, E.; Cao, A. Disposable versus reusable medical gowns: A performance comparison. *Am. J. Infect. Control* **2020**, *49*, 536–570. [CrossRef]
15. Šterman, S.; Townsend, K.; Salter, E.; Harrigan, K. Surveying Healthcare Workers to Improve the Design, Wearer Experience and Sustainability of PPE Isolation Gowns. *Stroj. Vestn.—J. Mech. Eng.* **2022**, *68*, 252–264. [CrossRef]
16. Sakač, N.; Madunić-Čačić, D.; Marković, D.; Ventura, B.D.; Velotta, R.; Ptiček Siročić, A.; Matasović, B.; Sermek, N.; Đurin, B.; Šarkanj, B.; et al. The 1,3-Dioctadecyl-1H-imidazol-3-ium Based Potentiometric Surfactant Sensor for Detecting Cationic Surfactants in Commercial Products. *Sensors* **2022**, *22*, 9141. [CrossRef]
17. Alygizakis, N.; Galani, A.; Rousis, N.I.; Aalizadeh, R.; Dimopoulos, M.A.; Thomaidis, N.S. Change in the chemical content of untreated wastewater of Athens, Greece under COVID-19 pandemic. *Sci. Total Environ.* **2021**, *799*, 149230. [CrossRef] [PubMed]
18. Hassani, S.; Henni, L.; Sidali, A.; Naitbouda, A.; Khereddine, A.; Dergham, D.; Lekoui, F. Effect of washing on quality, breathability performance and reusability of disposable face masks. *J. Med. Eng. Technol.* **2022**, *46*, 345–353. [CrossRef] [PubMed]

19. Cornelio, A.; Zanoletti, A.; Federici, S.; Ciacci, L.; Depero, L.-E.; Bontempi, E. Environmental Impact of Surgical Masks Consumption in Italy Due to COVID-19 Pandemic. *Materials* **2022**, *15*, 2046. [CrossRef]
20. Zorko, D.J.; Gertsman, S.; O'Hearn, K.; Timmerman, N.; Ambu-Ali, N.; Dinh, T.; Sampson, M.; Sikora, L.; McNally, J.D.; Choong, K. Decontamination interventions for the reuse of surgical mask personal protective equipment: A systematic review. *J. Hosp. Infect.* **2020**, *106*, 283–294. [CrossRef]
21. Remic, K.; Erjavec, A.; Volmajer Valh, J.; Šterman, S. Public Handling of Protective Masks from Use to Disposal and Recycling Options to New Products. *Strojniški Vestnik—J. Mech. Eng.* **2022**, *68*, 281–289. [CrossRef]
22. Erjavec, A.; Pohl, O.; Fras Zemljič, L.; Volmajer Valh, J. Significant Fragmentation of Disposable Surgical Masks—Enormous Source for Problematic Micro/Nanoplastics. *Pollut. Environ. Sustain.* **2022**, *14*, 12625. [CrossRef]
23. Prata, J.C.; Silva, A.L.P.; Duarte, A.C.; Rocha-Santos, T. Disposable over Reusable Face Masks: Public Safety or Environmental Disaster? *Environments* **2021**, *8*, 31. [CrossRef]
24. Pereira-Ávila, F.M.V.; Lam, S.C.; Góes, F.G.B.; Gir, E.; Pereira-Caldeira, N.M.V.; Teles, S.A.; Caetano, K.A.A.; Goulart, M.D.C.E.L.; Bazilio, T.R.; Silva, A.C.D.O.E. Factors associated with the use and reuse of face masks among Brazilian individuals during the COVID-19 pandemic. *Rev. Lat. Am. Enfermagem.* **2020**, *28*, e3360. [CrossRef] [PubMed]
25. Farzaneh, S.; Shirinbayan, M. Processing and Quality Control of Masks: A Review. *Polymers* **2022**, *14*, 291. [CrossRef] [PubMed]
26. Parac-Osterman, Đ. *Osnove o boji i Sustavi Vrednovanja*; Tekstilno-tehnološki fakultet: Zagreb, Hrvatska, 2013; ISBN 978-953-7105-11-2. (in Croatian)
27. Puebla, C. *Whiteness Assessment: A Primer Concepts, Determination and Control of Perceived Whiteness*; Technical Guide 2006; Axiphos GmbH: Lörrach, Germany, 2006.
28. Frece, J.; Markov, K. *Uvod u mikrobiologiju i fizikalno kemijsku analizu voda*; Inštitut za sanitarno inženirstvo: Vista, Slovenija, 2015; ISBN 978-961-92846-5-0. (in Croatian)
29. Application Bulletin 233/4 e. Potentiometric Determination of Anionic and Cationic Surfactants with Surfactant Electrodes, Metrohm, Switzerland. Available online: https://www.metrohm.com/en/applications/ab-application-bulletins/ab-233.html (accessed on 10 December 2022).
30. Application Bulletin 230/2 e. Potentiometric Determination of Nonionic Surfactants Based on Polyoxyethylene Adducts Using the NIO Electrode, Metrohm, Switzerland. Available online: https://www.metrohm.com/en_us/applications/ab-application-bulletins/ab-230.html (accessed on 10 December 2022).
31. Desalegn, A.; Gashaw, A.; Nalankilli, G. Impact of Sunlight Exposure to different Dyed fabrics on Color fastness to Washing. *Int. J. Mod. Sci. Technol.* **2017**, *2*, 360–365.
32. Xiao, S.; Yuan, Z.; Huang, Y. Disinfectants against SARS-CoV-2: A Review. *Viruses* **2022**, *14*, 1721. [CrossRef]
33. Dekanić, T.; Tarbuk, A.; Flinčec Grgac, S. Određivanje sposobnosti upravljanja vlagom vodoodbojne pamučne tkanine kondenzirane pri niskoj temperaturi. *Tekstil* **2018**, *67*, 176–188. (in Croatian).
34. Vujasinović, E.; Dragčević, K. Sorpcijska svojstva materijala za izradu jedara. *Tekstil* **2012**, *61*, 180–188. (in Croatian).
35. Fuk, B. Kuda s jednokratnim medicinskim maskama nakon upotrebe. *Sigurnost* **2020**, *62*, 421–424. (in Croatian).
36. Toharska, M. Neural Model of the Permeability Features of Woven Fabrics. *Text. Res. J.* **2004**, *74*, 1045–1049. [CrossRef]
37. Petrulyte, S.; Baktakyte, R. Istraživanje propusnosti zraka frotirnih tkanina u ovisnosti o primijenjenom postupku dorade. *Tekstil* **2008**, *57*, 21–27. (in Croatian).
38. Guo, J. The Effects of Household Fabrics Softeners on the Thermal Comfort and Flammability of Cotton and Polyester Fabrics. Master Thesis, Virginia Tech, Blacksburg, VA, USA, 29 April 2003.
39. Regent, A. Zaštine maske za lice-medicinske i maske za građanstvo. *Sigurnost* **2020**, *62*, 413–415. (in Croatian).
40. Wong, C.; Badri, K. Chemical Analyses of Palm Kernel Oil-Based Polyurethane Prepolymer. *Mater. Sci. Appl.* **2012**, *6*, 78–86. [CrossRef]
41. Pušić, T.; Soljačić, I. Kontrola kvalitete pranja prema RAL-GZ 992. *Tekstil* **2008**, *57*, 296–302.
42. Zupan, J. *Kemometrija in Obdelava Experimentalnih Podatkov*; Inštitut Nove revije, Zavod za humanistiko and National Institute of Chemistry: Ljubljana, Slovenia, 2009; pp. 154–167. ISBN 978-961-92463-3-7.

Disclaimer/Publisher's Note: The statements, opinions and data contained in all publications are solely those of the individual author(s) and contributor(s) and not of MDPI and/or the editor(s). MDPI and/or the editor(s) disclaim responsibility for any injury to people or property resulting from any ideas, methods, instructions or products referred to in the content.